Digital Transmission Engineering

Christoph Lange · Andreas Ahrens

Digital Transmission Engineering

Fundamentals and Techniques of Digital
Baseband Transmission

 Springer

Christoph Lange
Fachbereich 1: Ingenieurwissenschaften –
Energie und Information Hochschule für
Technik und Wirtschaft Berlin, University of
Applied Sciences
Berlin, Germany

Andreas Ahrens
Fakultät für Ingenieurwissenschaften
Hochschule Wismar, University of Applied
Sciences: Technology, Business and Design
Wismar, Germany

ISBN 978-3-658-46788-3   ISBN 978-3-658-46789-0   (eBook)
https://doi.org/10.1007/978-3-658-46789-0

Translation from the German language edition: "Übertragungstechnik" by Christoph Lange and Andreas Ahrens, © Der/die Herausgeber bzw. der/die Autor(en), exklusiv lizenziert an Springer Fachmedien Wiesbaden GmbH, ein Teil von Springer Nature 2023. Published by Springer Fachmedien Wiesbaden. All Rights Reserved.

This book is a translation of the original German edition "Übertragungstechnik" by Christoph Lange and Andreas Ahrens, published by Springer Fachmedien Wiesbaden GmbH in 2023. The translation was done with the help of an artificial intelligence machine translation tool. A subsequent human revision was done primarily in terms of content, so that the book will read stylistically differently from a conventional translation. Springer Nature works continuously to further the development of tools for the production of books and on the related technologies to support the authors.

This Springer imprint is published by the registered company Springer Fachmedien Wiesbaden GmbH, part of Springer Nature.
The registered company address is: Abraham-Lincoln-Str. 46, 65189 Wiesbaden, Germany

If disposing of this product, please recycle the paper.

In memory of

Prof. Dr.-Ing. habil. Rainer Kohlschmidt

and

Prof. Dr.-Ing. Reiner Rockmann

Preface

Motivation

In a multitude of areas of modern life, networking via communication networks has become indispensable. For example, it enables access to information that is largely independent of time and location on a scale that was utopian just a few decades ago. In today's information society, the processing, storage, and transmission of information therefore play a key role. These tasks are core areas of *communications engineering* and related fields in electrical engineering and computer science.

An important technical component is information transmission, which is tied to physical signals. This transmission of information using signals is a crucial task of *telecommunication technology* as a subfield of communications engineering. Communications engineering itself has evolved over long periods of time, starting from the dissemination of information using sound, light, and smoke signals in early times, through telegraphy and analog telecommunications technology in the past centuries, to modern messaging or communication technology, which is almost exclusively based on the exchange and processing of digital data. Thus, transmission technology is subject to continuous technical development, and its performance depends on the specific technology and techniques available; however, it is largely based on comparatively long-lasting insights, principles, and procedures.

The present book is an introductory presentation on communication technology. It is primarily aimed at students of technical disciplines who are seeking an introduction to the basics of communications engineering and, in particular, transmission technology, and who want to acquire a basic understanding of their principles and procedures. On this basis, methods of communication technology can be assessed in their performance and options for improvements and further developments can be independently developed. This book provides a solid foundation for a possible later technical deepening in communication technical special disciplines and techniques, which can take place in further studies or in professional practice.

Knowledge of engineering mathematics and the basics of electrical engineering are assumed. Knowledge of signal and system theory is also advantageous for reading; the necessary knowledge from this area is also presented compactly in a separate chapter and in a corresponding appendix.

The contents of this book are based on the lectures given by our academic teachers Prof. Dr.-Ing. habil. Rainer Kohlschmidt [Kohlschmidt, R.] and Prof. Dr.-Ing. Reiner Rockmann [Rockmann, R.] at the University of Rostock and previously at the Ingenieurhochschule für Seefahrt Warnemünde/Wustrow, an Engineering School for Maritime Navigation. Some of the teaching materials developed and used over many years in this context are, for example, [Kohlschmidt, 1983 a, Kohlschmidt, 1983 b, Kohlschmidt, 1998]. Their professional perspectives and modes of presentation were shaped by their engineering studies and subsequent scientific work in the fields of communications engineering, signal and system theory, high-frequency technology, and theoretical electrical engineering at the Technical University of Ilmenau (R. Kohlschmidt) and at the University of Rostock (R. Rockmann). The individuals who significantly influenced these working and research groups have presented their perspectives and insights in some of their own, now classic, works, for example, [Philippow, 1963, Philippow, 1992, Kress, 1977, Kress, 1979] and [Lange, 1965, Lange, 1983, Lange, 1988].

Subsequently, we ourselves as scientific assistants and staff have conveyed and represented these contents in teaching sessions at the University of Rostock - initially in exercises and laboratory practicals, later in lectures with accompanying exercises. After each of our own appointments at the Hochschule Wismar, a University of Applied Sciences for Technology, Business and Design (A. Ahrens) and at the Hochschule für Technik und Wirtschaft Berlin, a University of Applied Sciences for Technology and Economics (C. Lange), these contents are reflected in our own teaching sessions and lectures in the field of communications engineering. Our own views, ways of thinking and representations in communication technology, which have found their expression in this book, are shaped by our aforementioned esteemed academic teachers - to whom we would like to express our heartfelt thanks, albeit posthumously, at this point.

The aim of this book is to convey an understanding of basic principles and mechanisms that are common to the methods of information and signal transmission in communications engineering, and which will remain valid and enduring regardless of actual technologies, components, or systems available at a certain point in time. Therefore, it is our endeavor with this book to find a balance between a potentially very theory-loaded presentation on the one hand, and a potential short-livedness owed to the state of the art and practical implementation possibilities on the other hand.

Structure of the Book

In Chap. 1 signal and system theoretical basics are presented and repeated as far as they are necessary for the understanding of the further chapters of this book. They are usually conveyed in detail in preceding lectures. This chapter also serves to introduce the used terminology as well as mathematical designations and notations.

Chapter 2 provides the fundamentals of baseband transmission for non-distorting channels with additive white Gaussian-distributed noise as disturbance. Starting from a physically intuitive representation, the individual elements of such a transmission path are introduced and their effects on the useful signal and disturbance are examined. Quality criteria for assessing the quality of digital signal transmission are defined and applied. Using mathematical-analytical descriptions, influencing variables are identified and optimized. Numerical examples conclude the chapter and illustrate the effects that each of the previously introduced and discussed parameters have on the transmission.

In Chap. 3 the transmission channel *copper cable* is introduced as an example of a linearly distorting transmission channel. Starting from transmission line theory considerations, a system-theoretical model is developed that describes the distorting properties of copper two-wire lines as a transmission channel. In addition, crosstalk is also considered as a disturbance effect and described using system theory.

Chapter 4 contains the application of the fundamentals of baseband transmission introduced in Chap. 2 to the signal transmission over the copper cables described in Chap. 3. The considerations are expanded to include the now necessary compensation of the influences of the transmission channel on the useful signal using the example of the copper cable transmission channel. Methods of equalization are introduced and necessary optimizations of parameters are examined. Again, detailed numerical examples are provided.

Chapter 5 contains a brief summary as well as an outlook on current topics and further developments in the field of communication technology. These lead to the necessity to expand the knowledge acquired with this book: For this purpose, a number of textbooks and specialized works are available for the eager learner as well as for the seasoned practitioner. Some of these are referred to in the individual chapters.

Each chapter contains concluding exercises for practice, the solutions to which, along with the solution path, are provided in the respective chapter.

In some appendices, important relationships of signal and system theory, summaries of transformations, and other necessary mathematical functions, formulas, and relationships are presented: These are largely based on auxiliary sheets that have been created for lectures over the years and have proven to be useful aids and compact reference works in teaching practice.

Acknowledgements

At this point, the authors would like to thank all those who have contributed to the creation of this book.

Special thanks are primarily due to the families of the authors, who have made possible and continuously supported their respective professional careers, and who, through interest and encouragement, have contributed to the success of this book and its completion.

We would also like to express our heartfelt thanks to everyone who has supported and kindly accompanied us on our professional journey, initially after our studies at the Institute for Communications Engineering and Information Electronics at the University of Rostock, and later in the technology innovation areas of a telecommunications network operator in Berlin and at the Hochschule für Technik und Wirtschaft Berlin (C. Lange), as well as at the Hochschule Wismar (A. Ahrens). All these influences have contributed in one way or another to the success of this book.

Mr. Johannes F. Lange (M. Sc.) is thanked for discussions and support with some derivations and calculations.

Thanks are due to the publisher for their willingness to publish this book, as well as for their support and consistently excellent collaboration in its design and completion.

<table>
<tr><td>Berlin</td><td style="text-align:right">Christoph Lange</td></tr>
<tr><td>Wismar</td><td style="text-align:right">Andreas Ahrens</td></tr>
<tr><td>in April 2023</td><td></td></tr>
</table>

Bibliography for the Preface

[Kohlschmidt, R.] *Entry of "Rainer Kohlschmidt" in the Catalogus Professorum Rostochiensium*. University of Rostock. http://purl.uni-rostock.de/cpr/00001567. Version: 2018. – accessed on 15.03.2023

[Rockmann, R.] *Entry of "Reiner Rockmann" in the Catalogus Professorum Rostochiensium*. University of Rostock. http://purl.uni-rostock.de/cpr/00002046. Version: 2018. – accessed on 15.03.2023

[Kohlschmidt, 1983 a] KOHLSCHMIDT, R.: *Grundlagen analoger Systeme und Schaltungstechnik*. Teil I. Rostock: Ingenieurhochschule für Seefahrt Warnemünde/Wustrow, 1983. – Knowledge storage booklet

[Kohlschmidt, 1983 b] KOHLSCHMIDT, R.: *Grundlagen analoger Systeme und Schaltungstechnik*. Teil II. Rostock: Ingenieurhochschule für Seefahrt Warnemünde/Wustrow, 1983. – Knowledge storage booklet

[Kohlschmidt, 1998] KOHLSCHMIDT, R.: *Schaltungen und Baugruppen der Nachrichtentechnik*. Rostock : University of Rostock, 1998. – Study material

[Philippow, 1963] PHILIPPOW, E. (Ed.): *Taschenbuch Elektrotechnik*. Vol. 1–6. Berlin : Verlag Technik, since 1963

[Philippow, 1992] PHILIPPOW, E.: *Grundlagen der Elektrotechnik*. 9th, rev. ed. Berlin : Verlag Technik, 1992

[Kress, 1977] KRESS, D.: *Theoretische Grundlagen der Signal- und Informationsübertragung*. Berlin : Akademie-Verlag, 1977

[Kress, 1979] KRESS, D.: *Theoretische Grundlagen der Übertragung digitaler Signale*. Berlin : Akademie-Verlag, 1979

[Lange, 1965] LANGE, F. H.: *Signale und Systeme*. Bd. 1: Spektrale Darstellung; Bd. 2: Gesteuerte elektronische Systeme; Bd. 3: Regellose Vorgänge. Berlin : Verlag Technik, since 1965

[Lange, 1983] LANGE, F. H.: *Störfestigkeit in der Nachrichten- und Meßtechnik*. Berlin : Verlag Technik, 1983

[Lange, 1988] LANGE, F. H.: *Meßstochastik und Störsicherheit*. Berlin : Akademie-Verlag, 1988

Contents

About the Authors

Prof. Dr.-Ing. Christoph Lange teaches in the field of communications engineering and communication networks at the Hochschule für Technik und Wirtschaft Berlin, a University of Applied Sciences for Technology and Economics, in Berlin. Previously, after studying electrical engineering with a specialization in communications engineering and earning a doctorate in the field of transmission engineering at the University of Rostock, he worked for many years in various technology innovation areas of a large telecommunications network operator in Berlin.

Prof. Dr.-Ing. habil. Andreas Ahrens teaches in the field of communications engineering as well as signal and system theory at the Hochschule Wismar, a University of Applied Sciences for Technology, Business and Design. He completed a degree in electrical engineering with a focus on communications engineering, as well as a doctorate and habilitation in the field of communication transmission technology at the University of Rostock. He also spent time abroad at the University of Southampton and worked as a private lecturer at the University of Rostock.

Signal and System Theoretical Basics

1

Abstract

Signals can be found in very different forms as carriers of information. Large amounts of data must be processed in such a way that they can be efficiently transmitted over a given transmission channel or stored on a storage medium. To do this, it is necessary to describe signals in the time and frequency domain. Signals can have deterministic or random (or stochastic) characteristics; moreover, they can be analog or digital. The processing and linking of signals for information transmission and storage is done with the help of suitable systems. Problems of information and signal transmission can therefore usually be traced back to questions that concern the evaluation and transformation of signals by systems. This chapter provides an overview of the possibilities for describing signals and systems and presents the basics necessary for understanding telecommunications contexts, which are dealt with in the following chapters, using the time and frequency domain representation including the necessary transformations.

1.1 Introduction and Basics

In practical systems, signals occur in the form of voltages and currents as physically measurable quantities. These signals are primarily time functions that play a central role as transmission and reception quantities, for example, in communications engineering.

© The Editor(s) (if applicable) and The Author(s), under exclusive license to Springer
Fachmedien Wiesbaden GmbH, part of Springer Nature 2025
C. Lange and A. Ahrens, *Digital Transmission Engineering*,
https://doi.org/10.1007/978-3-658-46789-0_1

1

A distinction is made between deterministic and random (or stochastic) signals. While deterministic signals can be lawfully described by an algebraic expression, stochastic or random signals can only be described using the tools of stochastic[1]—for example, through correlation functions or distribution functions. An example of a deterministic signal is a harmonic oscillation, whereas noise is an example of a random signal.

Furthermore, signals can be defined as functions of continuous or discrete variables—with a finite or infinite set of values. If signals have an infinite set of values and are functions of a continuous time variable t, they are referred to as *analog* signals; these are thus time- and value-continuous. *Digital* signals, on the other hand, are functions of a discrete time variable and have a finite set of values; they are time- and value-discrete [30, 42]. For analog signals, the notation

$$u(t), x(t), h(t), \ldots \qquad \text{with} \quad t \in \mathbb{R} \tag{1.1}$$

should be used [42]. $\mathbb{R}$ describes the set of real numbers. Time-discrete signals, which can have either a continuous or a discrete range of values, are defined as follows:

$$x(i), y(k), z(l), \ldots \qquad \text{with} \quad i, k, l \in \mathbb{Z}. \tag{1.2}$$

$\mathbb{Z}$ describes the set of integers. Furthermore, the following should apply:

$$x(i) \in \mathbb{R} \qquad \text{or} \qquad x(i) \in \mathbb{C} \qquad \text{with} \quad i \in \mathbb{Z}. \tag{1.3}$$

The set of complex numbers is described by $\mathbb{C}$. The emphasis of a discrete range of values is expressed by the symbolism

$$\hat{x}(i) \in \mathbb{Z}, \qquad \text{with} \quad i \in \mathbb{Z}. \tag{1.4}$$

The notation $\hat{x}(i)$ denotes a digital signal [42].

For the investigation of communication technology relationships, it is often useful to consider time-discrete signals as sequences of numbers. For such finite sequences, a vector notation can be advantageously used. If $x(i)$ is limited to the interval $i_1 \leq i \leq i_2$ in time, the vector

$$\boldsymbol{x} = \left(x_{i_1}, x_{i_1+1}, x_{i_1+2}, \ldots, x_{i_2} \right)^{\mathrm{T}} \tag{1.5}$$

can be used [42]. The superscript $(\ldots)^{\mathrm{T}}$ denotes the transposition of a vector, so that the vector $\boldsymbol{x}$ is a column vector [75].

---

[1] Stochastics is a branch of mathematics that includes probability theory and statistics.

1.2 Deterministic and Stochastic Signals

1.2.1 Deterministic Signals

Periodic Time Functions and Fourier Analysis
One of the most well-known determined signals is the harmonic oscillation in the form

$$u(t) = U_0 \cos(\omega_0 t + \phi_0) = U_0 \cos(2\pi f_0 t + \phi_0), \qquad (1.6)$$

which is illustrated in Fig. 1.1. The value U_0 is referred to as amplitude or peak value, $\omega_0 = 2\pi f_0$ as angular frequency, f_0 as frequency and ϕ_0 as zero phase angle. Such signals $u(t)$ with the property

$$u(t + T) = u(t + 2T) = u(t) \qquad \text{for all } t \in \mathbb{R} \qquad (1.7)$$

are also referred to as periodic with the period T.

Furthermore, numerous signals with a harmonic characteristic show a time dependency in such a way that the amplitude is not constant over time, but decreases or increases with time. Such signals can be analytically described as follows:

$$u(t) = U_0 \, e^{-\sigma_0 t} \cos(\omega_0 t + \phi_0). \qquad (1.8)$$

For $\sigma_0 > 0$ we speak of a damped or decaying oscillation, for $\sigma_0 = 0$ of an undamped oscillation and for $\sigma_0 < 0$ of a growing or resonating oscillation. Figure 1.2 illustrates the resulting signal progression for a damped oscillation.

In addition, in many applications, two or more signals occur simultaneously and act on a system. The impact can be either additive or multiplicative.

Example 1.1 (Superposition) The sum of two harmonic oscillations is referred to as *superposition* when the frequencies are different from each other (e.g., $f_1 \gg f_2$ or $f_1 \ll f_2$) and it is referred to as *beating* in the special case when the frequencies are almost the same, i.e., $f_1 \approx f_2$. The following transformation shows which process

Fig. 1.1 Time signal $u(t)$ ($f_0 = 1\,\text{kHz}, U_0 = 1\,\text{V}, \phi_0 = 0$)

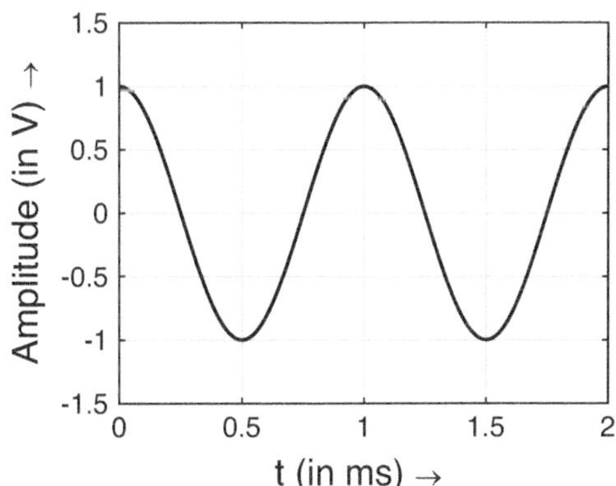

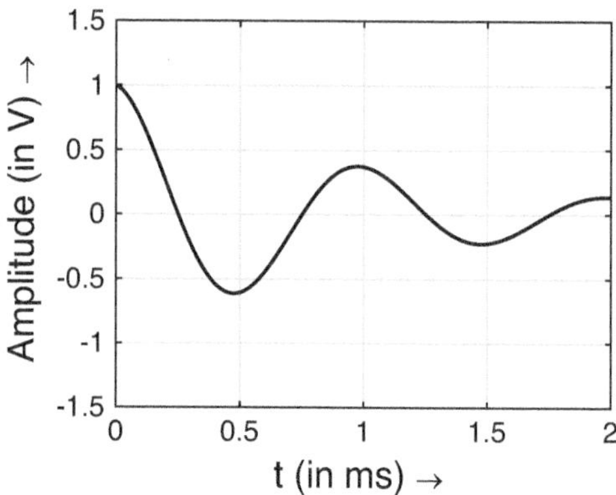

Fig. 1.2 Damped oscillation $u(t)$ according to (1.8) ($f_0 = 1\,\text{kHz}$, $U_0 = 1\,\text{V}$, $\phi_0 = 0$, $\sigma_0 = 10^3\,\text{s}^{-1}$)

occurs during the superposition of two harmonic oscillations $u_1(t) = U_1\,\sin(\omega_1\,t)$ and $u_2(t) = U_2\,\sin(\omega_2\,t)$ with identical amplitude, i.e. $U_1 = U_2 = U_0$, results in [39]. With

$$u(t) = u_1(t) + u_2(t) = U_0\,\sin(\omega_1\,t) + U_0\,\sin(\omega_2\,t) \tag{1.9}$$

the application of the addition theorem

$$\sin(x) + \sin(y) = 2\,\sin\left(\frac{x+y}{2}\right)\cos\left(\frac{x-y}{2}\right) \tag{1.10}$$

leads to

$$u(t) = 2\,U_0\,\sin\left(\frac{(\omega_1 + \omega_2)}{2}\,t\right)\cos\left(\frac{(\omega_1 - \omega_2)}{2}\,t\right). \tag{1.11}$$

Figures 1.3 and 1.4 illustrate the progression of the signals $u_1(t)$ and $u_2(t)$, while the resulting sum signal for the case $f_2 = 2f_1$ is shown in Fig. 1.5. For $\omega_1 = \omega_2 = \omega_0$ it follows with

$$\cos\left(\frac{(\omega_1 - \omega_2)}{2}\,t\right) = 1 \tag{1.12}$$

the result

$$u(t) = 2\,U_0\,\sin\left(\frac{(\omega_1 + \omega_2)}{2}\,t\right) = 2\,U_0\,\sin(\omega_0\,t). \tag{1.13}$$

Fig. 1.3 Time signal $u_1(t)$ ($f_1 = 1\,\text{kHz}, U_1 = 1\,\text{V}$)

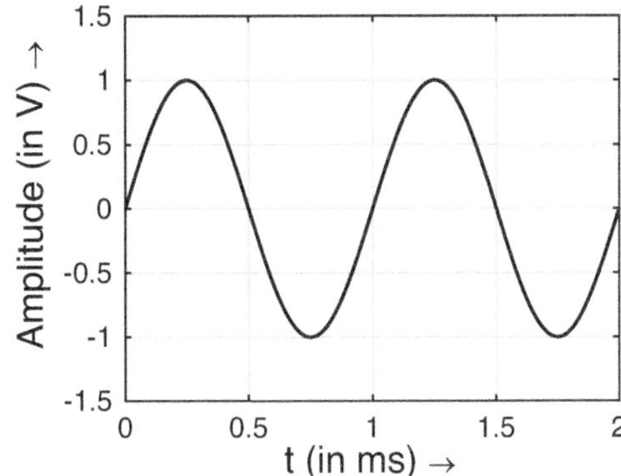

Fig. 1.4 Time signal $u_2(t)$ ($f_2 = 2\,\text{kHz}, U_2 = 1\,\text{V}$)

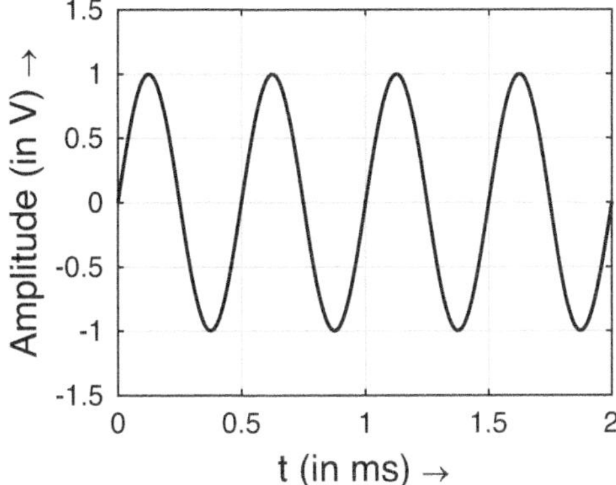

Fig. 1.5 Sum signal $u_1(t) + u_2(t)$ for $f_1 = 1\,\text{kHz}$, $f_2 = 2\,\text{kHz}$ and $U_0 = U_1 = U_2 = 1\,\text{V}$

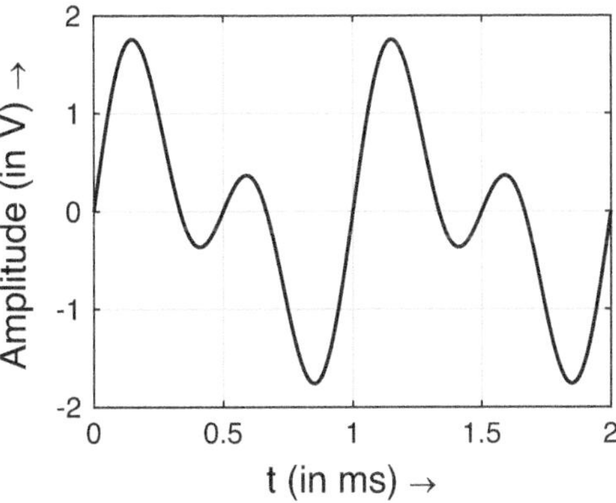

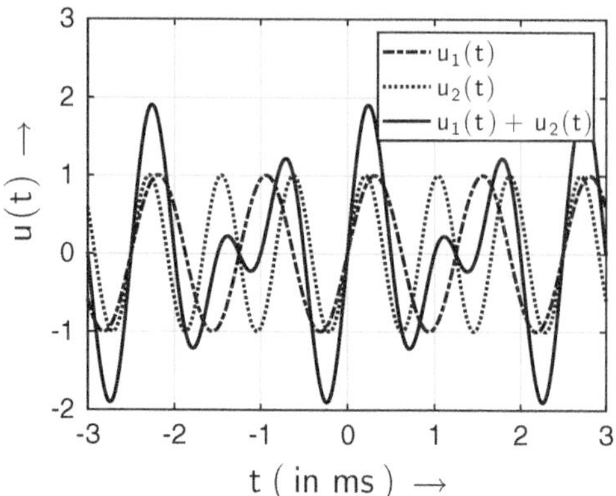

Fig. 1.6 Illustration of the signals with $f_1 = 0.8\,\text{kHz}$, $f_2 = 1.2\,\text{kHz}$ and $U_0 = U_1 = U_2 = 1\,\text{V}$

In this case, a harmonic oscillation with the frequency

$$\frac{(\omega_1 + \omega_2)}{2} = \omega_0 \tag{1.14}$$

and the amplitude $2\,U_0$ appears. For $\omega_1 \approx \omega_2$ a beat is obtained. The amplitude of the sum signal shows a periodically increasing and decreasing amplitude (see Fig. 1.6). $\square$

By combining such harmonic oscillations, one can generate periodic time events of any form. The mathematical basis for this is the theory of *Fourier series* [39].

It should be noted at this point that the sum of two periodic signals generally does not yield a periodic signal. For this to happen, the common period must be an integral multiple of the two given periods $T_1 = 1/f_1$ and $T_2 = 1/f_2$ [39]. This is only the case when the frequencies f_1 and f_2 are in a rational relationship to each other, which is exploited by the theory of Fourier series [30].

Example 1.2 (Sum of two periodic signals) The two frequencies f_1 and f_2 are in a rational relationship if this can be formed from two whole numbers. Thus, the rational ratio

$$\frac{f_1}{f_2} = \frac{2}{3} \tag{1.15}$$

leads to a periodic sum signal, while the exemplary choice of the frequency ratio

$$\frac{f_1}{f_2} = \frac{2.1}{3} \tag{1.16}$$

as a non-rational number does not yield a periodic sum signal.

Fig. 1.7 illustrates the resulting periodic sum signal when the two individual signals are overlaid additively, where the frequencies are in a rational ratio. In contrast, in

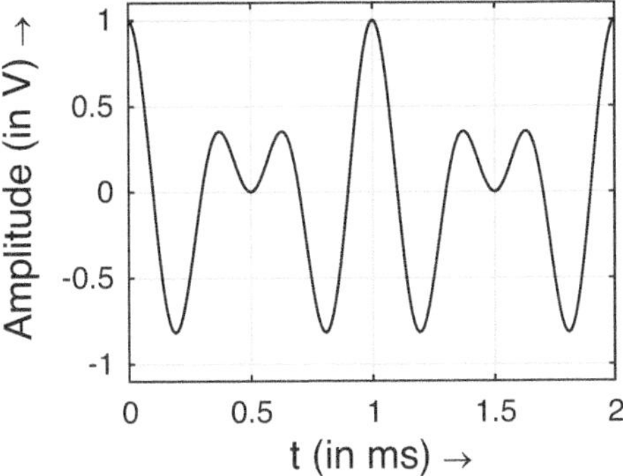

Fig. 1.7 Periodic time signal $u(t)$ ($f_1 = 2\,\text{kHz}, U_1 = 0.5\,\text{V}, \phi_1 = 0, f_2 = 3\,\text{kHz}, U_2 = 0.5\,\text{V}, \phi_2 = 0$)

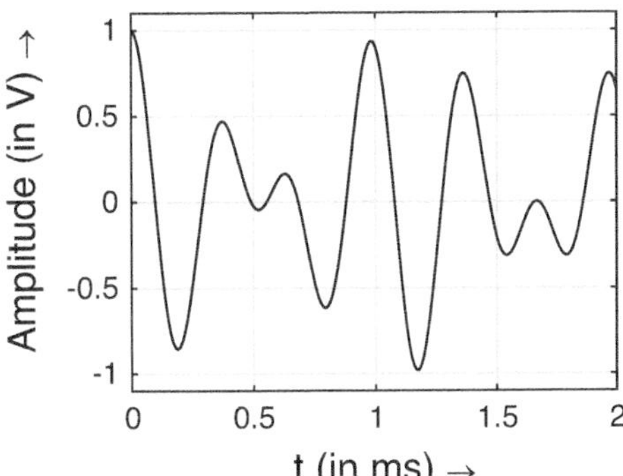

Fig. 1.8 Non-periodic time signal $u(t)$ ($f_1 = 2.1\,\text{kHz}, U_1 = 0.5\,\text{V}, \phi_1 = 0, f_2 = 3\,\text{kHz}, U_2 = 0.5\,\text{V}, \phi_2 = 0$)

Fig. 1.8, it can be seen that the property of periodicity is lost when there is no rational ratio. □

In the 19th century, J. Fourier[2] claimed that any periodic signal can be decomposed into harmonic components, which differ in amplitude, phase, and frequency, with the frequency of the components always being a multiple of the fundamental frequency. Consequently, any real-valued T-periodic function $u(t)$ can be expressed as a linear combination of sine and cosine oscillations:

$$u(t) = \frac{a_0}{2} + \sum_{n=1}^{\infty} \Big(a_n \cos(2\pi\,(nf_0)\,t) + b_n \sin(2\pi\,(nf_0)\,t) \Big). \tag{1.17}$$

The coefficients a_n and b_n are called Fourier coefficients and possess the unit of the time signal, $u(t)$. If $u(t)$ represents a voltage signal, then the Fourier coefficients have the

---

[2] Jean Baptiste Joseph Fourier (1768–1830), French mathematician and physicist.

unit Volt. The parameter f_0 characterizes the fundamental frequency and is equal to the inverse of the period duration T, i.e.,[3]

$$f_0 = \frac{\omega_0}{2\pi} = \frac{1}{T}. \tag{1.18}$$

Furthermore, the coefficient $a_0/2$ describes the DC component of the signal $u(t)$. In the case of a signal without a DC component, the coefficient $a_0/2$ equals zero. The partial oscillation for $n = 1$, i.e.

$$a_1 \cos(2\pi f_0 t) + b_1 \sin(2\pi f_0 t), \tag{1.19}$$

is called the fundamental oscillation (or 1st harmonic). The partial oscillation for $n > 1$, i.e.

$$a_n \cos(n\, 2\pi f_0 t) + b_n \sin(n\, 2\pi f_0 t), \tag{1.20}$$

is called the nth harmonic [39]. The equations for determining the Fourier coefficients are (see also [39])

$$\frac{a_0}{2} = \frac{1}{T} \int_{t_0}^{t_0+T} x(t)\, \mathrm{d}t, \tag{1.21}$$

$$a_n = \frac{2}{T} \int_{t_0}^{t_0+T} x(t)\, \cos(n\, \omega_0 t)\, \mathrm{d}t = \frac{2}{T} \int_{t_0}^{t_0+T} x(t)\, \cos(n\, 2\pi f_0 t)\, \mathrm{d}t, \tag{1.22}$$

$$b_n = \frac{2}{T} \int_{t_0}^{t_0+T} x(t)\, \sin(n\, \omega_0 t)\, \mathrm{d}t = \frac{2}{T} \int_{t_0}^{t_0+T} x(t)\, \sin(n\, 2\pi f_0 t)\, \mathrm{d}t. \tag{1.23}$$

The basis for obtaining the Fourier coefficients is the minimization of the square error between the original signal $u(t)$ and the signal reconstructed via the coefficients $a_0/2$, a_n and b_n $u_R(t)$ (see (1.17)) accordingly

$$\int_T [u(t) - u_R(t)]^2 \mathrm{d}t \longrightarrow \min. \tag{1.24}$$

It should be noted at this point that the Fourier coefficients a_n and b_n cannot be determined by measurement. Derived quantities that can be measured are the magnitude spectrum c_n and the phase spectrum φ_n. Practically, the magnitude and phase spectrum c_n and φ_n can be determined via the approach

---

[3] For the definition of frequency, see Appendix G.

$$u_R(t) = \frac{a_0}{2} + \sum_{n=1}^{\infty} c_n \cos\left(n\,\omega_0\,t - \varphi_n\right). \tag{1.25}$$

With the help of the addition theorem

$$\cos(\alpha - \beta) = \cos(\alpha)\cos(\beta) + \sin(\alpha)\sin(\beta) \tag{1.26}$$

the representation

$$u_R(t) = \frac{a_0}{2} + \sum_{n=1}^{\infty} \left(c_n \cos(\varphi_n)\cos(n\,\omega_0\,t) + c_n \sin(\varphi_n)\sin(n\,\omega_0\,t) \right) \tag{1.27}$$

can be found. The Fourier coefficients a_n and b_n result in

$$a_n = c_n \cos(\varphi_n) \quad \text{and} \quad b_n = c_n \sin(\varphi_n). \tag{1.28}$$

The quantities c_n and φ_n are measurable—for example, with a spectrum analyzer. The magnitude spectrum is real and results in

$$c_n = \sqrt{a_n^2 + b_n^2}. \tag{1.29}$$

and the phase spectrum is also real and can be determined by the expression

$$\varphi_n = \arctan\left(\frac{b_n}{a_n}\right). \tag{1.30}$$

The representation of the Fourier coefficients as a function of frequency is referred to as the *spectrum* of the periodic signal. It contains the fundamental frequency and its multiples, the harmonics, whose size is determined by the Fourier coefficients. Since in the spectrum of periodic time signals only multiples of the fundamental frequency are expected to have spectral components, the resulting *line spectrum* is a characteristic feature of periodic signals.

Example 1.3 (Fourier coefficients of a periodic time signal) Thus, the DC-free periodic sum signal shown in Fig. 1.5, i.e. $a_0/2 = 0$, in the form

$$u(t) = U_0 \sin(\omega_1\,t) + U_0 \sin(2\,\omega_1\,t) \tag{1.31}$$

with the fundamental frequency $f_1 = 1\,\text{kHz}$, the frequency $f_2 = 2f_1 = 2\,\text{kHz}$ and $U_0 = 1\,\text{V}$ the line spectrum indicated in Fig. 1.9. The Fourier coefficients different from zero are $b_1 = U_0 = 1\,\text{V}$ and $b_2 = U_0 = 1\,\text{V}$. $\square$

The additive combination of two harmonic time functions with given frequency, amplitude, and phase does not lead to the creation of new frequencies, as shown in Fig. 1.9.

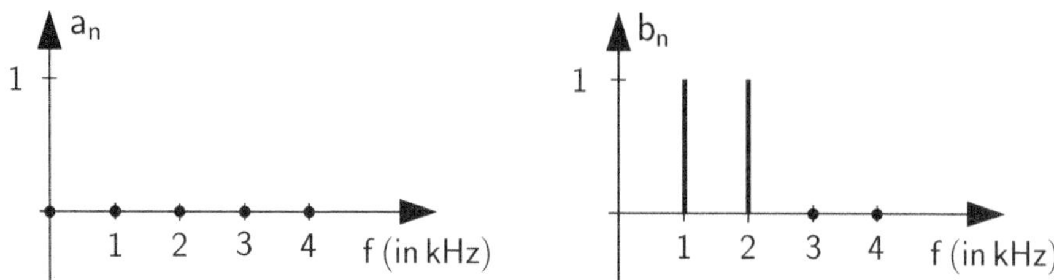

Fig. 1.9 Fourier coefficients of the periodic time signal shown in Fig. 1.5

Since the periodic time signal can theoretically be recalculated without error from the discrete values of the Fourier coefficients via Fourier synthesis, all the information of the periodic signal must be contained in the Fourier coefficients; the properties of a periodic time signal are expressed by the Fourier coefficients: Thus, the spectrum of a periodic time signal only has discrete values in the frequency range and is therefore referred to as a *line spectrum*.

Example 1.4 (Fourier coefficients of a periodic time signal with DC component) Now, continuing from the previous example, the DC-free periodic sum signal shown in Fig. 1.5 is extended by a DC component in the form

$$u(t) = U_0 + U_0 \sin(\omega_1 t) + U_0 \sin(2\,\omega_1 t) \tag{1.32}$$

with the fundamental frequency $f_1 = 1\,\text{kHz}$, the frequency $f_2 = 2f_1 = 2\,\text{kHz}$ and $U_0 = 1\,\text{V}$, the line spectrum shown in Fig. 1.9 is created. The Fourier coefficients different from zero are $b_1 = U_0 = 1\,\text{V}$ and $b_2 = U_0 = 1\,\text{V}$ and for the DC component, we get $a_0/2 = U_0 = 1\,\text{V}$. Since the DC component is frequency-independent, an additional spectral line is now created at $f = 0\,\text{Hz}$, which becomes visible in the representation of the a_n (see Fig. 1.10). □

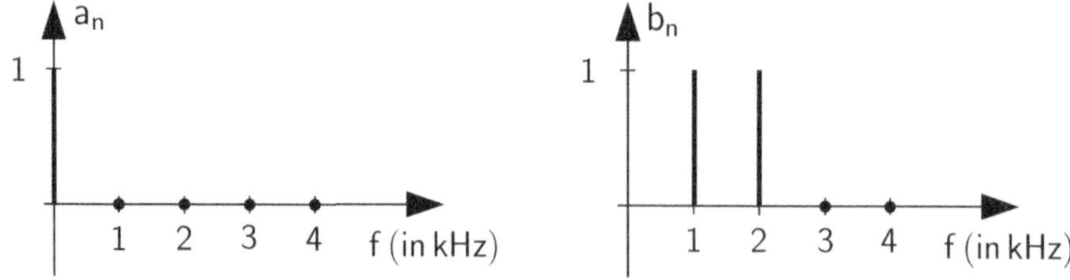

Fig. 1.10 Fourier coefficients of the periodic time signal shown in Fig. 1.5 with additional DC component

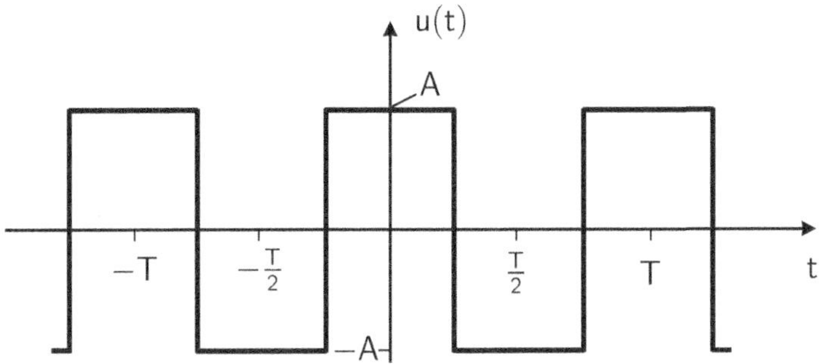

Fig. 1.11 Rectangle pulse sequence

Example 1.5 (Fourier coefficients of a periodic rectangular pulse sequence) The Fourier analysis of the DC-free (i.e. $a_0/2 = 0$) and even (i.e. $u(t) = u(-t)$) signal shown in Fig. 1.11 leads to the following course for the Fourier coefficients[4]

$$a_n = 2A\,\mathrm{si}\!\left(n\frac{\pi}{2}\right) \quad n = 1, 2, 3, \ldots \tag{1.33}$$

$$b_n = 0 \qquad\qquad n = 1, 2, 3, \ldots. \tag{1.34}$$

Insert: Derivation of the Fourier Coefficients of the Periodic Rectangular Pulse Sequence

The Fourier coefficients of the periodic rectangular pulse sequence shown in Fig. 1.11 are calculated step by step in the following.

Using the relationship

$$\frac{a_0}{2} = \frac{1}{T}\int_{t_0}^{t_0+T} x(t)\,\mathrm{d}t \tag{1.35}$$

the coefficient

$$\frac{a_0}{2} = 0, \tag{1.36}$$

is calculated over a whole period T, which can be arbitrarily located on the time axis: The total area under this curve is always zero, as positive and negative parts

---

[4] The sinc function is defined as $\mathrm{si}(x) = \sin(x)/x$.

compensate each other. The signal $u(t)$ is DC-free, which also appears plausible from the illustration (Fig. 1.11).

The coefficients a_n are determined by the relationship

$$a_n = \frac{2}{T} \int\limits_{t_0}^{t_0+T} x(t)\,\cos(n\,\omega_0\,t)\,\mathrm{d}t \tag{1.37}$$

and initially one obtains

$$a_n = \frac{2}{T} \int\limits_{-T/4}^{+T/4} A\,\cos(n\,\omega_0\,t)\,\mathrm{d}t - \frac{2}{T} \int\limits_{T/4}^{3/4\,T} A\,\cos(n\,\omega_0\,t)\,\mathrm{d}t$$

$$= \frac{2A}{Tn\,\omega_0}\,\sin(n\,\omega_0\,t)\,\Big|_{-T/4}^{+T/4} - \frac{2A}{Tn\,\omega_0}\,\sin(n\,\omega_0\,t)\,\Big|_{T/4}^{3/4\,T}$$

$$= \frac{2A}{Tn\,\omega_0}\left[\underbrace{\sin\left(n\,\omega_0\,\frac{T}{4}\right) - \sin\left(n\,\omega_0\left(-\frac{T}{4}\right)\right)}_{=\,\sin\left(n\,\omega_0\,\frac{T}{4}\right),\ \text{since odd function}} + \dots \right.$$

$$\left. \dots - \sin\left(n\,\omega_0\,\frac{3T}{4}\right) + \sin\left(n\,\omega_0\,\frac{T}{4}\right)\right]$$

$$= \frac{2A}{Tn\,\omega_0}\left[3\sin\left(n\,\omega_0\,\frac{T}{4}\right) - \sin\left(n\,\omega_0\,\frac{3T}{4}\right)\right].$$

If simplified according to

$$\omega_0\,T = \frac{2\pi}{T}\,T = 2\pi \tag{1.38}$$

it results in

$$n\,\omega_0\,\frac{T}{4} = n\cdot\frac{2\pi}{T}\cdot\frac{T}{4} = n\cdot\frac{\pi}{2}$$

$$n\,\omega_0\,\frac{3T}{4} = n\cdot\frac{2\pi}{T}\cdot\frac{3T}{4} = n\cdot\frac{3\pi}{2}$$

and the Fourier coefficients result

$$a_n = \frac{A}{n\,\pi}\left[3\sin\left(n\,\frac{\pi}{2}\right) - \sin\left(n\,\frac{3\pi}{2}\right)\right].$$

n	1	2	3	4
a_n	$\frac{4A}{\pi}$	0	$-\frac{4A}{3\pi}$	0

$$\tag{1.39}$$

A further transformation results in

$$\sin\left(\frac{3}{2}\,n\,\pi\right) = \sin\left(\frac{3}{2}\,n\,\pi \pm n \cdot 2\pi\right) = \sin\left(-n\frac{\pi}{2}\right) = -\sin\left(n\frac{\pi}{2}\right) \tag{1.40}$$

so that

$$\begin{aligned}
a_n &= \frac{A}{n\,\pi}\left[3\sin\left(n\frac{\pi}{2}\right) - \sin\left(n\frac{3\pi}{2}\right)\right] \\
&= \frac{A}{n\,\pi}\left[3\sin\left(n\frac{\pi}{2}\right) - \left(-\sin\left(n\frac{\pi}{2}\right)\right)\right] \\
&= \frac{4A}{n\,\pi}\sin\left(n\frac{\pi}{2}\right) = \frac{2A}{n\frac{\pi}{2}}\sin\left(n\frac{\pi}{2}\right)
\end{aligned}$$

follows.

Thus, with $\mathrm{si}(x) = \sin(x)/x$ one obtains the representation given above

$$a_n = 2A\,\mathrm{si}\left(n\frac{\pi}{2}\right). \tag{1.41}$$

The coefficients b_n are calculated by the relationship

$$b_n = \frac{2}{T}\int_{t_0}^{t_0+T} x(t)\,\sin(n\,\omega_0\,t)\,\mathrm{d}t \tag{1.42}$$

and since integration is again carried out over whole periods T that can be arbitrarily placed in time, one obtains

$$b_n = 0. \tag{1.43}$$

This can also be explained by the fact that the b_n represent the odd components of a function $u(t)$, but the given periodic rectangular pulse sequence is an even function.

The progression of the Fourier coefficients is illustrated in Fig. 1.12. The following applies:

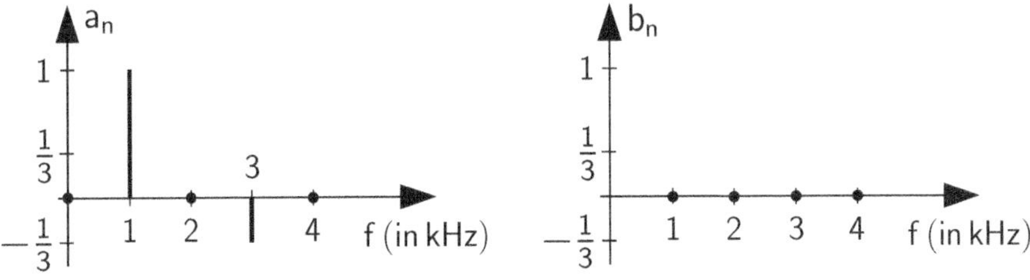

Fig. 1.12 Fourier coefficients (excerpt) of the time signal shown in Fig. 1.11 ($A = \pi/4\,\mathrm{V}$, $f_0 = 1/T = 1\,\mathrm{kHz}$)

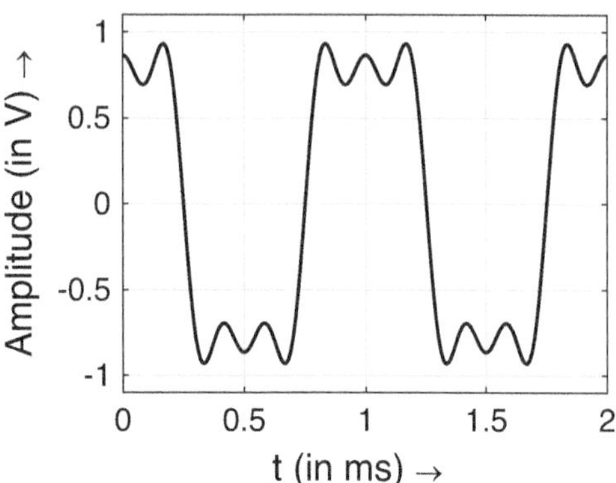

Fig. 1.13 Reconstructed time signal using the fundamental oscillation, the third and fifth harmonics ($A = \pi/4\,\mathrm{V}$, $T = 1/f_0 = 1\,\mathrm{ms}$)

$$u(t) = 2A \sum_{n=1}^{\infty} \mathrm{si}\left(n\,\frac{\pi}{2}\right) \cos(2\pi\,(nf_0)\,t). \tag{1.44}$$

It should be noted at this point that the representation according to (1.44) is not needed for the analysis of signal and system theoretical basics in many cases. Often, a representation corresponding to the time signal according to Fig. 1.12 is sought.

The reconstructed time signal using the fundamental oscillation, the third and fifth harmonics is shown in Fig. 1.13. □

The basis for the Fourier series is the approximation according to the equation

$$\frac{1}{T} \int_{t_0}^{t_0+T} (u(t) - u_\mathrm{R}(t))^2 \, \mathrm{d}t \to \mathrm{Min}. \tag{1.45}$$

with the minimization of the squares of the errors. It turns out that the Fourier series is poorly suited for approximating signals with jump or discontinuity points. At the points of discontinuity, an overshoot is noticeable, which does not decrease even when a higher number of Fourier coefficients are included: This property is referred to as *Gibbs's phenomenon*.[5] Figure 1.14 illustrates the Gibbs's phenomenon using the example of the rectangular pulse sequence.

Example 1.6 (Magnitude and phase spectrum of a rectangular pulse sequence) The following table reflects the determined values for the real Fourier coefficients a_n and b_n of the analyzed rectangular pulse sequence according to Fig. 1.11. □

n	1	2	3	4
a_n	$\frac{4A}{\pi}$	0	$-\frac{4A}{3\pi}$	0

---

[5] Josiah Willard Gibbs (1839–1903), American mathematician, physicist, and chemist

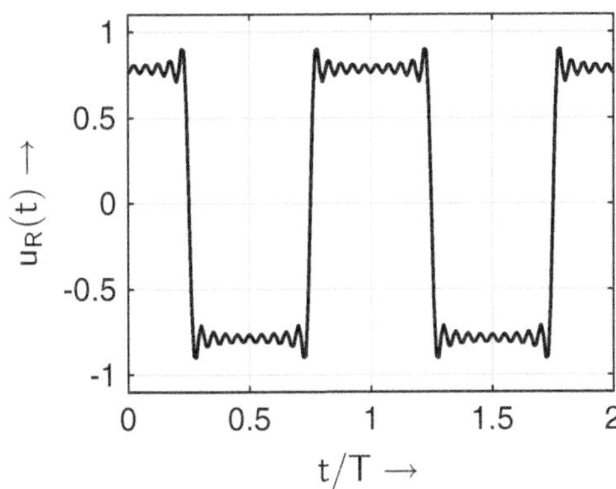

Fig. 1.14 To illustrate the Gibbs's phenomenon

With these, the rectangular pulse sequence can be represented as

$$u_R(t) = \frac{4A}{\pi} \cos(\omega_0 t) - \frac{4A}{3\pi} \cos(3\,\omega_0 t) + \dots \tag{1.46}$$

With the definition for the magnitude and phase spectrum in the form

$$c_n = \sqrt{a_n^2 + b_n^2} \qquad \text{and} \qquad \varphi_n = \arctan\left(\frac{b_n}{a_n}\right) \tag{1.47}$$

the magnitude and phase spectrum of the periodic rectangular pulse sequence shown in the following table is obtained. The value φ_n shows in the quadrants where the point (a_n, b_n) is located.

n	1	2	3	4
c_n	$\frac{4A}{\pi}$	0	$\frac{4A}{3\pi}$	0
φ_n	0	—	π	—

The table takes into account the fact that with a magnitude spectrum with the value $c_n = 0$ the value of the phase spectrum φ_n is not of interest. With

$$u_R(t) = \sum_{n=1}^{\infty} c_n \cos(n\,\omega_0\,t - \varphi_n) \tag{1.48}$$

the determined values for the magnitude and phase spectrum result in

$$u_R(t) = \frac{4A}{\pi} \cos(\omega_0\,t) + \frac{4A}{3\pi} \cos(3\,\omega_0\,t - \pi) + \dots \tag{1.49}$$

With $\cos(\alpha - \pi) = -\cos(\alpha)$ the result follows

$$u_R(t) = \frac{4A}{\pi} \cos(\omega_0\,t) - \frac{4A}{3\pi} \cos(3\,\omega_0\,t) + \dots \tag{1.50}$$

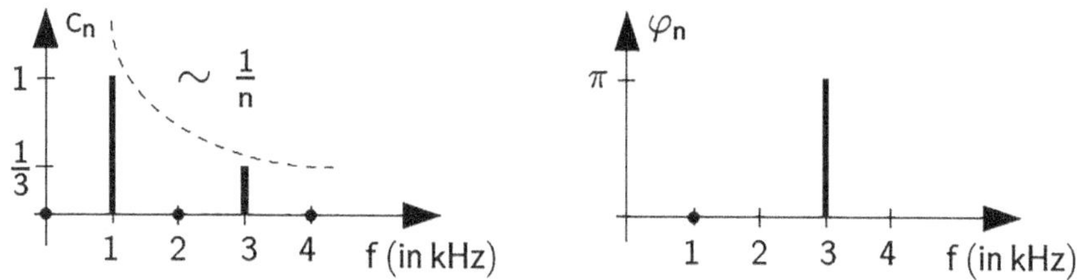

Fig. 1.15 Magnitude and phase spectrum of the rectangular pulse sequence shown in Fig. 1.11 ($A = \pi/4\,\text{V}$, $f_0 = 1/T = 1\,\text{kHz}$)

Figure 1.15 illustrates the resulting magnitude and phase spectrum of the rectangular pulse sequence shown in Fig. 1.11. $\square$

Characteristic Time Functions

Particularly important and useful is the familiarity with a series of non-periodic (aperiodic) signals, such as the rectangular pulse, the step function, and the ramp function.

Step Function The *step function* depicted in Fig. 1.16 can be analytically described as follows

$$u(t) = A \cdot 1(t). \tag{1.51}$$

The function $1(t)$ is referred to as the *unit step function*: It describes a jump of amplitude (jump height) 1 at the time $t = 0$ and is dimensionless. Since this function $1(t)$ is unitless, the step function $u(t)$ will take on the unit of the factor A. If the factor A describes a voltage value, the step function $u(t)$ has the unit Volt according to (1.51).

Ramp Function The *ramp function* depicted in Fig. 1.17 can be analytically described as follows

$$u(t) = \frac{A}{T_0} t \cdot 1(t) \tag{1.52}$$

and has the unit Volt.

Fig. 1.16 Step function

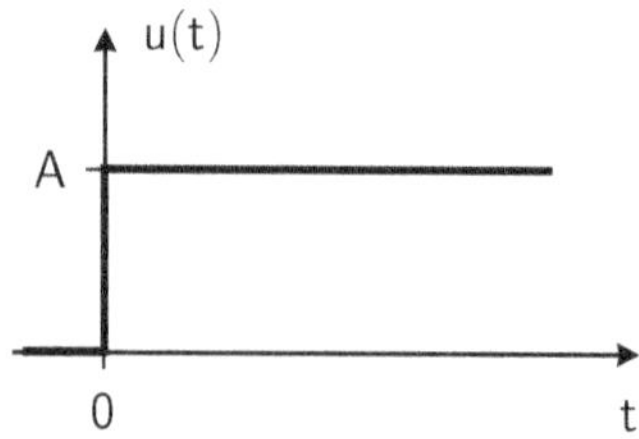

Fig. 1.17 Ramp function

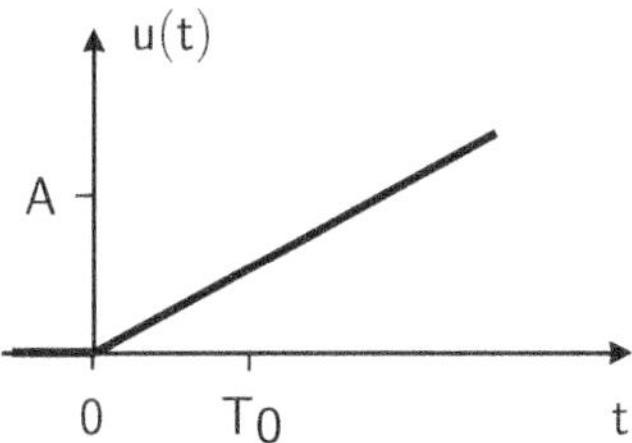

Rectangular Pulse   In addition to the step and ramp functions, the *rectangular pulse* has great importance in communications engineering (see Fig. 1.18), which can be derived from the knowledge of the step function to

$$u(t) = A \cdot 1(t) - A \cdot 1(t - T) \tag{1.53}$$

Dirac Impulse   The signal that arises when applying the limit transition $T \to 0$, i.e., reducing the pulse width T of the rectangular pulse $u(t)$ shown in Fig. 1.18 while keeping the pulse area $A\,T$ constant, for example, with $A = 1/T$ (see Fig. 1.19), is called a *Dirac impulse* or *Dirac delta function* (see Fig. 1.20):[6]

$$\lim_{T \to 0} u(t) = \delta(t). \tag{1.54}$$

The smaller the pulse duration T is chosen in Fig. 1.19, the larger the amplitude of the pulse must become in order for the pulse area $A\,T$ to remain constant.

The attempt to define the Dirac impulse leads to a function that is zero at all times $t \neq 0$ and grows beyond all limits at the time $t = 0$ [18, 30] (see Fig. 1.20), i.e.,

$$u(t) = \delta(t) = \begin{cases} \infty & t = 0 \\ 0 & \text{otherwise} \end{cases}. \tag{1.55}$$

Fig. 1.18 Rectangular pulse

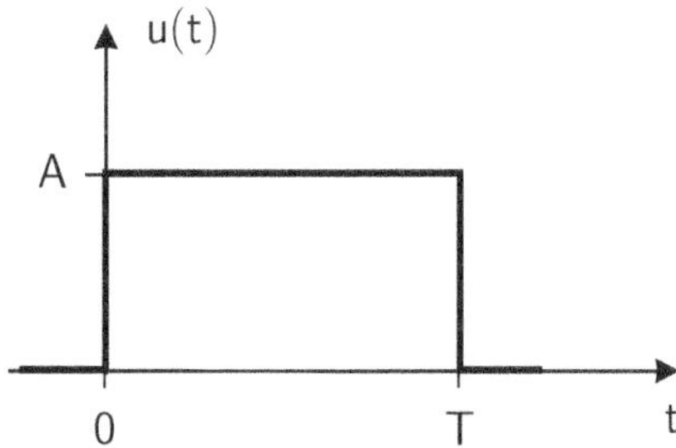

---

[6] Paul Adrien Maurice Dirac (1902–1984), British physicist.

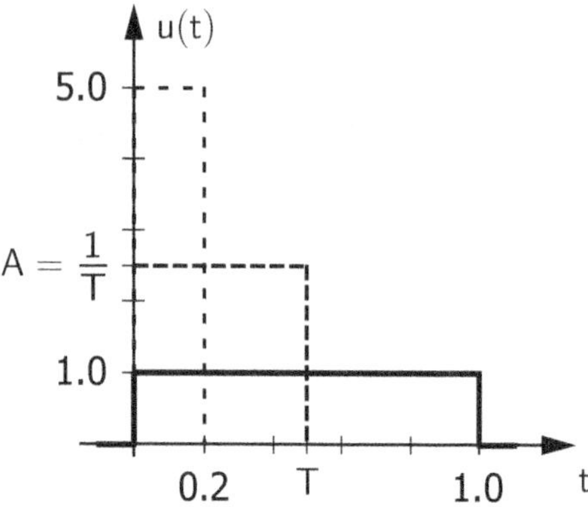

Fig. 1.19 Definition of the Dirac impulse via rectangular pulses of the same area

The Dirac impulse cannot be described by its immediate properties, but only by its effect on other functions, such as through the *selective property*. The relationship

$$\int_{-\infty}^{+\infty} u(t)\,\delta(t)\,\mathrm{d}t = u(0) \tag{1.56}$$

signals that the integral over the product of a function $u(t)$ with a Dirac impulse $\delta(t)$ suppresses—or masks—all function values of the function $u(t)$ for times $t \neq 0$ and only retains the value $u(0)$ of the function $u(t)$ at the point $t = 0$: In this case, the function value $u(0)$ is extracted from the time signal. However, if the function value at the point $t = t_0$ is to be extracted from a function $u(t)$, the following applies

$$\int_{-\infty}^{+\infty} u(t)\,\delta(t - t_0)\,\mathrm{d}t = u(t_0). \tag{1.57}$$

The Dirac impulse is usually represented by a vertical arrow at the time t_0 of its occurrence; this is illustrated in Fig 1.20 for the example $t_0 = 0$.

Energy and Power Signals Signals can also be classified according to their *energy* or their *power*. The energy of a real signal $u(t)$ is defined as

Fig. 1.20 Dirac impulse $\delta(t)$

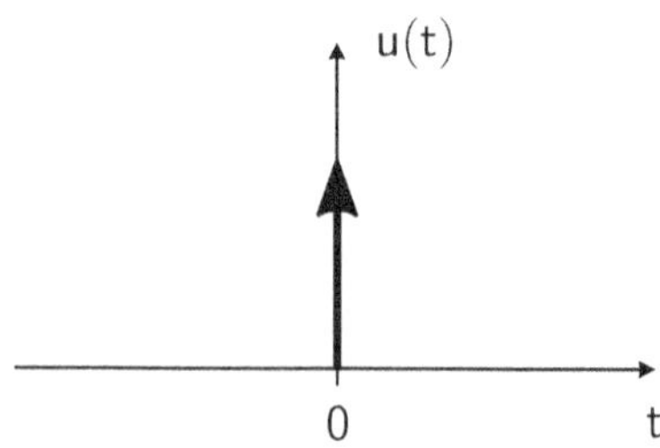

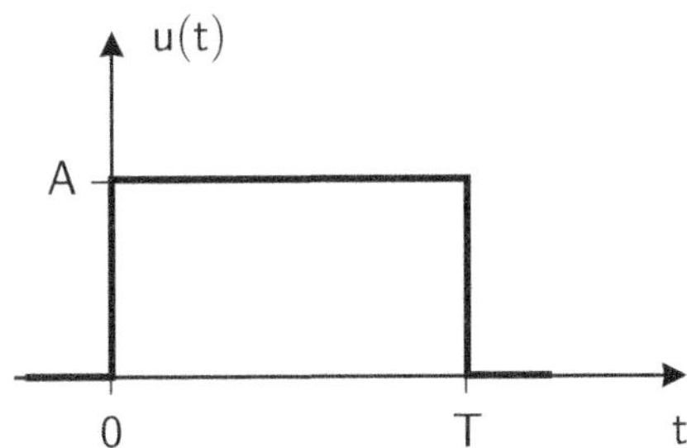

Fig. 1.21 The rectangular pulse as an example of an energy signal

$$E_s = \int\limits_{-\infty}^{+\infty} u^2(t)\, \mathrm{d}t. \tag{1.58}$$

A signal is referred to as an *energy signal* if $E_s < \infty$ is. The energy has the unit $\mathrm{V}^2\,\mathrm{s}$, if $u(t)$ is a voltage signal with the unit Volt. In this case, the signal $u(t)$ is referred to as energy-limited with the energy E_s. An example of an energy signal is the rectangular pulse (see Fig. 1.21).

The average *power* of a real signal $u(t)$ is defined as

$$P_s = \lim_{T \to \infty} \frac{1}{2T} \int\limits_{-T}^{+T} u^2(t)\, \mathrm{d}t \tag{1.59}$$

Signals whose power P_s is finite and not equal to zero are called *power signals*. The power is given in the unit V^2 when $u(t)$ is a voltage signal with the unit Volt. An example of a power signal is a harmonic sinusoidal oscillation (see Fig. 1.22 with a sinusoidal oscillation as an example of a harmonic oscillation). All physically interpretable signals have finite power; it applies $P_s < \infty$.

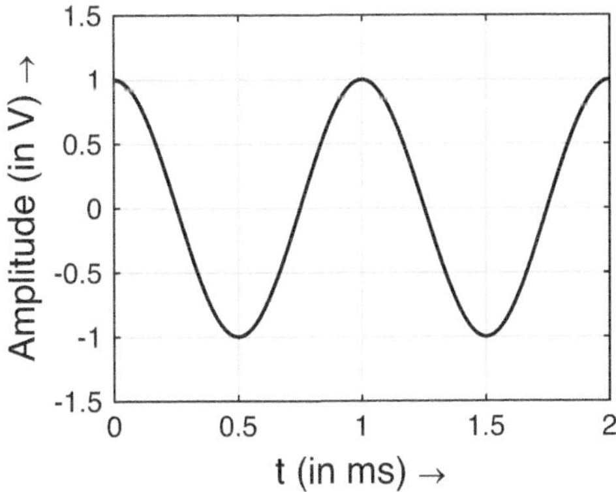

Fig. 1.22 Harmonic oscillation as an example of a power signal

Physical Power and Signal or System Theoretical Power It remains to be noted at this point that the expression power in signal and system theory is usually not used in the physical sense. Instead, the square of an amplitude value (in Volt) is referred to as signal or system theoretical power (in Volt^2). In this case, at a real resistance of $1\,\Omega$, the signal theoretical power is numerically—but not dimensionally—identical to the physical power (measured in Watt) (see also e.g. [30, 54]).

The advantage of such an approach is that it allows one to describe signals and, for example, examine and optimize transmission systems with signal and system theoretical means, regardless of a specific technology, which is each associated with a special resistance R (e.g. the characteristic impedance of a certain type of line).

The previous considerations for the analysis of periodic signals focused exclusively on positive frequencies. However, the generalized Fourier analysis, introduced by the mathematician Norbert Wiener, also allows the inclusion of negative frequencies. The following can be stated:

- technically meaningful (and measurable): positive frequencies f, f_0,
- theoretical approach: also negative frequencies $-f$, $-f_0$ are allowed.

Thus, a cosine oscillation always includes two frequencies, the positive at f_0 and the negative at $-f_0$. This fact can be used for a generalized frequency analysis and leads to a Fourier series representation in complex form, see Appendix G.

Fourier Transformation

In principle, all deterministic and stochastic signals can be equally described in a

- Time or original domain (independent variable t) or in a
- Spectral or image domain (independent variable f, ω)

Computational rules for the forward transformation, i.e., the transition from the original domain to the spectral domain (marked by the symbol $\circ\!\!-\!\!\bullet$), and the inverse transformation, i.e., the transition from the image domain to the original domain (marked by the symbol $\bullet\!\!-\!\!\circ$), are realized by functional transformations [16, 30].

While the Fourier series is limited to periodic time signals, the *Fourier Transformation* offers the possibility to examine aperiodic signals for their *frequency* composition.

Definition The Fourier transformation of an aperiodic time signal $u(t)$ provides its spectrum $U(f)$

$$U(f) = \mathcal{F}\{u(t)\} = \int\limits_{-\infty}^{+\infty} u(t)\, e^{-j2\pi f t}\, dt, \tag{1.60}$$

which is generally complex-valued. The inverse Fourier transformation is defined over

$$u(t) = \mathcal{F}^{-1}\{U(f)\} = \int_{-\infty}^{+\infty} U(f)\, e^{j2\pi ft}\, df. \tag{1.61}$$

$U(f)$ and $u(t)$ form a so-called *transformation pair.* $U(f)$ is the Fourier transform of $u(t)$ and $u(t)$ is in turn the inverse Fourier transform or the Fourier inverse transform of $U(f)$. In shorthand, this is clarified by the transformation symbol:

$$u(t) \quad \circ\!\!-\!\!\bullet \quad U(f). \tag{1.62}$$

Let $u(t)$ again be a one- or two-sided limited deterministic signal with

$$u(t)\ (\text{in V}), \tag{1.63}$$

i.e., $u(t)$ is a voltage. The Fourier transformation leads to the amplitude spectral density

$$U(f) = \int_{-\infty}^{+\infty} u(t)\, e^{-j2\pi ft}\, dt \quad \left(\text{in } \frac{\text{V}}{\text{Hz}} \text{ or Vs}\right). \tag{1.64}$$

and is measured in Volt per Hertz bandwidth. Here, Volt is the unit of the temporal voltage course $u(t)$ and the multiplicatively added unit s $= 1/\text{Hz}$ originates from the dt of the Fourier integral.

If the period of a periodic function is extended to infinity, while maintaining the shape of the signal, the Fourier series becomes the Fourier integral.

The spectral representations consist of discrete lines (line spectrum, see Fourier series representation) for periodic time functions and are continuous for one-time (aperiodic) processes.

The Fourier series provides a tool that decomposes periodic signals into their frequency components. Here it is shown that the period duration of the time signal T determines the distance of the spectral lines in the frequency range via the approach

$$f_0 = \frac{1}{T} \tag{1.65}$$

Now, if the pulse width is kept constant for a given periodic signal (see Fig. 1.23) and the period duration T is increased (theoretically to infinity), the result is that the distance of the spectral lines must decrease, specifically in the limit transition for $T \to \infty$ to $f_0 \to 0$.[7] Thus, aperiodic signals have a continuous spectrum.

---

[7] In Fig. 1.23, T denotes the period duration of the periodic time signal and τ the pulse width. In the following, the common designation T is used for the pulse duration or the pulse width in the case of aperiodic time signals.

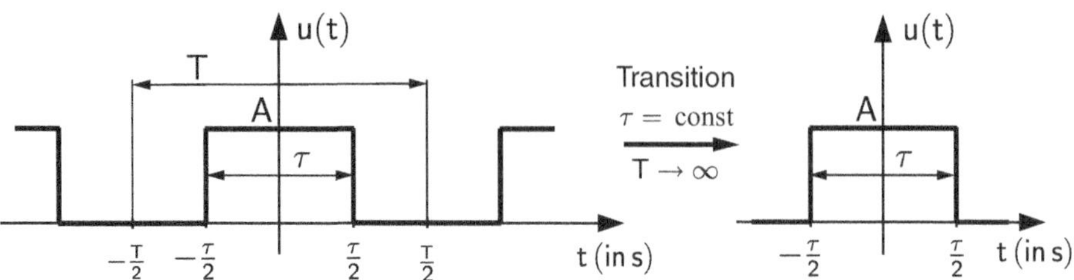

Fig. 1.23 Conversion of a rectangular pulse sequence (periodic) into a rectangular pulse (aperiodic)

The Fourier transform of the signal $u(t)$, i.e., $U(f)$, is also called the *spectrum* of a signal $u(t)$. This generally complex-valued function of the real variable f can be divided into a magnitude and phase spectrum for illustrative representation. The following applies

$$U(f) = |U(f)|\ e^{j\varphi(f)}, \tag{1.66}$$

with $|U(f)|$ as a description for the magnitude spectrum (also referred to as *amplitude response*) and $\varphi(f)$ as a description for the phase spectrum (also referred to as *phase response*).

Some important correspondences of the Fourier transformation for further considerations are compiled in Appendix C in Table C.1. There you will also find the transformation equations with the frequency variable ω, which can be used equally with the frequency variable f chosen here.

Application of the Fourier Transformation for Exemplary Signals In the following, the amplitude density spectra of some exemplary but typical time signals are calculated to demonstrate the application of the Fourier transformation and to show important resulting relationships.

Example 1.7 (amplitude spectral density of the rectangular pulse) As an example, the amplitude spectral density of the rectangular pulse according to Fig. 1.24 is analyzed here. Using the approach

$$U(f) = \int_{-T/2}^{+T/2} A\ e^{-j2\pi f t}\ dt \tag{1.67}$$

the amplitude spectral density $U(f)$ of the rectangular pulse is obtained as

$$U(f) = \frac{A}{\pi f}\ \sin(\pi f T). \tag{1.68}$$

The integration limits of $\pm T/2$ in (1.67) take into account that $u(t)$ is identically zero outside the interval from $-T/2$ to $T/2$ (see Fig. 1.24).

Fig. 1.24 Rectangular pulse

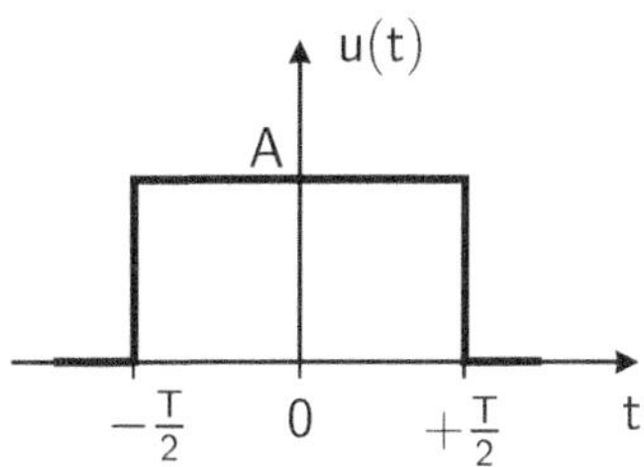

Fig. 1.25 Sinc function
$\mathrm{si}(x) = \sin{(x)}/x$

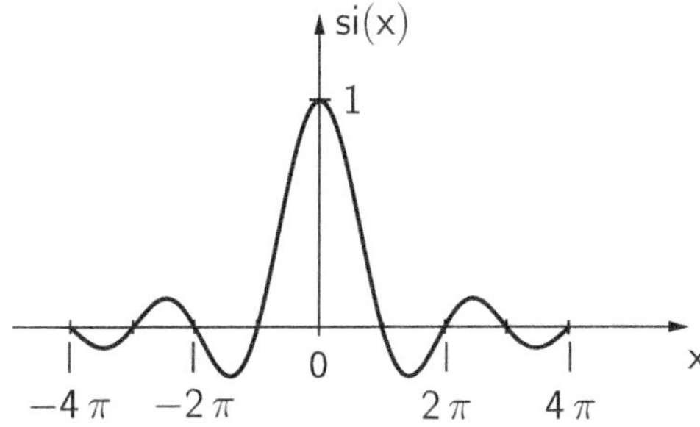

The sinc function is a mathematical function defined as follows:

$$\mathrm{si}(ax) = \frac{\sin(ax)}{ax}. \tag{1.69}$$

Using L'Hospital's rule [11, 74], it follows that $\mathrm{si}(0) = 1$. Fig. 1.25 illustrates the course of the sinc function.

If the numerator and denominator of the factor in (1.68) are each expanded by T, one can write with (1.69) for the spectral function of the rectangular pulse, i.e. the amplitude spectral density, as follows

$$U(f) = A\,T\,\mathrm{si}(\pi f T). \tag{1.70}$$

Insert: Derivation of the Spectrum of the Rectangular Signal
The Fourier transformation for calculating the spectrum of the rectangular signal is presented in detail below to clarify the calculation path:

$$U(f) = \mathcal{F}\{u(t)\} = \int_{-\infty}^{+\infty} u(t)\, e^{-j2\pi f t}\, dt = \int_{-T/2}^{+T/2} A\, e^{-j2\pi f t}\, dt$$

$$U(f) = A \int_{-T/2}^{+T/2} (\cos(2\pi f t) - j\sin(2\pi f t))\, dt$$

$$U(f) = A \int_{-T/2}^{+T/2} \cos(2\pi f t)\, dt - A j \underbrace{\int_{-T/2}^{+T/2} \sin(2\pi f t)\, dt}_{=\,0,\,\text{since odd function}}$$

$$U(f) = A \int_{-T/2}^{+T/2} \cos(2\pi f t)\, dt = 2 \cdot A \int_{0}^{T/2} \cos(2\pi f t)\, dt.$$

Using (see Appendix F.2 or e.g. [11, 74])

$$\int \cos ax\, dx = \frac{1}{a}\sin ax + C \tag{1.71}$$

the result of the integration is obtained as

$$U(f) = \frac{2A}{2\pi f}\left[\sin(2\pi f t)\right]_{0}^{T/2} \tag{1.72}$$

and after considering the integral limits

$$U(f) = \frac{A}{\pi f}\sin\left(2\pi f\,\frac{T}{2}\right). \tag{1.73}$$

Expanding the prefactor with T leads to

$$U(f) = \frac{AT}{\pi f T}\sin(\pi f T) \tag{1.74}$$

and with the definition of the sinc function

$$\mathrm{si}(ax) = \frac{\sin(ax)}{ax} \tag{1.75}$$

finally, for the spectral amplitude density of a single rectangle pulse

$$U(f) = A T\,\mathrm{si}(\pi f T). \tag{1.76}$$

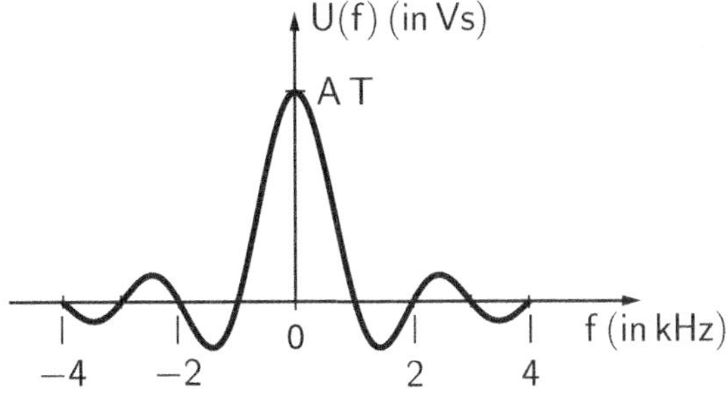

Fig. 1.26 Amplitude spectral density of the rectangular pulse ($T = 1$ ms and $A = 1$ V)

The resulting curve for $U(f)$ is shown in Fig. 1.26 and has the following properties:

- The maximum of $U(f)$ is at the frequency $f = 0$ and has the value $A T$ (area of the rectangle pulse in the time domain).
- The function $U(f)$ of the amplitude density spectrum has zeros at the frequencies $f_n = n/T$, with $n = \pm 1, \pm 2, \pm 3, \ldots$, i.e. $U(f = f_n) = 0$.

In the considered example case with a pulse duration of $T = 1$ ms, the spectrum has zeros at multiples of 1 kHz. $\square$

When using the Fourier transformation, one often tries to rely as much as possible on known correspondences to avoid the often complex evaluation of the Fourier integral. In doing so, one often encounters time-shifted signals, as shown in Fig. 1.27. Such time-shifted signals can be transformed into the image or spectral domain using the time shift theorem of the Fourier transformation, which will be illustrated in the following.

The Fourier transformation for calculating the spectrum of a time-shifted rectangular signal $u_1(t)$ is presented in detail below to clarify the calculation process and to establish connections to the time shift theorem of the Fourier transformation. The Fourier transformation of the time signal $u_1(t)$ leads to the spectrum

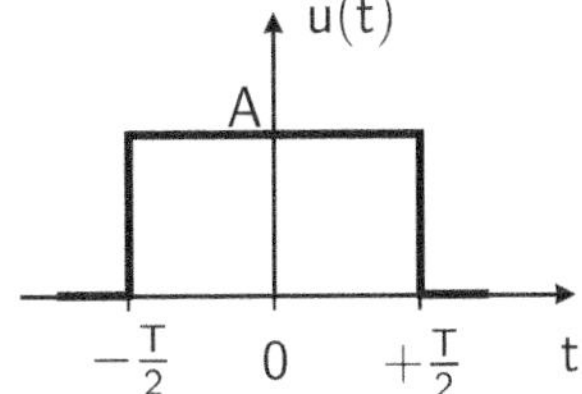

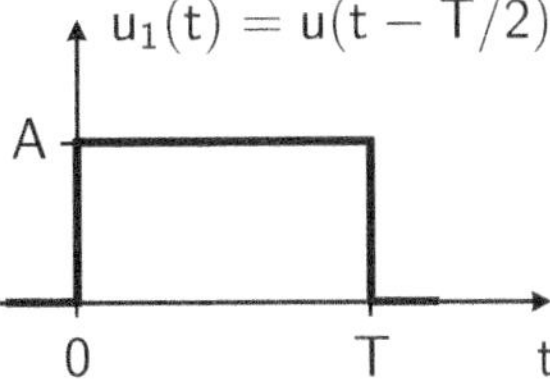

Fig. 1.27 Rectangular pulse and shifted rectangular pulse

$$U_1(f) = \mathcal{F}\{u_1(t)\} = \int_{-\infty}^{+\infty} u_1(t)\, e^{-j2\pi f t}\, dt = \int_{0}^{T} A\, e^{-j2\pi f t}\, dt$$

$$U_1(f) = -\frac{A}{j2\pi f}\, e^{-j2\pi f t}\, \Big|_{0}^{T} = -\frac{A}{j2\pi f}\left(e^{-j2\pi f T} - 1\right)$$

$$U_1(f) = \frac{A}{j2\pi f}\left(1 - e^{-j2\pi f T}\right)$$

$$U_1(f) = \frac{A}{j2\pi f}\, e^{-j2\pi f \frac{T}{2}}\left(e^{j2\pi f \frac{T}{2}} - e^{-j2\pi f \frac{T}{2}}\right).$$

Here, with

$$\sin(x) = \frac{1}{2j}\left(e^{jx} - e^{-jx}\right) \qquad \text{and} \qquad 2\pi f\,\frac{T}{2} = \pi f T \tag{1.77}$$

the intermediate result

$$U_1(f) = \frac{A}{\pi f}\, e^{-j\pi f T}\, \sin(\pi f T). \tag{1.78}$$

is obtained. Furthermore, with $\mathrm{si}(x) = \sin(x)/x$ the representation

$$U_1(f) = A\,T\, e^{-j\pi f T}\, \mathrm{si}(\pi f T). \tag{1.79}$$

is obtained. It turns out that the time shift does not affect the magnitude spectrum but only causes a change in the phase spectrum. With

$$\left| e^{-j\pi f T} \right| = 1 \tag{1.80}$$

the magnitude spectrum of the time-shifted signal $u_1(t)$ follows

$$|U_1(f)| = A\,T\, |\mathrm{si}(\pi f T)| = |U(f)|, \tag{1.81}$$

which matches the magnitude spectrum of the non-shifted pulse $u(t)$ (see also (1.76)). This relationship is also known under the term *time shift theorem* of the Fourier transformation and can be formulated as

$$u(t) \quad \circ\!\!-\!\!\bullet \quad U(f) \tag{1.82}$$

$$u(t - t_o) \quad \circ\!\!-\!\!\bullet \quad U(f) \cdot e^{-j2\pi f t_o} \tag{1.83}$$

(see also Appendix C).

To come as close as possible to physical reality, all determined signals in this book are understood as voltages $u(t)$ (in V). Through the Fourier transformation, the amplitude spectral density

$$U(f) = \int_{-\infty}^{+\infty} u(t)\, e^{-j2\pi ft}\, dt \quad \left(\text{in } \frac{V}{Hz} \text{ or } Vs \right), \tag{1.84}$$

measured in Volt per Hertz bandwidth is obtained.

Two other important characteristic functions are the time and frequency impulse. The function

$$u(t) = A \cdot \delta(t) \tag{1.85}$$

is referred to as a Dirac impulse in the time domain. Here, A is the area of the impulse or the impulse integral [31]. With the definition

$$\int_{-\infty}^{+\infty} \delta(t)\, dt = 1 \quad \longrightarrow \quad \delta(t) \quad \left(\text{in } \frac{1}{s} \text{ or } Hz \right) \tag{1.86}$$

it follows that the time impulse has the unit $(1/s)$. The value A—interpretable as the area of the impulse function—must then have the unit V s or V/Hz so that $u(t)$ in (1.85) has the required dimension of a voltage (in V).

Example 1.8 (amplitude spectral density of the Dirac pulse in the time domain) The spectrum of an exemplary time pulse of

$$u(t) = A \cdot \delta(t) = 1\,\frac{V}{Hz} \cdot \delta(t) \tag{1.87}$$

is thus a constant ($A = 1$ V/Hz) in the frequency domain (see Fig. 1.28) of

$$U(f) = A = 1\,\frac{V}{Hz}. \tag{1.88}$$

$\square$

The function

$$U(f) = B \cdot \delta(f) \tag{1.89}$$

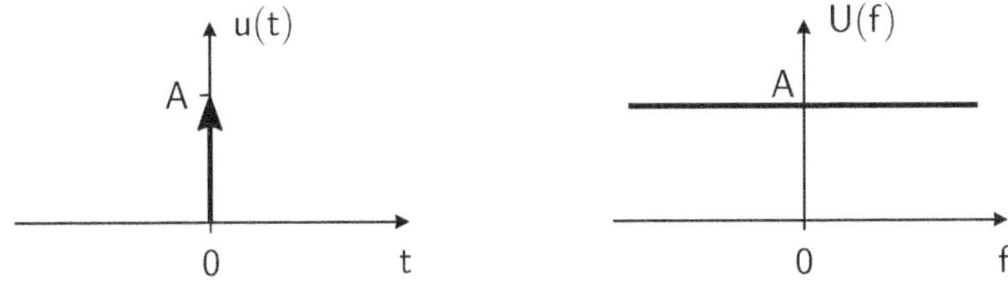

Fig. 1.28 Time pulse with associated description in the image domain

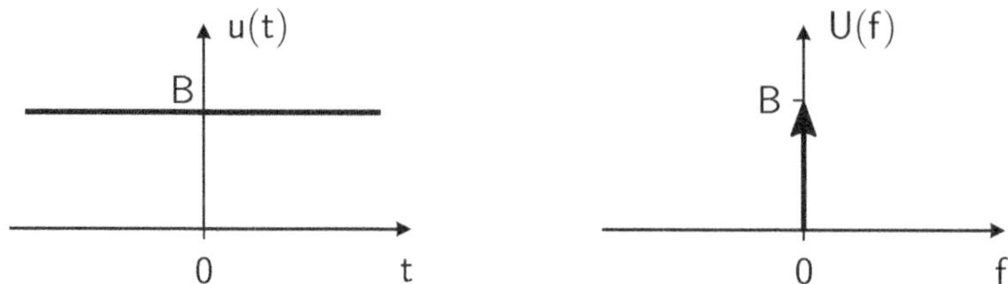

Fig. 1.29 Pulse in the frequency domain with associated description in the time domain

is referred to as a pulse in the frequency domain (analogous to the time pulse). Here, B is the area of the pulse (the pulse integral [31]). With the definition

$$\int_{-\infty}^{+\infty} \delta(f)\,\mathrm{d}f = 1 \quad \longrightarrow \quad \delta(f) \quad \left(\text{in } \frac{1}{\text{Hz}} \text{ or s}\right) \tag{1.90}$$

it follows that the frequency pulse has the unit (1/Hz). The value B (interpretable as the area of the pulse in the frequency domain) must then have the unit $\text{V} = \text{V/Hz} \cdot \text{Hz}$ so that $U(f)$ in (1.89) has the required dimension of an amplitude density (V/Hz).

Example 1.9 (amplitude spectral density of a direct current) The spectrum of an exemplary DC signal (direct current) of

$$u(t) = B = 1\,\text{V} \tag{1.91}$$

is a Dirac impulse in the frequency domain (see Fig. 1.29):

$$U(f) = B \cdot \delta(f) = 1\,\text{V} \cdot \delta(f). \tag{1.92}$$

The amplitude spectral density has the unit $\text{V/Hz} = \text{Vs}$. Thus, a DC signal leads to a spectral component at $f = 0\,\text{Hz}$. $\square$

Example 1.10 (Function values at zero and areas) From the Fourier transformation (1.60), the value of the amplitude density spectrum $U(f)$ at the point $f = 0$

$$U(f = 0) = \int_{-\infty}^{+\infty} u(t)\,\mathrm{e}^{-\mathrm{j}2\pi \cdot 0 \cdot t}\,\mathrm{d}t = \int_{-\infty}^{+\infty} u(t)\,\mathrm{d}t. \tag{1.93}$$

Thus, the value at the point $f = 0$ in the frequency domain corresponds to the area $\int_{-\infty}^{+\infty} u(t)\,\mathrm{d}t$ of the signal in the time domain (see Fig. 1.30).

Similarly, the value of the signal at $t = 0$ in the time domain can be determined via the inverse Fourier transformation (1.61) as

$$u(t = 0) = \int_{-\infty}^{+\infty} U(f)\,\mathrm{e}^{\mathrm{j}2\pi f \cdot 0}\,\mathrm{d}f = \int_{-\infty}^{+\infty} U(f)\,\mathrm{d}f \tag{1.94}$$

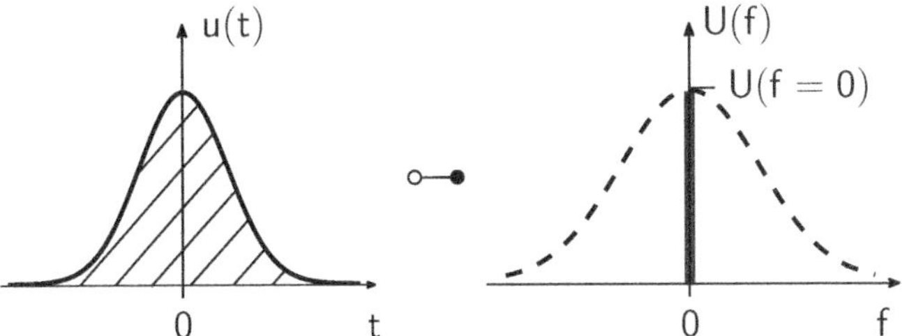

Fig. 1.30 On the equality of the value at $f = 0$ and the area of the time signal

Thus, the value of a time signal at the point $t = 0$ corresponds to the area $\int_{-\infty}^{+\infty} U(f)\,\mathrm{d}f$ of the amplitude density spectrum in the frequency domain (see Fig. 1.31).

In general, it can be stated that the value at zero in one domain always corresponds to the area under the curve in the other domain. Knowledge of these relationships can be helpful for many calculations and estimates. $\square$

Amplitude Density Spectra of Even and Odd Signals So far, in the analysis of the frequency composition of time signals, it has already become clear that even time signals have a real spectrum (see amplitude density spectrum of the symmetric rectangular pulse). Subsequent considerations are intended to further specify and generalize this fact.

Basically, all signals $u(t)$ can be decomposed into an even component $u_g(t)$ and an odd component $u_u(t)$ [45]. The even component can be calculated using

$$u_g(t) = \frac{1}{2}(u(t) + u(-t)) \tag{1.95}$$

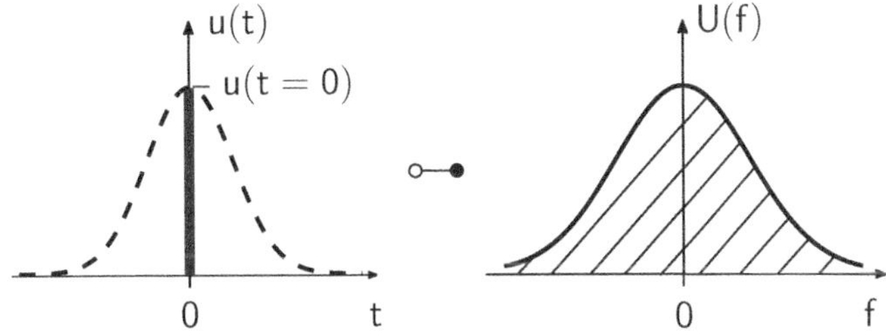

Fig. 1.31 On the equality of the value at $t = 0$ and the area of the amplitude density spectrum

Fig. 1.32 Time signal $u(t)$

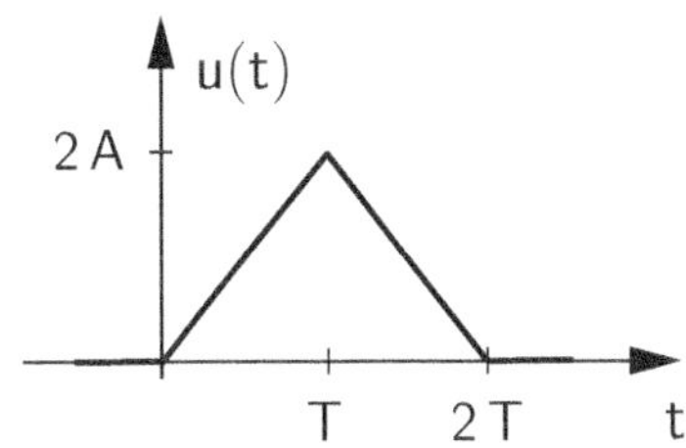

Fig. 1.33 Time signal $u(-t)$

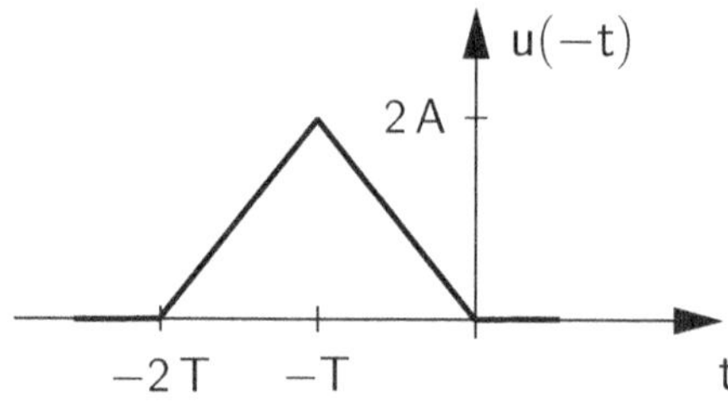

and for the odd component, the following applies

$$u_{\mathrm{u}}(t) = \frac{1}{2}(u(t) - u(-t)) \tag{1.96}$$

Furthermore, $u_{\mathrm{g}}(t) = u_{\mathrm{g}}(-t)$ and $u_{\mathrm{u}}(t) = -u_{\mathrm{u}}(-t)$ apply. It can be easily verified that the additive combination of $u_{\mathrm{g}}(t)$ and $u_{\mathrm{u}}(t)$ leads back to the time signal $u(t)$, since

$$u(t) = u_{\mathrm{g}}(t) + u_{\mathrm{u}}(t). \tag{1.97}$$

applies. Subsequently, for the time signal shown in Fig. 1.32, a decomposition into its even component $u_{\mathrm{g}}(t)$ and its odd component $u_{\mathrm{u}}(t)$ is sought. The decomposition of the time signal $u(t)$ into its even component $u_{\mathrm{g}}(t)$ and its odd component $u_{\mathrm{u}}(t)$ requires the

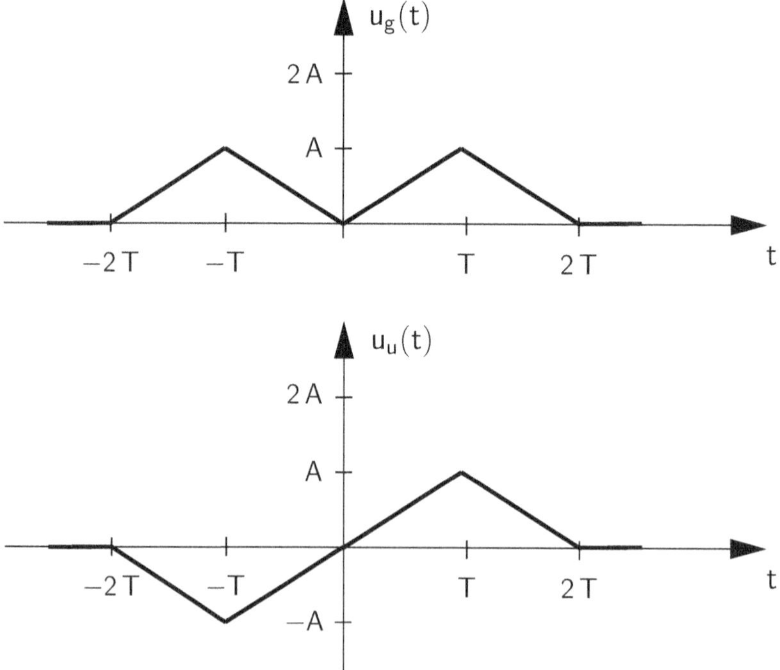

Fig. 1.34 Decomposition of $u(t)$ into its even component $u_{\mathrm{g}}(t)$ and its odd component $u_{\mathrm{u}}(t)$

determination of the signal $u(-t)$, which corresponds to a reflection on the y-axis (see also Fig. 1.33) [45].

The result of the decomposition of $u(t)$ into an even component $u_g(t)$ and an odd component $u_u(t)$ is illustrated in Fig. 1.34. The property of the time signal being *even* or *odd* also affects the spectrum, which is linked to the time signal via the Fourier transformation. It holds that

$$U(f) = \mathcal{F}\{u(t)\} = \int_{-\infty}^{+\infty} u(t)\, e^{-j2\pi f t}\, dt. \tag{1.98}$$

With $u(t) = u_g(t) + u_u(t)$ it follows first for the Fourier transformation of $u_g(t)$

$$U_g(f) = \int_{-\infty}^{+\infty} u_g(t)\, e^{-j2\pi f t}\, dt, \tag{1.99}$$

which can be written as

$$U_g(f) = \int_{-\infty}^{+\infty} u_g(t)\, \cos(2\pi f t)\, dt - j \int_{-\infty}^{+\infty} u_g(t)\, \sin(2\pi f t)\, dt. \tag{1.100}$$

The integral

$$\int_{-\infty}^{+\infty} u_g(t)\, \sin(2\pi f t)\, dt = 0 \tag{1.101}$$

gives the value zero, since here the integral of a product consisting of an even function and an odd function is calculated. As a result, for the spectrum of the even time signal $u_g(t)$ it can be written

$$U_g(f) = \int_{-\infty}^{+\infty} u_g(t)\, \cos(2\pi f t)\, dt. \tag{1.102}$$

It turns out that this spectrum is real. Accordingly, the spectrum of the odd time signal $u_u(t)$ can be formulated as

$$U_u(f) = \int_{-\infty}^{+\infty} u_u(t)\, e^{-j2\pi f t}\, dt \tag{1.103}$$

which can be simplified with the Euler relationship $e^{jx} = \cos x + j \sin x$ to

$$U_u(f) = \int_{-\infty}^{+\infty} u_u(t)\, \cos(2\pi f t)\, dt - j \int_{-\infty}^{+\infty} u_u(t)\, \sin(2\pi f t)\, dt. \tag{1.104}$$

The integral now results via

$$\int_{-\infty}^{+\infty} u_{\mathrm{u}}(t) \cos\left(2\,\pi f\,t\right) \mathrm{d}t = 0 \tag{1.105}$$

in the value zero, since here the integral of a product consisting of an odd function with an even function is calculated. As a result, the spectrum of the odd time signal $u_{\mathrm{u}}(t)$ can be written as

$$U_{\mathrm{u}}(f) = -\mathrm{j}\int_{-\infty}^{+\infty} u_{\mathrm{u}}(t) \sin\left(2\,\pi f\,t\right) \mathrm{d}t\,. \tag{1.106}$$

It turns out that this spectrum is imaginary.

With the help of the superposition theorem, the relationship

$$u(t) = u_{\mathrm{g}}(t) + u_{\mathrm{u}}(t) \quad \circ\!\!-\!\!\bullet \quad U(f) = U_{\mathrm{g}}(f) + U_{\mathrm{u}}(f) \tag{1.107}$$

can be derived as a result for the spectrum of the time signal $u(t)$.

1.2.2   Stochastic Signals

In fact, noise signals, which partly originate in the devices themselves or penetrate into the transmission paths (e.g., noise of resistors and semiconductors, interference from foreign transmitters, impulse interference from power supply systems or unwanted music), always occur in communications engineering in addition to these basic signals and generally the deterministic signals. Signals whose characteristic is not forecastable—and therefore not predictable or calculable—are referred to as non-deterministic, stochastic, statistical or random signals or random signals. An example of a stochastic signal $n(t)$ is given in Fig. 1.35.

Fig. 1.35 Representation of a stochastic signal

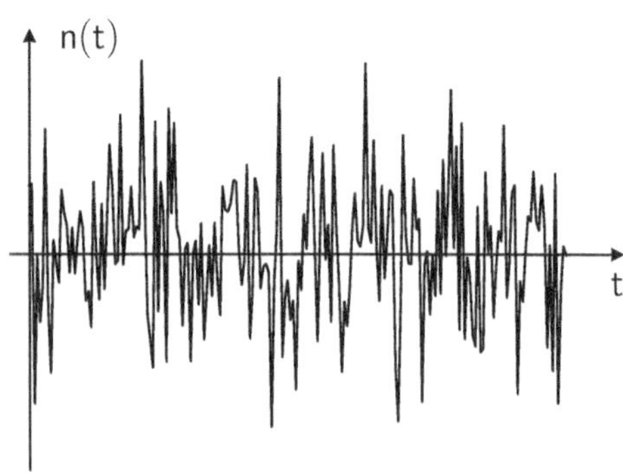

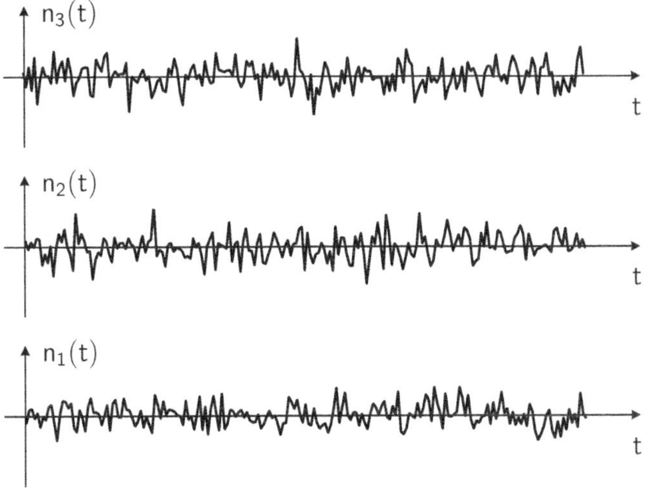

Fig. 1.36 Realizations of an ergodic random process

A process that generates random signals is called a random process. This can be imagined as any number of processes occurring simultaneously at time t (see Fig. 1.36).[8] The subsequent considerations in this book focus on *stationary, ergodic* processes, i.e., the statistical parameters describing the process should be independent of the choice of the point in time at which they are determined (stationarity) and a randomly selected realization or sample function of the process is sufficient to determine the statistical parameters (ergodicity). Here, stationarity is a prerequisite for ergodicity. Assuming that the realizations generated in Fig. 1.36 originate from an ergodic random process, a realization of the random process is sufficient to determine stochastic characteristics (such as mean or variance).

Even if the individual signal form cannot be described mathematically-analytically algebraically, deterministic characteristics can still be used to describe stochastic signals using probability theory and statistics (stochastics), such as the mean or variance. For real, ergodic signals, the linear mean (DC component, first-order moment) is calculated as

$$\overline{n(t)} = \lim_{T \to \infty} \frac{1}{2T} \int\limits_{-T}^{+T} n(t)\,\mathrm{d}t. \tag{1.108}$$

---

[8] The individual, simultaneous subprocesses are also referred to as *realizations* or *sample functions* of the stochastic process. In the chosen description, they are also voltages and can therefore be denoted as $u(t)$. To emphasize the random character of these sample functions, $n(t)$ was chosen as the mathematical notation (for *noise*), which is common in communication engineering literature. For the calculation of general characteristics, $u(t)$ can still be used as a (general) designation.

The square mean, i.e., the square of the effective value or the average power, is given by

$$U_R^2 = \overline{n^2(t)} = \lim_{T \to \infty} \frac{1}{2T} \int_{-T}^{+T} n^2(t)\, dt \tag{1.109}$$

and is given in V^2, provided that $n(t)$ appears in the form of a voltage (in V). This quantity (in V^2) is proportional to the physical power (in W) at a real and constant resistance (e.g., [29]).

Random variables and signals can be uniquely described by their *probability density-* or *distribution function.* If we denote the probability density of a random variable or a stochastic process $u(t)$ with $p(u)$, then the linear mean can be calculated as

$$\overline{u(t)} = m_1 = u_m = \int_{-\infty}^{+\infty} u\, p(u)\, du, \tag{1.110}$$

the quadratic mean as

$$\overline{u^2(t)} = m_2 = \int_{-\infty}^{+\infty} u^2\, p(u)\, du \tag{1.111}$$

and the variance as

$$\sigma^2 = \int_{-\infty}^{+\infty} (u - m_1)^2\, p(u)\, du = m_2 - m_1^2 \tag{1.112}$$

For zero-mean random signals ($m_1 = 0$), the variance σ^2 and the quadratic mean m_2 coincide.

Since the probability density function $p(u)$ is a deterministic function (and not a random function), the characteristics of linear mean, quadratic mean, and variance of the random time function $u(t)$ can be calculated with the given integral relationships (1.110), (1.111), and (1.112)—and they are therefore themselves deterministic quantities or functions.

It should be noted that the relationships in (1.108) and (1.109) are practically not or only approximately evaluable, since the time functions $n(t)$, over which integration is performed, are random functions. Nevertheless, the characteristics deterministic with them are determined quantities, as they can be calculated by other means as expected value (1.110) and quadratic mean (1.111) with the help of an integration over the determined function *probability density*, e.g., [74]. The relationships (1.108) and (1.109) can be used for an approximate experimental determination of the quantities.

Relationships between temporally adjacent values of stochastic signals can be expressed in the form of a measure for linear dependence, i.e., as a correlation function [29]. The temporal distance τ is the independent variable of the correlation function. In the case of the *autocorrelation function* (ACF), pairs of values from the same random function are considered. In contrast to the ACF, the *cross-correlation function* (CCF) represents a measure of the mutual dependence of two values from different functions. For a real, ergodic random process $n(t)$, the autocorrelation function is given by

$$\psi(\tau) = \lim_{T \to \infty} \frac{1}{2T} \int_{-T}^{+T} n(t)\, n(t + \tau)\, dt. \tag{1.113}$$

The autocorrelation function is used to describe random signals, but itself has a deterministic character.

Comparing (1.113) with (1.109), it becomes apparent that the square mean value U_R^2, i.e. the (root mean square)^2, for $\tau = 0$ is calculated as

$$U_R^2 = \psi(0) \tag{1.114}$$

At a real and constant resistance, the quantity U_R^2 is again proportional to the physical power in signal theory.

If one wants to determine the frequency composition of these signals, the Fourier transform of the autocorrelation function or the cross-correlation function must be determined. The Fourier transform of the ACF leads via

$$\Psi(f) = \int_{-\infty}^{+\infty} \psi(\tau)\, e^{-j 2\pi f \tau}\, d\tau \tag{1.115}$$

to the spectral power density $\Psi(f)$, which is also referred to as power spectral density (PSD). This relationship is referred to as the theorem of CHINTSCHIN[9] and WIENER[10] [29].

The power spectral density characterizes the distribution of the power of a random process over the frequency, as random signals are power signals.[11] The integration over the power spectral density, i.e.

$$U_R^2 = \psi(0) = \int_{-\infty}^{+\infty} \Psi(f)\, df \tag{1.116}$$

results in a measure for the power of the random process (see Fig. 1.37). The spectral power density is measured in V^2/Hz if $n(t)$ appeared in the form of a voltage. Thus, U_R^2 corresponds to the mean square of the signal (in V^2). If one works with a system-the-

---

[9] Alexander Yakovlevich Chintschin (1894–1959), Soviet mathematician (also: Khintchine or Khinchin)

[10] Norbert Wiener (1894–1964), American mathematician and philosopher

[11] In contrast, energy signals are described by an energy density spectrum which indicates the distribution of energy over the frequency.

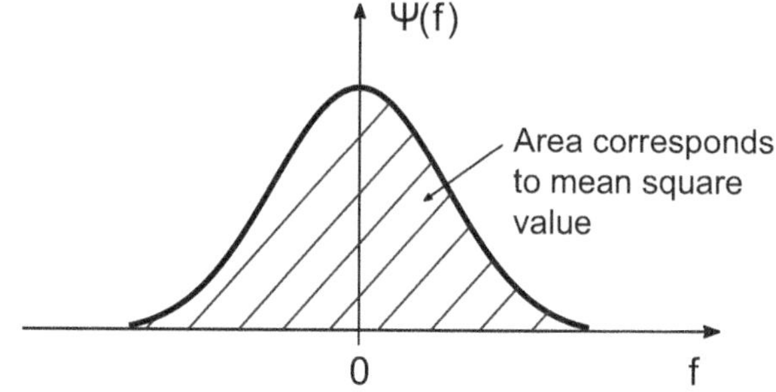

Fig. 1.37 The area under the power spectral density (squared amplitude spectral density) of a signal corresponds to the mean square of the signal (in V^2)

oretical power in the dimension $(\text{Amplitude})^2$, the power density is also referred to as squared amplitude density.

Many disturbances, such as amplifier noise, can be described by a random process with a power spectral density that is approximately constant over a wide frequency range. Such processes are often approximated by an idealized stochastic process whose power spectral density is constant for all frequencies and thus frequency-independent [18]. Such an idealized random process, in which almost all frequencies are equally represented, is also referred to as *white noise*—in analogy to white light, whose spectrum contains a multitude of wavelengths or frequencies: The effect of the white appearing light is caused by their superposition. The autocorrelation function of white noise can be represented using the already introduced Dirac impulse or Dirac delta function (Fig. 1.38):

$$\psi(\tau) = \Psi_0\, \delta(\tau). \tag{1.117}$$

This means that the values contained in a white noise signal $n(t)$ at different times t are uncorrelated [18]. From the Fourier transform of the autocorrelation function for white noise, it follows for the power spectral density (Fig. 1.38)

$$\Psi(f) = \mathcal{F}\{\psi(\tau)\} = \Psi_0. \tag{1.118}$$

Therefore, the following must apply for the mean square value and thus the power

$$U_R^2 = \psi(0) = \int_{-\infty}^{+\infty} \Psi_0\, df \longrightarrow \infty. \tag{1.119}$$

White noise is therefore an idealization (or an idealized model process) that cannot be physically realized, because processes with infinite power cannot be generated in physical reality.

For the description of communication technology contexts, it is often assumed that the disturbance amplitudes are Gaussian distributed and that the spectral noise power

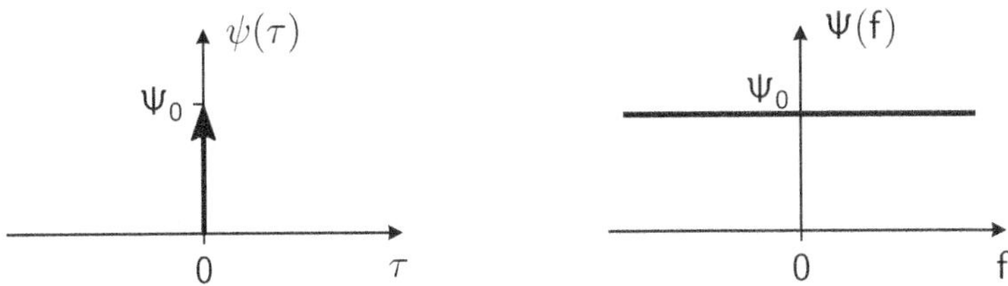

Fig. 1.38 Relationship between autocorrelation function and power spectral density in white noise

density of the disturbance is a constant. The assumption of a Gaussian distribution[12] is based on the validity of the central limit theorem [25]. However, it should be noted at this point that assuming a constant spectral power density, e.g. of the noise, the amplitudes are not automatically always Gaussian distributed. Also, the Gaussian distribution of the noise amplitudes does not imply the presence of a constant noise power spectral density.

1.2.3 On Spectral Characteristic Functions of Deterministic and Random Signals

In order to make the discussed spectral functions more illustrative and to clearly show the necessary difference in the treatment of deterministic and random signals, two thought experiments should be considered.

Amplitude Density Spectra of Deterministic Signals

To determine the spectral function of a deterministic voltage signal $u(t)$ in the image or frequency domain, this signal undergoes a thought experiment through a tunable bandpass (BP) of bandwidth $B = 1\,\mathrm{Hz}$. The passband is successively shifted by 1 Hz; the voltmeter registers the voltage present in the frequency range passed by the bandpass. The measurement arrangement is shown in Fig. 1.39 (left). The representation of the measurement result in the diagram in Fig. 1.39 (right) is as

$$U(f) = \frac{U_\mathrm{M}}{1\,\mathrm{Hz\ bandwidth}}, \tag{1.120}$$

thus as amplitude (U_M) related to a bandwidth (1 Hz)—and thus as *amplitude density*. The amplitude density is therefore given in the unit $\mathrm{V/Hz} = \mathrm{Vs}$.

---

[12] Johann Carl Friedrich Gauss (1777–1855), German mathematician, statistician, astronomer, geodesist, electrical engineer, and physicist.

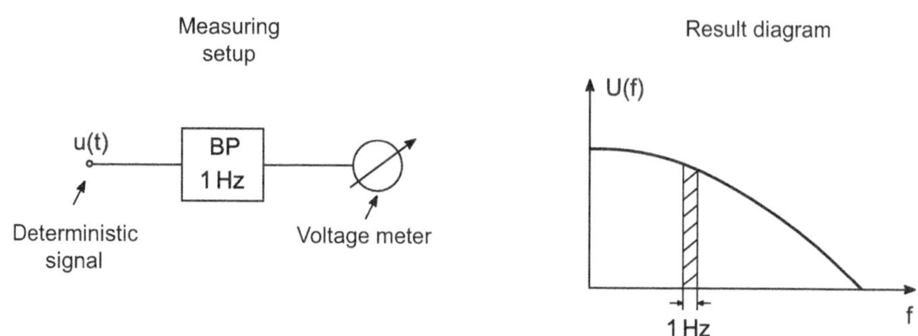

Fig. 1.39 Thought experiment for determining amplitude density in a determined signal

A bandpass of bandwidth $B = 1\,$Hz is often difficult to realize in practice. Then, for example, a bandpass of bandwidth $B = 10\,$kHz can be used and the measurement value is divided by the bandwidth B of the actually used bandpass: In this way, one practically obtains an approximation for the amplitude density to be determined.

The mathematical relationship is given by the Fourier transformation: $u(t) \;\circ\!\!-\!\!\bullet\; U(f)$ or $U(f) = \mathcal{F}\{u(t)\}$ and $u(t) = \mathcal{F}^{-1}\{U(f)\}$.

Power Density Spectra of Random Signals

The thought experiment for determining the spectral function of a random signal is similar to the procedure for a deterministic signal—but with one important change: Since the amplitude of a signal cannot be measured in a random signal—but only the power or the effective value—a power meter is used as a measuring device (Fig. 1.40, left). This measures the power present in the frequency range passed through by the bandpass. The representation in the diagram (in Fig. 1.40, right) is now as

$$\Psi(f) = \frac{P_\mathrm{M}}{1\,\text{Hz bandwidth}}, \tag{1.121}$$

thus as power (P_M) related to a bandwidth (1 Hz)—and thus as *power density*. The power density is therefore physically specified in the unit $\mathrm{W/Hz} = \mathrm{Ws}$.

At a real, constant resistance R, the *physical* power P (in W) is proportional to the square of the voltage U via the relationship $P = U^2/R$. In signal and system theory as well as in communications engineering applications, it is often advantageous to use directly $P = U^2$ (in V^2) as so-called *signal-* or *system-theoretical* power: The system-theoretical power is numerically equal to the physical power at a resistance of $R = 1\,\Omega$. Since in communications engineering, ratios of powers—e.g. signal-to-noise ratios—are

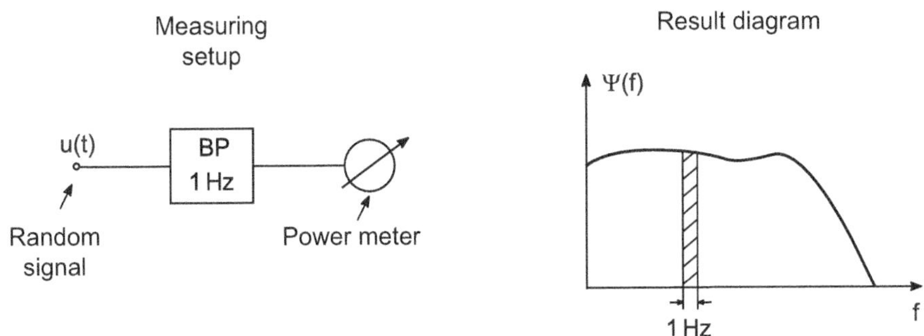

Fig. 1.40 Thought experiment for determining a power density with a random signal

of interest and significant for system optimization, this method can simplify the calculation and perform it independently of a specific technology. If a physical system needs to be dimensioned after optimization, the calculation of the physical power is necessary and easily possible with knowledge of the specific resistance R.

When working with the system-theoretical power, the "power density at $R = 1\,\Omega$" in relation to the frequency f (in W/Hz) is identical to the "square amplitude density" in relation to the frequency f (in V^2/Hz). Both terms are often used synonymously. Thus, *spectral power density* or simply *power density, power spectral density* and *square amplitude density* refer to identical spectral functions.

The mathematical relationship is again given by the Fourier transformation, now between the autocorrelation function of the random signal and the power density:
$\psi(\tau) \; \circ\!\!-\!\!\bullet \; \Psi(f)$ or $\Psi(f) = \mathcal{F}\{\psi(\tau)\}$ and $\psi(\tau) = \mathcal{F}^{-1}\{\Psi(f)\}$.

Parseval's Theorem
With the introduction of the Fourier transformation, aperiodic time signals $u(t)$ can be described in the frequency domain as

$$U(f) = \mathcal{F}\{u(t)\} = \int_{-\infty}^{+\infty} u(t)\; e^{-j2\pi f t}\, dt. \tag{1.122}$$

The inverse Fourier transformation yields the time signal corresponding to a spectrum from the frequency domain

$$u(t) = \mathcal{F}^{-1}\{U(f)\} = \int_{-\infty}^{+\infty} U(f)\; e^{j2\pi f t}\, df. \tag{1.123}$$

With PARSEVAL'S THEOREM[13], the energy of a time signal can now be determined both in the time domain and over the frequency domain. The signal energy (in V^2s) applies

$$E_s = \int\limits_{-\infty}^{+\infty} u^2(t)\, dt = \int\limits_{-\infty}^{+\infty} |U(f)|^2\, df. \tag{1.124}$$

With the correlation function for energy signals in the form (the superscript E is to indicate an energy signal)

$$\psi^E(\tau) = \int\limits_{-\infty}^{+\infty} u(t)\, u(t+\tau)\, dt. \tag{1.125}$$

from this, for $\tau = 0$, the signal energy can also be determined. A comparison of (1.125) with (1.124) shows for $\tau = 0$ a match: The evaluation of the autocorrelation function can also be used for the determination of the signal energy. Since the determination of the autocorrelation function can be expressed via the convolution

$$\psi^E(\tau) = u(\tau) * u(-\tau). \tag{1.126}$$

The transformation of the expression into the spectral domain leads to the energy density spectrum

$$\Psi^E(f) = \mathcal{F}\{\psi^E(\tau)\} = U(f) \cdot U^*(f) = |U(f)|^2 \tag{1.127}$$

and the final integration—analogous to the evaluation of the power density spectrum for power signals—to the energy of a time signal

$$E_s = \int\limits_{-\infty}^{+\infty} u^2(t)\, dt = \int\limits_{-\infty}^{+\infty} \Psi^E(f)\, df = \int\limits_{-\infty}^{+\infty} |U(f)|^2\, df. \tag{1.128}$$

Thus, the signal energy can be determined from

- the integral over the square of the time signal,
- the autocorrelation function at the point $\tau = 0$ or
- the integral over the energy density spectrum.

Insert: Derivation of the Relationship Between the Autocorrelation Function and the Energy Density Spectrum

The operation described in (1.126) can be expressed by the convolution integral

---

[13] Marc-Antoine Parseval des Chênes (1755–1836), French mathematician.

$$\psi^E(\tau) = \int\limits_{-\infty}^{+\infty} u(t)\, u(t+\tau)\, \mathrm{d}t \tag{1.129}$$

which in the frequency domain leads to a multiplication of the corresponding amplitude density spectra. Based on the approach that every time signal can be decomposed into an even and an odd component ($u_g(\tau)$ and $u_u(\tau)$), the following relationships in the frequency domain can be derived. With

$$u(\tau) = u_g(\tau) + u_u(\tau) \quad \circ\!\!-\!\!\bullet \quad U(f) = U_g(f) + U_u(f) \tag{1.130}$$

and

$$u(-\tau) = u_g(-\tau) + u_u(-\tau) = u_g(\tau) - u_u(\tau) \tag{1.131}$$

it initially follows for $u(-\tau) = u_g(\tau) - u_u(\tau)$, which can be represented in the frequency domain as

$$u(-\tau) = u_g(\tau) - u_u(\tau) \quad \circ\!\!-\!\!\bullet \quad U^*(f) = U_g(f) - U_u(f) \tag{1.132}$$

This results in the frequency domain for the energy density spectrum

$$\Psi^E(f) = \mathcal{F}\{\psi^E(\tau)\} = U(f) \cdot U^*(f) = |U(f)|^2, \tag{1.133}$$

which describes the distribution of a signal's energy over frequency.

1.3 Characterization of Systems

1.3.1 Basics

In communication systems, signals can be information-carrying functions that are altered or transformed by a system. Systems can be components of the transmitter and receiver or even the transmission channel itself. The task of system theory is to describe system arrangements that influence, change and link input signals according to certain laws, which depend on the properties of the system. Possibilities for describing a system are given by:

- Response functions to characteristic time functions (impulse response, step response),
- System properties in the frequency domain (transfer functions, attenuation, phase response) and the
- Response to random processes with predefined statistical properties.

Such a transmission system can be understood as a mathematical model of a technical arrangement that is subjected to input signals and reacts in a desired manner with output signals. Essential for a system is its transmission characteristic, i.e., how an input signal is influenced by the system and how it is present at its output after this evaluation or transformation by the system.

1.3.2   System Description in the Time and Frequency Domain

A system that is linear and whose properties are independent of temporal shifts of the input signal is referred to as a linear time-invariant system or briefly as an LTI system (linear time-invariant system). A system is linear if the superposition theorem is applicable. In addition, such systems possess the properties of *stability* and *causality*.

Forms of System Description

In terms of their transmission properties, LTI systems are fully described by the weighting function $g(t)$ or the transfer function $G(f)$—as the Fourier transform $G(f) = \mathcal{F}\{g(t)\}$ of the weighting function $g(t)$.

The transfer function is used in the form

$$G(f) = \frac{U_2(f)}{U_1(f)} \tag{1.134}$$

(see Fig. 1.41); it is therefore dimensionless. Derived from the generally complex transfer function $G(f)$ are the *magnitude characteristic* $|G(f)|$ (amplitude spectrum) and the *angle or phase characteristic* $\varphi(f)$ (phase spectrum) via

$$G(f) = |G(f)|\, e^{j\,\varphi(f)}. \tag{1.135}$$

In addition to the transfer function $G(f)$, as a complete system characteristic, in the form $G(f) = |G(f)|\, e^{j\,\varphi(f)}$, the representation derived from classical network theory

$$G(f) = e^{-(\alpha(f)+j\,\beta(f))} = e^{-\alpha(f)} \cdot e^{-j\,\beta(f)} \tag{1.136}$$

with $|G(f)| = e^{-\alpha(f)}$ and the parameters *attenuation* $\alpha(f) = -\ln(|G(f)|)$ and *phase* $\beta(f) = -\varphi(f)$ is known.

To illustrate the frequency composition of signals and systems, in addition to the linear representation, the logarithmic representation has proven to be a suitable tool. Thus, the representation of

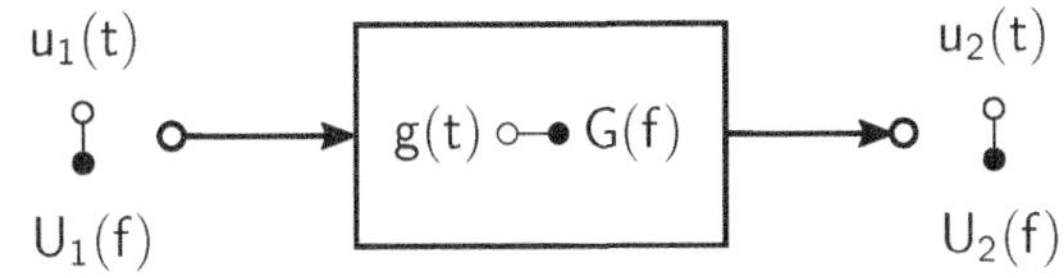

Fig. 1.41 For the definition of a system with determined input and output variables

$$u_1(t) \qquad \qquad \boxed{g(t) \circ\!\!-\!\!\bullet\, G(f)} \qquad \qquad u_2(t)$$
$$U_1(f) \qquad \qquad \qquad \qquad \qquad \qquad U_2(f)$$

$$20\lg|G(f)| = 20\log_{10}|G(f)| \qquad \text{and} \qquad \varphi(f) \tag{1.137}$$

with a logarithmic frequency axis is referred to as a Bode diagram.[14] This takes advantage of the fact that suitable scales—such as the logarithmic representation—result in additive relationships between the overall system and the individual elementary systems in terms of the course of magnitude and phase (see Fig. 1.48).

Through the Fourier transformation, the transfer function is linked with the weighting function

$$g(t) = \int_{-\infty}^{+\infty} G(f)\, e^{j2\pi ft} df \qquad \left(\text{in Hz or } \frac{1}{s}\right) \tag{1.138}$$

which has the unit $1/s$.

Relative Level and Attenuation Specifications

Usually, for relative specifications of levels and attenuations, the auxiliary unit dB for Dezibel is used, named after A. G. Bell.[15] The Dezibel serves to indicate the use of the decimal logarithm of a ratio of two similar power or energy quantities, e.g.:

$$L(\text{in dB}) = 10\lg\left(\frac{P_2}{P_1}\right) \tag{1.139}$$

When using amplitude quantities, e.g. voltages, one obtains (with $P_1 = U_1^2/R$ and $P_2 = U_2^2/R$)

$$L(\text{in dB}) = 10\lg\left(\frac{U_2^2/R}{U_1^2/R}\right) = 10\lg\left(\frac{U_2^2}{U_1^2}\right) = 10\lg\left(\frac{U_2}{U_1}\right)^2$$

$$L(\text{in dB}) = 20\lg\left(\frac{U_2}{U_1}\right)$$

---

[14] Hendrik Wade Bode (1905–1982), US-American mathematician and physicist
[15] Alexander Graham Bell (1847–1922), British, later US-American inventor and entrepreneur

Another auxiliary unit related to the Dezibel and widely used in the past for electrical lines is Np for Neper, named after J. NAPIER.[16] The Neper is used to denote the use of the natural logarithm of a ratio of two similar field sizes, e.g. voltages:

$$L(\text{in Np}) = \ln\left(\frac{U_2}{U_1}\right) \tag{1.140}$$

The conversion factor between dB and Np is $20\lg e = 8.686$.

Example 1.11 (Conversion of 1 Np to dB) A ratio U_2/U_1, which corresponds to 1 Np, is to be converted into dB. Starting from

$$L(\text{in Np}) = \ln\left(\frac{U_2}{U_1}\right) = 1\,\text{Np} \tag{1.141}$$

one initially obtains

$$1\,\text{Np} = \ln\left(\frac{U_2}{U_1}\right) \qquad \Big| \quad e^{(\dots)}$$

$$e^1 = \frac{U_2}{U_1}$$

and the specification in dB reads

$$L(\text{in dB}) = 20\lg\left(\frac{U_2}{U_1}\right) = 20\lg(e) \approx 8.686\,\text{dB} \tag{1.142}$$

$\square$

Example 1.12 (Conversion of 1 dB to Np) A ratio U_2/U_1, which corresponds to 1 dB, is to be converted into Np. Starting from

$$L(\text{in dB}) = 20\lg\left(\frac{U_2}{U_1}\right) = 1\,\text{dB} \tag{1.143}$$

one initially obtains

$$1\,\text{dB} = 20\lg\left(\frac{U_2}{U_1}\right) \qquad \Big| \quad : 20$$

$$\frac{1}{20} = \lg\left(\frac{U_2}{U_1}\right) \qquad \Big| \quad 10^{(\dots)}$$

$$10^{1/20} = \frac{U_2}{U_1}$$

---

[16] John Napier (Latinized Neper), (1550–1617), Scottish mathematician, natural philosopher, and theologian

and the specification in Np is

$$L(\text{in Np}) = \ln\left(\frac{U_2}{U_1}\right) = \ln\left(10^{1/20}\right) = \frac{1}{20}\ln 10 \approx 0.115\,\text{Np} \qquad (1.144)$$

$\square$

The conversion between both specifications is therefore:

- $1\,\text{dB} = 0.115\,\text{Np}$,
- $1\,\text{Np} = 8.686\,\text{dB}$.

Both the Dezibel and the Neper are not real units of measurement, as they each represent a ratio of similar quantities—and thus a dimensionless quantity. They indicate the base of the logarithm used (decimal or natural logarithm). Both (pseudo-) units are used, with the dB being much more common and widespread than the Np.

Application of System Description Forms to Exemplary Systems
In the following examples, a selection of the system description forms introduced in this section is applied to specific exemplary systems that play an important role in communications engineering and that will be revisited and utilized in the following chapters.

Example 1.13 (Amplitude and phase response of an ideal low-pass filter) Fig. 1.42 illustrates the frequency and phase response of an ideal low-pass filter. The inverse Fourier transformation

$$g(t) = \int\limits_{-\frac{1}{2T}}^{+\frac{1}{2T}} A\,T\,e^{j2\pi ft}\,df \qquad (1.145)$$

results in the weighting function $g(t)$ to

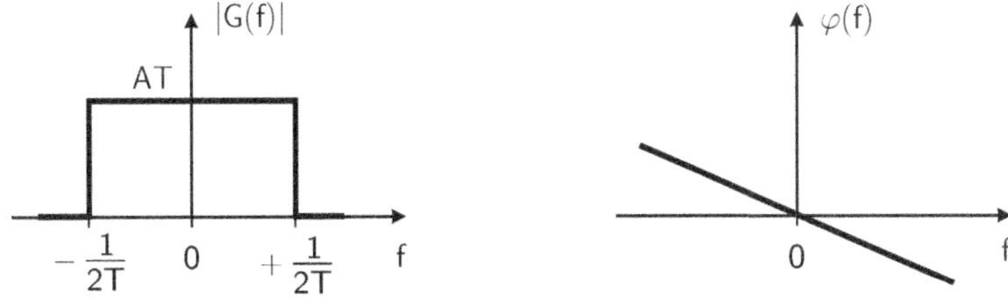

Fig. 1.42 Amplitude and phase response of an ideal low-pass filter

$$g(t) = A \operatorname{si}\left(\pi \, \frac{t}{T}\right). \tag{1.146}$$

Insert: Derivation of the Weighting Function of the Ideal Low Pass Filter

The inverse Fourier transformation

$$g(t) = \int_{-\frac{1}{2T}}^{+\frac{1}{2T}} A\, T \, e^{j 2\pi f t} \, df \tag{1.147}$$

of the transfer function of the ideal low-pass filter leads after performing the integration to

$$g(t) = \frac{A\,T}{j 2\pi t} \; e^{j 2\pi f t} \Bigg|_{-\frac{1}{2T}}^{+\frac{1}{2T}}. \tag{1.148}$$

Inserting the integration limits results in

$$g(t) = \frac{A\,T}{j 2\pi t}\left(e^{j \frac{2\pi t}{2T}} - e^{-j \frac{2\pi t}{2T}}\right). \tag{1.149}$$

This expression can be simplified to

$$g(t) = \frac{A\,T}{j 2\pi t}\left(e^{j \frac{\pi t}{T}} - e^{-j \frac{\pi t}{T}}\right) \tag{1.150}$$

which can be written with $\sin(x) = \frac{1}{2j}(e^{jx} - e^{-jx})$ as

$$g(t) = \frac{A\,T}{\pi t}\, \sin\left(\pi \, \frac{t}{T}\right) \tag{1.151}$$

and with $\operatorname{si}(ax) = \frac{\sin(ax)}{ax}$ leads to the result

$$g(t) = A \operatorname{si}\left(\pi \, \frac{t}{T}\right). \tag{1.152}$$

If one also considers a constant delay τ_0, which an input signal requires for the system run-through, then the phase response of an ideal low pass, as indicated in Fig. 1.43, can be represented. In this case, the phase response results in

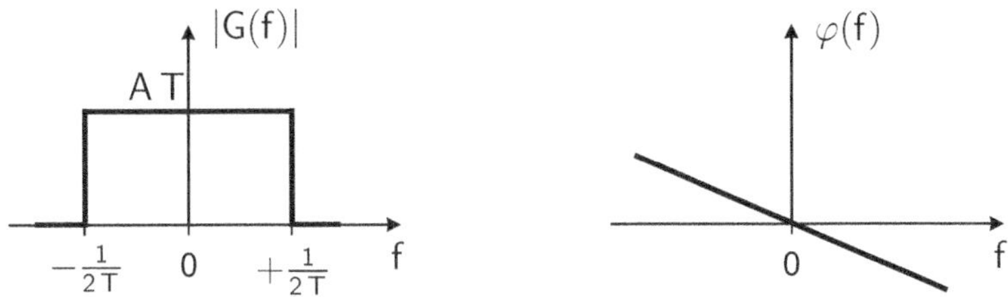

Fig. 1.43 Amplitude and phase response of an ideal low pass with linear phase dependence

$$\varphi(f) = -2\pi f \tau_0. \tag{1.153}$$

Through the inverse Fourier transformation

$$g(t) = \int_{-\frac{1}{2T}}^{+\frac{1}{2T}} A\,T\,e^{-j2\pi f\,\tau_0}\,e^{j2\pi ft}\,df \tag{1.154}$$

the weighting function $g(t)$ results in

$$g(t) = A\,\mathrm{si}\left(\pi\,\frac{(t-\tau_0)}{T}\right). \tag{1.155}$$

$\square$

The demand for a constant group delay τ_g, i.e.,

$$\tau_g = -\frac{1}{2\pi}\,\frac{d\,\varphi(f)}{df} = \text{constant}, \tag{1.156}$$

with a constant magnitude characteristic in the considered frequency interval, i.e. $|G(f)| = 1$, is a prerequisite for the realization of distortion-free transmission. If the system is subjected to an input signal $u_1(t)$ in this case, it experiences only a time shift during system throughput, regardless of the spectral composition of the input signal—provided all frequency components of the input signal are within the passband of the filter. The output signal in this case results in

$$u_2(t) = u_1(t - \tau_g) \tag{1.157}$$

and is only time-shifted by τ_g compared to the input time function. The system shown in Fig. 1.43 has a group delay of $\tau_g = \tau_0$.

In contrast, real systems will have a more gradual transition from the passband to the stopband.

Example 1.14 (First order low pass) For a first order low pass, the transfer function results in

$$G(f) = \frac{1}{1 + j\frac{f}{f_g}}. \tag{1.158}$$

The value f_g describes the 3-dB cutoff frequency. At this point, the magnitude of the transfer function has dropped to $1/\sqrt{2}$ of the maximum value. Consequently, it must then hold:

$$|G(f = f_g)| = \frac{1}{\sqrt{2}}. \tag{1.159}$$

The bandwidth of a given signal shape can be calculated from the lowest and highest contained spectral component. In filter technology, the range is referred to as *bandwidth*, at whose cutoff frequencies the voltage amplitude has changed by the factor $\sqrt{2}$ (corresponds to a change by 3 dB) compared to the maximum or minimum of the amplitude.

The Bode diagram corresponding to (1.137) for the first order low pass is shown in Fig. 1.44: The upper part shows the amplitude response and the lower part shows the phase response, each with a logarithmically divided frequency axis.

The logarithmic amplitude response indicated in the upper part of Fig. 1.44 is often represented in a *straight line approximation* for better clarity (see also Fig. 1.45): Up to the cutoff frequency $f_g = 10\,\text{kHz}$ this would be a horizontal line at 0 dB and from this cutoff frequency a line falling with a slope of $-20\,\text{dB/Dekade}$. The term *decade* refers to a power of ten, so a decade is, for example, $10^4 \ldots 10^5$ or $10^5 \ldots 10^6$. $\qquad\square$

Fig. 1.44 Bode diagram of the first order low pass (cutoff frequency: $f_g = 10\,\text{kHz}$)

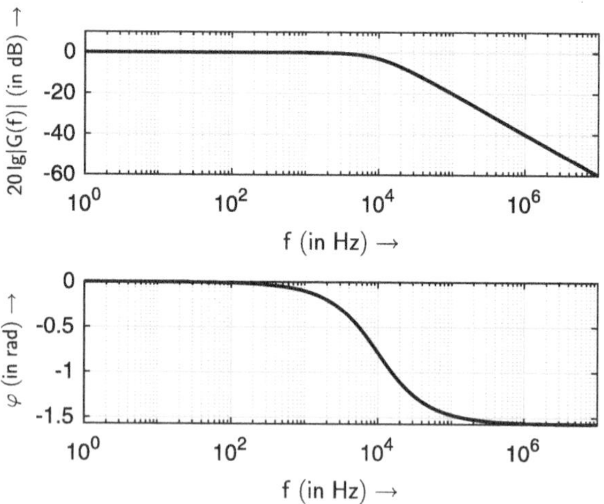

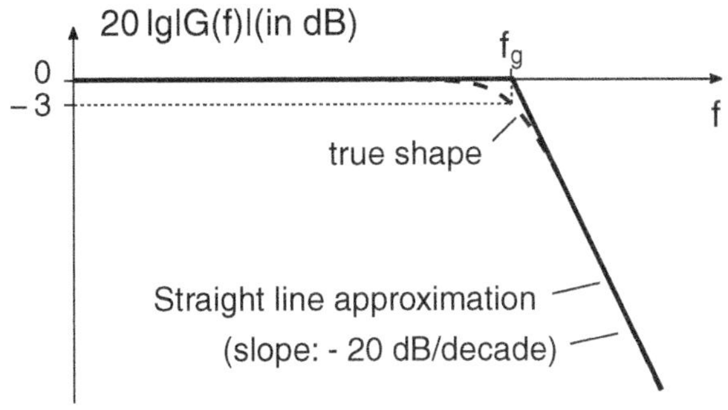

Fig. 1.45 Amplitude response of a first order low pass and straight line approximation (frequency axis: logarithmically divided)

Example 1.15 (Cosine roll-off low-pass) Often is used in transmission engineering for $G(f)$ a cosine roll-off transfer function

$$G(f) = \begin{cases} 1 & \text{for} \quad f \leq \frac{1-r}{2T} \\ \frac{1}{2}\left[1 + \sin\left(\frac{\pi\,(1-2\,f\,T)}{2\,r}\right)\right] & \text{for} \quad \frac{1-r}{2T} \leq f \leq \frac{1+r}{2T} \\ 0 & \text{for} \quad f > \frac{1+r}{2T} \end{cases} \tag{1.160}$$

with the roll-off factor r aimed for (see Fig. 1.46); for justification and application see Chap. 2.

Through the inverse Fourier transformation, the weighting function is obtained (see Fig. 1.47) to

$$g(t) = \frac{1}{T}\,\text{si}\left(\pi\,\frac{t}{T}\right)\frac{\cos\left(r\,\pi\,t/T\right)}{1 - 4(r\,t/T)^2}. \tag{1.161}$$

$\square$

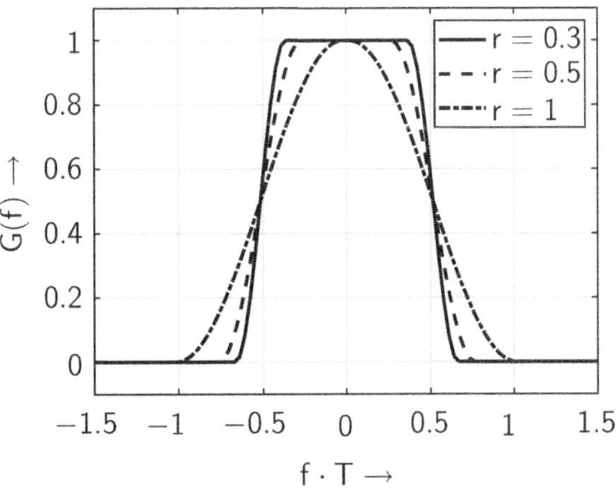

Fig. 1.46 Cosine roll-off spectrum for various roll-off factors

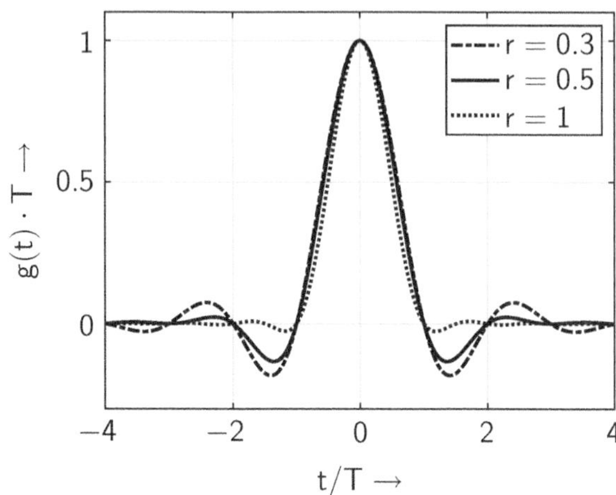

Fig. 1.47 Cosine roll-off weighting functions for various roll-off factors r

Serial Concatenation of Systems

From the serial concatenation of individual elementary systems

$$G(f) = G_1(f) \cdot G_2(f) \tag{1.162}$$

it follows when choosing the logarithmic representation with

$$G(f) = |G(f)|\, e^{j\,\varphi(f)}, \tag{1.163}$$

that by applying logarithmic laws, an additive accumulation of the individual magnitude and phase characteristics to the respective magnitude or phase response of the overall system is possible. For

$$G_1(f) = |G_1(f)|\, e^{j\,\varphi_1(f)} \tag{1.164}$$

and

$$G_2(f) = |G_2(f)|\, e^{j\,\varphi_2(f)} \tag{1.165}$$

this results in:

$$20\log_{10}|G(f)| = 20\log_{10}|G_1(f)| + 20\log_{10}|G_2(f)|. \tag{1.166}$$

In addition, the following additive relationship can be derived for the phase progression

$$\varphi(f) = \varphi_1(f) + \varphi_2(f). \tag{1.167}$$

On Calculation in the Original and the Image Domain

Figure 1.49 exemplifies the possibilities of calculation in the original and image domain, based on [16]. While calculation in the original or time domain is often possible but

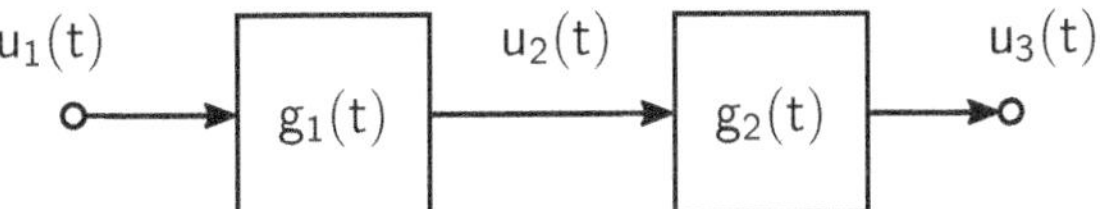

Fig. 1.48 Serial concatenation of elementary systems to build arbitrarily complex transmission systems

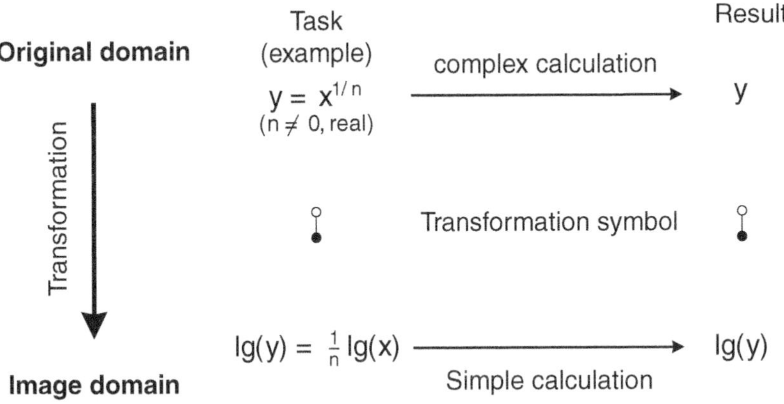

Fig. 1.49 On possibilities in the calculation in the original and image domain

time-consuming, calculation in the image domain offers advantages in many cases. For example, operations of infinitesimal calculus in the time domain are mapped onto algebraic operations in the image domain, which are often much easier to handle computationally. For the description of the basic principles of signal and system theory, the image domain represents the spectral domain, which directly (e.g., via Fourier transformation) or indirectly (e.g., via Laplace transformation) provides information about the frequency composition of signals.

The previous explanations have already shown that the transition to the image domain can offer advantages in calculation and the intuitive interpretation of signal properties. Figure 1.49 illustrates the potential of calculation in the image domain: For example, it would be relatively complex and time-consuming to determine the value of the function $y = x^{1/n}$ in the original domain, especially for large numbers x and n. The transition to the image domain is achieved here by applying a logarithmic function: It results in the expression $\log_{10}(y) = \lg(y) = 1/n \cdot \lg(x)$ in the image domain, which can be calculated with manageable effort. The subsequent back transformation leads via $y = 10^a$ with $a = 1/n \cdot \lg(x)$ to the desired result y.

1.4 Transmission of Signals over Systems

1.4.1 Deterministic Signals

Since signals in communication technology contexts typically represent information, it is often desired or necessary in many applications to transmit signals—and thus information—over spatial distances (e.g., between locations), to store (transmit over a temporal distance), or to link with other information. To do this, relationships between signals must be established and these relations have to be described mathematically [18]. This task is performed by the system, which establishes relationships between several signals. In many cases, signals are given to systems and the corresponding system response is sought. This can happen both in the frequency domain and in the time domain.

Description of the Transformation of Signals through Systems in the Time and the Spectral Domain

The input-output relationship of the system shown in Fig. 1.50 in the spectral domain reads

$$U_2(f) = G(f) \cdot U_1(f). \tag{1.168}$$

Since $G(f)$ is dimensionless, the amplitude density present at the input of the system $U_1(f)$ (in V/Hz) also produces an amplitude density $U_2(f)$ (in V/Hz) at the output.

In the time domain, the combination of the input signal $u_1(t)$ with the system's weighting function can be described for linear, time-invariant systems via the convolution integral. The system response results in

$$u_2(t) = \int_{-\infty}^{+\infty} u_1(\tau)\, g(t - \tau)\, \mathrm{d}\tau. \tag{1.169}$$

Alternatively, the following shorthand notation is used for the convolution of two signals

$$u_2(t) = u_1(t) * g(t). \tag{1.170}$$

Since the convolution operation is dimension-bearing, the voltage present at the input $u_1(t)$ (in V) is mapped into an output voltage $u_2(t)$ (in V): $g(t)$ (in 1/s) multiplied by the unit (s) from the $\mathrm{d}\tau$ of the convolution integral cancels out and the dimension of

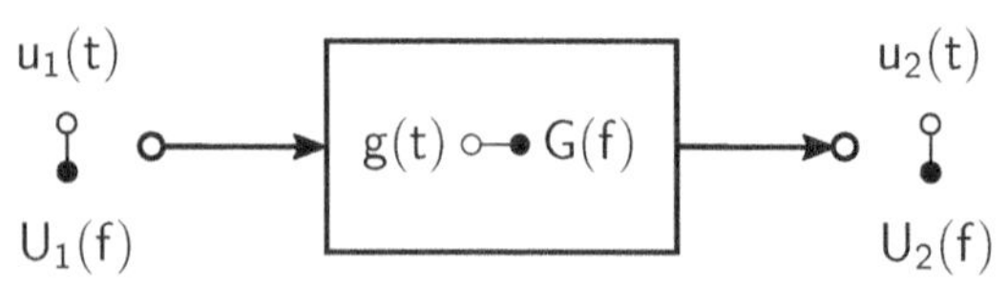

Fig. 1.50 For the definition of a system with deterministic input and output variables

$u_1(t)$ (in V) remains for the result $u_2(t)$ (in V). In convolution, unlike in the calculation of the correlation function, the commutative law applies:

$$u_2(t) = u_1(t) * g(t) = g(t) * u_1(t). \tag{1.171}$$

The execution of the convolution operation requires the following steps:

- Temporal reflection,
- (Temporal) shift,
- Calculation of the integral of the product $u_1(\tau) \cdot g(t - \tau)$.

Insert: Conversion of the Convolution Integral into the Spectral Domain

After the combination of two signals in the time domain was introduced via convolution, the question arises whether the convolution operation can be converted into an equivalent structure in the image or spectral domain. This resulted in the convolution integral

$$u_2(t) = \int_{-\infty}^{+\infty} u_1(\tau)\, g(t - \tau)\, d\tau, \tag{1.172}$$

which can also be expressed by the shorthand notation $u_2(t) = u_1(t) * g(t)$. The Fourier transformation of the signal $u_2(t)$ leads to the amplitude spectral density

$$U_2(f) = \int_{-\infty}^{+\infty} u_2(t)\, e^{-j2\pi ft}\, dt. \tag{1.173}$$

Substituting $u_2(t)$ into (1.173) results via

$$U_2(f) = \int_{-\infty}^{+\infty} \underbrace{\int_{-\infty}^{+\infty} u_1(\tau)\, g(t - \tau)\, d\tau}_{u_2(t)}\; e^{-j2\pi ft}\, dt \tag{1.174}$$

in the relationship

$$U_2(f) = \int_{-\infty}^{+\infty} \int_{-\infty}^{+\infty} u_1(\tau)\, g(t - \tau)\, e^{-j2\pi ft}\, dt\, d\tau. \tag{1.175}$$

The expression can be transformed into

$$U_2(f) = \int_{-\infty}^{+\infty} u_1(\tau) \underbrace{\int_{-\infty}^{+\infty} g(t - \tau)\, e^{-j2\pi ft}\, dt}_{\mathcal{F}\{g(t-\tau)\}=G(f)\cdot e^{-j2\pi f\tau}}\, d\tau \tag{1.176}$$

and simplified by applying the time shift theorem of the Fourier transformation to

$$U_2(f) = \int_{-\infty}^{+\infty} u_1(\tau) \cdot G(f) \cdot e^{-j2\pi f\tau}\, d\tau .$$

(1.177)

This can be written as

$$U_2(f) = G(f) \underbrace{\int_{-\infty}^{+\infty} u_1(\tau)\, e^{-j2\pi f\tau}\, d\tau}_{\mathcal{F}\{u_1(\tau)\}=U_1(f)} .$$

(1.178)

This shows that the combination of two signals in the time domain via the convolution integral in the frequency domain leads to the multiplication of the corresponding spectra. The amplitude spectral density of the input signal is multiplicatively linked with the transfer function, so that

$$U_2(f) = U_1(f) \cdot G(f)$$

(1.179)

is obtained.

Example 1.16 (Convolution of two signals) The convolution operation in the time domain is to be illustrated using an example (see also Fig. 1.51).

The step-by-step execution of the convolution integral results in:

1. Area for $0 \le t \le T_1$
   The value of the convolution integral in the 1st area (see Fig. 1.52) results in

$$u(t) = \int_0^t U_1 \cdot \frac{1}{T} \cdot d\tau = U_1 \cdot \frac{1}{T} \cdot t.$$

(1.180)

2. Area for $T_1 \le t \le T_2$
   The value of the convolution integral in the 2nd area (see Fig. 1.53) results in

$$u(t) = \int_0^{T_1} U_1 \cdot \frac{1}{T} \cdot d\tau = U_1 \cdot \frac{1}{T} \cdot T_1.$$

(1.181)

3. Area for $T_2 \le t \le T_1 + T_2$
   The value of the convolution integral in the 3rd area (see Fig. 1.54) results in

$$u(t) = \int_{t-T_2}^{T_1} U_1 \cdot \frac{1}{T} \cdot d\tau = U_1 \cdot \frac{1}{T} \cdot (T_1 - (t - T_2)).$$

(1.182)

Finally, the result shown in Fig. 1.55 is obtained.

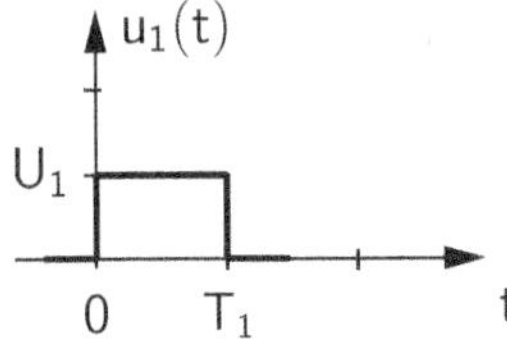
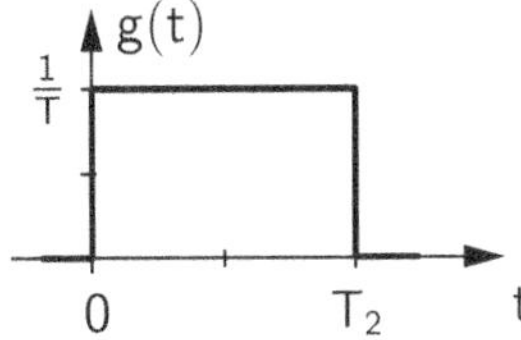

Fig. 1.51 Input signal $u_1(t)$ and weighting function $g(t)$ of the system

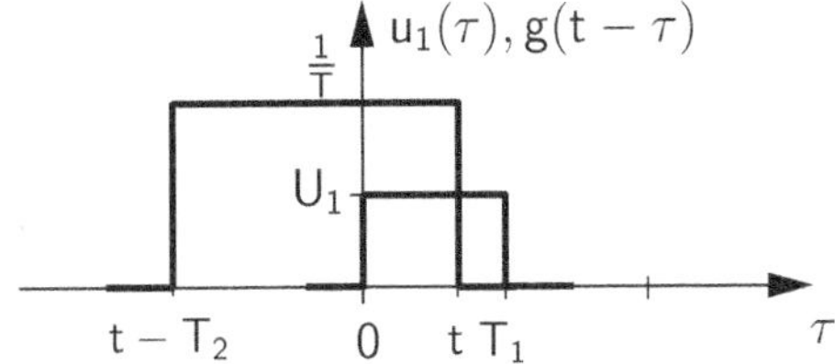
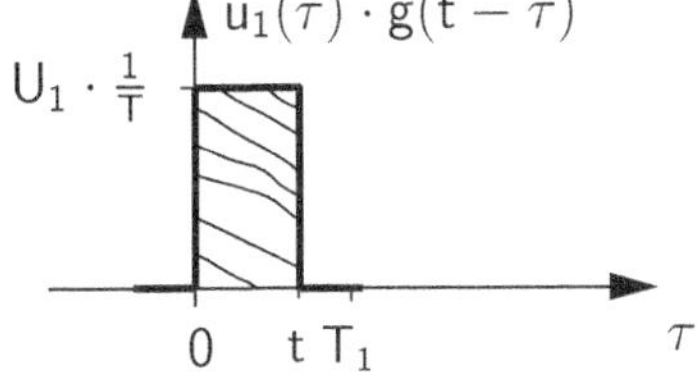

Fig. 1.52 Evaluation of the product $u_1(\tau) \cdot g(t - \tau)$ for the time shift $0 \le t \le T_1$ (first area)

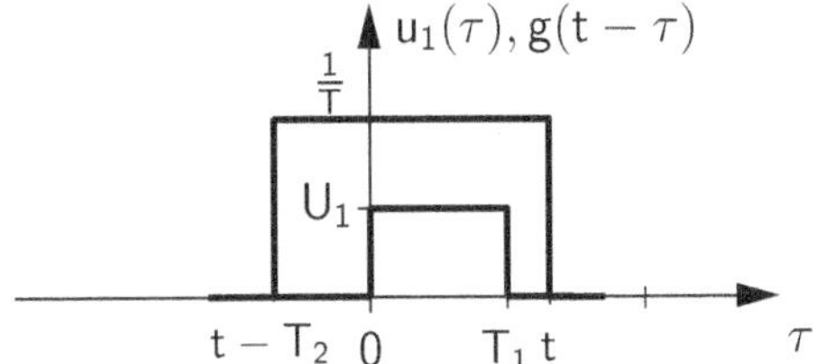
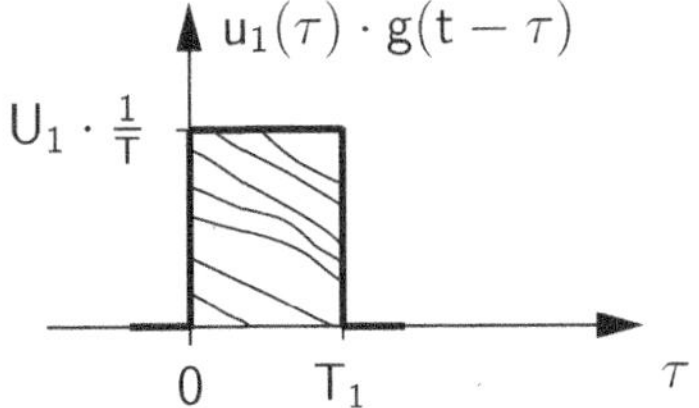

Fig. 1.53 Evaluation of the product $u_1(\tau) \cdot u_2(t - \tau)$ for the time shift $T_1 \le t \le T_2$ (second area)

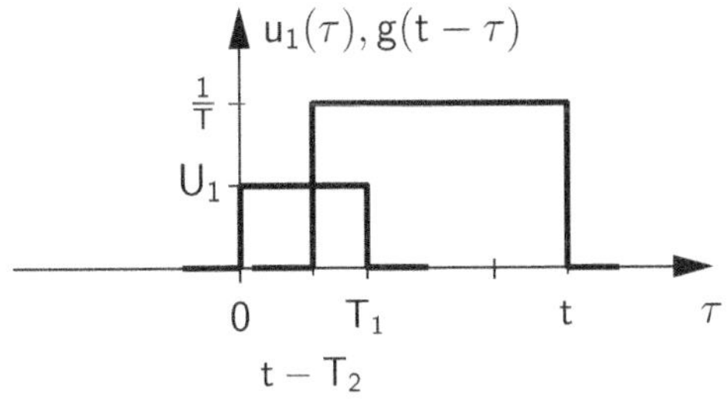
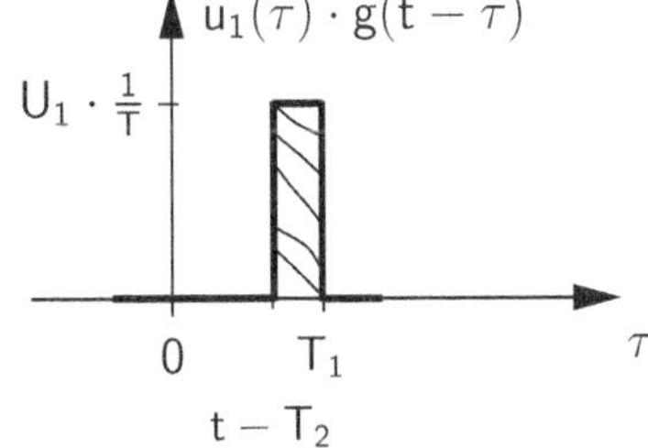

Fig. 1.54 Evaluation of the product $u_1(\tau) \cdot u_2(t - \tau)$ for the time shift $T_2 \le t \le T_1 + T_2$ (third area)

It should be noted that the weighting function $g(t)$ as system characteristic with dimensionless transfer function $G(f)$ has the dimension $(\text{time})^{-1}$; on the other hand, the impulse response as the system's reaction to an impulse at the input has the dimension of the input signal (here thus in the unit V). In this respect, the terms *weighting function* and *impulse response* are to be distinguished and not to be used synonymously.

If the system $g(t)$ is subjected to a Dirac impulse

$$u_1(t) = A\,\delta(t) \tag{1.183}$$

as an input signal, the Dirac impulse response is obtained

$$u_2(t) = u_1(t) * g(t) \tag{1.184}$$

$$u_2(t) = A\,\delta(t) * g(t) \tag{1.185}$$

$$u_2(t) = A \cdot g(t) \tag{1.186}$$

as a signal at the output of the system. The weighting function and the Dirac impulse response differ only by the factor A; however, this is dimensionally loaded.

Insert: Input-Output Relationship for Linear, Time-Invariant Systems: Derivation of the Convolution Formula

An attempt will now be made to derive the input-output relationship for linear, time-invariant systems using the example of the impulse response. The input-output relationship is defined as

$$u_2(t) = u_1(t) * g(t) = \int_{-\infty}^{+\infty} u_1(\tau) \cdot g(t - \tau)\,\mathrm{d}\tau \tag{1.187}$$

If the system $g(t)$ is subjected to an impulse

$$u_1(t) = A\,\delta(t) \tag{1.188}$$

as an input signal, the impulse response is obtained

$$u_2(t) = u_1(t) * g(t) \tag{1.189}$$

$$u_2(t) = A\,\delta(t) * g(t) \tag{1.190}$$

$$u_2(t) = A \cdot g(t) \tag{1.191}$$

as a signal at the output of the system. A linear, time-invariant system responds to an input pulse $u_1(t) = A\,\delta(t)$ with the output signal $u_2(t) = A\,g(t)$ (impulse response). The fact that an input signal $u_1(t) = A\,\delta(t)$ is mapped to the output signal $u_2(t) = A\,g(t)$ can now be used to extend this relationship to any input signals $u_1(t)$.

With the help of the selective property of the Dirac impulse, any input signal $u_1(t)$ can be represented as follows:

$$u_1(t) = \int\limits_{-\infty}^{+\infty} u_1(\tau) \cdot \delta(t-\tau)\,\mathrm{d}\tau = u_1(t) \cdot \underbrace{\int\limits_{-\infty}^{+\infty} \delta(t-\tau)\,\mathrm{d}\tau}_{=1} = u_1(t). \tag{1.192}$$

This equation can be intuitively understood as the superposition of infinitely many shifted Dirac impulses $\delta(t-\tau)$ with the weight $u_1(\tau)$. Each individual Dirac impulse $\delta(t-\tau)$ (here with the scaling factor $A=1$) then leads to the impulse response $g(t-\tau)$. The output signal $u_2(t)$ then results again from the superposition of infinitely many impulse responses $g(t-\tau)$ with the weights $u_1(\tau)$. Thus, the output signal can be written as

$$u_2(t) = \int\limits_{-\infty}^{+\infty} u_1(\tau) \cdot g(t-\tau)\,\mathrm{d}\tau. \tag{1.193}$$

For the purpose of controlling the convolution integral, an input signal $u_1(t) = A\,\delta(t)$ is now applied to the system. This results in

$$u_2(t) = \int\limits_{-\infty}^{+\infty} A\,\delta(\tau) \cdot g(t-\tau)\,\mathrm{d}\tau. \tag{1.194}$$

With the selective property of the Dirac impulse (which occurs at the time $\tau = 0$), the convolution integral can be simplified to

$$u_2(t) = A\,g(t) \underbrace{\int\limits_{-\infty}^{+\infty} \delta(\tau)\,\mathrm{d}\tau}_{=1} \tag{1.195}$$

and the well-known result of the impulse response at the output is obtained:

$$u_2(t) = A\,g(t). \tag{1.196}$$

Sampling of Time Signals

In signal processing, signals often exist in a discrete-time form. These discrete-time signals are generated by the *sampling* of a continuous-time signal. In contrast to the continuous signal (Fig. 1.1), the discrete-time signal is only defined at discrete points in time with a fixed interval T_a on the time axis. The value T_a denotes the sampling period and its

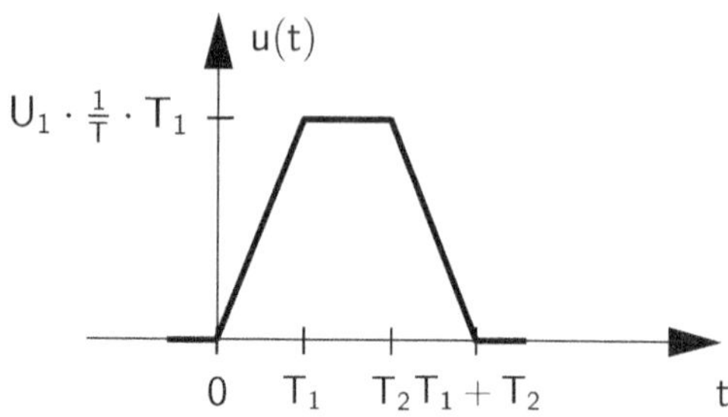

Fig. 1.55 Result $u_2(t)$ of the convolution operation

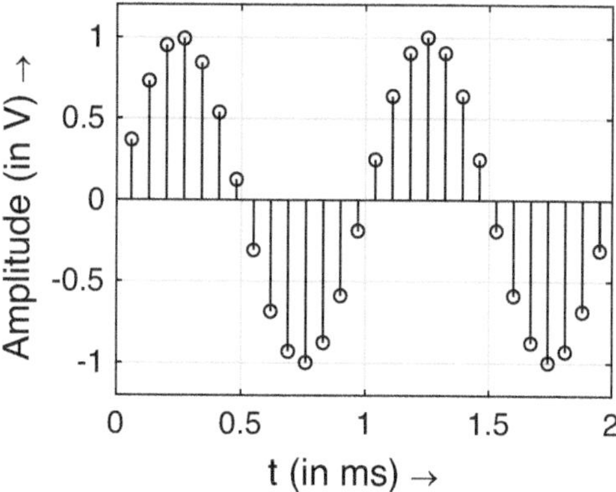

Fig. 1.56 Representation of a discrete-time signal

reciprocal is called the sampling frequency or sampling rate $f_a = 1/T_a$. An example of a discrete-time signal is given in Fig. 1.56.

A mathematical model for the generation of a discrete-time signal can be found in Fig. 1.57.

The multiplication of the signal $u(t)$ with a sequence of Dirac impulses leads via the approach

$$u_a(t) = u(t) \cdot T_a \sum_{k=-\infty}^{\infty} \delta(t - k\,T_a) = T_a \sum_{k=-\infty}^{\infty} u(k\,T_a)\,\delta(t - k\,T_a) \qquad (1.197)$$

to a continuous-time representation of a signal sampled with the sampling frequency $f_a = 1/T_a$. The sampling values $u(k\,T_a)$ of the continuous signal appear as weights of the Dirac impulses. The scaling factor T_a ensures that the ideally sampled signal $u_a(t)$ also has the dimension of a voltage and thus the unit Volt.

Figure 1.58 illustrates the creation of a discrete-time signal. Discrete-time (discrete) signals are only defined at discrete, usually equidistant, points in time $k\,T_a$. Here, k is to be understood as an independent discrete time variable.

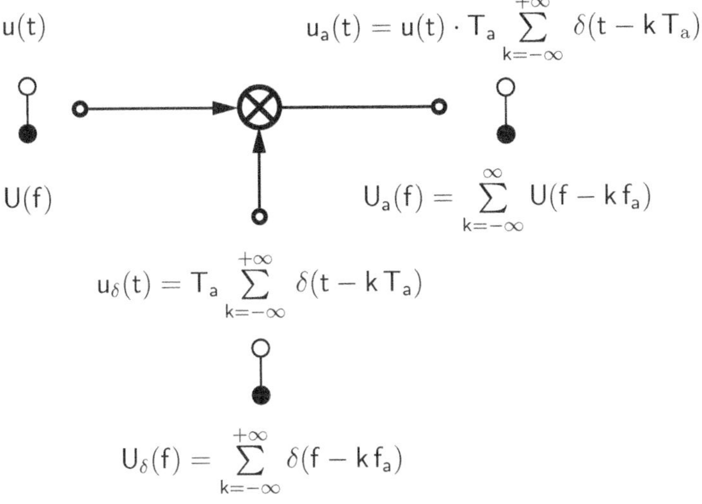

Fig. 1.57 Structure of an ideal sampling unit and its description in the time and spectral domain

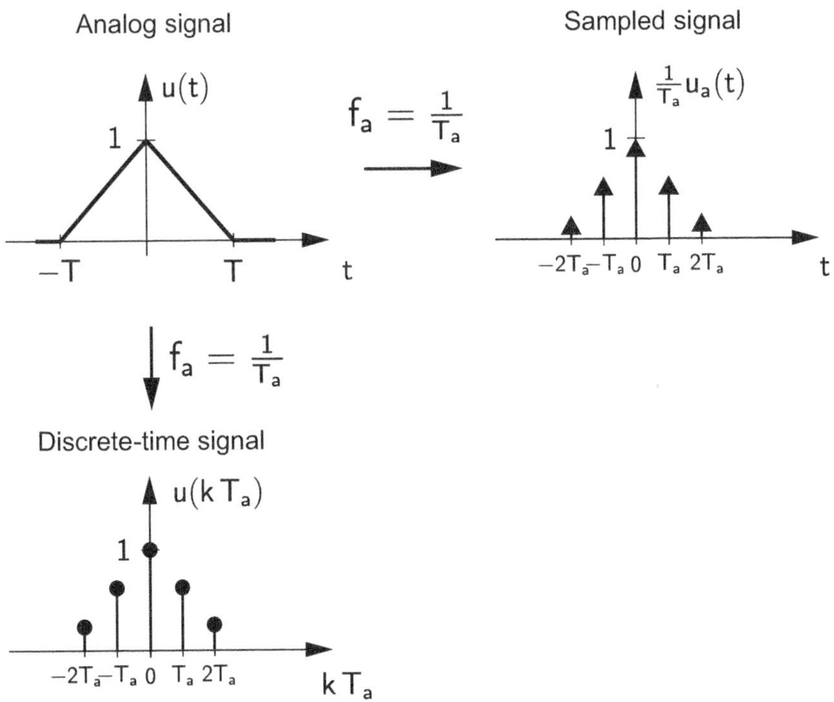

Fig. 1.58 Representation forms of discrete-time signals

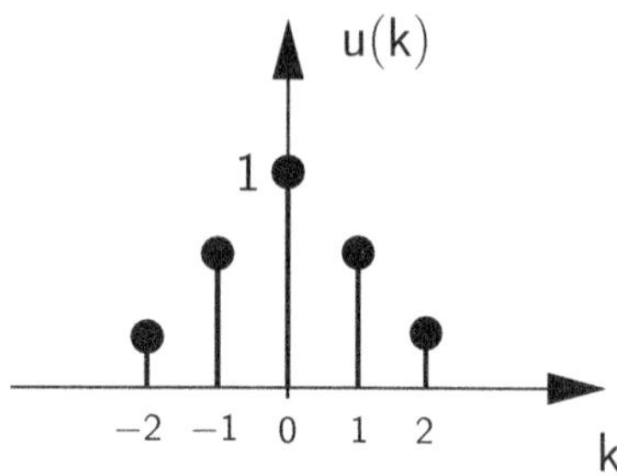

Fig. 1.59 Representation of a discrete-time signal

Often, the specification of T_a is omitted and discrete-time signals are represented as in Fig. 1.59.

Mathematical Model of Sampling The sampling represents an idealized process that triggers samples (sampling values) from the continuous-time signal at the interval of the sampling period T_a. With the definition introduced here

$$u_\mathrm{a}(t) = u(t) \cdot T_\mathrm{a} \sum_{k=-\infty}^{+\infty} \delta(t - k\,T_\mathrm{a}) \qquad (1.198)$$

the ideally sampled signal $u_\mathrm{a}(t)$ has the unit Volt, provided $u(t)$ was in the unit Volt (see, for example, [30]). Thus, the system functions introduced so far, such as amplitude density (in Vs), can be used unchanged for description. In contrast, in the literature, the ideally sampled signal is often also represented by

$$u_\mathrm{a}(t) = u(t) \sum_{k=-\infty}^{+\infty} \delta(t - k\,T_\mathrm{a}) \qquad (1.199)$$

(see, for example, [26, 48]), which is also suitable as a model to extract samples (sampling values) from the continuous-time signal $u(t)$ at the interval of the sampling period T_a. However, in this case, the ideally sampled signal $u_\mathrm{a}(t)$ has the unit V/s, which leads to a conflict in dimensions with the description of signals and systems introduced so far.

In order to capture the representations commonly used in the literature on the one hand, and not to abandon the correctness of the dimension (or the units) of the sampled signal on the other hand, this book will use the notation

$$u_\mathrm{a}(t) = u(t) \cdot C \sum_{k=-\infty}^{+\infty} \delta(t - k\,T_\mathrm{a}) \qquad (1.200)$$

as a mathematical model for sampling. The sampling period T_a—which always also causes an amplitude scaling of the sampled signal in relation to the analog signal—is replaced by constant C, which has the dimension of a time (with the unit s) and whose magnitude is set to 1 in the following.

On the Significance of the Constant C The function

$$u(t) = A \cdot \delta(t) \tag{1.201}$$

is referred to as a pulse in the time domain. Here, A is the area of the pulse. The time pulse has the unit (1/s) (see, for example, Appendix A). The value A—interpretable as the area of the pulse function—must then have the unit Vs or V/Hz so that $u(t)$ in (1.201) has the required dimension of a voltage (in V). The following applies

$$[\text{time function}] = [\text{Area of the Dirac impulse}] \cdot [\text{Dirac impulse}]$$
$$u(t) \quad = \quad A \quad \cdot \quad \delta(t) \quad .$$

In the context of sampling, different definitions for the definition of the area of the time pulse have been established. With the definition of the Dirac pulse sequence in the form

$$u_\delta(t) = C \sum_{k=-\infty}^{+\infty} \delta(t - k T_a) \tag{1.202}$$

the ideally sampled signal results in

$$u_a(t) = u(t) \cdot u_\delta(t) = C \sum_{k=-\infty}^{+\infty} u(k T_a) \delta(t - k T_a). \tag{1.203}$$

Figures 1.60 and 1.61 illustrate the significance of the constant C during sampling. Here, the value $C \cdot u(k T_a)$ is to be understood as the area of the Dirac pulse and in the case of $C = T_a$ takes the value $T_a \cdot u(k T_a)$ (in Vs) and in the case of $C = 1$ s takes the value $1 \cdot u(k T_a)$ (in Vs) on.

In the further course of the book, the constant C is set to the value $C = 1$ s, which has the advantage that the weights of the Dirac impulses in the ideally sampled signal correspond to the sampling values themselves, i.e., $u(k T_a)$, (without violating the units in the subsequent functional transformations). However, it should be noted that the constant C is dimensioned.

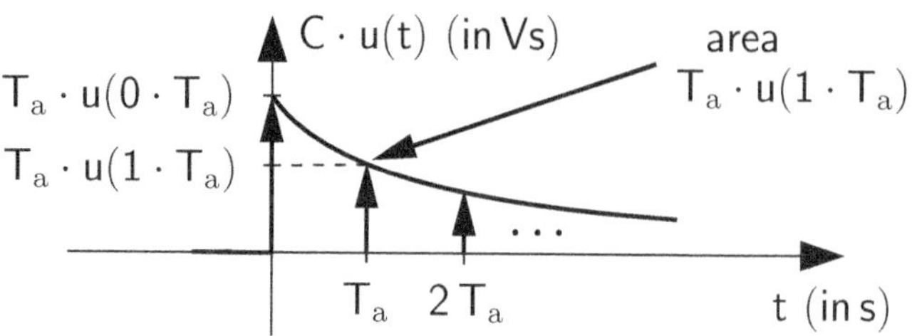

Fig. 1.60 On the significance of the constant $C = T_a$ within the sampling process

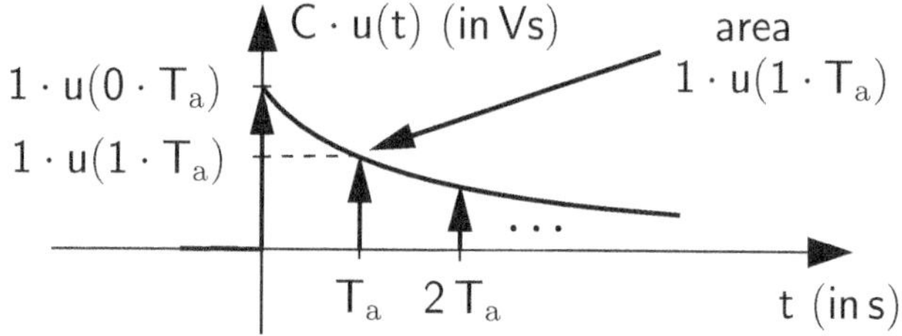

Fig. 1.61 On the significance of the constants $C = 1$ s within the sampling process

Key statement:

In Fig. 1.61, the area $C \cdot u(k\,T_a)$ with $C = 1$ s numerically (but not in terms of units) corresponds to the sampling value $u(k\,T_a)$, while in Fig. 1.60 the area $C \cdot u(k\,T_a)$ with $C = T_a$ also does not numerically correspond to the sampling value $u(k\,T_a)$.

Even though the signal-theoretical description should come as close as possible to practice (e.g., by choosing a suitable basis (time)), the specific choice of the time constants C has no effect on the signal-theoretical relationships. The discretization of a time signal leads to a periodicity of the spectrum of the signal to be sampled.

Spectrum of Sampled Signals The sampling of a time signal with the sampling period T_a corresponds to a periodicity with the period $f_a = 1/T_a$ in the spectral range.

Figures 1.62 and 1.63 illustrate the periodicity in the frequency range associated with sampling.

For the analysis of the frequency composition of continuous-time signals, the Fourier and Laplace transformations are available. The Laplace transformation offers the advantage that it is defined for a much larger class of signals and can avoid convergence problems when evaluating the Fourier integral for a large class of time functions. For the analysis of the frequency composition of discrete-time signals, the Z transformation or the discrete Fourier transformation (FTD, DFT) are available [16, 48]. The FTD (Fourier transform for time-discrete signals or time-discrete Fourier transform) offers the possibility to examine discrete-time signals for their frequency composition. The resulting spectrum

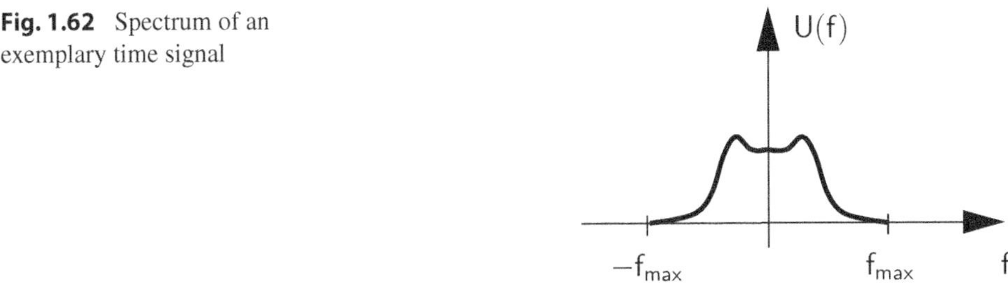

Fig. 1.62 Spectrum of an exemplary time signal

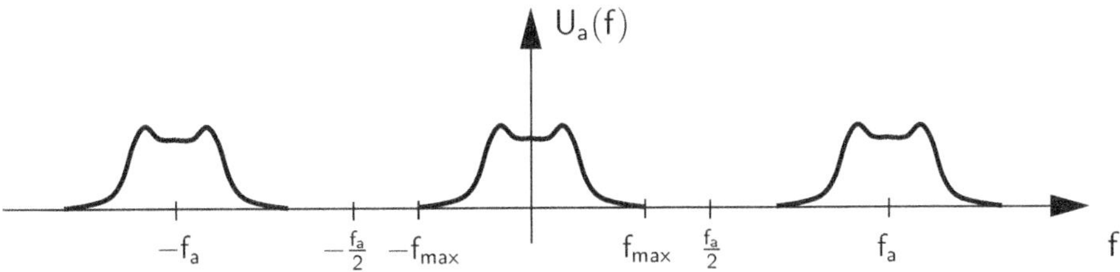

Fig. 1.63 Spectrum of an exemplary sampled time signal

is a continuous periodic function in f. Individual spectral components can be calculated using the DFT (discrete Fourier transform), which is the most commonly used method for analyzing the frequency composition of discrete-time signals. Here, a discrete-time signal—assuming periodic continuation in the frequency range—is mapped onto a discrete, periodic frequency spectrum.

Furthermore, it should be noted at this point that power signals can also have a deterministic character and can differ in their properties from energy signals, although both types of signals can be lawfully described by an algebraic expression.

While deterministic power signals (e.g. $u(t) = U_0 \sin(\omega_1 t)$) can be described by a power spectral density, i.e., a spectral function that provides information about the frequency distribution of a signal's power, energy signals (e.g. $u(t) = \text{rect}(t/T)$) are described by an energy density spectrum, i.e., a spectral function that provides information about the frequency distribution of a signal's energy.

1.4.2 Stochastic Signals

The input-output relationship for random variables (see Fig. 1.64) in the frequency range reads

$$\Psi_2(f) = |G(f)|^2 \cdot \Psi_1(f). \tag{1.204}$$

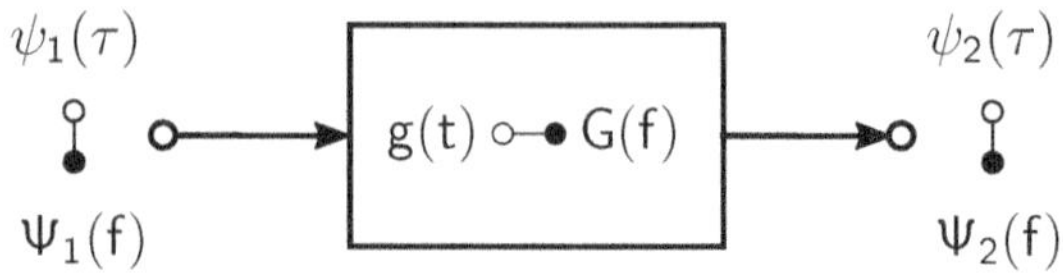

Fig. 1.64 For the definition of a system with random input and output variables

Since $G(f)$ is dimensionless (and thus also $|G(f)|^2$), calls the power density (or square amplitude density) present at the system's input $\Psi_1(f)$ (in V^2/Hz) also a power density $\Psi_2(f)$ (in V^2/Hz) at the output. In the time domain, the input-output relationship is defined by the convolution

$$\psi_2(\tau) = \psi_g(\tau) * \psi_1(\tau) = \int\limits_{-\infty}^{+\infty} \psi_g(t) \cdot \psi_1(t - \tau)\, dt \tag{1.205}$$

with

$$\psi_g(\tau) = \int\limits_{-\infty}^{+\infty} g(t) \cdot g(t + \tau)\, dt \tag{1.206}$$

as the autocorrelation function (ACF) of the weighting function $g(t)$. In (1.206), the unit of an $g(t)$ (in $1/s$) cancels out against the unit (s) of the dt from the convolution integral, leaving the dimension of $g(t)$ (in $1/s$) for the result $\psi_g(\tau)$ (in $1/s$). Thus, in (1.205) $\psi_g(\tau)$ has the unit $(1/s)$, which cancels out against the unit (s) of the dt from the convolution integral, leaving as the unit for the result $\psi_2(\tau)$ in (1.205) the unit of the autocorrelation function.$\psi_1(\tau)$ of the input signal (in V^2).

For the analysis of the interplay of input signal (i.e., random signal), system, output signal (i.e., random signal) according to Fig. 1.64, the consideration will be limited exclusively to stationary and ergodic random processes with Gaussian distribution. When passing through LTI systems, this Gaussian distribution is preserved, and it is particularly possible to calculate the resulting error probability in the transmission of digital signals for such Gaussian-distributed noise signals.

A derived, refined model for random processes results when white noise is subjected to filtering. In this case, a band-limited white noise generally arises: This is also known as *colored* noise—also in reference to light properties that make light appear colored when only a few spectral components are present or in the extreme case a single light wavelength is observed.

Example 1.17 (Filtering of white noise) As is well known, the Fourier transformation of white noise leads to the power spectral density

$$\psi(\tau) = \Psi_0\,\delta(\tau) \circ\!\!-\!\!\bullet \quad \Psi(f) = \Psi_0. \tag{1.207}$$

If this white random process is subjected to a filtering with the weighting function (see Fig. 1.65)

$$g(t) = \begin{cases} \dfrac{1}{T} & \text{for} \quad |t| < \frac{T}{2} \\[2mm] 0 & \text{otherwise} \end{cases} \tag{1.208}$$

and the associated autocorrelation function

$$\psi_g(\tau) = \begin{cases} \dfrac{1}{T}\left(1 - \dfrac{|\tau|}{T}\right) & \text{for} \quad |\tau| < T \\[2mm] 0 & \text{otherwise} \end{cases} \tag{1.209}$$

(see Fig. 1.66), the power spectral density results

$$\Psi_2(f) = |G(f)|^2 \cdot \Psi_1(f) = \Psi_0\,\mathrm{si}^2(\pi f T). \tag{1.210}$$

Through the inverse Fourier transformation, $\psi_2(\tau)$ can be determined to

$$\psi_2(\tau) = \begin{cases} \dfrac{\Psi_0}{T}\left(1 - \dfrac{|\tau|}{T}\right) & \text{for} \quad |\tau| < T \\[2mm] 0 & \text{otherwise} \end{cases} \tag{1.211}$$

The mean square value (i.e., the power) of the random process after passing through the filter with the weighting function $g(t)$ results in

$$U_{\mathrm{R}}^2 = \psi_2(0) = \int_{-\infty}^{\infty} \Psi_2(f)\,\mathrm{d}f = \Psi_0 \int_{-\infty}^{\infty} \mathrm{si}^2(\pi f T)\,\mathrm{d}f = \frac{\Psi_0}{T} < \infty. \tag{1.212}$$

Fig. 1.65 Weighting function $g(t)$

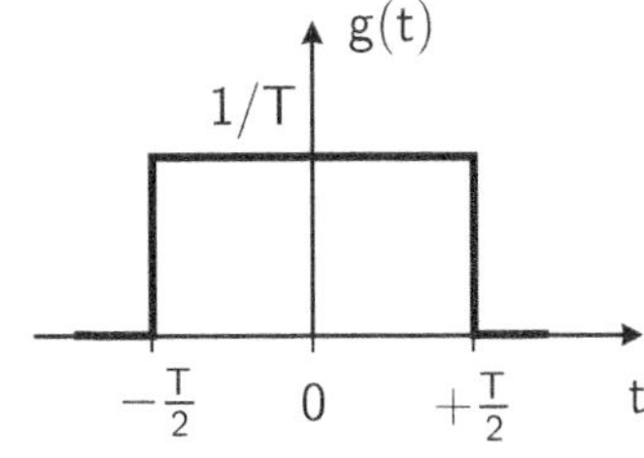

Fig. 1.66 Autocorrelation function $\psi_g(t)$

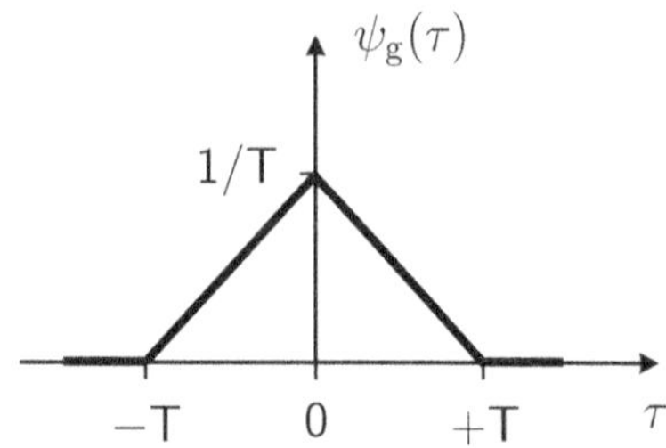

Insert: Derivation of the Power Density Spectrum when Filtering White Noise With a Sinc Low-Pass Filter

If this white random process is subjected to filtering with the weighting function (see Fig. 1.65)

$$g(t) = \begin{cases} \dfrac{1}{T} & \text{for} \quad |t| < \dfrac{T}{2} \\ 0 & \text{otherwise} \end{cases} \tag{1.213}$$

the corresponding autocorrelation function can be calculated using the approach

$$\psi_g(\tau) = \int_{-\infty}^{+\infty} g(t) \cdot g(t + \tau)\, \mathrm{d}t . \tag{1.214}$$

The autocorrelation function can be calculated using convolution. Here applies

$$\psi_g(\tau) = g(\tau) * g(-\tau). \tag{1.215}$$

The conversion of the time domain operation into the spectral domain results in

$$\Psi_g(f) = |G(f)|^2. \tag{1.216}$$

With the weighting function $g(t)$ in the form

$$g(t) = \begin{cases} \dfrac{1}{T} & \text{for} \quad |t| < \dfrac{T}{2} \\ 0 & \text{otherwise} \end{cases} \tag{1.217}$$

follows for the transfer function (in reference to the previously derived amplitude spectral density of a rectangular pulse)

$$G(f) = \mathrm{si}(\pi f T). \tag{1.218}$$

Since this low-pass filter in the frequency domain follows a sinc function $\mathrm{si}(\pi f T)$, it is also referred to as a *sinc low-pass filter*. With

$$\Psi_g(f) = |G(f)|^2 = \text{si}^2(\pi f T) \tag{1.219}$$

the desired power density spectrum at the filter output results

$$\Psi_2(f) = \Psi_0 \cdot |G(f)|^2 = \Psi_0\,\text{si}^2(\pi f T). \tag{1.220}$$

$\square$

Random signals can be described via correlation and distribution functions. Zero-mean, Gaussian-distributed noise signals exhibit the probability density function

$$p(u) = \frac{1}{U_R\sqrt{2\pi}} \cdot e^{-\frac{u^2}{2U_R^2}} \tag{1.221}$$

where U_R indicates the root mean square—or the dispersion—of the disturbance (see Fig. 1.67).

In communications engineering, the probability is often sought that the disturbance amplitude U exceeds or falls below a given threshold U_E.

The probability $P_{U_E} = P(U \geq U_E)$ given the knowledge of the probability density function $p(u)$ over (see Fig. 1.68)

$$P_{U_E} = \int_{U_E}^{\infty} p(u)\mathrm{d}u \tag{1.222}$$

can be calculated. With the help of the complementary Gaussian error function $\text{erfc}(x)$ in the form (see Appendix E)

$$\text{erfc}(x) = \frac{2}{\sqrt{\pi}} \int_{x}^{\infty} e^{-y^2}\,\mathrm{d}y \tag{1.223}$$

the probability $P_{U_E} = P(U \geq U_E)$ that the disturbance amplitude exceeds the threshold U_E is given by

$$P_{U_E} = \frac{1}{2}\text{erfc}\left(\frac{U_E}{\sqrt{2}\,U_R}\right) = \frac{1}{2}\left[1 - \text{erf}\left(\frac{U_E}{\sqrt{2}\,U_R}\right)\right]. \tag{1.224}$$

Fig. 1.67 Probability density function of a zero-mean, Gaussian-distributed noise signal

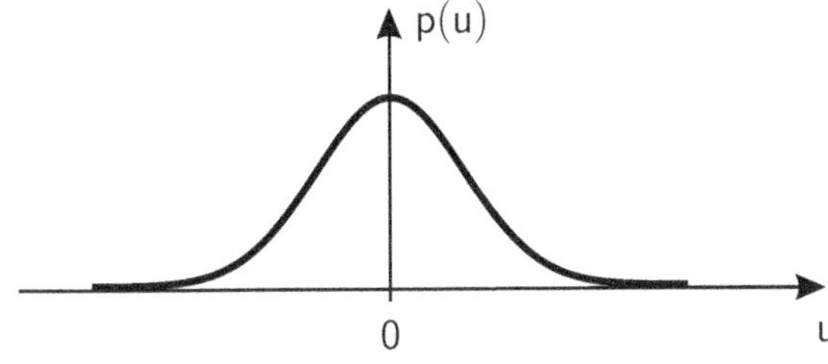

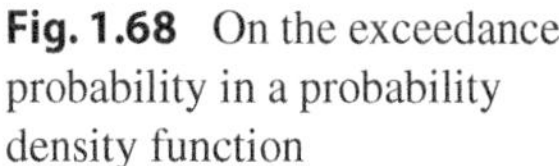

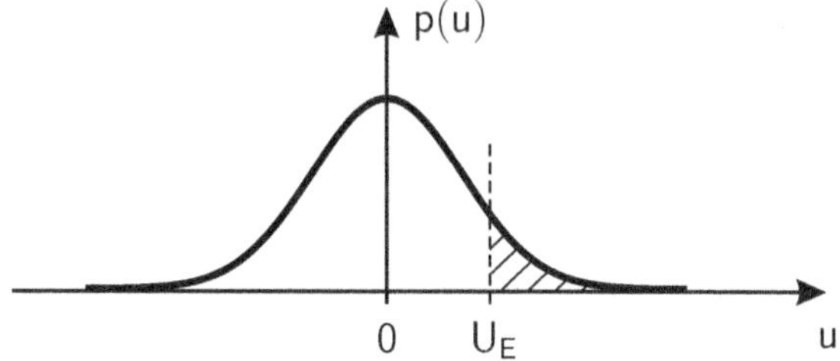

Fig. 1.68 On the exceedance probability in a probability density function

A detailed calculation of this relationship is found in the derivation of the error probability in Chap. 2.

1.5   Further Functional Transformations

In the preceding sections of this chapter, it was already mentioned that in addition to the discussed Fourier transformation, there are other functional transformations that establish a connection between an original domain (time domain) and an image domain (frequency domain).

Two more such transformations, important for this textbook, will be briefly introduced below:

- The *Laplace transformation* is an extension of the Fourier transformation and allows a much larger class of time functions (than the Fourier transformation itself) to be transformed into an image domain—the frequency domain.
- The *Z-transformation* allows time-discrete signals to be transformed into an image domain. It is based on the Laplace transformation.

1.5.1   Laplace Transformation

With the introduction of a complex frequency in the form $p = \sigma + j\omega$ in the Fourier-Transformation[17] convergence difficulties that lead to the Fourier integral not being able to be evaluated are avoided. In this book, we work with the one-sided Laplace transformation, which is defined for time signals that are limited to the time interval $0 \leq t < \infty$

Such signals have great practical significance, as a large number of processes in technology have the property that they begin or began at a certain point in time: This point in time is arbitrarily set to $t = 0$ [16]. The Laplace transformation also has a two-sided

---

[17] Pierre-Simon Laplace (1749–1827), French mathematician and astronomer.

definition of this functional transformation, see e.g. [18], which is also referred to as complex or general Fourier transformation [16], as it represents a generalized Fourier transformation with a complex frequency.

The Laplace transformation represents an extension or a generalization of the Fourier transformation and allows the transformation of a much larger class of time functions or signals into an spectral domain than that it is possible with the Fourier transformation. However, it should be noted that if the Fourier transform is missing for the signals, no statement can be made about their frequency composition. Since the sufficient condition for the existence of the Fourier integral and thus the Fourier transformation in the form

$$\lim_{t \to \infty} x(t) = 0 \tag{1.225}$$

is not met for a large number of signals such as step, ramp, cosine and sine functions, the parameter σ is introduced to avoid convergence difficulties and the Fourier transform belonging to the newly emerging signal $x(t)\, e^{-\sigma t}$ is determined. The factor $e^{-\sigma t}$ forces the convergence of the Fourier integral for a larger class of functions, namely those that increase or rise less with time than the *convergence-enforcing factor* $e^{-\sigma t}$ decreases with increasing time progress. In contrast to the Fourier transformation, the Laplace transformation thus also allows exponentially rising time functions.

In this case, the approach

$$X(j\omega) = X(\omega) = \int_0^\infty x(t)\, e^{-j\omega t}\, dt \tag{1.226}$$

can be transformed into the form

$$X(j\omega) = X(\omega) = \int_0^\infty x(t)\, e^{-\sigma t}\, e^{-j\omega t}\, dt = \int_0^\infty x(t)\, e^{-(\sigma + j\omega) t}\, dt. \tag{1.227}$$

With the complex frequency variable $p = \sigma + j\omega$, this results in the Laplace transform for one-sided limited time signals

$$X(p) = \int_0^\infty x(t)\, e^{-p t}\, dt. \tag{1.228}$$

The complex frequency variable $p = \sigma + j\omega$ is a pure computational quantity that cannot be measured.

The inverse Laplace transformation is

$$x(t) = \mathscr{L}^{-1}\{X(p)\} = \frac{1}{2\pi j} \int_{\sigma - j\infty}^{\sigma + j\infty} X(p)\, e^{p t}\, dp. \tag{1.229}$$

Important laws and calculation rules of the Laplace transformation as well as a selection of common correspondences are compiled in Appendix D.

Since in this book, in connection with the Laplace transformation, only one-sided time functions with the property

$$x(t) = 0 \qquad \text{für} \quad t < 0$$

are considered, the addition $1(t)$ is omitted—provided there is no risk of confusion. In this respect, the notations

$$x(t) \qquad \text{und} \qquad x(t) \cdot 1(t)$$

and

$$x(t - t_0) \qquad \text{und} \qquad x(t - t_0) \cdot 1(t - t_0)$$

are used interchangeably.

Convergence Area

Since both functional transformations (Laplace and Fourier transformation) are related, it is of interest when Fourier and Laplace transformations can be used interchangeably: This statement can be determined by analyzing the convergence area. The convergence area describes the range of values of p for which the definition equation $X(p)$ can be evaluated for a given signal $x(t)$ and the Laplace integral (1.228) converges.

Both transformations can be used completely interchangeably when the range of values of p completely includes the imaginary axis. In this case, the Fourier transform of a given time signal $x(t)$ can be derived from the Laplace transform by evaluating the complex frequency variable $p = \sigma + j\omega$ for $\sigma = 0$.

Figure 1.69 illustrates this relationship. In case a), both transformations can be used completely equally, while in case b) a Laplace transform can be specified, but no Fourier transform. This also means that for the time signal in case b), no Fourier transform exists and therefore no frequency composition for the signal $x(t)$ can be found.

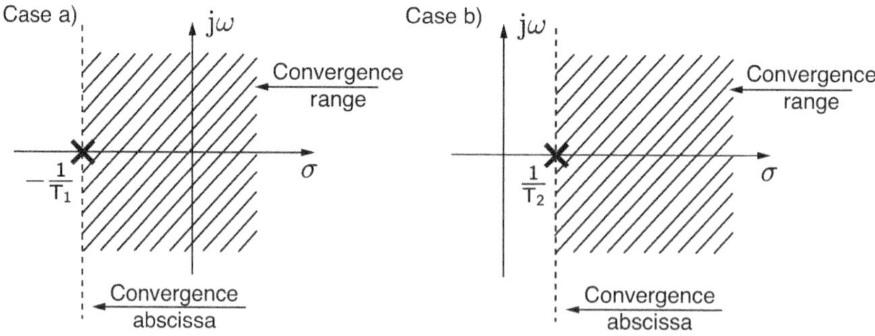

Fig. 1.69 For the definition of the convergence area

Signals where the imaginary axis limits the convergence area are often referred to as marginally stable systems and require separate consideration.

For determining the convergence area, it is helpful to transform a given fractional-rational function into the following form consisting of poles, zeros, and a constant

$$X(p) = K \, \frac{\displaystyle\prod_{\nu=0}^{N} (p - p_{0\,\nu})}{\displaystyle\prod_{\mu=0}^{M} (p - p_{\infty\,\mu})}. \tag{1.230}$$

Here, $p_{0\,\nu}$ are the zeros of the numerator polynomial and $p_{\infty\,\mu}$ are the zeros of the denominator polynomial, which are also referred to as *pole positions* or simply as *poles*. It is common for the zeros to be represented as 'o' and the pole positions as 'x' in the pole-zero diagram. Figure 1.70 illustrates the pole-zero representation for a system with a simple real pole (first-order low pass) with the transfer function

$$G(p) = \frac{1}{1 + p\,T_{\mathrm{g}}}. \tag{1.231}$$

The transfer function of this system was given as a Fourier transform in (1.158) and shown in Figs. 1.44 and 1.45.

Example 1.18 (Laplace transformation of a time function) The time function

$$x(t) = \mathrm{e}^{at} \cdot 1(t) \tag{1.232}$$

is given. In Fig. 1.71, the basic course of this time function for three important cases $a > 0$, $a = 0$ and $a < 0$ is shown.

For the Laplace transform of the function $x(t)$ applies

$$X(p) = \mathscr{L}\{x(t)\} = \int_0^\infty \mathrm{e}^{at} \cdot \mathrm{e}^{-pt}\,\mathrm{d}t = \int_0^\infty \mathrm{e}^{-(p-a)t}\,\mathrm{d}t$$

$$= \frac{1}{-(p-a)}\mathrm{e}^{-(p-a)t}\Big|_0^\infty = 0 - \frac{1}{-(p-a)}$$

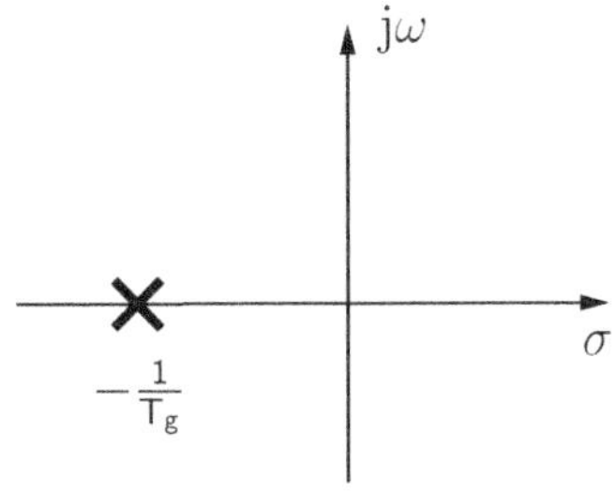

Fig. 1.70 Pole-zero diagram for a first-order low pass filter

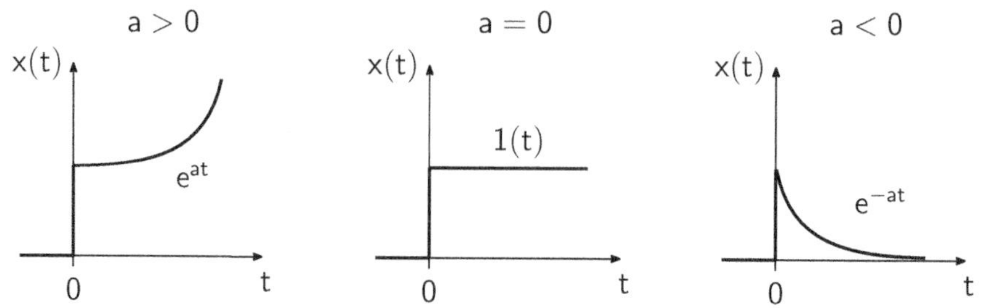

Fig. 1.71 Basic course of the time function $x(t) = e^{at} \cdot 1(t)$ for different a

and it results (for $\mathrm{Re}\{p\} = \sigma > a$)

$$X(p) = \mathscr{L}\{x(t)\} = \frac{1}{p - a}. \tag{1.233}$$

Results for special cases, which are significant in technical practice, are obtained for

$$a = 0 \quad \Longrightarrow \quad \mathscr{L}\{1(t)\} = \frac{1}{p}$$

$$a \to -a \quad \Longrightarrow \quad \mathscr{L}\{e^{-at}\} = \frac{1}{p + a}$$

$$a = \pm j\omega \quad \Longrightarrow \quad \mathscr{L}\{e^{\pm j\omega t}\} = \frac{1}{p \mp j\omega}.$$

Example 1.19 (Laplace transformation of a Dirac pulse) If one transforms a Dirac impulse $x(t) = \delta(t)$ using the Laplace transformation, the result is

$$X(p) = \mathscr{L}\{x(t)\} = \int_0^\infty \delta(t) \cdot e^{-pt}\,dt = \int_0^\infty \delta(t) \cdot e^{-p0}\,dt = \int_0^\infty \delta(t)\,dt = 1. \tag{1.234}$$

A Dirac impulse in the time domain corresponds to a constant in the image domain:

$$x(t) = \delta(t) \quad \circ\!\!-\!\!\bullet \quad X(p) = 1. \tag{1.235}$$

Since here the entire p domain forms the convergence domain, the transition $p = j\omega$ can be made and the Fourier transform can be determined as

$$x(t) = \delta(t) \quad \circ\!\!-\!\!\bullet \quad X(j\omega) = 1.$$

(1.236)

$\square$

On the Relationship between Fourier and Laplace Transformations

Since the Fourier transformation (1.60) does not exist for time-increasing functions, the Laplace transformation (1.228) was introduced and defined using a convergence-enforcing factor.

For the application of the Fourier and Laplace transformations, the question often arises as to when both transformations are equivalent and can be applied equally. This will be illustrated after the discussion on the convergence domain using a thought experiment.

The frequency response or the transfer function of a given system $G(f)$—e.g., a filter or a broadband amplifier—is to be measured. The arrangement is shown in Fig. 1.72: A tunable frequency generator is provided as a source, which generates sinusoidal or cosine-shaped—thus time-harmonic—voltage signals with adjustable frequency and constant amplitude. The system with the frequency response to be determined is connected downstream of this source, and a measuring device (voltage meter) is connected to it, which records the voltage at the output of the system. A special feature is the two potentiometers, which are switched into the setup before and after the unknown system: Depending on the setting angle α of the common actuator, they provide an evaluation of the respective signal with a real factor v_1 (before the system) and v_2 (after the system) and thus simulate the factor $\mathrm{e}^{-\sigma t}$ of the Laplace transformation. They are coupled in such a way that for every point in time $v_1 \cdot v_2 = \text{const.} = 1$ applies, e.g., $v_1 \sim \alpha$ and $v_2 \sim 1/\alpha$, see Fig. 1.73, upper part. An exemplary relationship between the gain factors v_1 and v_2 with σ could be $v_1 = \mathrm{e}^{-\sigma t}$ and $v_2 = \mathrm{e}^{\sigma t}$.

If the frequency response at the output is now measured by tuning the input frequency, two cases can arise:

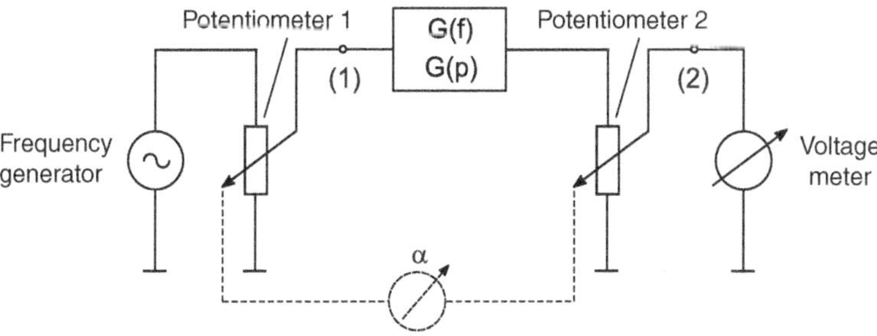

Fig. 1.72 Arrangement of a thought experiment on the equivalence of Fourier and Laplace transformations

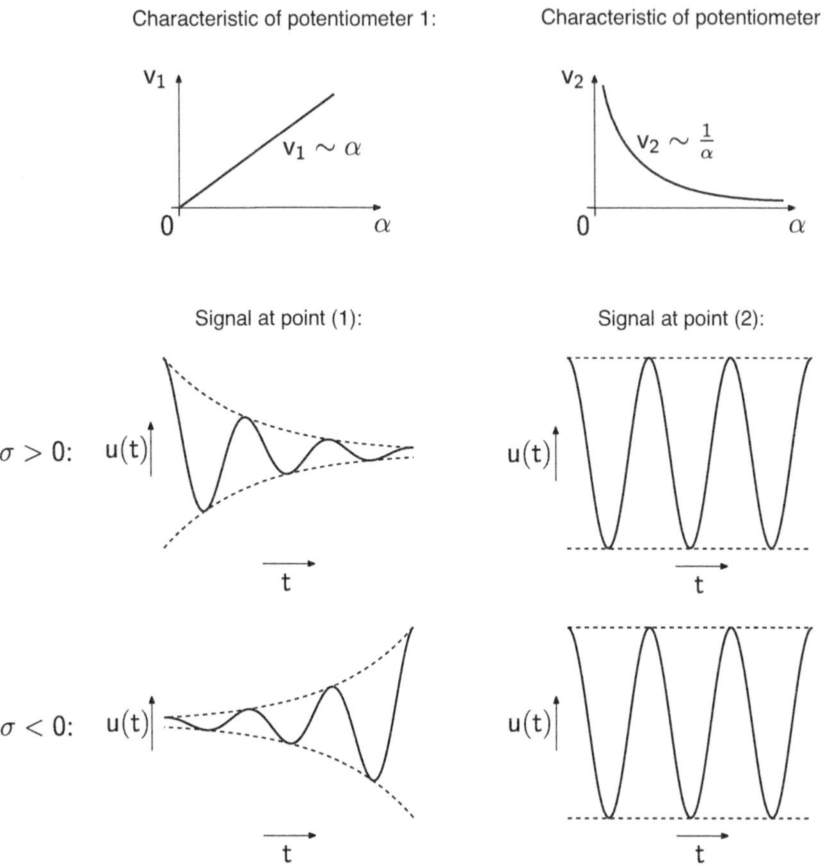

Fig. 1.73 For the thought experiment: Characteristics of the potentiometers and resulting signals at the points (1) and (2)

1. The amplitude of the sinusoidal or cosine-shaped time function remains constant during the measurement: $v_1 = v_2 = 1$. This corresponds to the Fourier transformation ($\sigma = 0$).
2. During the measurement, the control knob for both potentiometers is operated: At point (1), a sinusoidal or cosine-shaped signal is obtained, which either rises exponentially ($\sigma < 0$) or decays ($\sigma > 0$). At point (2), a sinusoidal or cosine-shaped signal with constant amplitude is obtained in both cases, since the rise or decay caused by v_1 is reversed by v_2, see Fig. 1.73, lower part.

The measuring device "notices" nothing of these changes by the potentiometers in the second case of the thought experiment or more generally of the change introduced by the convergence-enforcing factor in the Laplace transformation: The measured transfer functions are the same in both cases and thus Fourier and Laplace transformations—where both are defined—are equivalent and lead to identical results.

Therefore, in practical handling, one can choose which way a transfer function of a system should be determined, whereby boundary conditions regarding the convergence area must be observed.

1.5.2 Z-Transformation

The Z-Transformation is the discrete-time counterpart to the Laplace transformation and thus a tool for describing discrete-time signals in the image domain. A simple way to generate discrete-time signals from analog signals is the multiplication of a continuous-time signal $x(t)$ with a pulse comb or a periodic Dirac pulse sequence[18]

$$x_\delta(t) = \sum_{k=-\infty}^{+\infty} \delta(t - k\,T_\mathrm{a}).$$ (1.237)

The parameter T_a is also referred to as the sampling period. The inverse of the sampling period characterizes the sampling frequency f_a

$$f_\mathrm{a} = \frac{1}{T_\mathrm{a}}.$$ (1.238)

The sampling frequency plays a crucial role when a continuous-time signal is to be accurately recovered from a discrete-time signal. The result of sampling a continuous-time signal $x(t)$ with a Dirac pulse comb $x_\delta(t)$ results in an ideally sampled signal (using the selective property of the Dirac pulse) in the form (Fig. 1.74)

$$x_\mathrm{a}(t) = \sum_{k=-\infty}^{+\infty} x[k\,T_\mathrm{a}] \cdot \delta(t - k\,T_\mathrm{a}).$$ (1.239)

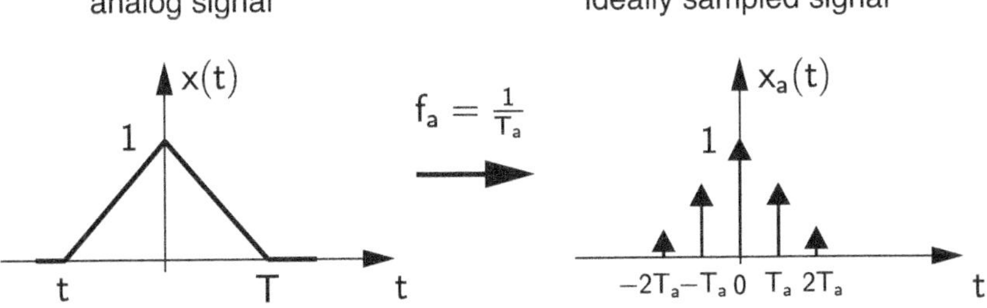

Fig. 1.74 For generating sequences from analog signals

---

[18] In the following, the simplified representation with $C = 1$ is used for sampling.

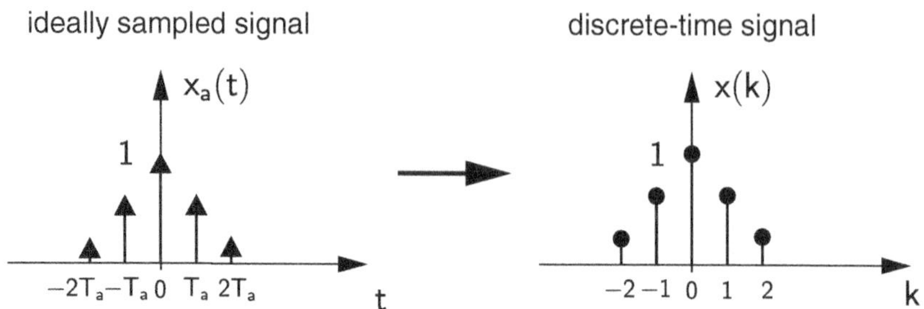

Fig. 1.75 Representation of discrete-time signals

The ideally sampled signal represents the time sequence (sampling sequence) represented by the sampling values $x[k] := x[k\,T_a]$. For simplification, it is common to write only $x(k)$ since the equidistant step size T_a is a self-evident attribute (Fig. 1.75). In the following, square brackets are used for better distinguishability when an element of the time sequence (sampling sequence) is to be analyzed, e.g., $x[2]$ and round brackets when the entire time sequence is described, e.g., $x(k)$.

Furthermore, in system theory, it is common to work with causal signals, i.e., signals that start at $t = 0$. Thus, the ideally sampled signal can be described as

$$x_a(t) = \sum_{k=0}^{\infty} x[k] \cdot \delta(t - k\,T_a) \ . \tag{1.240}$$

By applying the time shift theorem of the Laplace transformation in the form

$$x(t - t_0) \cdot 1(t - t_0) \ \circ\!\!-\!\!\bullet \ X(p) \cdot e^{-p\,t_0} \tag{1.241}$$

with $x(t) \cdot 1(t) \ \circ\!\!-\!\!\bullet \ X(p)$.

The Laplace transform of the one-sided (limited to non-negative times) and ideally sampled signal $x_a(t)$ results in

$$X_a(p) = \sum_{k=0}^{\infty} x[k] \cdot e^{-p\,k\,T_a}. \tag{1.242}$$

With the independent image variable

$$z = e^{p\,T_a} \tag{1.243}$$

the Z-transform of the discrete-time signal $x(k)$ yields

$$X(z) = \sum_{k=0}^{\infty} x[k] \cdot z^{-k} \, . \tag{1.244}$$

It turns out that $X(z)$ is a special series with descending powers of z (Laurent series). It can be written as

$$X(z) = x[0] + x[1] \cdot z^{-1} + x[2] \cdot z^{-2} + x[3] \cdot z^{-3} + \dots \, . \tag{1.245}$$

The factor z^{-k} in (1.244) can be interpreted as an operator that shifts a process by k cycles to the right. This allows a relatively clear transition from the knowledge of system characteristic functions weight sequence and Z-transfer function to system realizations— which is usually much more complicated in the case of continuous systems [16].

Example 1.20 (Z-Transformation of the unit step sequence) The unit step sequence is given as

$$x(k\,T_{\mathrm{a}}) = 1(k) = \{1, 1, 1, 1, \dots\} \tag{1.246}$$

for $k = 0, 1, 2, \dots$ [7]. The Z-Transformation results in

$$X(z) = \mathcal{Z}\{1(k)\} = 1 + \frac{1}{z} + \frac{1}{z^2} + \frac{1}{z^3} + \dots = \sum_{k=0}^{\infty} z^{-k} \tag{1.247}$$

and a transformation initially leads to

$$\frac{1}{z} \cdot X(z) = \frac{1}{z} \cdot \sum_{k=0}^{\infty} z^{-k}$$

$$\frac{1}{z} \cdot X(z) = z^{-1} \cdot \sum_{k=0}^{\infty} z^{-k} = \sum_{k=0}^{\infty} z^{-k-1} = \sum_{k=0}^{\infty} z^{-(k+1)}$$

$$\frac{1}{z} \cdot X(z) = \sum_{k=1}^{\infty} z^{-k} = \underbrace{\sum_{k=0}^{\infty} z^{-k}}_{X(z)} - z^{-0} = \sum_{k=0}^{\infty} z^{-k} - 1$$

and further results in

$$\frac{1}{z} \cdot X(z) = X(z) - 1$$

$$X(z) = z \cdot X(z) - z$$

$$z = z \cdot X(z) - X(z) = X(z) \cdot (z - 1)$$

and thus finally as a result

$$X(z) = \mathcal{Z}\{1(k)\} = \frac{z}{z-1}. \tag{1.248}$$

$\square$

The original function corresponding to a given Z-Transform is obtained via the inverse transformation:

$$x(k) = \mathcal{Z}^{-1}\{X(z)\}. \tag{1.249}$$

The inverse Z-Transformation is defined via the inverse integral

$$x(k) = \frac{1}{2\pi\mathrm{j}} \oint X(z) \cdot z^{k-1}\, \mathrm{d}z, \tag{1.250}$$

which is to be extended over a closed curve that runs within the convergence area and includes all singularities of $X(z)$ [7]. This type of inverse transformation can be applied to any function $X(z)$; the evaluation can be done using residue calculation, but the computational effort is often high. Other methods for the practical execution of the Z-inverse transformation are partial fraction decomposition followed by the use of correspondence tables, series development, and the application of a recursion formula [7].

The distribution of poles and zeros plays a crucial role in the convergence behavior of discrete-time systems. Figure 1.76 illustrates a possible distribution of poles and zeros of a discrete-time system. Stable systems are characterized by the fact that poles (marked with $\times$) lie within the unit circle. Zeros, as with time-continuous systems, are marked with the symbol $\bigcirc$. Systems whose poles all lie at the origin are also referred to as FIR systems (finite impulse response) or as systems with finite impulse response length. Thus, the system given in case a) can be described by the transfer function

$$G(z) = \frac{z + \frac{1}{2}}{z} = 1 + \frac{1}{2}z^{-1} \tag{1.251}$$

and in the time domain by the weighting function

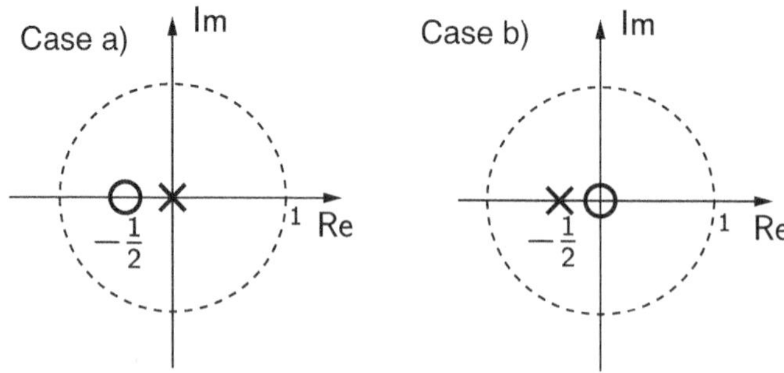

Fig. 1.76 Pole-zero plot of given discrete-time systems

$$g(k) = \delta(k) + \frac{1}{2}\,\delta(k-1)\,. \tag{1.252}$$

The symbol $\delta(k)$ defines a discrete Dirac impulse with the property

$$\delta(k) = \begin{cases} 1 & \text{for } k = 0 \\ 0 & \text{for } k \neq 0 \end{cases}. \tag{1.253}$$

If a discrete system $g(k)$ is subjected to an input signal $u_1(k)$ (see also Fig. 1.77), the output signal $u_2(k)$ in the image domain results from

$$G(z) = \frac{U_2(z)}{U_1(z)} \qquad \text{to} \qquad U_2(z) = G(z) \cdot U_1(z). \tag{1.254}$$

With $G(z)$, defined in (1.251), it follows

$$U_2(z) = G(z) \cdot U_1(z) = \frac{z + 1/2}{z} \cdot U_1(z) = \left(1 + \frac{1}{2} z^{-1}\right) \cdot U_1(z), \tag{1.255}$$

which in the time domain leads to

$$u_2(k) = u_1(k) + \frac{1}{2} \cdot u_1(k-1)\,. \tag{1.256}$$

Figure 1.78 illustrates the resulting system structure.

In contrast, the pole-zero configuration given in case b) describes an IIR system (infinite impulse response). Such systems are also known as systems with infinite impulse response length.

In this case, the system given in case b) can be described by the transfer function

Fig. 1.77 For the definition of a system with discrete input and output variables

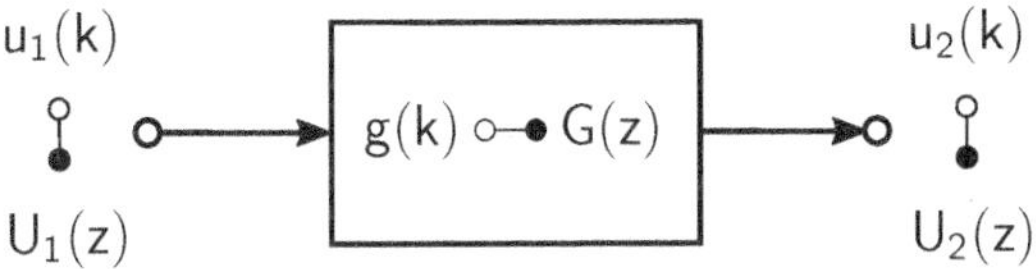

Fig. 1.78 Exemplary discrete-time model of the given FIR transmission system $G(z)$

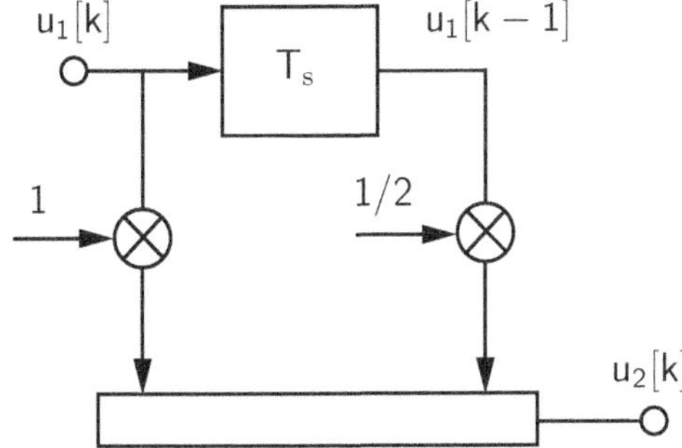

$$G(z) = \frac{z}{z + 1/2} \tag{1.257}$$

Now, if a discrete system $g(k)$ is subjected to an input signal $u_1(k)$, the output signal $u_2(k)$ in the image domain results from

$$G(z) = \frac{U_2(z)}{U_1(z)} \qquad \text{to} \qquad U_2(z) = G(z) \cdot U_1(z). \tag{1.258}$$

With $G(z)$, defined in (1.257), it follows

$$U_2(z) = G(z) \cdot U_1(z) = \frac{z}{z + 1/2} \cdot U_1(z) = \frac{1}{1 + 1/2\,z^{-1}} \cdot U_1(z). \tag{1.259}$$

By solving this equation, it follows

$$U_1(z) = \left(1 + \frac{1}{2}\,z^{-1}\right) \cdot U_2(z) = U_2(z) + \frac{1}{2}\,z^{-1} \cdot U_2(z), \tag{1.260}$$

which can also be written as

$$U_2(z) = U_1(z) - \frac{1}{2}\,z^{-1} \cdot U_2(z)\,. \tag{1.261}$$

The back transformation into the time domain leads to

$$u_2(k) = u_1(k) - \frac{1}{2} \cdot u_2(k - 1). \tag{1.262}$$

If such a system is now subjected to a discrete Dirac impulse to determine the Dirac impulse response, it has an infinite impulse response length, unlike the FIR system. In this case, for $u_2(k)$ the sequence follows

$$u_2(k) = (1, -0.5, 0.25, -0.125, \ldots). \tag{1.263}$$

Detailed presentations on the Z transformation as well as, for example, correspondence tables and important theorems can be found in [7, 68, 74].

1.6   Summary and Bibliographic Notes

Signals as carriers of information play an important role in many everyday devices and systems. Examples can be found in a variety of ways, such as in the transmission, storage, and processing of speech, music, or video images, which can initially be analog or digitized.

The transport of this generally very large amount of data requires a comprehensive knowledge of signal and system properties. In this chapter, basic relationships of signal and system theory were presented. Both deterministic and random signals were considered and typical examples were given, which play an important role especially in communications engineering. These include the description of signals and systems in the time and frequency domain, with suitable and frequently used transformations for continuous and discrete signals and systems being introduced.

Signal and system theory plays an important role in communications engineering and beyond in a variety of scientific branches and specialist areas of electrical engineering and computer science, e.g. in automation and control technology, in computer engineering and in adjacent areas. It thus forms a universal basis for the description, understanding, and design of a variety of technical processes and devices.

The essential basics of signal and system theory presented in this section are summarized compactly in Appendix A.

For a more in-depth study of signal and system theory, reference is made to the extensive literature. Considerations on this complex of topics can be found, for example, in the textbooks [18, 30] as well as [54] and [49]. Classic works on signal and system theory are [36–38] and [45]. An interactive approach to signal and system theory is offered by [27] with the possibility of conducting your own computer-aided experiments. The Z-Transformation is extensively treated in [7]. A more in-depth consideration of Fourier, Laplace, and Z-Transformation can be found in [68], a compact summary of signal description and the necessary functional transformations is offered by [16]. In [47], for example, further application possibilities of signal theory are presented. The description and processing of stochastic signals can be found in, e.g. [10, 20, 73]. All of the references mentioned here are in German.

1.7 Problems

Problem 1.1 (Fourier Series)
Given are the (real) Fourier coefficients of a DC-free signal $u(t)$ according to Fig. 1.79.
The following are sought:

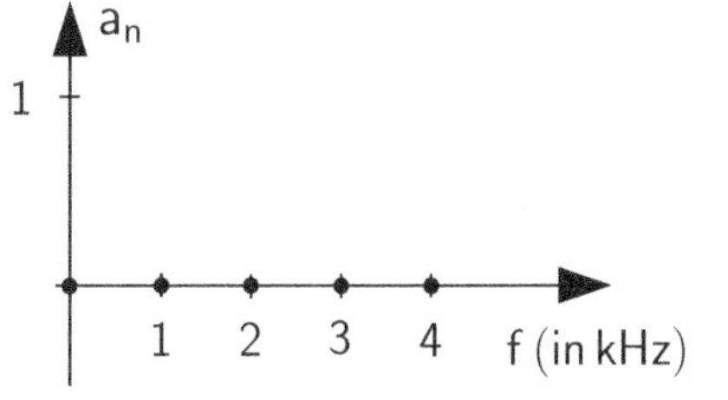

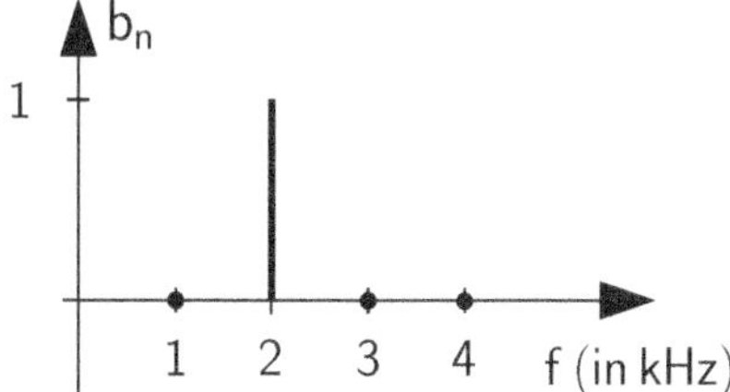

Fig. 1.79 Fourier coefficients Problem 1.1

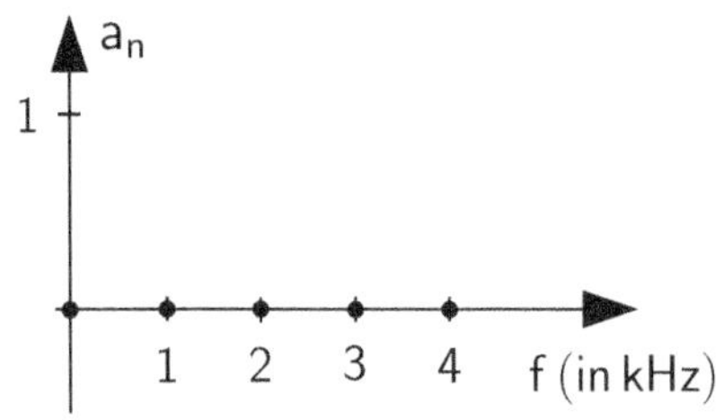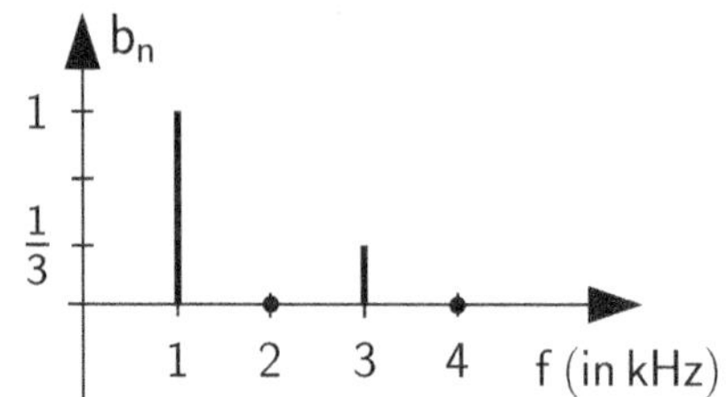

Fig. 1.80 Fourier coefficients Problem 1.2

- $u(t)$ analytically and graphically as well as
- the period of the time signal $u(t)$.

Problem 1.2 (Fourier Series)
Given are the (real) Fourier coefficients of a DC-free signal $u(t)$ according to Fig. 1.80.
   The following are sought:

- $u(t)$ analytically and graphically as well as
- the period of the time signal $u(t)$.

Problem 1.3 (Fourier and Laplace Transformation)
Given is the function $u(t)$ outlined in Fig. 1.81.
- Calculate the amplitude spectral density of this function $u(t)$ using the time shift theorem!

Fig. 1.81 Time function $u(t)$
for Problem 1.3

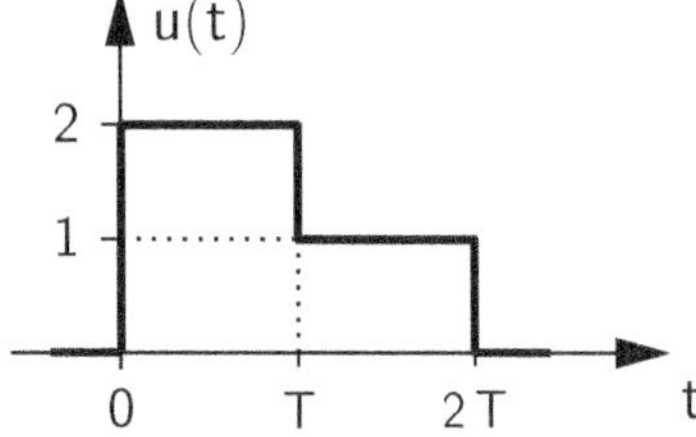

Fig. 1.82 Time function $u(t)$
for Problem 1.4

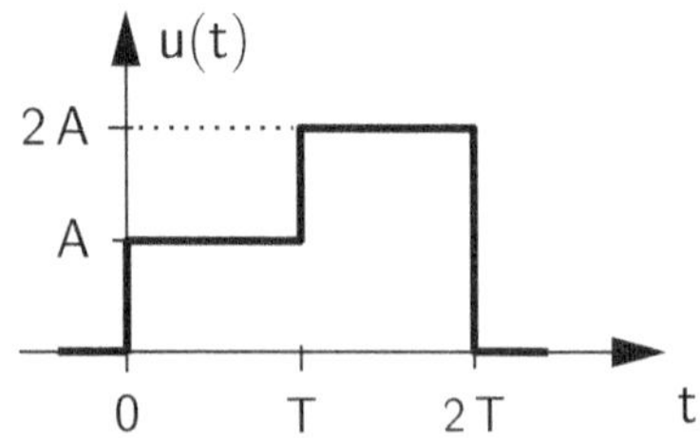

Fig. 1.83 Time function $u(t)$
for Problem 1.5

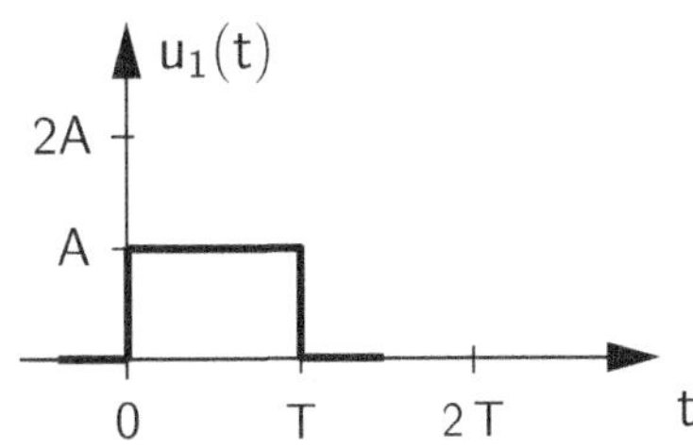

Problem 1.4 (Laplace Transformation)
Given is the function $u(t)$ outlined in Fig. 1.82.

a) Calculate the Laplace transform $U(p)$!
b) Can a Fourier transform $U(\omega)$ be specified for the function $u(t)$?

Problem 1.5 (System response in the time domain)
A system responds to the input signal $u_1(t) = A \cdot 1(t)$ (step function with the amplitude
$A = 1\,\mathrm{V}$) with the step response $u_2(t)$ in the form

$$u_2(t) = A \ e^{-t/T_1} \ 1(t). \tag{1.264}$$

- Calculate the system response to the input pulse $u_1(t)$ outlined in Fig. 1.83!
- Sketch the determined result!

1.8   Solutions of the Problems

Solution of Problem 1.1 The analytical description for the periodic signal $u(t)$ is to be
determined. The property that the signal must be periodic is derived from the given line
spectrum. First, the period of the signal can be determined from the distance between the
spectral lines in the spectral domain. Here applies

$$f_0 = \frac{1}{T} \tag{1.265}$$

with the period T of the time signal $x(t)$. It can be derived from the representations in
Fig. 1.79 that the period of the time signal $T = 1\,\mathrm{ms}$ is ($f_0 = 1/T = 1\,\mathrm{kHz}$). Furthermore, it shows that the time signal has no even components, as the coefficients a_n have
no components different from zero.

Taking into account that the coefficient b_1 is associated with the frequency f_0 and the
coefficient b_2 is associated with the frequency $2f_0$, the time signal is obtained

$$u(t) = 1 \cdot \sin\left(2\,\pi\,2f_0\,t\right). \tag{1.266}$$

Figure 1.84 illustrates the time signal resulting in this case.

Solution of Problem 1.2 The analytical description for the periodic signal $u(t)$ is to be determined. First, the period of the signal can be determined from the distance between the spectral lines in the image area. Here applies

$$f_0 = \frac{1}{T} \tag{1.267}$$

with the period T of the time signal $u(t)$. It can be derived from the representations in Fig. 1.80 that the period of the time signal $T = 1\,\mathrm{ms}$ is ($f_0 = 1/T = 1\,\mathrm{kHz}$). Furthermore, it shows that the time signal contains no even components, as the coefficients a_n have no components different from zero.

If one now additionally considers that the coefficient b_1 is linked with the frequency f_0 and the coefficient b_2 with the frequency $2f_0$, one obtains the time signal

$$u(t) = 1 \cdot \sin\left(2\pi f_0\, t\right) + \frac{1}{3} \cdot \sin\left(2\pi\,(3f_0)\, t\right). \tag{1.268}$$

Fig. 1.85 shows the resulting time signal $u(t)$.

Fig. 1.84 Time signal $u(t)$ for Problem 1.1

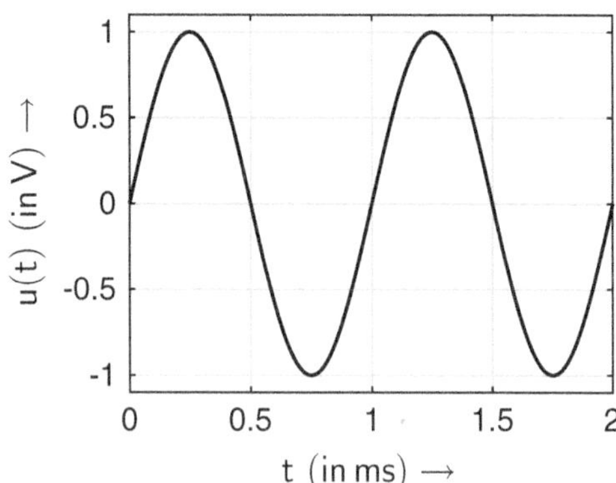

Fig. 1.85 Time signal $u(t)$ for Problem 1.2

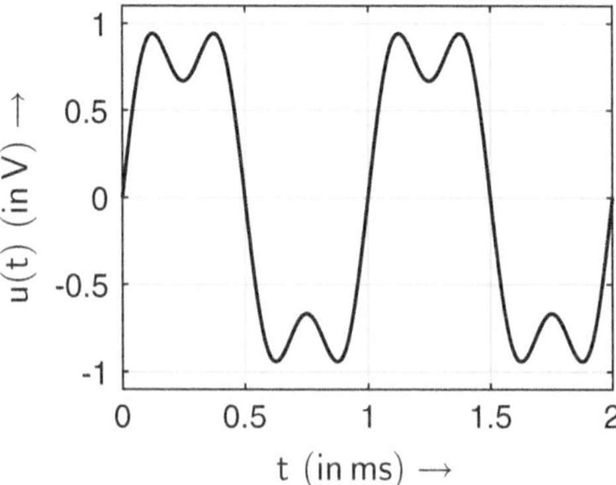

Solution of Problem 1.3 The signal given as a sketch in the task can be mathematically described as

$$u(t) = 2 \cdot 1(t) - 1 \cdot 1(t - T) - 1 \cdot 1(t - 2T) \tag{1.269}$$

The amplitude spectral density is the Fourier transform $U(\omega) = \mathcal{F}\{u(t)\}$ of the given time signal. This can either be calculated directly by evaluating the Fourier integral or via the Laplace transformation $U(p) = \mathcal{L}\{u(t)\}$ and evaluation of the Laplace transform at the point $p = j\omega$, provided that the $j\omega$ axis lies in the convergence area and thus the Fourier transform exists. This calculation method, which initially appears to be a detour, is often less complex when applying the correspondences than the direct one and will be used here.

For the determination of the Laplace transform $U(p) = \mathcal{L}\{u(t)\}$ one needs the correspondence for the unit step function

$$1(t) \quad \circ\!\!-\!\!\bullet \quad \frac{1}{p} \tag{1.270}$$

and the time shift theorem

$$\mathcal{L}\{u(t - b)\} = e^{-bp}\,\mathcal{L}\{u(t)\} \tag{1.271}$$

Both relationships can be taken from Appendix D, for example.

The correspondences for the unit step functions shifted by T and $2T$ result in

$$1(t - T) \quad \circ\!\!-\!\!\bullet \quad \frac{1}{p}e^{-pT} \tag{1.272}$$

and

$$1(t - 2T) \quad \circ\!\!-\!\!\bullet \quad \frac{1}{p}e^{-2pT}. \tag{1.273}$$

By the additive composition of the individual transformed terms, one obtains

$$U(p) = \mathcal{L}\{u(t)\} = \frac{2}{p} - \frac{1}{p}e^{-pT} - \frac{1}{p}e^{-2pT} \tag{1.274}$$

as a solution for the overall Laplace transform

$$U(p) = \frac{1}{p}\left(2 - e^{-pT} - e^{-2pT}\right). \tag{1.275}$$

The convergence area extends over all p, i.e., the Fourier transform of $u(t)$ exists, as in particular the $j\omega$ axis lies in the convergence area.

With $p = j\omega$ follows from (1.275) the Fourier transform

$$U(\omega) = \mathcal{F}\{u(t)\} = \frac{1}{j\omega}\left(2 - e^{-j\omega T} - e^{-2j\omega T}\right). \tag{1.276}$$

Solution of Problem 1.4 The Laplace transform and—if possible—the Fourier transform of a given time signal $u(t)$ should be specified. The signal given as a sketch in the task can be analytically written as

$$u(t) = A \cdot 1(t) + A \cdot 1(t - T) - 2A \cdot 1(t - 2T) \tag{1.277}$$

a) For the determination of the Laplace transform $U(p) = \mathcal{L}\{u(t)\}$ one again needs the correspondence for the unit step function

$$1(t) \quad \circ\!\!-\!\!\bullet \quad \frac{1}{p} \tag{1.278}$$

as well as the time shift theorem

$$\mathcal{L}\{u(t - b)\} = e^{-bp}\,\mathcal{L}\{u(t)\} \tag{1.279}$$

Both relationships can be taken from Appendix D, for example. This gives the correspondences for the unit step functions shifted by T and $2T$

$$1(t - T) \quad \circ\!\!-\!\!\bullet \quad \frac{1}{p}e^{-pT} \tag{1.280}$$

and

$$1(t - 2T) \quad \circ\!\!-\!\!\bullet \quad \frac{1}{p}e^{-2pT} \tag{1.281}$$

and through the additive composition of the individual transformed terms

$$U(p) = \mathcal{L}\{u(t)\} = \frac{A}{p} + \frac{A}{p}e^{-pT} - \frac{2A}{p}e^{-2pT} \tag{1.282}$$

as a solution for the overall Laplace transform

$$U(p) = \frac{A}{p}\left(1 + e^{-pT} - 2\,e^{-2pT}\right). \tag{1.283}$$

The Laplace transform $U(p)$ has a pole at $p_{\infty 1} = 0$ and a zero at $p_{0 1} = 0$.

b) The convergence range extends over all p, i.e., the Fourier transform of $u(t)$ exists, as in particular the $j\omega$-axis lies in the convergence range. With $p = j\omega$ follows from (1.283) the Fourier transform

$$U(\omega) = \mathcal{F}\{u(t)\} = \frac{A}{j\omega}\left(1 + e^{-j\omega T} - 2\,e^{-2j\omega T}\right). \tag{1.284}$$

Solution of Problem 1.5 It is to represent the system response to the input pulse outlined in the task $u_1(t)$. With the given values, one can initially recognize that the step response $u_2(t) = h(t)$ (response of the system to an input step) is given. If one now wants to represent the system response to the given input pulse $u_1(t)$ (rectangle pulse), the weighting function $g(t)$ or the transfer function $G(p)$ must first be determined. Since in many cases the calculation in the image or spectral range has advantages, the system response $u_2(t)$ should be determined via the image range, i.e., via the determination of the transfer function $G(p)$ and subsequently the Laplace transform $U_2(p)$ associated with the system response $u_2(t)$.

For the determination of the Laplace transform $G(p) = \mathcal{L}\{g(t)\}$ one needs the correspondence for the unit step function

$$1(t) \quad \circ\!\!-\!\!\bullet \quad \frac{1}{p} \tag{1.285}$$

and for the time-limited exponential function

$$e^{-t/T_1}\, 1(t) \quad \circ\!\!-\!\!\bullet \quad \frac{1}{p + \frac{1}{T_1}} \tag{1.286}$$

as well as the time shift theorem

$$\mathcal{L}\{x(t - b) \cdot 1(t - b)\} = e^{-bp}\,\mathcal{L}\{x(t) \cdot 1(t)\}. \tag{1.287}$$

These re

lationships can be taken from Appendix D, for example.

Thus, for the input signal $u_1(t) = A \cdot 1(t)$ the Laplace transform

$$U_1(p) = A \cdot \frac{1}{p} \tag{1.288}$$

and for the step response $u_2(t) = h(t)$ the Laplace transform

$$U_2(p) = \frac{A}{p + \frac{1}{T_1}}. \tag{1.289}$$

From the quotient

$$G(p) = \frac{U_2(p)}{U_1(p)} = \frac{p}{p + \frac{1}{T_1}} \tag{1.290}$$

the transfer function $G(p)$ can be determined, which forms the basis for determining the system response to the given input pulse (rectangle). The input signal $u_1(t)$ can be represented as

$$u_1(t) = A \cdot 1(t) - A \cdot 1(t - T) \tag{1.291}$$

using the unit step function. For the unit step function shifted by T, considering the time shift theorem, the correspondence

$$1(t - T) \quad \circ\!\!-\!\!\bullet \quad \frac{1}{p} e^{-pT}, \tag{1.292}$$

applies, and thus for the input signal $u_1(t)$ the Laplace transform

$$U_1(p) = \frac{A}{p} - \frac{A}{p} e^{-pT} \tag{1.293}$$

can be derived, which can be simplified to

$$U_1(p) = \frac{A}{p} \left(1 - e^{-pT} \right). \tag{1.294}$$

The system response $u_2(t)$ can be determined via the image domain and the approach

$$G(p) = \frac{U_2(p)}{U_1(p)}. \tag{1.295}$$

It results in

$$U_2(p) = U_1(p) \cdot G(p). \tag{1.296}$$

With the determined values for $U_1(p)$ and $G(p)$ results in the Laplace transform $U_2(p)$ of the system response to

$$U_2(p) = \frac{A}{p} \left(1 - e^{-pT} \right) \cdot \frac{p}{p + \frac{1}{T_1}}, \tag{1.297}$$

which simplifies to

$$U_2(p) = \frac{A}{p + \frac{1}{T_1}} \left(1 - e^{-pT} \right) \tag{1.298}$$

The inverse transformation leads to the system response $u_2(t)$. With the correspondences

$$\frac{1}{p + \frac{1}{T_1}} \quad \bullet\!\!-\!\!\circ \quad e^{-t/T_1} \, 1(t) \tag{1.299}$$

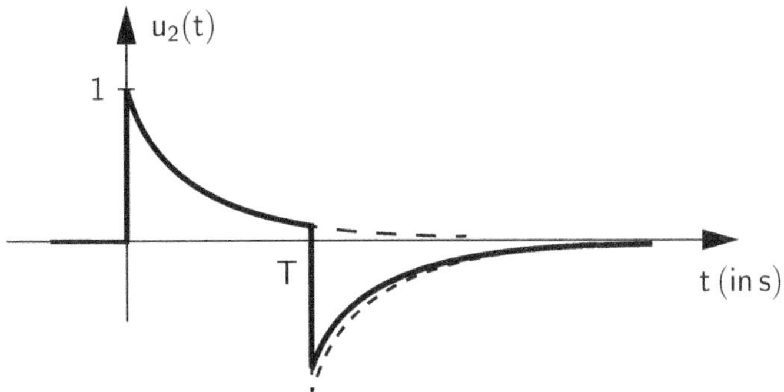

Fig. 1.86   System response $u_2(t)$

and

$$U(p)\,\mathrm{e}^{-p\,t_0} \quad\bullet\!\!-\!\!\circ\quad u(t-t_0)\cdot 1(t-t_0) \tag{1.300}$$

the time signals for the partial terms result

$$\frac{1}{p+\frac{1}{T_1}} \quad\bullet\!\!-\!\!\circ\quad \mathrm{e}^{-t/T_1}\,1(t) \tag{1.301}$$

and

$$\frac{1}{p+\frac{1}{T_1}}\cdot\mathrm{e}^{-pT} \quad\bullet\!\!-\!\!\circ\quad \mathrm{e}^{-\frac{(t-T)}{T_1}}\,1(t-T)\ . \tag{1.302}$$

Finally, the system response $u_2(t)$ is obtained by the additive composition of the individual inverse transformed terms to

$$u_2(t) = A\,\mathrm{e}^{-t/T_1}\,1(t) - A\,\mathrm{e}^{-\frac{(t-T)}{T_1}}\,1(t-T). \tag{1.303}$$

Fig. 1.86 illustrates the system response $u_2(t)$ over the additive overlay of the determined partial terms of the time signals for $A = 1\,\mathrm{V}$.

Basics of Baseband Transmission

2

Abstract

The fundamental problem or core task of communication technology is to transmit a given flow of information with minimal energy and maximum speed over a disturbed channel with minimal probability of error. This chapter introduces this basic problem of communication—in short: the reliable and efficient transmission of information over disturbed channels. It explains how to design suitable methods for baseband transmission to reliably transmit information over disturbed channels. This includes the transmission over distortion-free channels that are disturbed by noise. Such channels are usually referred to as AWGN channels (Additive White Gaussian Noise). Initially, using the example of binary transmission, it is explained in detail how the useful signal and disturbance are evaluated on the transmission path by the individual components of the transmission system. Quality criteria are defined that are suitable for capturing the transmission quality. Essential influencing variables, dependencies, and resulting optimization options are identified and discussed. Subsequently, some of the insights gained are generalized and important criteria are derived that characterize high-performance transmission systems. Building on this, some improved and extended concepts are presented that enable the optimization of transmission systems and thus the improvement of the performance capability. Detailed numerical examples serve to illustrate the general insights gained and the principles developed.

© The Editor(s) (if applicable) and The Author(s), under exclusive license to Springer   91
Fachmedien Wiesbaden GmbH, part of Springer Nature 2025
C. Lange and A. Ahrens, *Digital Transmission Engineering*,
https://doi.org/10.1007/978-3-658-46789-0_2

2.1 Basic Model of Binary Baseband Transmission

The following introduces and illustrates the basic problem of communication technology using the example of binary information transmission. Binary transmission is of great practical importance, as often either digital information elements in the form of bits and thus binary, i.e., two-valued, signals are present, e.g., in computer communication, or binary digital signals are generated by converting analog signals before transmission, e.g., in modern telephony, audio or video transmission.[1] Moreover, this representation allows a physically intuitive discussion of important properties of digital signal transmission using signal and system theoretical means, without having to use theoretically complicated methods that are needed for advanced transmission methods, but can obscure the core in an introductory consideration, e.g., through mathematical complexity.

2.1.1 Model and Components of the Baseband Transmission System

Physically Intuitive Representation

In Fig. 2.1, the basic model of binary message transmission with its components is shown as a block diagram. A baseband transmission is considered, i.e., the transmission takes place in the original frequency range of the transmit signal without a shift into other frequency ranges, as is necessary, for example, in radio communication. In practice, baseband systems can be used for transmission over electrical lines, as the baseband signals are low-pass signals and electrical wire pairs generally allow direct baseband transmission due to their low-pass characteristics. The block diagram was therefore designed for the baseband transmission to be discussed here.

The goal of information or message transmission is to transmit the data sequence present on the sending side as error-free as possible to the receiving side. The data sequence is a sequence of bits, i.e., a random sequence of $'0'$ and $'1'$, which is either directly available—e.g., when transmitting files—or is generated by an analog-digital conversion—e.g., in audio and video signal transmission and processing.

The *transmitter* has the task of converting the bit sequence to be transmitted, the transmission data sequence, into a physical signal—e.g., into a voltage—in such a way that the transmit signal is suitable for transmission over the given channel in terms of transmission power, amplitude range, and occupied bandwidth. At the output of the transmitter, the transmit signal $u_1(t)$ is present as a voltage, which contains the information-carrying data sequence. As an example, a unipolar transmission with the two voltage

---

[1] The conversion of analog signals into a binary digital signal or a bit sequence can be done by a so-called *Pulse Code Modulation (PCM)* with the steps *sampling, quantization* and *coding*, see e.g., [42, 48]. In the further course of this book, it is assumed that a bit sequence is already available that is to be transmitted.

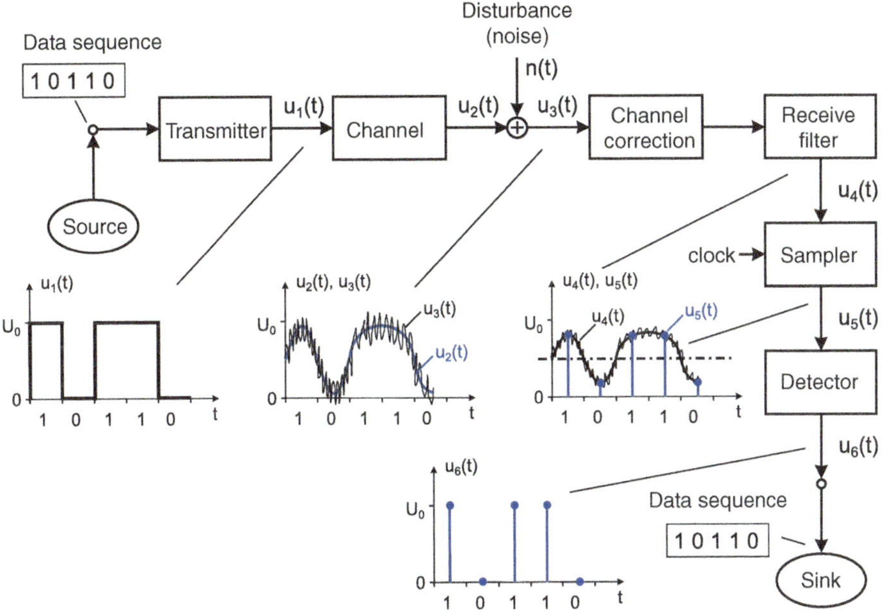

Fig. 2.1 Transmission system for the basic model of binary message transmission in baseband

states 0 and U_0 with NRZ rectangular pulses (Non Return to Zero)[2] is considered: If a '1' is present in the data sequence, the voltage state U_0 is sent and if a '0' is present, the voltage state 0 V is sent. This type of transmit signal allows a physically intuitive representation of the basic principles of communication technology and can also be easily generated in practice by switching voltages. Alternative and advanced concepts of transmit signal shaping will be discussed later in this chapter. The way in which an element of the transmission data sequence, i.e., a single bit, is physically represented as a pulse is referred to as the *basic transmission pulse*.

During transmission over the *channel*, the transmitted signal is generally deformed or distorted, and disturbances are superimposed on it, which can mostly be modeled as random signals in the form of noise and are additively added to the useful signal. The signal deformation or distortion caused by the channel can be understood as a filtering of the signal $u_1(t)$. At the channel output, the signal $u_2(t)$ deformed or distorted by the channel is initially present, to which the disturbance signal $n(t)$ is additively superimposed as noise.

---

[2] In this type of transmit signal, the respective voltage state is maintained for the entire symbol duration or clock period: If a symbol '1' with the amplitude U_0 is sent, the voltage does not return to the voltage 0 V during the duration of this symbol—hence the name *Non Return to Zero, NRZ* of this pulse type. The same applies analogously for the '0' symbols. The term *unipolar* means that only signals with voltages of one polarity, usually positive or non-negative voltage amplitudes, are used for transmission.

At the receiver input, the distorted and disturbed signal $u_3(t) = u_2(t) + n(t)$ is present. Practically important transmission channels are copper cables, radio channels, and fiber optic cables, which generally cause signal distortions through different physical mechanisms, such as frequency dependence, and additionally cause disturbances through noise.[3]

A *channel correction* compensates for the distorting influence of the channel on the transmitted signal and is often referred to as *equalization* or *equalizer*, as it essentially reverses the *dis*tortions caused by the channel, i.e., it *equalizes*. The channel correction is also a filtering and it can be realized in various ways, e.g., by means of suitable electronic circuits or by algorithms of digital signal processing.

The *receive filter* has the task of shaping the receiving pulse in such a way that it is as suitable as possible for the subsequent sampling and recognition and also to limit the bandwidth of the disturbance and thus the noise power. At the output of the receive filter, the possibly equalized and disturbed signal $u_4(t)$ evaluated by the receive filter appears, with the disturbance being limited in its effect.

The *sampler* takes samples from the received signal at the sampling times. For this purpose, a clock information is supplied to the sampler, which can be obtained from the received signal itself or which is additionally provided, e.g., via a control channel. The sampling times are generally determined significantly by the pulse width and thus by the rate at which information is transmitted, as well as by the type of transmit pulse shaping in interaction with the receive filtering.

In the *decision maker (detector)*, a decision is made regarding a threshold as to whether the sample value is greater or smaller than the decision or detection threshold. The decision maker is also referred to as a *detector*. At the output of the decision maker, depending on the type of practical implementation, the voltage $u_6(t)$ with the decided amplitudes at the sampling times or directly the received or estimated data sequence is present.

In Fig. 2.1, signal waveforms are drawn in based on an exemplary data sequence for the individual points in the baseband transmission system to illustrate how signals can look like. In the example, the waveforms are shown in such a way that all bits of the data sequence are correctly decided and recognized. This does not always have to be the case, depending on the specific influence of the channel and the intensity of the noise disturbance. How this can be captured theoretically and mathematically is the subject of the following sections.

Simplifications and Signal and System Theoretical Description

In order to be able to work out essential basic relationships in the transmission of digital signals, mathematical descriptions are introduced and some simplifying assumptions are made:

---

[3] If a transmission over radio channels is considered, for example, the transmitter in Fig. 2.1 would include a modulation to shift the transmitted signal into the given radio frequency range, and a corresponding frequency shift back, the demodulation, would have to be provided in the receiver. Then, in particular, the example signals shown in Fig. 2.1 for baseband transmission would have different characteristics.

- The basic transmission pulse $x_s(t)$ is an NRZ-rectangle impulse of the symbol duration or pulse width T_s and the amplitude U_0.
- It is assumed that the channel does not cause any distortions of the signal, i.e. the channel transfer function is set as $G_k(f) = 1$ or $G_k(\omega) = 1$—corresponding to a weighting function $g_k(t) = \delta(t)$ of the channel.

Insert: AWGN Channel

The received signal at the output of a distortion-free, noise-disturbed channel is

$$u_3(t) = D\,u_1(t - t_0) + n(t) \tag{2.1}$$

with the transmission factor D, the basic delay time t_0 (group delay) and the noise component $n(t)$. This results in the transfer function of the block *Channel* to

$$G_k(f) = D \cdot e^{-j2\pi f t_0} \quad \circ\!\!-\!\!\bullet \quad g_k(t) = D \cdot \delta(t - t_0) \,. \tag{2.2}$$

Often, for simplification, a conceptual time shift is made to compensate for the basic delay time t_0 and the transmission factor is chosen to be $D = 1$, so that in the case of an AWGN channel the condition

$$u_3(t) = u_1(t) + n(t) \tag{2.3}$$

applies.

- If the channel does not cause any distortions, these do not need to be equalized, so that for the transfer function of the channel correction—the equalizer—in Fig. 2.1 also $G_{\text{korr}}(f) = 1$ or $G_{\text{korr}}(\omega) = 1$ applies.
- The receive filter is a first-order low-pass filter with the 3-dB -cutoff frequency f_g or the time constant $T_g = 1/\omega_g = 1/(2\pi f_g)$.
- The clock for the sampling in the receiver is assumed to be ideal and known.

This leaves the simplified baseband transmission system shown in Fig. 2.2.[4] The individual elements are mathematically described in the following and an analysis is gradually introduced using the tools of signal and system theory.

It is assumed that only linear systems are used in the transmission path in Fig. 2.2. From a practical point of view, this is an assumption that only approximately applies, as,

---

[4] Since the channel in the simplified transmission system does not cause any deformation of the transmit signal $u_1(t)$, with reference to Fig. 2.1 $u_2(t) = u_1(t)$ applies and therefore $u_2(t)$ in Fig. 2.2 is not separately entered. In a practical system, a basic delay t_0 would have to be taken into account, so that $u_2(t) = u_1(t - t_0)$ would apply: This basic delay is set to $t_0 = 0$ for further system analysis, without restricting generality.

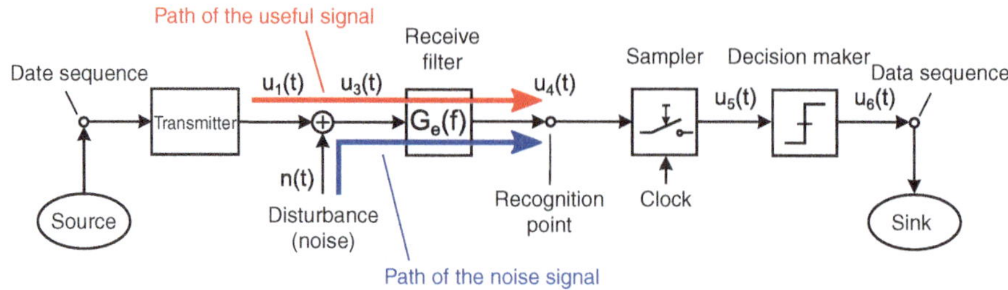

Fig. 2.2 Simplified transmission system for baseband transmission

for example, real circuits always have a limited dynamic range and often contain electronic components that, strictly speaking, have nonlinear characteristics—e.g. transistors. However, linear systems can often be assumed approximatively, as real systems can be considered as linear systems with small error. The superposition theorem then applies and the signal and system theoretical treatment is simplified: Signal and disturbance can be considered and analyzed separately from each other during their passage and their evaluation by the individual systems in the transmission path.

Under these conditions, the transmit signal passes through the receive filter $G_e(f)$ up to the detection point (marked in Fig. 2.2 as *Path of the useful signal*), and the disturbance is also evaluated by the receive filter $G_e(f)$ on the way to the detection point (marked in Fig. 2.2 as *Path of the noise signal*).

In the following sections, the path and evaluation of the useful signal as well as the path and evaluation of the noise signal for the given case (NRZ transmit signal, first-order receiver low-pass filter) are described in detail and mathematically captured using the tools of signal and system theory.

Transmit signal The transmit signal $u_1(t)$ is created by overlaying the data sequence with a single pulse, which is referred to as the basic transmit pulse $x_s(t)$. This basic transmit pulse is an NRZ-rectangle impulse of width T_s and amplitude U_0. The NRZ basic transmit pulse can be specified in the time and frequency domain as

$$x_s(t) = \begin{cases} U_0 & \text{for} \quad |t| \le \frac{T_s}{2} \\ 0 & \text{for} \quad |t| > \frac{T_s}{2} \end{cases} \quad \circ\!\!-\!\!\bullet \quad X_s(f) = U_0\, T_s\, \text{si}(\pi f\, T_s) \tag{2.4}$$

(or $X_s(\omega) = U_0\, T_s\, \text{si}(\omega\, T_s/2)$ with the frequency variable ω), where the link between time and frequency domain is obtained via the Fourier transformation (see Chap. 1 and Appendix A). Fig. 2.3 shows the basic transmit pulse in the time and frequency domain.

The frequency at which information-carrying symbols follow each other and are transmitted is $f_T = 1/T_s$. It is referred to as *symbol rate, symbol sequence frequency, symbol clock frequency* or *clock frequency*. Since in the basic model considered here, only binary symbols are transmitted initially, these symbols are simultaneously bits and

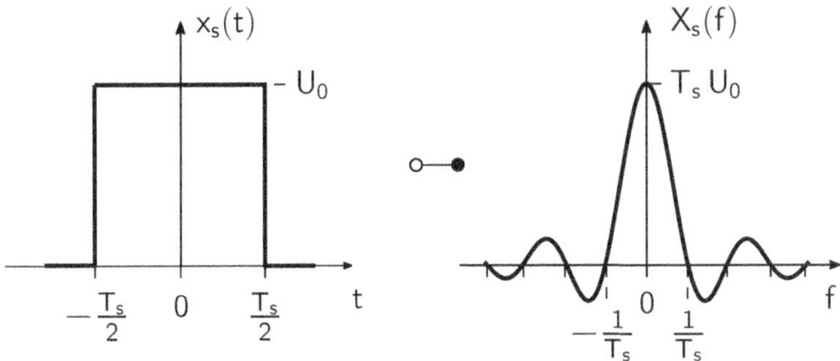

Fig. 2.3 NRZ basic transmit pulse in the time and frequency domain

the clock frequency here corresponds to the bit rate or bit sequence frequency f_B, with which bits follow each other: $f_T = f_B = 1/T_s$. In multi-level transmission systems discussed later in this chapter, bit and symbol rate must be distinguished.

Important characteristics of the transmit signal, e.g. for the design of transmitter components, are the *amplitude range* and the *average transmit power:*

- The amplitude range is the voltage interval in which the transmit signal lies; it ranges from A_{min} to A_{max}. The components of the transmitter must be designed for this amplitude range and work within it, for example, as linearly and distortion-free as possible. For the unipolar NRZ transmit signal $u_1(t)$, $A_{min} = 0$ and $A_{max} = U_0$ apply.
- The average transmit power can be calculated with reference to Fig. 2.1 and 2.3 according to

$$P_s = \overline{u_1^2(t)} = \lim_{T \to \infty} \frac{1}{2T} \int_{-T}^{+T} u_1^2(t)\, dt \tag{2.5}$$

for the random, binary and unipolar NRZ transmit signal $u_1(t)$ with the two equally distributed amplitude levels 0 and U_0: It results in

$$P_s = \frac{1}{2T_s} \left[\int_0^{T_s} U_0^2\, dt + \int_{T_s}^{2T_s} 0^2\, dt \right] = \frac{1}{2T_s} \left[U_0^2 \cdot T_s + 0^2 \cdot T_s \right] \tag{2.6}$$

and one obtains

$$P_s = \frac{U_0^2}{2}. \tag{2.7}$$

Often, in communications engineering for optimization purposes, a numerical value of $P_s = 1\,\text{V}^2$ (system-theoretical power) is chosen for reasons of comparability.

Circuit   Pole-zero diagram

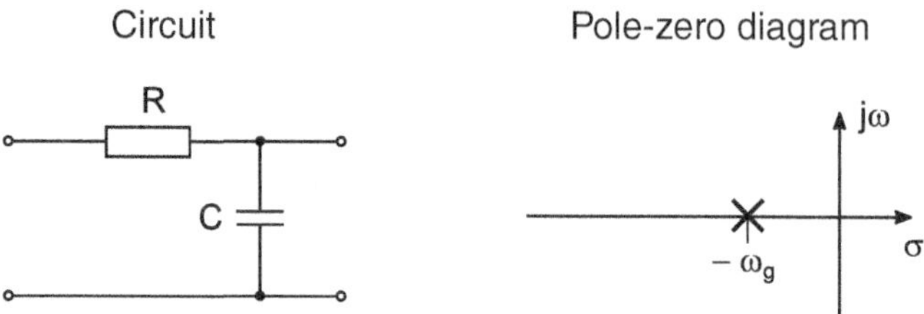

Fig. 2.4 Circuit and pole-zero diagram of a first-order RC low-pass filter

Receive Filter The receive filter is essentially freely selectable and is here, according to the previously made assumptions, a first-order RC low-pass filter [30]:

$$G_e(p) = \frac{1}{1 + p\,T_g} \quad \bullet\!\!-\!\!\circ \quad g_e(t) = \frac{1}{T_g}\,e^{-\frac{t}{T_g}} \cdot 1(t) \text{ with } T_g = \frac{1}{\omega_g} = \frac{1}{2\pi\,f_g}. \tag{2.8}$$

The cutoff frequency f_g is a 3-dB cutoff frequency, i.e., at the frequency $f = f_g$, the magnitude of the transfer function from (2.8) has dropped to a value of $1/\sqrt{2} \approx 0.7071$ or by $-20\lg(1/\sqrt{2}) = 3\,\text{dB}$ compared to the value at the frequency $f = 0$.

For the first-order RC low-pass filter, one obtains for $\sigma = 0$ and thus $p = \sigma + j\omega \Longrightarrow j\omega = j2\pi f$ as transfer function with the frequency variable ω initially

$$G_e(p) = \frac{1}{1 + pT_g} \quad \Longrightarrow \quad G_e(\omega) = \frac{1}{1 + j\omega T_g} = \frac{1}{1 + j\frac{\omega}{\omega_g}} \tag{2.9}$$

and with the frequency variable f:

$$G_e(p) = \frac{1}{1 + pT_g} \quad \Longrightarrow \quad G_e(f) = \frac{1}{1 + j\frac{2\pi f}{2\pi f_g}} = \frac{1}{1 + j\frac{f}{f_g}}. \tag{2.10}$$

In Fig. 2.4, a realization of the receive filter as a circuit and the associated pole-zero diagram are given. The time constant of the receive filter is determined by the elements R and C: $T_g = RC$. Hence the name RC low-pass filter. It is *first order,* since at the cutoff frequency $\omega_g = 1/T_g$ there is *one* pole (see Chap. 1).

Since the imaginary axis lies within the convergence range of the Laplace integral (see Chap. 1), the transition $p = \sigma + j\omega \quad \Longrightarrow \quad p = j\omega$ for $\sigma = 0$ can be made and the Fourier transform $G_e(\omega)$ or $G_e(f)$ corresponding to the Laplace transform $G_e(p)$ can be given.

2.1.2 Signal Path and Evaluation of the Useful Signal

System for Evaluating the Useful Signal
The transmission system according to Fig. 2.1 is characterized by the linearity of the individual functional blocks. The useful signal and disturbance, whose paths are illus-

trated when passing through the transmission system in Fig. 2.2, can therefore—as already mentioned—be treated separately. In this section, the path of the useful signal through the transmission system shown in Fig. 2.2 is examined in detail and the effects calculated that the systems located in the transmission path have on the useful signal. On this basis, a suitable size is defined that describes the quality or quality of the useful signal at the detection point.

For the useful signal, the block diagram shown in Fig. 2.5 remains: The data sequence to be sent in Fig. 2.1 is denoted by $a(k)$, where it consists of a random sequence of '0' and '1'; k is an integer and describes that an element of the data sequence is sent in each symbol clock, i.e. every $k \cdot T_s$. The transmit signal $u_s(t)$ is created from the superposition of the basic transmission pulses $x_s(t)$ evaluated with the random data sequence $a(k)$ from the elements $\{0, 1\}$ and temporally spaced T_s—this transmit signal corresponds to $u_1(t)$ in the baseband basic model in Fig. 2.1. Since the channel and channel correction, according to the prerequisites for the simplified transmission system, do not cause any deformations of the useful signal, only the receive filter $G_e(f)$ or $G_e(p)$ remains as a system that influences and deforms the useful signal. If only the useful signal is considered, the received useful signal $u_e(t)$ is present at the output of the receive filter: This signal is the transmit signal $u_s(t)$ evaluated by the receive filter. It corresponds in Fig. 2.1 to the signal $u_4(t)$—however *without* the superimposed and contained noise component. The definition of the received useful signal $u_e(t)$ is therefore useful, since the useful signal and disturbance can be treated separately. In a real transmission system, this undisturbed received useful signal cannot be observed, it is a pure—and yet useful—computational quantity.

Received Useful Signal and Received Basic Pulse
The received useful signal—similar to the transmitted signal—can be represented as an overlay of basic pulses, which are evaluated with the random data sequence $a(k)$ from the elements $\{0, 1\}$, and temporally successive at intervals T_s. However, the evaluation is now done with the *received basic pulse* $x_e(t)$. The *received basic pulse* is created when a single transmitted basic pulse $x_s(t)$ is evaluated by the receiving filter $G_e(f)$ or $G_e(p)$ •—○ $g_e(t)$, i.e., by filtering the transmitted basic pulse through the receiving filter. This can be written in the time domain as convolution $x_e(t) = x_s(t) * g_e(t)$.

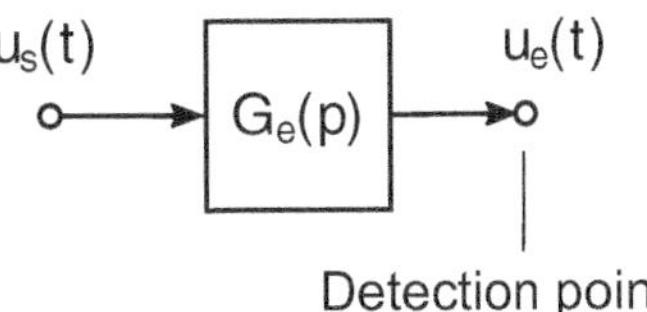

Fig. 2.5 Block diagram for evaluating the useful signal

Therefore, the received basic pulse is now calculated. This is advantageously done in the spectral or image domain using the Laplace transformation.[5] The NRZ transmitted basic pulse of duration T_s and amplitude U_0 is represented as causal

$$x_s(t) = \begin{cases} U_0 & \text{for} \qquad 0 < t < T_s \\ 0 & \text{Otherwise} \end{cases} \tag{2.11}$$

and can be formulated with $1(t)$ as a unit step function as

$$x_s(t) = U_0 \cdot [1(t) - 1(t - T_s)]. \tag{2.12}$$

The Laplace transform of the rectangular NRZ transmitted basic pulse of duration T_s with amplitude U_0 is (using the shift theorem of the Laplace transformation)

$$X_s(p) = \frac{U_0}{p} - \frac{U_0\,e^{-pT_s}}{p} = U_0 \left(\frac{1}{p} - \frac{e^{-pT_s}}{p} \right) = \frac{U_0}{p} \left(1 - e^{-pT_s}\right). \tag{2.13}$$

The Laplace transform of the received basic pulse is thus, after multiplication with $G_e(p)$, according to (2.8)

$$X_e(p) = X_s(p) \cdot G_e(p) = \frac{U_0}{p} \cdot \frac{1 - e^{-pT_s}}{1 + pT_g}. \tag{2.14}$$

By factoring out T_g in the denominator, one obtains

$$X_e(p) = \frac{U_0}{p} \cdot \frac{1 - e^{-pT_s}}{T_g\left(\frac{1}{T_g} + p\right)} = \frac{U_0}{T_g} \cdot \frac{1 - e^{-pT_s}}{p\left(\frac{1}{T_g} + p\right)}. \tag{2.15}$$

Using the inverse Laplace transformation of the thus rearranged Laplace transform of the received basic pulse

$$x_e(t) = \frac{U_0}{T_g} \cdot \mathscr{L}^{-1} \left\{ \frac{1 - e^{-pT_s}}{p\left(p + \frac{1}{T_g}\right)} \right\} \tag{2.16}$$

one obtains, with the help of the correspondence (see Appendix D or [11, 13, 74])

$$\frac{1}{p(p + a)} \quad\bullet\!\!-\!\!\circ\quad \frac{1}{a}\left(1 - e^{-at}\right) \cdot 1(t) \tag{2.17}$$

---

[5]It should be noted that when using the Laplace transformation, the time signals can only be analyzed for times $t \geq 0$, as the Laplace transformation used in this book is only defined for non-negative times. For this purpose, a causal representation of the NRZ rectangular pulse is used—from 0 to T_s—and the transmitted basic pulse is shifted by $T_s/2$; this does not change its magnitude spectrum (see Chap. 1).

the time function of the received basic pulse

$$x_e(t) = U_0\left[\left(1 - e^{-\frac{t}{T_g}}\right)\cdot 1(t) - \left(1 - e^{-\frac{t-T_s}{T_g}}\right)\cdot 1(t - T_s)\right].\qquad(2.18)$$

Fig. 2.6 shows this rectangular response $x_e(t)$ of a low-pass filter with a single real pole—i.e., the received basic pulse—at two exemplary receiving filter cutoff frequencies $f_g = 1/(2\pi T_g)$.

The spectral effect of the receiving filter is illustrated by the logarithmic amplitude responses of the transfer function $G_e(f)$ (Bode diagram) shown in Fig. 2.7: With a large filter time constant T_g or small cutoff frequency f_g, the usable frequency range is relatively strongly limited by the receiving filter, while with a small time constant T_g or large cutoff frequency f_g of the receiving filter, a comparatively large spectral passband is available. The drop of the magnitude frequency response above the cutoff frequency f_g occurs with 20 dB/Dekade, i.e., within a frequency range of 10^n Hz $\ldots$ 10^{n+1} Hz (n integer), the value of the magnitude frequency response decreases by 20 dB, e.g., in the range of 10 MHz $\ldots$ 100 MHz or 100 MHz $\ldots$ 1 GHz. This results in a descending straight line in the representation with a vertical axis that is logarithmically divided according to the decimal logarithm.

General Representation of the Evaluation of the Useful Signal

The previously described evaluation of the transmit signal $u_s(t)$ or the transmission basic pulse $x_s(t)$ by a receive filter $G_e(f)$ can be generalized and represented in an alternative way: To generate the transmission basic pulse, a Dirac impulse

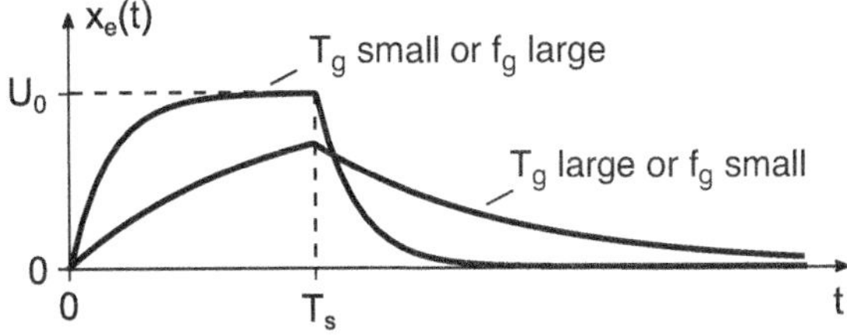

Fig. 2.6 Rectangular response of a low-pass filter with a single-real pole at high ($f_g = 1/T_s$) and low ($f_g = 0.2/T_s$) cutoff frequency

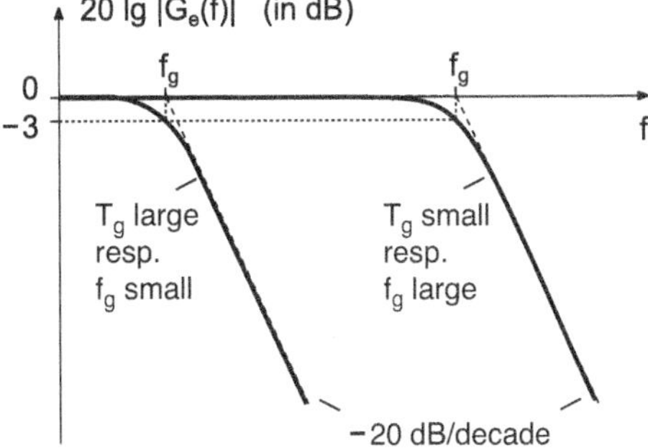

Fig. 2.7 Spectral effect of the receive filter $G_e(f)$

$x_q(t) = U_0\,T_s\,\delta(t) \circ\!\!-\!\!\bullet\, X_q(f) = U_0\,T_s$ is evaluated by the transmit filter with the weighting function $g_s(t) \circ\!\!-\!\!\bullet\, G_s(f)$, and the transmit basic pulse is obtained

$$x_s(t) = x_q(t) * g_s(t) = U_0\,T_s\,g_s(t) \quad\circ\!\!-\!\!\bullet\quad X_s(f) = U_0\,T_s\,G_s(f). \tag{2.19}$$

The receiving basic pulse is created after further filtering of this signal, now by the receiving filter $g_e(t) \circ\!\!-\!\!\bullet\, G_e(f)$, as

$$x_e(t) = x_q(t) * g_s(t) * g_e(t) \quad\circ\!\!-\!\!\bullet\quad X_e(f) = U_0\,T_s\,G_s(f)\,G_e(f). \tag{2.20}$$

The creation of the transmit signal $u_s(t)$ and the received useful signal $u_e(t)$ can be represented analogously, if not a single Dirac impulse, but a periodic Dirac pulse sequence is assumed as the input signal of the transmit filter, which emits Dirac pulses $U_0\,T_s\,\delta(t)$ at the interval of the symbol duration T_s, which are weighted—i.e., multiplied—with the dimensionless amplitude coefficients $a[k]$ of the data sequence, i.e., the source signal is

$$u_q(t) = U_0\,T_s\,\sum_{k=-\infty}^{+\infty} a[k]\cdot\delta(t - k\,T_s). \tag{2.21}$$

Fig. 2.8 shows the extension of the block diagram shown in Fig. 2.5 for the evaluation of the useful signal by the transmit filter $G_s(f)$: Both forms of description are equivalent, but the representation now introduced in Fig. 2.8 is more flexible, as various transmit filter functions $g_s(t) \circ\!\!-\!\!\bullet\, G_s(f)$ are allowed and can be captured.

For example, if a Dirac impulse $U_0\,T_s\,\delta(t)$ passes through a transmit filter

$$G_s(f) = \mathrm{si}(\pi f T_s) \quad\bullet\!\!-\!\!\circ\quad g_s(t) = \frac{1}{T_s}\,\mathrm{rect}\!\left(\frac{t}{T_s}\right), \tag{2.22}$$

the NRZ rectangular pulse (Fig. 2.3)

$$x_s(t) = U_0\,\mathrm{rect}\!\left(\frac{t}{T_s}\right) \quad\circ\!\!-\!\!\bullet\quad X_s(f) = U_0\,T_s\,\mathrm{si}(\pi f T_s). \tag{2.23}$$

is created as the transmit basic pulse. This has shown the equivalence of both forms of description using the example of the NRZ transmission basic pulse.

Fig. 2.8 Block diagram for evaluating the useful signal (general description form)

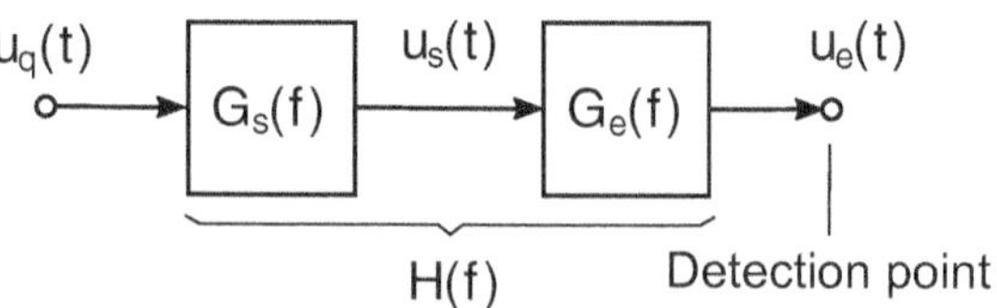

This form of representation allows different filter functions to be used as transmit filters $G_s(f)$, for example, to optimally account for the conditions on the transmission channel. Therefore, it will be revisited in the further parts of this chapter.

The transfer function of the transmission system from the input of the transmit filter $G_s(f)$ to the output of the receive filter $G_e(f)$ is denoted by $H(f) \bullet\!\!-\!\!\circ h(t)$ as the overall transfer function or resulting transfer function of the baseband system (see Fig. 2.8). Accordingly, $h(t)$ is the resulting overall weighting function, which is often also referred to as the overall impulse response. The functions $H(f)$ and $h(t)$ have proven useful for the analysis of the evaluation of the useful signal in transmission systems. For the overall transmission system, in the example considered here of a non-distorting transmission channel, which therefore only includes the transmission and receiving filters in the useful signal path, the relationship applies

$$h(t) = g_s(t) * g_e(t) \quad \circ\!\!-\!\!\bullet \quad H(f) = G_s(f) \cdot G_e(f). \tag{2.24}$$

If the useful signal path contains further components before the symbol clock sampling on the receiving side, e.g., a distorting channel and a channel correction (see, for example, Fig. 2.1), these are part of the overall transmission system and the overall transfer function $H(f)$ results from the multiplication of the transfer functions of all involved subsystems—the overall weighting function $h(t)$ is obtained via the convolution of the weighting functions of the subsystems.

Origin of the Received Useful Signal

The received useful signal $u_e(t)$ arises, as already mentioned, by the superposition of the basic received impulses evaluated with the individual elements $a[k] \in \{0, 1\}$ of the data sequence $a(k)$.

Using the example section of a data symbol sequence $a(k) = (1\,0\,1\,1\,0)$ already used in Fig. 2.1, the origin of the received useful signal is illustrated in Fig. 2.9: If you use the basic received impulse according to (2.18) and assume a sufficiently long sequence of $'0'$ symbols as history, the considered section of the received useful signal can be written as

$$u_e(t) = a[0] \cdot x_e(t) + a[1] \cdot x_e(t - T_s) + a[2] \cdot x_e(t - 2\,T_s) + \ldots$$
$$\ldots + a[3] \cdot x_e(t - 3\,T_s) + a[4] \cdot x_e(t - 4\,T_s) \tag{2.25}$$

(here with $k = 0, 1, 2, 3, 4$) or with the concrete data symbols as

$$u_e(t) = 1 \cdot x_e(t) + 0 \cdot x_e(t - T_s) + 1 \cdot x_e(t - 2\,T_s) + \ldots$$
$$\ldots + 1 \cdot x_e(t - 3\,T_s) + 0 \cdot x_e(t - 4\,T_s). \tag{2.26}$$

In Fig. 2.9, the resulting exemplary received useful signal $u_e(t)$ and the individual basic received impulses $x_e(t)$ shifted by multiples of the symbol duration T_s and evaluated with the data symbols $a[k]$ are shown, so that it becomes recognizable how the filtered useful signal at the output of the receive filter is created by superposition. In addition, a decision threshold is entered, which in the case of two-level, unipolar transmission with the

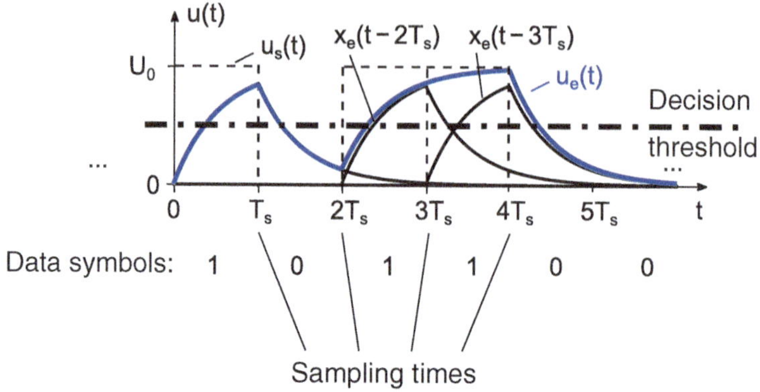

Fig. 2.9 On the formation of the received useful signal

transmission amplitudes 0 and U_0 and redundancy-free source with equally probable data symbols is at $U_E = U_0/2$.

Important criteria for obtaining sample values that have as large a difference as possible to the decision threshold—to allow reliable detection—are:

- Choice of the correct sampling time. In practice, clock synchronization plays a significant role (assumed to be ideal here).
- Setting the correct height of the decision threshold.

The sample values depend on the

- Type of filter (e.g., first-order low-pass, n-th order low-pass, other low-pass filter functions),
- Pulse duration (pulse width) T_s or the rate $1/T_s$ at which the pulses follow each other—i.e., the symbol or bit rate (or symbol or bit sequence frequency).

If one generalizes the previous consideration, the received useful signal at the output of the receiver filte for data symbol sequences of any length can be formulated as a sum

$$u_e(t) = \sum_{k=-\infty}^{+\infty} a[k] \cdot x_e(t - k\,T_s) \,. \tag{2.27}$$

The basic received impulses $x_e(t)$ have—depending on the receive filter cutoff frequency f_g—generally a much longer temporal extent than the pulse width T_s of the rectangular pulses used on the transmitter side: Therefore, there may be mutual influences in the received signal at neighboring sampling points. These are referred to as *intersymbol interference (ISI)*. They cause an influence of consecutively transmitted symbols at the sampling points.

The lower the cutoff frequency f_g of the receive filter, the more the received pulses are blurred and their temporal extent increases: As a result, the intersymbol disturbances become stronger and reliable detection becomes more difficult. The size of the intersymbol disturbances depends on the cutoff frequency of the receive filter.

From the standpoint of useful signal transmission, a cutoff frequency f_g of the receive filter as large as possible is therefore desirable, as then the received signal is little—or not at all—distorted and the intersymbol disturbances are minimal or disappear.

It is now of interest how, after this qualitative description of the situation regarding the useful signal at the detection point, the intersymbol disturbances and the associated distortions of the useful signal can be quantitatively recorded: A suitable method for this will be presented in the following.

Eye Diagram

A useful representation for determining the quality of the received useful signal at the output of the receiving filter is the *eye diagram.* Sections of the received useful signal of a certain duration—usually integer multiples of the symbol duration T_s—are written on top of each other at random data symbol sequence. This can be done practically with an oscilloscope that is triggered with the clock information (symbol clock frequency f_T) of the received signal. Often, signal sections of duration T_s or $2\,T_s$ are displayed as eye diagrams. The eye diagram thus consists of all superimposed sections of a digital signal, the duration of which is an integer multiple of the symbol duration T_s [65]. The eye diagram is a suitable means for evaluating the reception quality of the received useful signal. The term *eye diagram* comes from the fact that the graphical representation resembles an eye.

Fig. 2.10 shows some of the curves of the received useful signal that arise from various sections of the data symbol sequence $a(k)$. In particular, it can be seen that a curve that limits the eye diagram from the inside is created when a $'1'$ pulse is sent after a

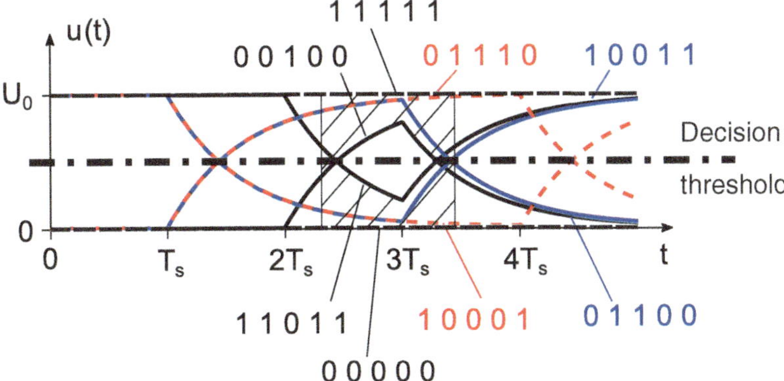

Fig. 2.10   For the construction of the eye diagram of the received useful signal

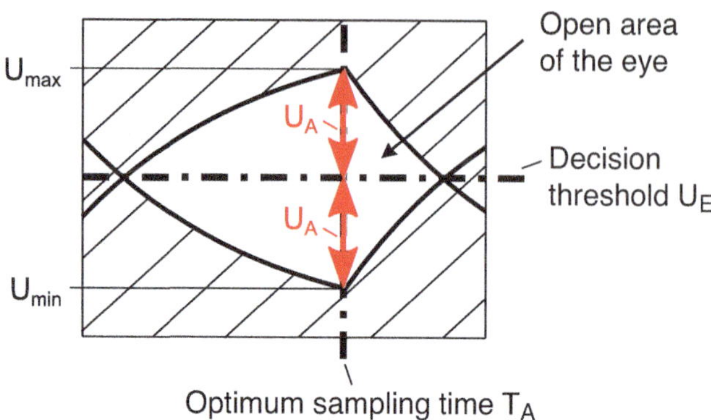

Fig. 2.11 Eye diagram (principle)

long sequence of ′0′ pulses and this single ′1′ pulse is followed by a long sequence of ′0′ pulses. The other inner boundary line of the eye diagram is created when a single ′1′ pulse is sent after a long sequence of ′0′ pulses, followed by a long sequence of ′1′ pulses.

The distance between the inner lines of the eye, the vertical eye opening, is of particular interest for assessing the transmission quality, as this is where the probability of a wrong decision in case of disturbed transmission is highest. The inner lines of the eye diagram thus represent the worst case *(worst case)*. The optimal sampling or detection time T_A should be chosen for maximum reliability of detection so that the vertical eye opening becomes maximum: At these times, the received signal is preferably sampled. This particularly interesting area is hatched in Fig. 2.10—and shown separately in Fig. 2.11.

Fig. 2.11 shows a schematic representation of the eye diagram in two-level transmission with important parameters. The decision threshold U_E is arranged in the middle between the two transmission amplitude levels. The optimal sampling time T_A is where the vertical eye opening is largest. As amplitudes, the values U_{max} or U_{min} are reached at the optimal sampling time for the inner eye lines: These values are generally—in the case of remaining intersymbol interference—not identical to the transmission amplitude levels. The *half* vertical eye opening U_A gives the value for the maximum permissible disturbance amplitude at the detection point, at which there is just no wrong decision: It is thus an important parameter for assessing the quality or the quality of the useful signal at the detection point and thus the overall signal transmission. The hatched areas are the regions where eye lines are displayed for different random symbol sequences, but these are not the inner lines of the eye diagram.

Calculation of the Vertical Eye Opening

The half vertical eye opening as a quality measure for the transmission of the useful signal up to the detection point is accessible to calculation. The useful sample value $x_e(t = T_A)$ of the received basic pulse as the main value is reduced in the worst case by sample values $x_e(t = T_A \pm T_s), x_e(t = T_A \pm 2T_s), \ldots$ of the received basic pulse, which originate from precursors and postcursors.

The *whole* vertical eye opening for the worst case is the distance between the two innermost lines of the eye diagram at the optimal sampling point [65]. It can be calculated by reducing the main value by the intersymbol interferences generally caused by the precursors and postcursors:

$$U_{AA} = x_e(T_A) - \sum_{\substack{k=-\infty \\ k \neq 0}}^{+\infty} |x_e(T_A - k\,T_s)|. \tag{2.28}$$

A rearrangement yields

$$U_{AA} = x_e(T_A) \underbrace{\left[1 - \frac{\sum_{\substack{k=-\infty \\ k \neq 0}}^{+\infty} |x_e(T_A - k\,T_s)|}{x_e(T_A)} \right]}_{k_{ISI}}, \tag{2.29}$$

and in compact form one obtains

$$U_{AA} = x_e(T_A) \cdot k_{ISI}. \tag{2.30}$$

The factor k_{ISI} describes the intersymbol interferences on the transmission path and is fixed with the overall frequency response $H(f)$ or the overall weighting function $h(t)$ of the transmission system between the transmitter filter input and the receive filter output. Intersymbol interferences can generally occur—depending on the type of components that determine the overall frequency response $H(f)$—both constructively and destructively. For the calculation of U_{AA}, only the components of the received basic pulse $|x_e(T_A - k\,T_s)|$ mit $k \neq 0$ that contribute to a reduction of $x_e(t = T_A)$ are to be considered, since only the vertical eye opening for the *worst* case is considered.

The *half* vertical eye opening for the worst case results with (2.28) to

$$U_A = \frac{1}{2} \left[x_e(T_A) - \sum_{\substack{k=-\infty \\ k \neq 0}}^{+\infty} |x_e(T_A - k\,T_s)| \right]. \tag{2.31}$$

Fig. 2.12 illustrates the calculation of the vertical eye opening according to (2.28) and (2.31) for the special case considered in this section, in which only pulse postcursors and no pulse precursors occur due to the shape of the received basic pulse. The optimal sampling time is at $T_A = T_s$ and the received basic pulse only has non-negative components: The number of ISI sample values depends on the length of the received basic pulse in

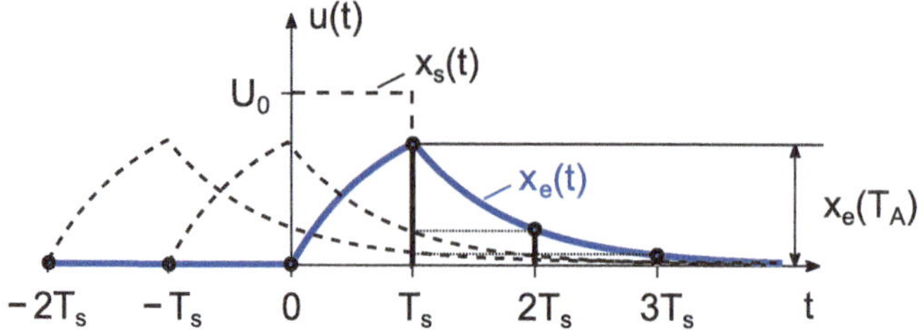

Fig. 2.12 On the calculation of the vertical eye opening

relation to the symbol duration T_s—and thus both on the clock frequency f_T of the transmission and on the cutoff frequency f_g of the receive filter.

In the special case considered here, the half vertical eye opening can also be calculated by determining the difference between the useful sample value or main value $x_e(T_s)$ and the decision threshold U_E: In this way, an analytical, not too complicated expression for the half vertical eye opening can be determined, which allows further mathematical evaluation. This will be shown in the following example calculation.

Calculation of the Half Vertical Eye Opening Depending on the Receive Filter Cutoff Frequency

The innermost boundary lines of the eye at the detection point arise when NRZ rectangular pulses are emitted and received filtering through a first-order RC low-pass filter, either after a long sequence of $'0'$ pulses a single $'1'$ pulse and then again $'0'$ pulses or vice versa after a long sequence of $'1'$ pulses a single $'0'$ pulse and then again $'1'$ pulses are sent or received (Fig. 2.13). With other transmission symbol sequences, the vertical eye opening is not further reduced by ISI contributions.

The optimal sampling point T_A is where the vertical eye opening is largest. Figure 2.6 shows that under the considered conditions the optimal sampling point is at $T_A = T_s$ (see also (2.18)). The optimal threshold for a unipolar, binary transmission with the uni-

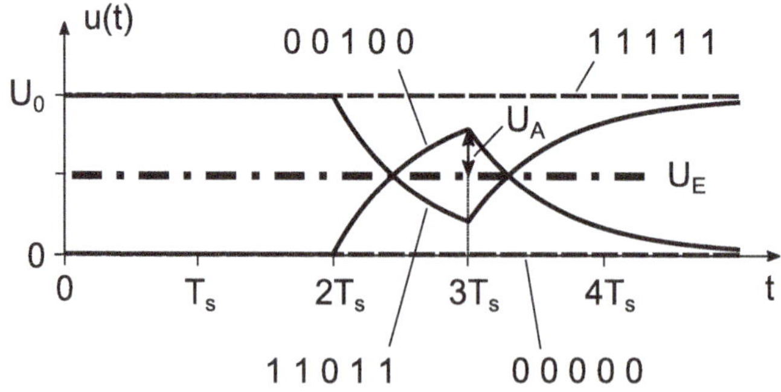

Fig. 2.13 On the calculation of the half vertical eye opening

formly distributed amplitude levels 0 and U_0 is at $U_E = U_0/0$. Under these conditions, the half vertical eye opening can be calculated after

$$U_A = x_e\,(T_A) - U_E = \left[U_0\left(1 - e^{-\frac{T_s}{T_g}}\right)\right] - \frac{U_0}{2}\,. \tag{2.32}$$

It results in

$$U_A = U_0 - \frac{U_0}{2} - U_0\,e^{-\frac{T_s}{T_g}} = \frac{U_0}{2} - U_0\,e^{-\frac{T_s}{T_g}} \tag{2.33}$$

and finally, the result is

$$U_A = U_0\left(\frac{1}{2} - e^{-\frac{T_s}{T_g}}\right) \tag{2.34}$$

or as a function of the receiving filter cutoff frequency f_g:

$$U_A = U_0\left(\frac{1}{2} - e^{-2\pi f_g T_s}\right). \tag{2.35}$$

For reliable detection of the transmitted data sequence in the receiver, the eye must be vertically open, i.e., for the half vertical eye opening (2.35) $U_A > 0$ must apply. According to (2.35), for small receiving filter cutoff frequencies f_g, the half vertical eye opening becomes negative if the received basic pulse at the optimal sampling point $x_e\,(T_A)$ is smaller than the decision threshold U_E (see Fig. 2.13): This area is not usefully usable for reliable message transmission and the half vertical eye opening U_A is set to zero for these areas of the receiving filter cutoff frequency.

Figure 2.14 shows the basic characteristic of the half vertical eye opening as a function of the receiving filter cutoff frequency: There is an area at small cutoff frequencies

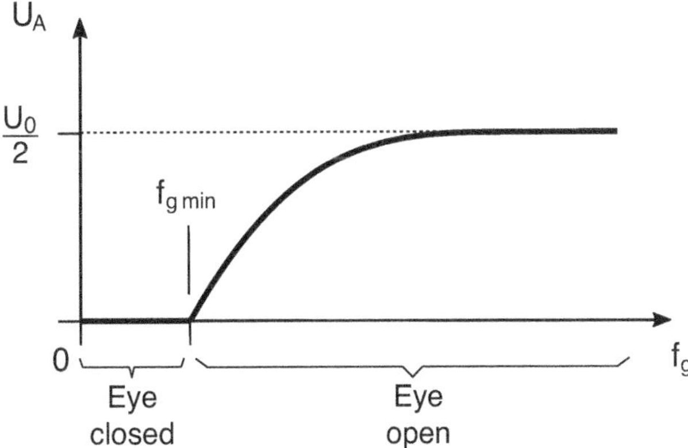

Fig. 2.14 Half vertical eye opening at the detection point as a function of the receive filter cutoff frequency f_g

f_g of the receiving filter, where the eye is closed and thus no detection is possible; the intersymbol interferences are too severe here. Above the cutoff frequency $f_{g\,min}$, at which the eye opens, the eye opening increases and tends towards a maximum value, which is given by the half-level transmit amplitude $U_0/2$; larger than $U_0/2$ the half vertical eye opening cannot become even with an infinitely running receiving filter cutoff frequency.

The receiving filter cutoff frequency $f_{g\,min}$ can be calculated as the zero of (2.35)

$$U_A = U_0 \left(\frac{1}{2} - e^{-2\pi f_g T_s} \right) = 0$$

$$\frac{1}{2} - e^{-2\pi f_g T_s} = 0$$

$$\frac{1}{2} = e^{-2\pi f_g T_s}$$

$$\ln \left(\frac{1}{2} \right) = -2\pi f_g T_s$$

$$-\ln 2 = -2\pi f_g T_s$$

$$f_g = \frac{\ln 2}{2\pi T_s},$$

so that as the receive filter cutoff frequency, at which the eye just opens vertically

$$f_{g\,min} = \frac{\ln 2}{2\pi T_s} \tag{2.36}$$

results. Thus, the half vertical eye opening can also be written as

$$U_A = \begin{cases} U_0 \left(\frac{1}{2} - e^{-2\pi f_g T_s} \right) & \text{for } f_g \geq \frac{\ln 2}{2\pi T_s} \\ [1ex] 0 & \text{for } f_g < \frac{\ln 2}{2\pi T_s} \end{cases} \tag{2.37}$$

to capture the areas where the eye is vertically open or closed.

Example 2.1 (Half vertical eye opening for the transmission of unipolar NRZ rectangular signals and receive filtering by a first order *RC* low-pass filter) The half vertical eye opening is to be calculated according to the general relationship (2.31) and the result is to be compared with the relationship obtained by another method (2.34) or (2.35) (or (2.37)) for this case.

The basic receive impulse $x_e(t)$ according to (2.18) is non-negative and only has trailing edges. The optimal sampling time is $T_A = T_s$ (see Fig. 2.6). According to (2.31), the result is

$$U_A = \frac{1}{2} \left[x_e(T_A) - \sum_{\substack{k=-\infty \\ k\neq 0}}^{+\infty} |x_e(T_A - k T_s)| \right] = \frac{1}{2} \left[x_e(T_s) - \sum_{k=1}^{\infty} x_e(T_s + k T_s) \right], \tag{2.38}$$

under the assumption that the basic receive impulse considered here and especially its symbol clock sampling values only take positive values and thus $|x_{\mathrm{e}}(k\,T_{\mathrm{s}})| = x_{\mathrm{e}}(k\,T_{\mathrm{s}})$ applies (see Fig. 2.12). The first of the symbol clock sampling values are, using (2.18):

$$x_{\mathrm{e}}(T_{\mathrm{s}}) = U_0\left(1 - e^{-\frac{T_{\mathrm{s}}}{T_{\mathrm{g}}}}\right)$$

$$x_{\mathrm{e}}(2T_{\mathrm{s}}) = U_0\left[\left(1 - e^{-\frac{2T_{\mathrm{s}}}{T_{\mathrm{g}}}}\right) - \left(1 - e^{-\frac{2T_{\mathrm{s}}-T_{\mathrm{s}}}{T_{\mathrm{g}}}}\right)\right]$$

$$= U_0\left[1 - e^{-\frac{2T_{\mathrm{s}}}{T_{\mathrm{g}}}} - 1 + e^{-\frac{2T_{\mathrm{s}}-T_{\mathrm{s}}}{T_{\mathrm{g}}}}\right]$$

$$= U_0\left[e^{-\frac{T_{\mathrm{s}}}{T_{\mathrm{g}}}} - e^{-\frac{2T_{\mathrm{s}}}{T_{\mathrm{g}}}}\right]$$

$$x_{\mathrm{e}}(3T_{\mathrm{s}}) = U_0\left[\left(1 - e^{-\frac{3T_{\mathrm{s}}}{T_{\mathrm{g}}}}\right) - \left(1 - e^{-\frac{3T_{\mathrm{s}}-T_{\mathrm{s}}}{T_{\mathrm{g}}}}\right)\right]$$

$$= U_0\left[1 - e^{-\frac{3T_{\mathrm{s}}}{T_{\mathrm{g}}}} - 1 + e^{-\frac{3T_{\mathrm{s}}-T_{\mathrm{s}}}{T_{\mathrm{g}}}}\right]$$

$$= U_0\left[e^{-\frac{2T_{\mathrm{s}}}{T_{\mathrm{g}}}} - e^{-\frac{3T_{\mathrm{s}}}{T_{\mathrm{g}}}}\right]$$

$$x_{\mathrm{e}}(4T_{\mathrm{s}}) = U_0\left[\left(1 - e^{-\frac{4T_{\mathrm{s}}}{T_{\mathrm{g}}}}\right) - \left(1 - e^{-\frac{4T_{\mathrm{s}}-T_{\mathrm{s}}}{T_{\mathrm{g}}}}\right)\right]$$

$$= U_0\left[1 - e^{-\frac{4T_{\mathrm{s}}}{T_{\mathrm{g}}}} - 1 + e^{-\frac{4T_{\mathrm{s}}-T_{\mathrm{s}}}{T_{\mathrm{g}}}}\right]$$

$$= U_0\left[e^{-\frac{3T_{\mathrm{s}}}{T_{\mathrm{g}}}} - e^{-\frac{4T_{\mathrm{s}}}{T_{\mathrm{g}}}}\right]$$

$$\vdots$$

If these symbol clock sampling values are inserted into (2.38), the result for the half vertical eye opening is initially

$$U_{\mathrm{A}} = x_{\mathrm{e}}(T_{\mathrm{s}}) - (x_{\mathrm{e}}(2T_{\mathrm{s}}) + x_{\mathrm{e}}(3T_{\mathrm{s}}) + x_{\mathrm{e}}(4T_{\mathrm{s}}) + \ldots)$$

$$U_{\mathrm{A}} = \frac{U_0}{2}\left[\left(1 - e^{-\frac{T_{\mathrm{s}}}{T_{\mathrm{g}}}}\right) - \left\{\left(e^{-\frac{T_{\mathrm{s}}}{T_{\mathrm{g}}}} - e^{-\frac{2T_{\mathrm{s}}}{T_{\mathrm{g}}}}\right) + \left(e^{-\frac{2T_{\mathrm{s}}}{T_{\mathrm{g}}}} - e^{-\frac{3T_{\mathrm{s}}}{T_{\mathrm{g}}}}\right) + \ldots\right.\right.$$

$$\left.\left.\ldots + \left(e^{-\frac{3T_{\mathrm{s}}}{T_{\mathrm{g}}}} - e^{-\frac{4T_{\mathrm{s}}}{T_{\mathrm{g}}}}\right) + \ldots\right\}\right].$$

It can be seen that consecutive summands each contain the same terms, which are associated with different signs, so that according to

$$U_{\mathrm{A}} = \frac{U_0}{2}\left[\left(1 - e^{-\frac{T_{\mathrm{s}}}{T_{\mathrm{g}}}}\right) - \left(e^{-\frac{T_{\mathrm{s}}}{T_{\mathrm{g}}}} - e^{-\frac{2T_{\mathrm{s}}}{T_{\mathrm{g}}}}\right) - \left(e^{-\frac{2T_{\mathrm{s}}}{T_{\mathrm{g}}}} - e^{-\frac{3T_{\mathrm{s}}}{T_{\mathrm{g}}}}\right) - \ldots\right.$$

$$\left.\ldots - \left(e^{-\frac{3T_{\mathrm{s}}}{T_{\mathrm{g}}}} - e^{-\frac{4T_{\mathrm{s}}}{T_{\mathrm{g}}}}\right) - \ldots\right]$$

and further

$$U_{\mathrm{A}} = \frac{U_0}{2}\left[\left(1 - \mathrm{e}^{-\frac{T_\mathrm{s}}{T_\mathrm{g}}}\right) - \mathrm{e}^{-\frac{T_\mathrm{s}}{T_\mathrm{g}}} + \mathrm{e}^{-\frac{2T_\mathrm{s}}{T_\mathrm{g}}} - \mathrm{e}^{-\frac{2T_\mathrm{s}}{T_\mathrm{g}}} + \mathrm{e}^{-\frac{3T_\mathrm{s}}{T_\mathrm{g}}} - \mathrm{e}^{-\frac{3T_\mathrm{s}}{T_\mathrm{g}}} + \mathrm{e}^{-\frac{4T_\mathrm{s}}{T_\mathrm{g}}} - \ldots\right] \quad (2.39)$$

almost all terms cancel each other out. For the last term in each case (here when the summation is terminated after $x_\mathrm{e}(4T_\mathrm{s})$: $\mathrm{e}^{-\frac{4T_\mathrm{s}}{T_\mathrm{g}}}$) it applies $\lim_{k\to\infty} \mathrm{e}^{-kT_\mathrm{s}/T_\mathrm{g}} = 0$ when the summation is terminated after a finite number of symbol clock sampling values. Only

$$U_{\mathrm{A}} = \frac{U_0}{2}\left[\left(1 - \mathrm{e}^{-\frac{T_\mathrm{s}}{T_\mathrm{g}}}\right) - \mathrm{e}^{-\frac{T_\mathrm{s}}{T_\mathrm{g}}} + \underbrace{\mathrm{e}^{-\frac{4T_\mathrm{s}}{T_\mathrm{g}}}}_{\approx 0}\right] = \frac{U_0}{2}\left[1 - \mathrm{e}^{-\frac{T_\mathrm{s}}{T_\mathrm{g}}} - \mathrm{e}^{-\frac{T_\mathrm{s}}{T_\mathrm{g}}}\right], \quad (2.40)$$

remains, so that

$$U_{\mathrm{A}} = \frac{U_0}{2}\left[1 - 2\,\mathrm{e}^{-\frac{T_\mathrm{s}}{T_\mathrm{g}}}\right] \quad (2.41)$$

or

$$U_{\mathrm{A}} = U_0\left[\frac{1}{2} - \mathrm{e}^{-\frac{T_\mathrm{s}}{T_\mathrm{g}}}\right] \quad (2.42)$$

results. If this result is compared with the result obtained by another method (2.34) or (2.35) (or (2.37)) for the half vertical eye opening, it becomes clear that both methods lead to identical formulations. $\qquad\square$

2.1.3   Noise Path and Evaluation of the Noise Disturbance

System for Evaluating the Noise Disturbance

For the evaluation of the disturbance, the transmission system shown in Fig. 2.15 remains: The noise as a disturbance is evaluated on the way to the detection point by the receive filter $G_\mathrm{e}(f)$ or $G_\mathrm{e}(p)$.

The disturbance is white, Gaussian noise with the spectral power density Ψ_0. Causes for the disturbance are the noise of the receiver input resistance and further external disturbance sources, which can often be modeled approximately as white, Gaussian noise.

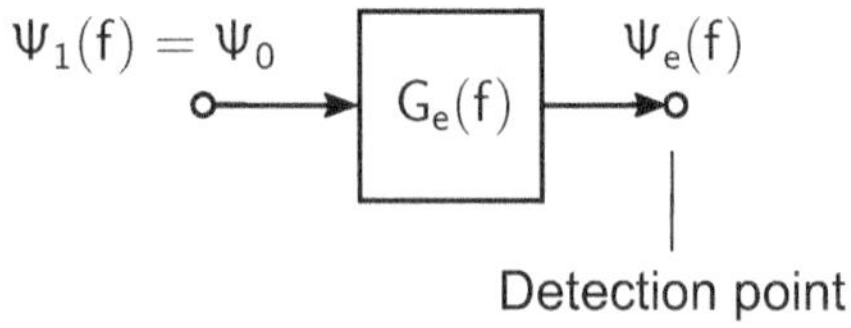

Fig. 2.15 Block diagram for the evaluation of the noise disturbance

Insert: Signal and System Theoretical Relationships for Random Signals
Some signal and system theoretical relationships and correlations for random signals are helpful for calculating noise characteristics. These will be compactly summarized in the following.

Noise Power and Noise Effective Value In Fig. 2.16, the power spectral density $\Psi(f)$ and autocorrelation function $\psi(\tau)$ of a random signal are schematically represented, which are connected by Fourier transformation, i.e., $\Psi(f) = \mathcal{F}\{\psi(\tau)\}$ applies.

The power P of the random signal is the area marked by hatching under the power spectral density and can be calculated as

$$P = \int_{-\infty}^{+\infty} \Psi(f)\, df = \frac{1}{2\pi} \int_{-\infty}^{+\infty} \Psi(\omega)\, d\omega \tag{2.43}$$

The autocorrelation function gives

$$\psi(\tau) = \mathcal{F}^{-1}\{\Psi(f)\} = \int_{-\infty}^{+\infty} \Psi(f)\, e^{j2\pi f \tau}\, df \tag{2.44}$$

and at the point $\tau = 0$ results in

$$\psi(\tau = 0) = \int_{-\infty}^{+\infty} \Psi(f)\, e^{j2\pi f \cdot 0}\, df = \underbrace{\int_{-\infty}^{+\infty} \Psi(f)\, df}_{P}. \tag{2.45}$$

The power of a random signal can be calculated over the area under the power density spectrum $\Psi(f)$ using integration and as a function value of the autocorrelation function $\psi(\tau)$ at the point $\tau = 0$:

$$P = \int_{-\infty}^{+\infty} \Psi(f)\, df = \psi(\tau = 0). \tag{2.46}$$

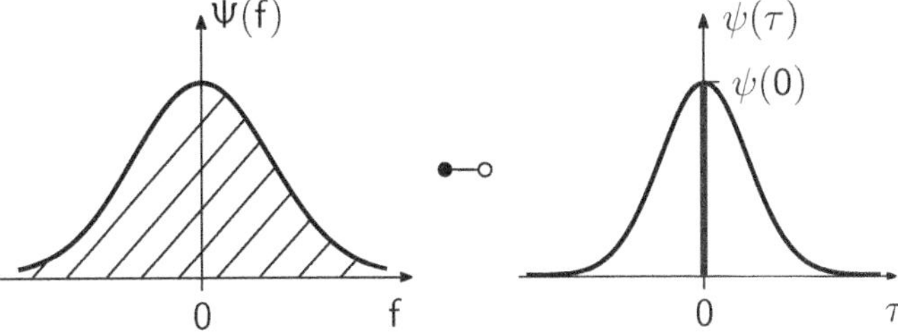

Fig. 2.16 Power spectral density and autocorrelation function of a random process

The effective value of the noise results in

$$u_{\text{eff}} = \sqrt{P} = U_{\text{R}}. \tag{2.47}$$

White Noise The *white noise* is a theoretical model of broadband noise. Power spectral density and autocorrelation function are shown in Fig. 2.17.
An important feature of white noise is that it has infinite power, as, for example, the area under the function of the power density spectrum $\Psi(f) = \Psi_0$ and thus the integral (2.43) becomes infinite.

Filtering of Noise If a noise signal with a power density $\Psi_1(f)$ is transmitted over a system $G(f)$ (see Fig. 2.18), the power spectral density at the system output becomes

$$\Psi_2(f) = \Psi_1(f) \cdot |G(f)|^2. \tag{2.48}$$

The noise power at the filter output is then

$$P_2 = \int_{-\infty}^{+\infty} \Psi_2(f)\,\mathrm{d}f = \int_{-\infty}^{+\infty} \Psi_1(f) \cdot |G(f)|^2\,\mathrm{d}f, \tag{2.49}$$

and for even power spectral density applies

$$P_2 = 2 \int_{0}^{\infty} \Psi_1(f) \cdot |G(f)|^2\,\mathrm{d}f. \tag{2.50}$$

For white noise at the filter input, whose power spectral density $\Psi_1(f) = \Psi_0$ is constant with respect to the frequency f, one obtains

$$P_2 = 2\,\Psi_0 \int_{0}^{\infty} |G(f)|^2\,\mathrm{d}f. \tag{2.51}$$

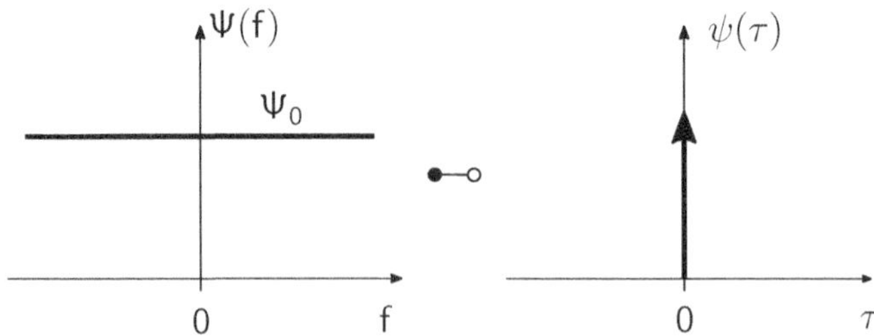

Fig. 2.17 Power spectral density and autocorrelation function of white noise

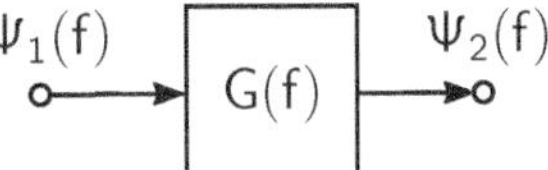

Fig. 2.18 Block diagram for noise filtering

Calculation of Noise Power

The power density of the noise signal at the detection point (see Fig. 2.15) is

$$\Psi_e(f) = \Psi_0 |G_e(f)|^2, \tag{2.52}$$

and the noise power as a suitable measure for the intensity of the disturbance at the detection point is obtained via integration

$$U_R^2 = \Psi_0 \int_{-\infty}^{+\infty} |G_e(f)|^2 \, df. \tag{2.53}$$

The noise power is determined by the area under $|G_e(f)|^2$, i.e., the shape of the transfer function $G_e(f)$ of the receiving filter—and thus its cutoff frequency—significantly influences the effective noise power at the detection point. With increasing cutoff frequency of the receiving filter, the noise power present at the detection point increases. Figure 2.19 illustrates the relationships in the filtering of white noise.

The hatched area below $\Psi_e(f)$ in Fig. 2.19 represents the noise power U_R^2, and it increases especially when the cutoff frequency f_g increases.

From the standpoint of the noise disturbance, a smallest possible cutoff frequency f_g of the receive filter is desirable, as then the noise power at the output of the receive filter is small and thus the disturbance in its effects at the detection point is minimized.

Calculation of the Noise Power at the Output of the Receive Filter Depending on its Cutoff Frequency

For the considered case of a first-order RC low-pass filter as a receive filter, the noise power at the detection point is now calculated. The starting point is the transfer function according to (2.9) or (2.10). The magnitude of the transfer function is

Fig. 2.19 For the filtering of noise at the receiver

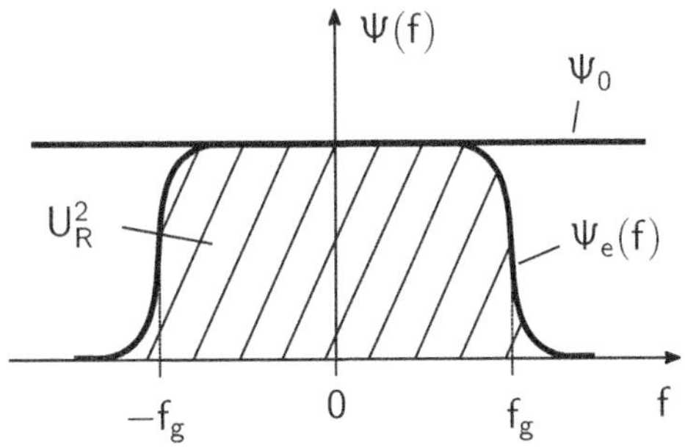

$$G_e(f) = \frac{1}{1 + j\frac{f}{f_g}} \implies |G_e(f)| = \frac{1}{\sqrt{1^2 + \left(\frac{f}{f_g}\right)^2}} \tag{2.54}$$

and the square of this magnitude of the transfer function is

$$|G_e(f)|^2 = \frac{1}{1 + \left(\frac{f}{f_g}\right)^2} = \frac{1}{1 + \frac{f^2}{f_g^2}} \implies |G_e(f)|^2 = \frac{f_g^2}{f_g^2 + f^2}. \tag{2.55}$$

The noise power is thus

$$U_R^2 = \Psi_0 \int_{-\infty}^{+\infty} \frac{f_g^2}{f_g^2 + f^2}\, df = 2\, \Psi_0 f_g^2 \int_0^{+\infty} \frac{1}{f_g^2 + f^2}\, df. \tag{2.56}$$

For the calculation of the integral, one uses

$$\int \frac{1}{a^2 + x^2}\, dx = \frac{1}{a} \arctan \frac{x}{a} + C \tag{2.57}$$

from Appendix F.2 or [11, 74] and obtains

$$\begin{aligned} U_R^2 &= 2\, \Psi_0 f_g^2 \left[\frac{1}{f_g} \arctan \frac{f}{f_g} \right]_0^{+\infty} \\ &= 2\, \Psi_0 \frac{f_g^2}{f_g} \left[\arctan \frac{\infty}{f_g} - \arctan \frac{0}{f_g} \right] \\ &= 2\, \Psi_0 f_g \left[\frac{\pi}{2} - 0 \right] = 2\, \Psi_0 f_g \frac{\pi}{2}. \end{aligned} \tag{2.58}$$

The result for the noise power at the detection point is

$$U_R^2 = \Psi_0\, \pi\, f_g. \tag{2.59}$$

The noise power is directly proportional to the cutoff frequency of the receive filter f_g and therefore increases linearly with increasing cutoff frequency of the receive filter (Fig. 2.20).

Fig. 2.20 Noise power at the detection point as a function of the receive filter cutoff frequency f_g

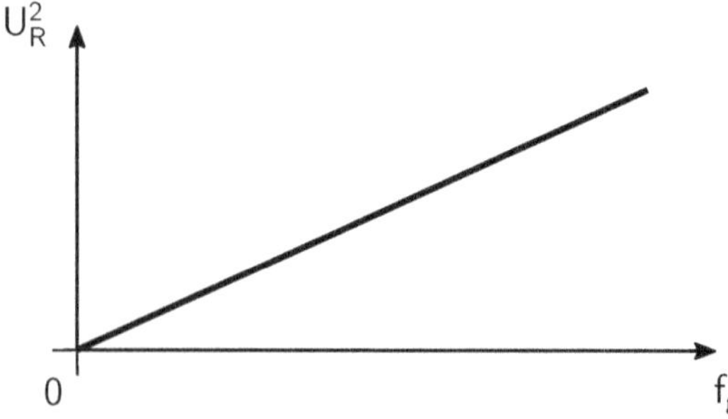

2.1.4   Quality of Transmission

If a transmission system is to be examined, dimensioned or optimized, a criterion for the quality or a measure for the quality of digital signal transmission—i.e., a quality or quality criterion—must be found and defined that is accessible to calculation or measurement [1]. Common and practical is the definition of a *signal-to-noise ratio* as a quotient, in which a useful signal power and a disturbance power are related at the output of the receiving filter. The larger this ratio is, the more reliable is the transmission.

As a measure of the useful signal power at the detection point, the square of the *half vertical eye opening* (2.31) is used here, and as a measure of the intensity of the disturbance—also at the detection point—the noise power or the square of the noise effective value (2.53) is used. If these two quantities are put into relation to each other, the result is the *signal-to-noise ratio*

$$\varrho = \frac{(\text{half vertical eye opening})^2}{(\text{noise effective value})^2} = \frac{U_A^2}{U_R^2}. \tag{2.60}$$

The signal-to-noise ratio according to (2.60) is the ratio of the available signal power as the square of the half vertical eye opening U_A for the worst case and the noise power U_R^2 at the detection point.

Since this quantity represents the signal-to-noise ratio at the detection point it is occasionally also referred to as the *detection* signal-to-noise ratio, in order to distinguish it from other definitions of the signal-to-noise ratio within the transmission system. A discussion of some signal-to-noise ratios used in the transmission of digital signals is carried out in a section later in this chapter to clarify mutual relationships.

If the signal-to-noise ratio is given in dB, the following applies

$$\varrho^* = 10 \cdot \lg \varrho \qquad (\text{in dB}). \tag{2.61}$$

The signal-to-noise ratio in dB according to (2.61) is often also referred to as *signal-to-noise distance* or *interference distance*.

Both the useful signal and the disturbance are evaluated by the receive filter: Thus, the cutoff frequency f_g of the receive filter according to Figs. 2.14 and 2.20 plays an important role for the achievable transmission quality. To summarize the influence of the receive filter cutoff frequency in the evaluation of useful signal and disturbance once again, the half vertical eye opening U_A and the noise effective value U_R are shown together in a diagram in Fig. 2.21: Below a cutoff frequency $f_{g\,min}$, the eye is closed; above $f_{g\,min}$, the half vertical eye opening increases monotonically and approaches a maximum value, which is determined by the half transmission amplitude. In contrast, the noise effective value U_R—and thus the noise power U_R^2—increases unlimitedly with increasing cutoff frequency f_g; the noise effective value $U_R = \sqrt{\Psi_0\,\pi\,f_g}$ is proportional to $\sqrt{f_g}$ according to (2.59), i.e. $U_R \sim \sqrt{f_g}$ or $U_R^2 \sim f_g$.

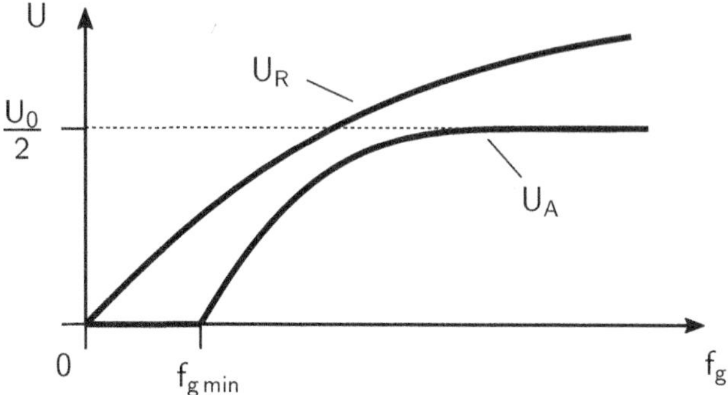

Fig. 2.21 On the role of the receive filter cutoff frequency f_{g}

As a result of this consideration of the role of the receive filter cutoff frequency with regard to transmission quality, the following problem arises:

- A large cutoff frequency f_{g} (or ω_{g}) of the receive filter causes a large vertical eye opening, but also a large noise power.
- A small cutoff frequency f_{g} (or ω_{g}) of the receive filter leads to a small vertical eye opening and causes a small noise power.

The signal-to-noise ratio is inversely dependent on the half vertical eye opening U_{A} and the noise effective value U_{R} or the noise power U_{R}^2 from the cutoff frequency f_{g} (or ω_{g}) of the receive filter: To achieve a transmission that is as powerful and reliable as possible under the given boundary conditions, an optimization of the receive filter cutoff frequency is possible and necessary, so that a maximum signal-to-noise ratio is achieved. For the dimensioning of the receive filter, the corresponding optimal cutoff frequency is set or, in a circuit design, the time constant $T_{\mathrm{g}} = R\,C$ is adjusted accordingly through the appropriate choice of component values R and C of the RC low-pass filter.

Figure 2.22 illustrates these relationships and shows the signal-to-noise ratio (2.60) as a function of the receive filter cutoff frequency f_{g}. It becomes clear that for the receive filter cutoff frequency f_{g}, there exists an optimum at which the signal-to-noise ratio exhibits a maximum ϱ_{max}: At very small receiving filter cutoff frequencies f_{g}, the eye is closed and therefore reliable detection and decision about received bits is not possible. After the eye has opened vertically above $f_{\mathrm{g\,min}}$, the signal-to-noise ratio ϱ increases with increasing cutoff frequency f_{g}, as the half vertical eye opening U_{A} (see Fig. 2.14) rapidly enlarges in this range, but the noise in its power U_{R}^2 still remains relatively small (see Fig. 2.20). With further increasing receive filter cutoff frequency f_{g}, a point is reached at which the increases U_{A}^2 and U_{R}^2 in dependence on f_{g} just balance out: Here, the signal-to-noise ratio reaches its maximum value ϱ_{max}, the receive filter cutoff frequency is

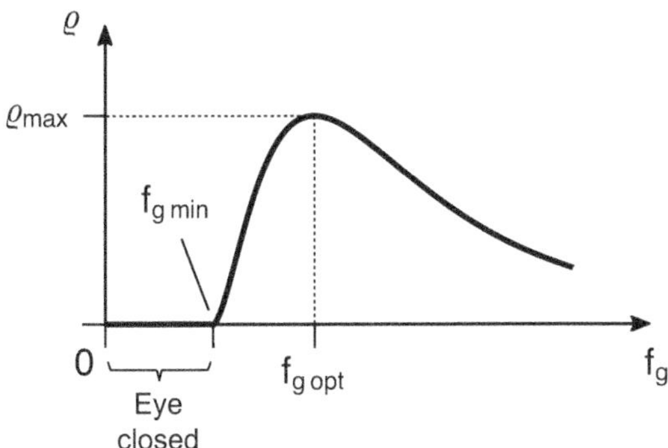

Fig. 2.22 Signal-to-noise ratio ϱ at the detection point as a function of the receive filter cutoff frequency f_g

therefore—with respect to a maximum signal-to-noise ratio—optimal and is therefore referred to as $f_{g\,opt}$. After this maximum, the signal-to-noise ratio decreases with further increasing receive filter cutoff frequency, as the half vertical eye opening approaches a maximum value, which is determined by the half-level transmit amplitude $U_0/2$, and the noise now predominates and continues to rise unrestrictedly.

The practical determination of the optimal cutoff frequency of the receive filter is generally only possible numerically, i.e., either through the numerical evaluation of mathematical relationships that describe the transmission path, or through computer simulation or experiment; only in special cases is an analytical determination of the optimal cutoff frequency and the resulting maximum signal-to-noise ratio possible.

2.1.5 Relationship Between Signal-to-Noise Ratio and Error Probability

Since ultimately bits are transmitted in a digital information or signal transmission and these are to be detected in the receiver, a quality criterion has the greatest significance that allows an assessment of the quality of this decision about received bits. Thus, the average bit error probability is the most suitable quality measure in principle, as it allows a statement about the probability with which a transmitted bit, transmitted over a transmission channel, is incorrectly recognized and decided in the receiver [1]. A nearly equivalent measure in practical transmission systems is the average bit error rate or the average bit error ratio, i.e., the number of bits incorrectly decided on average in the receiver relative to the number of transmitted bits (e.g., [24]).

Therefore, for the two-level, unipolar baseband transmission (see Fig. 2.1), the relationship between the signal-to-noise ratio (2.60) and the desired bit error probability is determined. A special feature to be mentioned is that the transmitted symbols in this examined transmission system are binary ('0' and '1') and thus directly represent bits. The error probability P_f calculated in the following is therefore immediately the desired bit error probability P_b. In general—especially in multi-level transmission—a distinction must be made between symbol and bit error probability (see the section on multi-level transmission later in this chapter).

First, the noise is considered again: In terms of information transmission, two essential characteristics of noise are important, some of which have already been used:

1. Spectral distribution: The noise is white, i.e., it has a constant power density (or square amplitude density) Ψ_0 (in V^2/Hz) in the considered frequency range.
2. Amplitude distribution: Electronic noise is assumed to be a random process (stochastic process) with an amplitude distribution that corresponds to a Gaussian or normal distribution.

This model for noise is appropriate and has proven itself in the description and analysis of transmission systems, although it is of course only an approximation of the real physical processes.

The probability density of the Gaussian or normal distribution is

$$p(u) = \frac{1}{\sigma\sqrt{2\pi}} \cdot e^{-\frac{(u-u_m)^2}{2\sigma^2}}, \tag{2.62}$$

where u_m is the linear mean or expected value and σ is the dispersion or standard deviation (σ^2: variance) of the random noise voltage [25]. The maximum value of the probability density $p(u)$ is at the position of the expected value u_m and is $p_{max} = \frac{1}{\sigma\sqrt{2\pi}}$. Figure 2.23 shows the probability density function of a Gaussian or normal distribution with its characteristics.

The parameters u_m and σ can be determined from temporal considerations: u_m corresponds to the DC component $\overline{u(t)}$ of the noise voltage $u(t)$ and σ^2 to its variance, which can be determined from its square mean value $\overline{u^2(t)}$ and the DC component $\overline{u(t)}$ according to $\sigma^2 = \overline{u^2(t)} - \left[\overline{u(t)}\right]^2$. The prerequisite for determining the parameters via temporal considerations is that the noise process is *stationary*[6] and *ergodic*[7].

If the Gaussian process is mean-free, then $u_m = 0$ applies: Then $\sigma^2 = U_R^2$ is the power of the noise and $\sigma = U_R$ is the effective noise value.

In communications engineering, disturbances, e.g., by noise, can often be modeled by mean-free Gaussian processes that are additively superimposed on the signal amplitudes (here: 0 and U_0).

---

[6] A random process $X(t)$ is called stationary if all mean values are independent of a shift in the observation time by any time t_0.

[7] A random process $X(t)$ is called ergodic if the time averages are the same for all sample functions, i.e., time average = ensemble average. Ergodic processes must always be stationary.

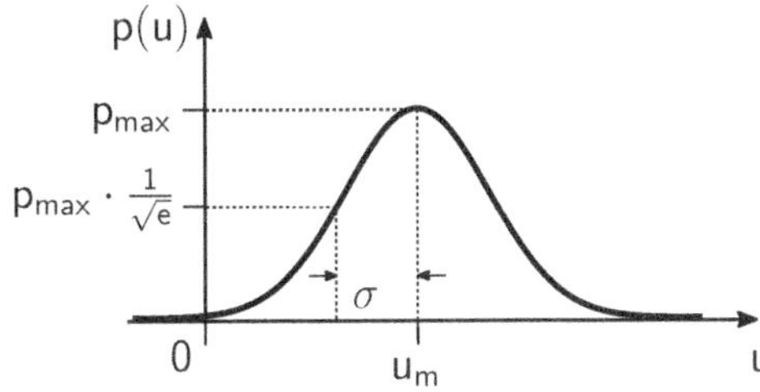

Fig. 2.23 Probability density of a Gaussian distribution

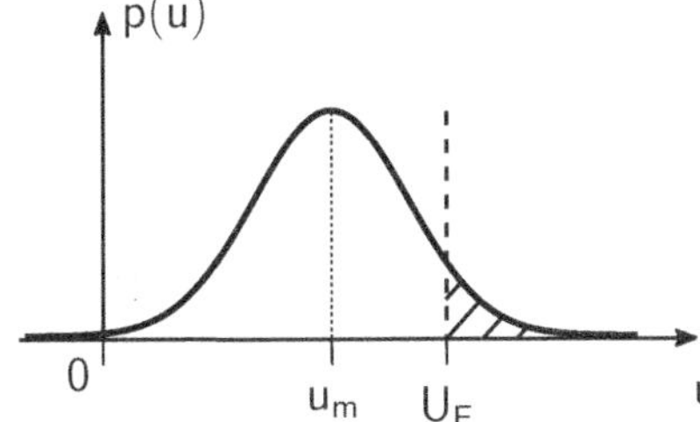

Fig. 2.24 On the calculation of the exceedance probability of a threshold

Insert: Exceedance Probability

For the calculation of the error probability, the probability is needed with which a threshold U_E is exceeded given the probability density function $p(u)$ of the noise. Therefore, this probability will be calculated and derived in detail. Figure 2.24 shows the probability density function $p(u)$ with the threshold U_E; the hatched area represents the sought probability and can be calculated as

$$P(u \geq U_E) = \int_{U_E}^{\infty} p(u)\, du \,. \tag{2.63}$$

Substituting (2.62) into (2.63) initially results in

$$P(u \geq U_E) = \frac{1}{\sigma \sqrt{2\pi}} \int_{U_E}^{\infty} e^{-\frac{(u-u_m)^2}{2\sigma^2}}\, du \,. \tag{2.64}$$

If the curve is shifted to the left by the expected value u_m, a modified probability density $p(u + u_m)$ is obtained with $u_m = 0$: The hatched area does not change (see Fig. 2.25), but the calculation is simplified. Therefore, it can be further written:

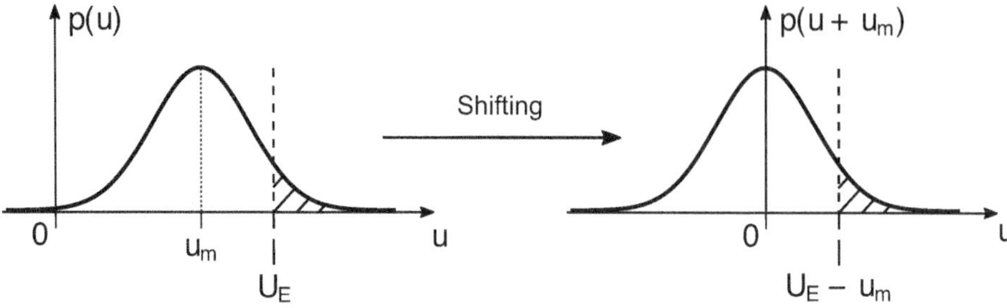

Fig. 2.25 On the shift of the probability density function in the calculation of the exceedance probability of a threshold

$$
P(u \geq U_{\mathrm{E}}) = \frac{1}{\sigma \sqrt{2\pi}} \int_{U_{\mathrm{E}}-u_{\mathrm{m}}}^{\infty} e^{-\frac{u^2}{2\sigma^2}}\, \mathrm{d}u
$$

$$
= \frac{1}{\sigma \sqrt{2\pi}} \int_{U_{\mathrm{E}}-u_{\mathrm{m}}}^{\infty} e^{-\left(\frac{u}{\sqrt{2}\sigma}\right)^2}\, \mathrm{d}u
$$

$$
= \frac{1}{\sigma \sqrt{2\pi}} \int_{\frac{U_{\mathrm{E}}-u_{\mathrm{m}}}{\sqrt{2}\sigma}}^{\infty} e^{-\left(\frac{u}{\sqrt{2}\sigma}\right)^2}\, \mathrm{d}\left(\frac{u}{\sqrt{2}\sigma}\right) \cdot \sqrt{2}\sigma
$$

$$
= \frac{\sqrt{2}\sigma}{\sigma \sqrt{2\pi}} \int_{\frac{U_{\mathrm{E}}-u_{\mathrm{m}}}{\sqrt{2}\sigma}}^{\infty} e^{-\left(\frac{u}{\sqrt{2}\sigma}\right)^2}\, \mathrm{d}\left(\frac{u}{\sqrt{2}\sigma}\right)
$$

$$
= \frac{1}{\sqrt{\pi}} \int_{\frac{U_{\mathrm{E}}-u_{\mathrm{m}}}{\sqrt{2}\sigma}}^{\infty} e^{-\left(\frac{u}{\sqrt{2}\sigma}\right)^2}\, \mathrm{d}\left(\frac{u}{\sqrt{2}\sigma}\right)
$$

$$
P(u \geq U_{\mathrm{E}}) = \frac{1}{2} \cdot \frac{2}{\sqrt{\pi}} \int_{\frac{U_{\mathrm{E}}-u_{\mathrm{m}}}{\sqrt{2}\sigma}}^{\infty} e^{-\left(\frac{u}{\sqrt{2}\sigma}\right)^2}\, \mathrm{d}\left(\frac{u}{\sqrt{2}\sigma}\right).
$$

With the definition of the complementary Gaussian error function (or complementary Gaussian error function, see Appendix E and e.g. [4, 66])

$$
\mathrm{erfc}(x) = \frac{2}{\sqrt{\pi}} \int_{x}^{\infty} e^{-y^2}\, \mathrm{d}y = 1 - \mathrm{erf}(x) \tag{2.65}
$$

and

$$\text{erf}(x) = \frac{2}{\sqrt{\pi}} \int_0^x e^{-y^2} \, dy \tag{2.66}$$

finally results—with $y = u/(\sqrt{2}\sigma)$—the probability that the threshold U_E is exceeded given the probability density function $p(u)$:

$$P(u \geq U_\text{E}) = \frac{1}{2} \text{erfc}\left(\frac{U_\text{E} - u_\text{m}}{\sqrt{2}\sigma}\right) = \frac{1}{2}\left[1 - \text{erf}\left(\frac{U_\text{E} - u_\text{m}}{\sqrt{2}\sigma}\right)\right]. \tag{2.67}$$

In the considered basic model of binary baseband transmission, '0' symbols (bits) are sent as voltage 0 V and '1' symbols (bits) as voltage U_0. Assuming initially that no inter-symbol interference is effective at the detection point (e.g., at a suitably high cutoff frequency f_g of the receive filter), Fig. 2.26 shows the present reception situation for two different noise powers. In the absence of the additive noise disturbance, under this assumption at the detection point (at the optimal sampling time), the amplitudes 0 or U_0 would result.

There are two possibilities for misdecisions in the receiver:

- If a '0' has been sent, the receiver incorrectly decides on '1'. Then the probability is to be calculated that the received voltage $u(t)$ at the sampling time is greater than the threshold U_E when '0' has been sent.
- If a '1' has been sent, the receiver incorrectly decides on '0'. Now the probability is to be calculated that the received voltage $u(t)$ at the sampling time is less than the threshold U_E when '1' has been sent.

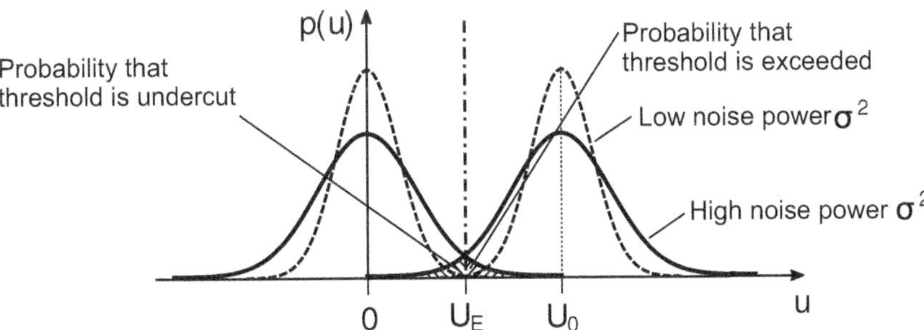

Fig. 2.26 For the calculation of the error probability

The overall probability that an error occurs in the decision at the receiver is the sum of both probabilities, each weighted with the occurrence probability for a transmitted '0' and a transmitted '1' (according to the law of total probability, e.g. [76])

$$P_\mathrm{f} = P(1) \cdot P(0|1) + P(0) \cdot P(1|0). \tag{2.68}$$

For further calculation, the practically important special case is assumed, that the transmitted symbols are equally likely, i.e. it applies

$$P(0) = P(1) = \frac{1}{2}. \tag{2.69}$$

Therefore, it is a symmetric binary source [42, 48].

With any cut-off frequency f_g of the receive filter, intersymbol interferences can generally occur and the eye is not fully open vertically. Figure 2.27 shows the conditions in this case using the eye diagram and the probability density functions with the characteristics: It becomes clear that now in the worst case (*worst case*) the values U_max and U_min—instead of U_0 and 0—limit the eye inward at the optimal sampling time. The mean values of the Gaussian processes thus shift from 0 to U_min and from U_0 to U_max.

For the calculation of the error probability, it is further assumed that the threshold is in the middle between the two transmission amplitude levels (0 and U_0) and thus under the given conditions also in the middle between the two useful signal amplitude levels (U_max and U_min) at the detection point, i.e. :

$$U_\mathrm{E} = \frac{U_0}{2} \qquad \mathrm{bzw.} \qquad U_\mathrm{E} = \frac{U_\mathrm{max} - U_\mathrm{min}}{2} + U_\mathrm{min}. \tag{2.70}$$

The total probability for a wrong decision on the receiving side simplifies with (2.69) to

$$P_\mathrm{f} = \frac{1}{2}[P(0|1) + P(1|0)]. \tag{2.71}$$

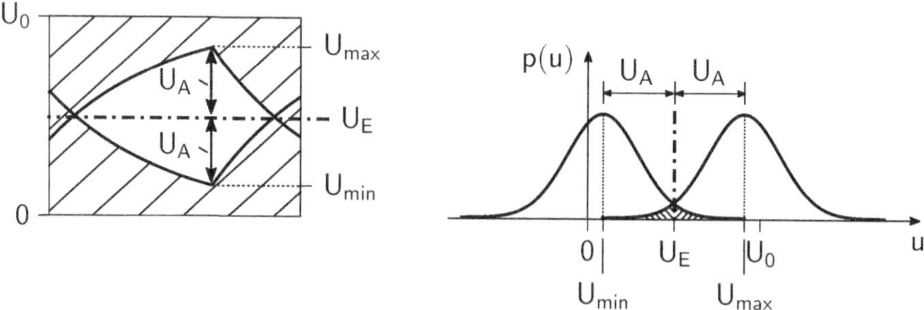

Fig. 2.27 For the calculation of the error probability in intersymbol interferences

The remaining probabilities for wrong decisions to be calculated are

$$P(0|1) = \int_{-\infty}^{U_E} p(u - U_{max})\, du \tag{2.72}$$

and

$$P(1|0) = \int_{U_E}^{\infty} p(u - U_{min})\, du. \tag{2.73}$$

The error probability obtained is

$$P_f = \frac{1}{2}\left[\int_{-\infty}^{U_E} p(u - U_{max})du + \int_{U_E}^{\infty} p(u - U_{min})du\right]. \tag{2.74}$$

Since the exceedance probability (2.67) has already been calculated, (2.74) can be advantageously rewritten with the identity

$$\int_{-\infty}^{U_E} p(u - U_{max})du = 1 - \int_{U_E}^{\infty} p(u - U_{max})du \tag{2.75}$$

(taking advantage of the normalization condition of probability density functions, e.g. [11, 76]) to

$$P_f = \frac{1}{2}\left[1 - \int_{U_E}^{\infty} p(u - U_{max})du + \int_{U_E}^{\infty} p(u - U_{min})du\right]. \tag{2.76}$$

and with (2.67) results

$$P_f = \frac{1}{2}\left[1 - \frac{1}{2}\left[1 - \mathrm{erf}\left(\frac{U_E - U_{max}}{\sqrt{2}\sigma}\right)\right] + \frac{1}{2}\left[1 - \mathrm{erf}\left(\frac{U_E - U_{min}}{\sqrt{2}\sigma}\right)\right]\right]. \tag{2.77}$$

Insert: Derivation of the Error Probability

According to Fig. 2.27, $U_E - U_{max} = -U_A$ or $U_{max} - U_E = U_A$ and $U_E - U_{min} = U_A$ apply. If you put this relationship into (2.77) and take into account the designation $\sigma = U_R$ for the noise effective value, you get

$$P_f = \frac{1}{2}\left[1 - \frac{1}{2}\left[1 - \mathrm{erf}\left(\frac{U_E - U_{max}}{\sqrt{2}\,U_R}\right)\right] + \frac{1}{2}\left[1 - \mathrm{erf}\left(\frac{U_E - U_{min}}{\sqrt{2}\,U_R}\right)\right]\right]$$

$$= \frac{1}{2}\left[1 - \frac{1}{2}\left[1 - \mathrm{erf}\left(\frac{-U_A}{\sqrt{2}\,U_R}\right)\right] + \frac{1}{2}\left[1 - \mathrm{erf}\left(\frac{U_A}{\sqrt{2}\,U_R}\right)\right]\right].$$

The function $\mathrm{erf}(x)$ is an odd function, i.e., the relationship $\mathrm{erf}(x) = -\mathrm{erf}(-x)$ applies (see Appendix E). If you take this into account, you get further

$$P_\mathrm{f} = \frac{1}{2}\left[1 - \frac{1}{2}\left[1 + \mathrm{erf}\left(\frac{U_\mathrm{A}}{\sqrt{2}\,U_\mathrm{R}}\right)\right] + \frac{1}{2}\left[1 - \mathrm{erf}\left(\frac{U_\mathrm{A}}{\sqrt{2}\,U_\mathrm{R}}\right)\right]\right]$$

$$= \frac{1}{2}\left[1 - \frac{1}{2} - \frac{1}{2}\,\mathrm{erf}\left(\frac{U_\mathrm{A}}{\sqrt{2}\,U_\mathrm{R}}\right) + \frac{1}{2} - \frac{1}{2}\,\mathrm{erf}\left(\frac{U_\mathrm{A}}{\sqrt{2}\,U_\mathrm{R}}\right)\right]$$

$$= \frac{1}{2}\left[1 - \frac{1}{2}\,\mathrm{erf}\left(\frac{U_\mathrm{A}}{\sqrt{2}\,U_\mathrm{R}}\right) - \frac{1}{2}\,\mathrm{erf}\left(\frac{U_\mathrm{A}}{\sqrt{2}\,U_\mathrm{R}}\right)\right]$$

$$= \frac{1}{2}\left[1 - \mathrm{erf}\left(\frac{U_\mathrm{A}}{\sqrt{2}\,U_\mathrm{R}}\right)\right].$$

With the definition of the signal-to-noise ratio (2.60), one gets

$$\varrho = \frac{U_\mathrm{A}^2}{U_\mathrm{R}^2} \qquad \Longrightarrow \qquad \frac{U_\mathrm{A}}{U_\mathrm{R}} = \sqrt{\varrho} \tag{2.78}$$

and the error probability is ultimately

$$P_\mathrm{f} = \frac{1}{2}\left[1 - \mathrm{erf}\left(\sqrt{\frac{\varrho}{2}}\right)\right] = \frac{1}{2}\,\mathrm{erfc}\left(\sqrt{\frac{\varrho}{2}}\right). \tag{2.79}$$

In the case of Gaussian, zero-mean noise as a disturbance, the error probability P_f as a function of the signal-to-noise ratio ϱ for binary transmission results in

$$P_\mathrm{f} = \frac{1}{2}\left[1 - \mathrm{erf}\left(\sqrt{\frac{\varrho}{2}}\right)\right] = \frac{1}{2}\,\mathrm{erfc}\left(\sqrt{\frac{\varrho}{2}}\right). \tag{2.80}$$

It should be mentioned again that the determined error probability P_f, which represents the probability that transmitted symbols are incorrectly interpreted in the receiver—and which is therefore often referred to as *symbol error probability*—in the case of binary transmission directly corresponds to the bit error probability P_b, since the binary symbols directly represent bits.

Figure 2.28 shows the relationship between error probability P_f and signal-to-noise ratio ϱ according to (2.80) in the diagram. On the horizontal axis, the signal-to-noise ratio ϱ can be plotted with logarithmic division of the axis or the signal-to-noise ratio ϱ^* (in dB). For an error probability of $P_\mathrm{f} = 10^{-10}$, for example, a signal-to-noise ratio $\varrho \approx 40$ or $\varrho^* = 10 \lg 40 \approx 16\,\mathrm{dB}$ is necessary.

The maximum error probability is $P_{\mathrm{f}\,\mathrm{max}} = 0.5$: Then, on average, every second transmitted bit is detected as erroneous in the receiver and there is maximum uncertainty on the receiving side about what has been transmitted. If $P_\mathrm{f} = 1$ were the maximum error probability, i.e. all bits were detected incorrectly on the receiving side, one would only

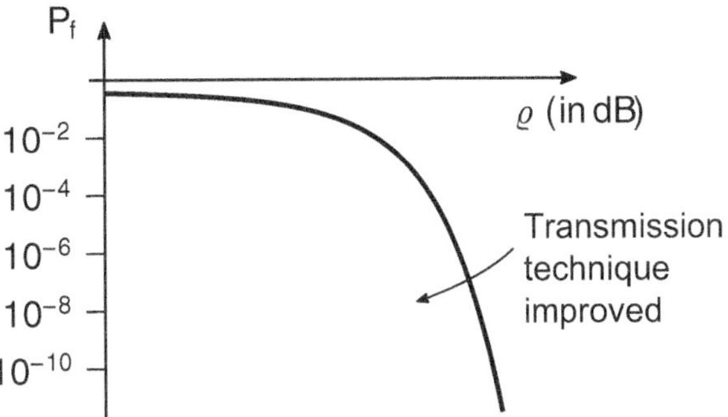

Fig. 2.28 Error probability P_f as a function of the signal-to-noise ratio ϱ

have to transmit with $P_f = 1$—thus maximally poor, i.e. with very little effort, e.g. in terms of transmission power or energy –, and invert all received bits in the receiver to achieve the ideal case of error-free transmission with $P_f = 0$.

The further to the left the curve $P_f(\varrho)$ shown in Fig. 2.28 is in the diagram, the better or more efficient is the associated transmission method: Then, for the same signal-to-noise ratio ϱ, a lower error probability P_f is achieved.

Taking into account that the signal-to-noise ratio ϱ in the considered transmission system is a function of the cutoff frequency of the receive filter f_g, the error probability is also a function of this cutoff frequency. Figure 2.29 shows the error probability P_f as a function of the receive filter cutoff frequency f_g with logarithmic division of the vertical axis.

The optimal cutoff frequency $f_{g\,opt}$ leads to the minimum error probability $P_{f\,min}$. Due to the monotonic relationship (2.80) between signal-to-noise ratio and error probability, optimizing the receive filter cutoff frequency for maximum signal-to-noise ratio and minimum error probability is equivalent. Often, therefore, optimization is done for maximum signal-to-noise ratio, as this is more accessible mathematically and computationally or in terms of measurement technology.

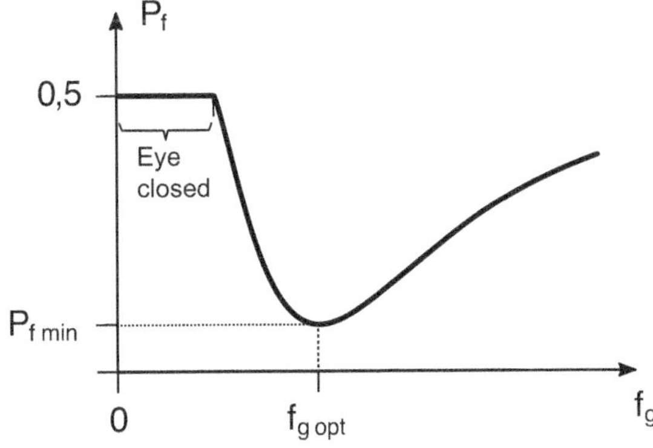

Fig. 2.29 Bit error probability P_f at the detection point as a function of the receive filter cutoff frequency f_g (logarithmic division of the vertical axis)

With the relationship (2.80), a monotonic relationship between signal-to-noise ratio ϱ and error probability P_f was established. The signal-to-noise ratio can therefore be used as a measure of quality for the investigation and optimization of digital transmission systems. When using the signal-to-noise ratio ϱ for the worst case *(worst case)* according to (2.60), an upper limit for the error probability [1] is obtained. This corresponds to the previous approach in this book.

The bit error probability was calculated here for the examined—two-level, unipolar—transmission system as a function of the signal-to-noise ratio ϱ. It is generally difficult to calculate directly for more complex transmission methods, as it depends on many properties of the specific transmission method and the transmission channel, as well as the boundary conditions.[8] Therefore, simulations [24, 64] are often used for the analysis and optimization of transmission systems, or alternative quality criteria are applied [29]. When using alternative quality criteria, it is essential that they have a clear, i.e., monotonic, relationship with the (bit) error probability as the most meaningful quality measure for digital transmission [1, 29]. This was shown here, and the insights and results obtained for the specific transmission system can be generalized, so that the signal-to-noise ratio ϱ can often be used as a quality criterion for the practical design and optimization of transmission systems.

2.1.6   Numerical Example: Baseband Transmission over a Distortion-Free Channel with Noise Disturbance

Introduction and Task
With the help of a numerical example, the procedure presented in the previous sections for describing and optimizing a baseband transmission system, which is operated over a distortion-free channel with pure noise disturbance, is illustrated using a specific example. The transmission system under consideration is shown in Fig. 2.30.

The transmitter of a baseband transmission system sends a binary, unipolar, uniformly distributed random sequence of rectangular NRZ pulses with amplitudes 0 and U_0 and pulse width T_s. The average transmission power P_s is given; from this, the transmission amplitude U_0 is determined. The influence of the channel consists solely of the additive addition of a noise disturbance. This disturbance is modeled as white, Gaussian noise with the spectral power density Ψ_0. The receiving filter $G_e(f)$ is a first-order low-pass filter with a real pole at the cutoff frequency f_g. This corresponds to the transmission system that was examined in the previous sections.

---

[8] The multitude of possible transmission methods, their components, and the resulting dependencies exceed the content and scope of this chapter. The interested reader is referred to [3, 23, 26, 62, 72] for further study.

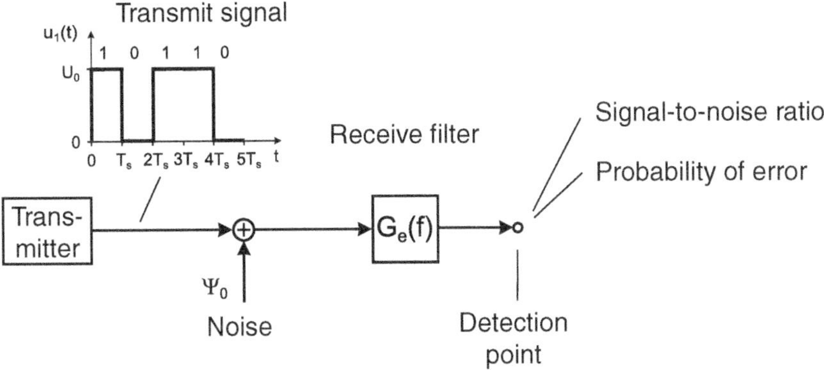

Fig. 2.30 Transmission system for the numerical example in unipolar, binary baseband transmission

The achievable quality of the transmission, i.e., the signal-to-noise ratio ϱ and the bit error probability P_b, which in the two-level transmission considered here coincides with the symbol error probability P_f, i.e. $P_b = P_f$, is to be calculated. For this, the individual necessary quantities (U_A, U_R^2) are to be calculated and possibly optimized (f_g).

Given Values
The following numerical values are given:

- Signal: Average transmit power $P_s = 1\,\mathrm{V}^2$, pulse duration $T_s = 1\,\mu s$, unipolar transmission,
- Disturbance: Noise power spectral density $\Psi_0 = 2 \cdot 10^{-8}\,\frac{\mathrm{V}^2}{\mathrm{Hz}}$.

Procedure and Results
From the given average transmission power P_s, the transmission amplitude is first calculated using (2.7) over

$$P_s = \frac{U_0^2}{2} \quad \Longrightarrow \quad U_0 = \sqrt{2P_s}. \qquad (2.81)$$

For $P_s = 1\,\mathrm{V}^2$, the transmission amplitude $U_0 = \sqrt{2}\,\mathrm{V} \approx 1.414\,\mathrm{V}$ is obtained.

The receive filter is predefined in its structure as a first-order RC low-pass filter with a single real pole, but its cutoff frequency f_g is freely selectable and thus optimizable. Therefore, the half vertical eye opening according to (2.35) and the noise power according to (2.59) are determined as a function of the receiving filter cutoff frequency. In Fig. 2.31, the half vertical eye opening and in Fig. 2.32, the noise power at the detection point, each as a function of the receiving filter cutoff frequency f_g, are shown. The typical curves are recognizable: The half vertical eye opening for the worst case

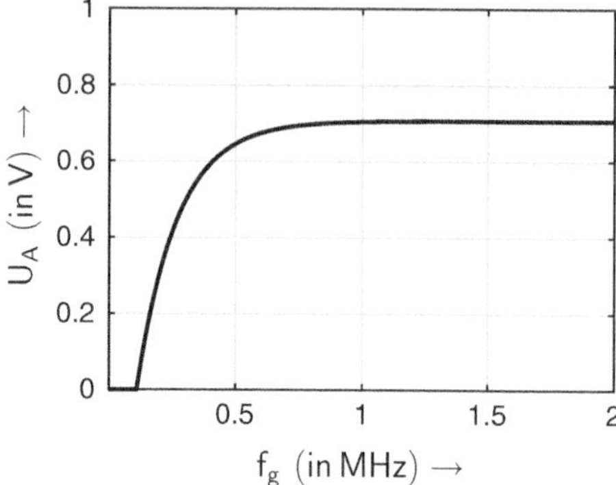

Fig. 2.31 Half vertical eye opening U_A as a function of the receive filter cutoff frequency f_g

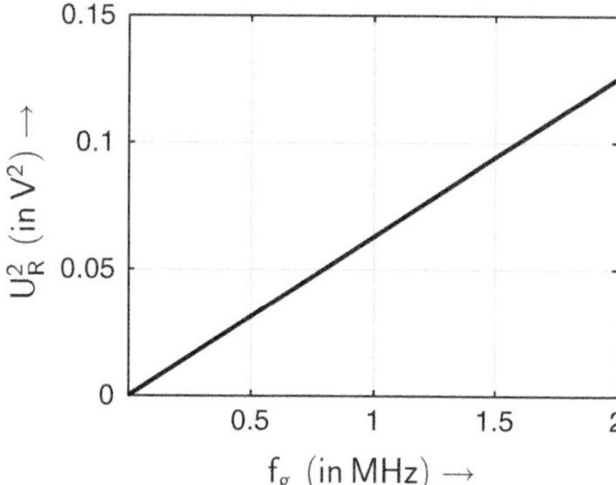

Fig. 2.32 Noise power U_R^2 as a function of the receive filter cutoff frequency f_g

(worst case) is initially zero when the eye is closed and tends towards a maximum of $U_0/2 = 1/\sqrt{2}\,\mathrm{V} \approx 0.707\,\mathrm{V}$ predetermined by the transmission amplitude U_0, while the noise power grows linearly and unlimitedly with increasing receiving filter cutoff frequency.

If the signal-to-noise ratio according to (2.60) and the bit error probability according to (2.80), also as a function of the receiving filter cutoff frequency f_g, are now calculated and shown in Figs. 2.33 and 2.34, it is possible to determine the optimal cutoff frequency $f_\mathrm{g\,opt}$ of the receiving filter. It is read from one of the diagrams where the signal-to-noise ratio becomes maximum—or the bit error probability minimum—and $f_\mathrm{g\,opt} = 394\,\mathrm{kHz}$ is obtained.

Numerical Results for Optimized Transmission System

Once the optimal cutoff frequency $f_\mathrm{g\,opt}$ of the receive filter has been determined, all other values for signal and noise at the detection point can be determined and the quality of transmission in an optimized transmission system can be specified. The individual

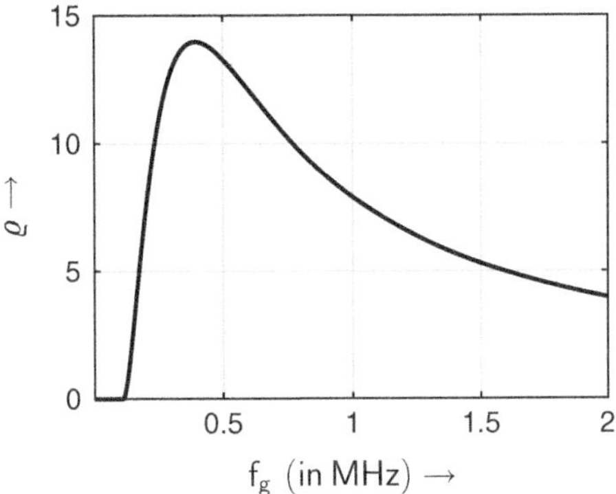

Fig. 2.33 Signal-to-noise ratio ϱ as a function of the receiving filter cutoff frequency f_{g}

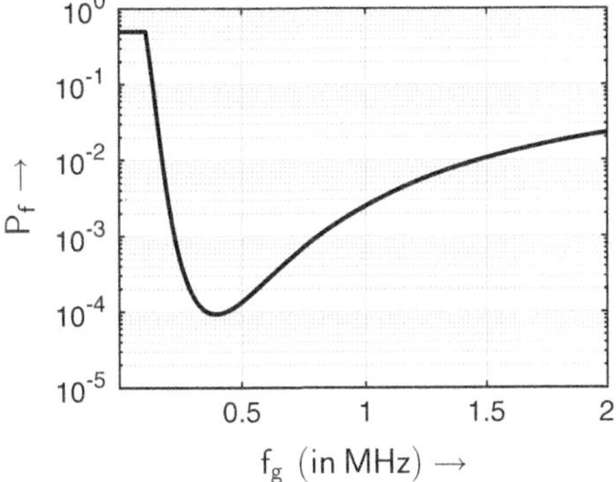

Fig. 2.34 Bit error probability P_{f} as a function of the receiving filter cutoff frequency f_{g}

relationships—for the half vertical eye opening (2.37), the noise power (2.59), the signal-to-noise ratio (2.60) and the error probability (2.80)—each formulated as functions of the cutoff frequency f_{g} of the receive filter or its time constant $T_{\mathrm{g}} = 1/(2\pi f_{\mathrm{g}})$, are evaluated at the point $f_{\mathrm{g}} = f_{\mathrm{g\,opt}}$. The numerical results are compiled in Table 2.1.

The maximum achievable signal-to-noise ratio is $\varrho_{\max} = 13.97$ or $\varrho^{*}_{\max} = 10\lg \varrho_{\max} = 11.45\,\mathrm{dB}$; it leads to a bit error probability of $P_{\mathrm{f\,min}} = 9.27 \cdot 10^{-5}$.

The optimal receive filter cutoff frequency $f_{\mathrm{g\,opt}}$ is below half the bit sequence frequency or bit rate $f_{\mathrm{B}} = 1/T_{\mathrm{s}}$ at $f_{\mathrm{g\,opt}} \cdot T_{\mathrm{s}} \approx 0.4$. As a result, the half vertical eye opening U_{A} has a value that is comparatively significantly below the maximum half vertical eye opening $U_0/2$: This maximum half vertical eye opening $(U_0/2)$ would be achieved at a (very) large receive filter cutoff frequency f_{g} (see, for example, Fig. 2.31). This would in turn cause a (strong) increase in noise power compared to $f_{\mathrm{g\,opt}}$ (see Fig. 2.32). All these factors together lead to the optimal cutoff frequency in the size that results in its optimization with respect to a maximum signal-to-noise ratio or a minimum bit error probability.

Table 2.1 Given values and numerical results at optimal receive filter cutoff frequency $f_{g\,opt}$ of the first order low pass filter

Size	Numerical value
Average transmit power	$P_s = 1\,V^2$
Transmit amplitude	$U_0 = 1.414\,V$
Noise power spectral density	$\Psi_0 = 2 \cdot 10^{-8}\,V^2/Hz$
Pulse duration (bit or symbol duration)	$T_s = 1\,\mu s$
Optimal receive filter cutoff frequency	$f_{g\,opt} = 394\,kHz$
Half vertical eye opening	$U_A = 0.588\,V$
Noise power	$U_R^2 = 0.025\,V^2$
Signal-to-noise ratio	$\varrho_{max} = 13.97$
Bit error probability	$P_{f\,min} = 9.27 \cdot 10^{-5}$

Some Diagrams for Optimized Transmission System

This section is intended to illustrate with graphical representations how some of the time and spectral functions look like in an optimized transmission system.

Figure 2.35 shows the amplitude frequency response of the receive filter $G_e(f)$ in the Bode diagram and Fig. 2.36 the eye diagram of the useful signal at the detection point—each for the case of the optimized cutoff frequency of the receiving low pass filter ($f_{g\,opt} = 394\,kHz$). It can be seen that the eye is not fully open vertically, as the band limitation of the receive filter starts relatively early and thus in the optimized transmission system the useful signal is deformed, but above all the noise can be relatively well limited in its power on the receiving side.

Fig. 2.35 Amplitude response of the receive filter (Bode diagram) at optimal cutoff frequency $f_{g\,opt}$

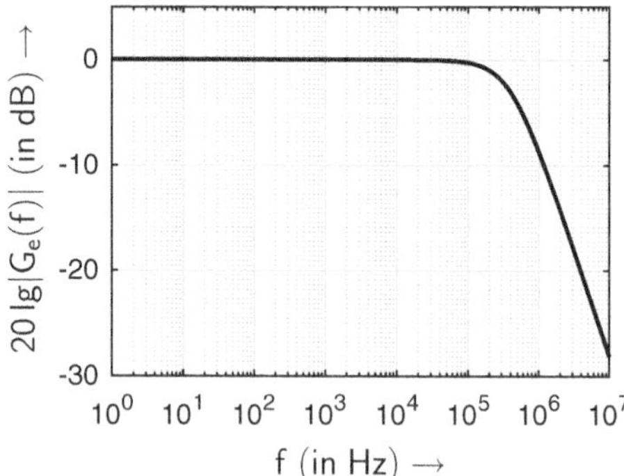

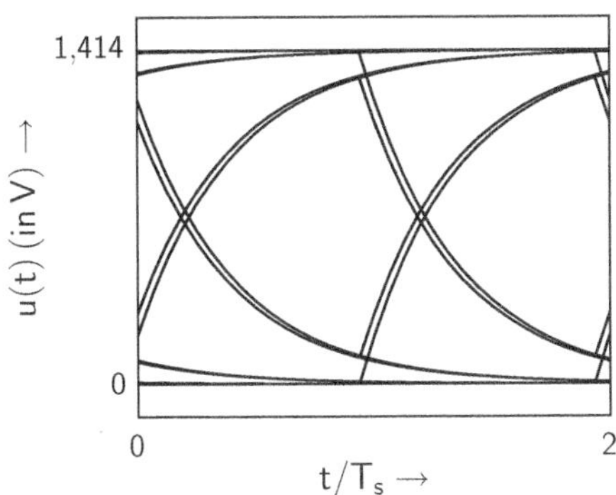

Fig. 2.36 Eye diagram of the useful signal at the detection point with optimal receive filter cutoff frequency $f_{g\,\mathrm{opt}}$

2.1.7   Insights into Transmission Engineering

Important insights that were developed in this section using the basic model of baseband transmission are:

- Disturbances are inevitable in message transmission: Every signal transmission is inherently disturbed.
- A receive filter is therefore always necessary in principle, the practical implementation can be done in various ways.
- Disturbances cause errors in threshold decision: The received bit sequence generally contains errors. A measure of the quality of the transmission is the bit error probability, which provides information about the probability with which transmitted bits are decided incorrectly in the receive under given boundary conditions.
- Due to monotonic functional relationships between bit error probability and signal-to-noise ratio, the practically much easier to handle signal-to-noise ratio can be used as a quality criterion or quality measure for assessing the reliability of a digital signal transmission.
- The signal-to-noise ratio at the detection point was defined as the ratio of the square of half the vertical eye opening and the noise power, both at the receive filter output, which in turn both depend on the type and cutoff frequency of the receive filtering.

An important task in transmission engineering is to minimize the bit error probability under given transmission conditions: For this purpose, some methods and procedures will be introduced in the further course of this chapter, which clarify basic relationships and enable an optimal design of baseband transmission components under given boundary conditions.

2.2 Improved Transmission Concepts in Baseband

2.2.1 Introduction and Motivation

There are a number of possibilities to improve the performance and quality of the transmission compared to the basic model of baseband transmission previously discussed. Some important ones will be introduced and discussed in this section. First, filters other than those previously used for pulse shaping and receive filtering can be used to improve the properties of the transmitted signal and the overall achievable transmission quality at the detection point. In this process, general criteria are derived that provide guidance for the design of transmit and receive filters, and optimal functions for the receive filter can be specified in terms of the signal-to-noise ratio or error probability. For example, the question arises whether the first-order RC low-pass filter used in the previous sections for receive filtering already leads to the maximum achievable signal-to-noise ratio ϱ: This question will be investigated in one of the following sections. Another possibility to make the transmission more powerful, especially with a limited given bandwidth, is multi-level transmission, which is also introduced in this section and examined in combination with the improved filter functions.

2.2.2 Alternative Transmit and Receive Filter Characteristics

So far, the transmission with an NRZ-rectangle impulse (see Fig. 2.3) as the basic transmit impulse has been extensively studied.

Advantages are:

- The maximum signal energy ($U_0^2 T_s$) lies in the time interval T_s. This can be advantageous for applications where only a limited transmission time is available (e.g., location applications, synchronization).
- Such impulses are practically easy to generate by switching voltages (flip-flops).

Disadvantages are:

- The basic transmit impulse has a comparatively wide spectrum, it decays with $1/\omega$ or $1/f$ with frequency.
- In the case of band limitation (e.g., by the transmission channel or by the receive filter), impulse distortions remain and intersymbol interferences result (e.g., through the RC low-pass filter).

The following discusses two alternative pulse shapes or filter functions that have gained significance in the transmission of digital signals, as they allow improvements in the quality of transmission: First, the *Gaussian* pulse or low-pass filter is discussed, followed by the *Root Raised Cosine* pulse or low-pass filter.

Gaussian Pulse and Gaussian Filter
Description in Time and Frequency Domain A *Gaussian* pulse exhibits the following characteristics in the time and frequency domain

$$x(t) = U_0\, e^{-(\ln 2)\cdot\left(\frac{t}{\tau}\right)^2} \quad \circ\!\!-\!\!\bullet \quad X(\omega) = U_0\, \frac{2\sqrt{\pi \ln 2}}{\omega_G}\, e^{-(\ln 2)\cdot\left(\frac{\omega}{\omega_G}\right)^2}. \tag{2.82}$$

(Fig. 2.37). Key parameters are the constant τ and the cutoff frequency ω_G or f_G, which are related via

$$\tau = \frac{2\ln 2}{\omega_G} = \frac{\ln 2}{\pi f_G} \tag{2.83}$$

The spectrum of the basic transmission pulse can also be written as a function of the frequency variable $f = \omega/(2\pi)$:

$$X(f) = U_0\, \frac{\sqrt{\ln 2}}{\sqrt{\pi} f_G}\, e^{-(\ln 2)\cdot\left(\frac{f}{f_G}\right)^2}. \tag{2.84}$$

The cutoff frequency f_G (or ω_G) in this definition is a 6-dB -cutoff frequency, i.e., at the frequency $f = f_G$, the magnitude of the spectrum of (2.84) has fallen to half or by $-20\lg(1/2) = 6\,\mathrm{dB}$ compared to the value at the frequency $f = 0$ (or $\omega = 0$).

The practical realization of a Gaussian transmission pulse can approximately be achieved via the impulse response of a Gaussian-low-pass filter with the transfer function

$$G(f) = e^{-\ln(2)\cdot\left(\frac{f}{f_G}\right)^2} \quad \bullet\!\!-\!\!\circ \quad g(t) = \sqrt{\frac{\pi}{\ln 2}}\, f_G \cdot e^{-\left(\frac{\pi}{\sqrt{\ln 2}}\right)^2\cdot t^2 f_G^2} \tag{2.85}$$

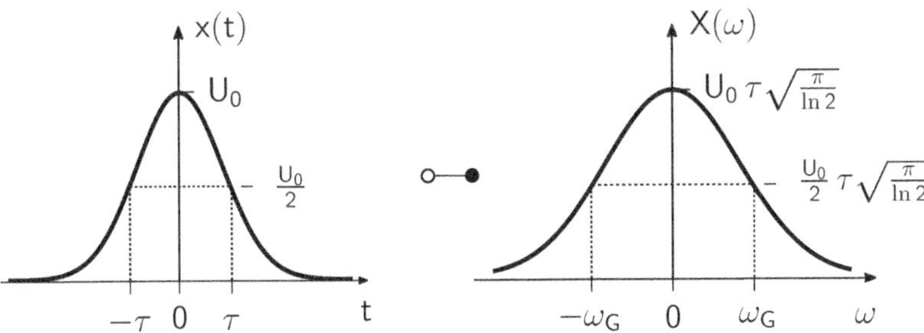

Fig. 2.37 Gaussian pulse in time and frequency domain

[29].

Characteristics of the Gaussian low-pass filter are:

- a similar principal functional course in the time and frequency domain with $\tau \sim 1/\omega_G$; this behavior is referred to as *self-reciprocal*[9]: In both domains, the function decays with e^{-x^2};
- a compromise between temporal pulse width and bandwidth: Maximum steep course in the frequency domain without overshooting in the time domain.

Series Connection of Transmit and Receive Filters For use in transmission systems, two combinations have proven to be practically significant:

a) Gaussian basic transmit impulse and Gaussian low-pass as receive filter,
b) NRZ-rectangle basic transmit impulse and Gaussian low-pass as receive filter.

Both variants are briefly discussed with reference to Fig. 2.8.

On a) *Gaussian basic transmit impulse and Gaussian low-pass as receive filter*

If one considers the series connection of two Gaussian low-pass filters $G_s(f)$ and $G_e(f)$, one obtains for the total transfer function of the cascade of transmit and receive filters

$$G_{\text{ges}}(f) = G_s(f) \cdot G_e(f) = e^{-(\ln 2)\cdot\left(\frac{f}{f_{Gs}}\right)^2} \cdot e^{-(\ln 2)\cdot\left(\frac{f}{f_{Ge}}\right)^2}$$

$$= e^{-(\ln 2)\left[\left(\frac{f}{f_{Gs}}\right)^2+\left(\frac{f}{f_{Ge}}\right)^2\right]} = e^{-(\ln 2)\cdot f^2\left[\left(\frac{1}{f_{Gs}}\right)^2+\left(\frac{1}{f_{Ge}}\right)^2\right]}.$$

From the form of the transfer function, it can be seen that this again corresponds to a Gaussian low-pass

$$G_{\text{ges}}(f) = e^{-(\ln 2)\cdot\left(\frac{f}{f_{G\,\text{ges}}}\right)^2}. \tag{2.86}$$

The cutoff frequency $f_{G\,\text{ges}}$ of this total transfer function $G_{\text{ges}}(f)$ is obtained via

$$\left(\frac{1}{f_{G\,\text{ges}}}\right)^2 = \left(\frac{1}{f_{Gs}}\right)^2 + \left(\frac{1}{f_{Ge}}\right)^2$$

$$\frac{1}{f_{G\,\text{ges}}} = \sqrt{\left(\frac{1}{f_{Gs}}\right)^2 + \left(\frac{1}{f_{Ge}}\right)^2}$$

---

[9] Other self-reciprocal signals include, for example, the hyperbolic cosine pulse and the Dirac impulse sequence, which each show the same principal functional dependencies on time t and frequency f (or ω) in both the time and frequency domain [30].

finally to

$$f_{G\,ges} = \frac{1}{\sqrt{\left(\frac{1}{f_{G\,s}}\right)^2 + \left(\frac{1}{f_{G\,e}}\right)^2}}.$$ (2.87)

This relationship (2.87) can be further transformed via

$$f_{G\,ges}^2 = \frac{1}{\frac{1}{f_{G\,s}^2} + \frac{1}{f_{G\,e}^2}} = \frac{1}{\frac{f_{G\,e}^2 + f_{G\,s}^2}{f_{G\,s}^2 \cdot f_{G\,e}^2}}$$

$$f_{G\,ges}^2 = \frac{f_{G\,s}^2 \cdot f_{G\,s}^2}{f_{G\,s}^2 + f_{G\,e}^2}$$

to

$$f_{G\,ges} = \frac{f_{G\,s} f_{G\,e}}{\sqrt{f_{G\,s}^2 + f_{G\,e}^2}}.$$ (2.88)

The series connection of two Gaussian low-pass filters is again a Gaussian low-pass filter with a changed cutoff frequency $f_{G\,ges}$; the cutoff frequency $f_{G\,ges}$ is smaller than $f_{G\,s}$ and $f_{G\,e}$. This means that the basic receive impulse present at the output of the receiving filter is also a Gaussian impulse (2.82), but with the—smaller—cutoff frequency $f_{G\,ges}$. The basic receive impulse therefore has a larger half-life period constant τ than the basic transmit impulse and therefore has a larger temporal extension than the basic transmit impulse. This consideration can be extended to the series connection of a larger number of Gaussian low-pass filters.

On b) *NRZ-rectangle basic transmission impulse and Gaussian low-pass as receive filter*
    If an NRZ-rectangle impulse (2.4) is filtered by a Gaussian receive low-pass with the cutoff frequency f_G, one obtains as the basic receive impulse at the detection point

$$x_e(t) = \frac{U_0}{2}\left[\mathrm{erf}\left(\frac{\pi f_G}{\sqrt{\ln(2)}}\left(t + \frac{T_s}{2}\right)\right) - \mathrm{erf}\left(\frac{\pi f_G}{\sqrt{\ln(2)}}\left(t - \frac{T_s}{2}\right)\right)\right].$$ (2.89)

Figure 2.38 exemplarily shows the rectangle response $x_e(t)$ of a Gaussian low-pass filter at two exemplary filter cutoff frequencies.

Fig. 2.38 Response to a rectangular impulse of a Gaussian low-pass filter at high ($f_G = 1/T_s$) and low ($f_G = 0.2/T_s$) cutoff frequency

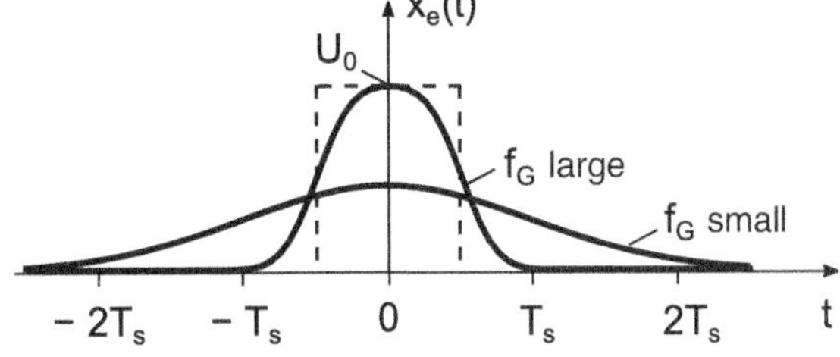

Insert: Derivation of the Response of a Gaussian Low-Pass Filter to a Rectangular Impulse

The response of a Gaussian low-pass filter to an NRZ-rectangle impulse symmetrically located at the time $t = 0$ with the amplitude U_0 and the impulse width T_s is to be derived. For this purpose, the response of the Gaussian low-pass filter with the weighting function

$$g(t) = \sqrt{\frac{\pi}{\ln 2}}\, f_G \cdot e^{-(\ln 2)\left(\frac{t}{\tau}\right)^2} \tag{2.90}$$

and the relationship

$$\tau = \frac{\ln 2}{\pi f_G} \tag{2.91}$$

between cutoff frequency f_G and half-life period constant τ to a unit jump is first calculated:

$$s_T(t) = \int_{-\infty}^{t} g(\theta)\, \mathrm{d}\theta = 1 - \int_{t}^{+\infty} g(\theta)\, \mathrm{d}\theta. \tag{2.92}$$

One obtains

$$s_T(t) = 1 - \sqrt{\frac{\pi}{\ln 2}}\, f_G \int_{t}^{\infty} e^{-\left(\sqrt{\ln 2}\,\frac{\theta}{\tau}\right)^2} \mathrm{d}\theta$$

$$= 1 - \sqrt{\frac{\pi}{\ln 2}}\, f_G \int_{t\cdot\frac{\sqrt{\ln 2}}{\tau}}^{\infty} e^{-\left(\sqrt{\ln 2}\,\frac{\theta}{\tau}\right)^2} \mathrm{d}\left(\theta\frac{\sqrt{\ln 2}}{\tau}\right) \cdot \frac{\tau}{\sqrt{\ln 2}}$$

$$= 1 - \sqrt{\frac{\pi}{\ln 2}}\, f_G \cdot \frac{\tau}{\sqrt{\ln 2}} \int_{t\cdot\frac{\sqrt{\ln 2}}{\tau}}^{\infty} e^{-\left(\sqrt{\ln 2}\,\frac{\theta}{\tau}\right)^2} \mathrm{d}\left(\theta\frac{\sqrt{\ln 2}}{\tau}\right)$$

and with (2.91) results

$$s_\mathsf{T} = 1 - \sqrt{\frac{\pi}{\ln 2}}\, f_\mathrm{G} \cdot \frac{\ln 2}{\pi f_\mathrm{G} \sqrt{\ln 2}} \int\limits_{t \cdot \frac{\pi f_\mathrm{G} \sqrt{\ln 2}}{\ln 2}}^{\infty} \mathrm{e}^{-\left(\sqrt{\ln 2}\,\frac{\theta}{\tau}\right)^2} \,\mathrm{d}\!\left(\theta\, \frac{\sqrt{\ln 2}}{\tau}\right)$$

$$= 1 - \frac{1}{\sqrt{\pi}} \int\limits_{t \cdot \frac{\pi f_\mathrm{G}}{\sqrt{\ln 2}}}^{\infty} \mathrm{e}^{-\left(\sqrt{\ln 2}\,\frac{\theta}{\tau}\right)^2} \,\mathrm{d}\!\left(\theta\, \frac{\sqrt{\ln 2}}{\tau}\right)$$

$$= 1 - \frac{1}{2}\frac{2}{\sqrt{\pi}} \int\limits_{\frac{t\,\pi f_\mathrm{G}}{\sqrt{\ln 2}}}^{\infty} \mathrm{e}^{-\left(\sqrt{\ln 2}\,\frac{\theta}{\tau}\right)^2} \,\mathrm{d}\!\left(\sqrt{\ln 2}\, \frac{\theta}{\tau}\right).$$

With the definition of the complementary Gaussian error function (see Appendix E)

$$\mathrm{erfc}(x) = \frac{2}{\sqrt{\pi}} \int\limits_{x}^{\infty} \mathrm{e}^{-y^2}\,\mathrm{d}y = 1 - \mathrm{erf}(x) \tag{2.93}$$

it results in

$$s_\mathsf{T} = 1 - \frac{1}{2}\,\mathrm{erfc}\!\left(\frac{t\,\pi f_\mathrm{G}}{\sqrt{\ln 2}}\right)$$

$$= 1 - \frac{1}{2}\left[1 - \mathrm{erf}\!\left(\frac{\pi f_\mathrm{G}\,t}{\sqrt{\ln 2}}\right)\right]$$

$$= 1 - \frac{1}{2} + \frac{1}{2}\,\mathrm{erf}\!\left(\frac{\pi f_\mathrm{G}\,t}{\sqrt{\ln 2}}\right)$$

$$= \frac{1}{2} + \frac{1}{2}\,\mathrm{erf}\!\left(\frac{\pi f_\mathrm{G}\,t}{\sqrt{\ln 2}}\right).$$

One obtains as the response of the Gaussian low-pass filter to a unit step function

$$s_\mathsf{T} = \frac{1}{2}\left[\mathrm{erf}\!\left(\frac{\pi f_\mathrm{G}\,t}{\sqrt{\ln(2)}}\right) + 1\right]. \tag{2.94}$$

The response of a Gaussian low-pass filter with the cutoff frequency f_G to an NRZ-rectangle impulse symmetrically located at the time $t = 0$ with the amplitude U_0 and the impulse width T_s results via

$$x_\mathrm{e}(t) = U_0\left[s_\mathsf{T}\!\left(t + \frac{T_\mathrm{s}}{2}\right) - s_\mathsf{T}\!\left(t - \frac{T_\mathrm{s}}{2}\right)\right] \tag{2.95}$$

to

$$x_\mathrm{e}(t) = \frac{U_0}{2}\left[\mathrm{erf}\!\left(\frac{\pi f_\mathrm{G}}{\sqrt{\ln(2)}}\left(t + \frac{T_\mathrm{s}}{2}\right)\right) - \mathrm{erf}\!\left(\frac{\pi f_\mathrm{G}}{\sqrt{\ln(2)}}\left(t - \frac{T_\mathrm{s}}{2}\right)\right)\right]. \tag{2.96}$$

Filtering of White Noise by a Gaussian Low-Pass Filter For the noise power at the detection point, one obtains according to (2.53) with (2.85) as the receive filter transfer function $G_\mathrm{e}(f)$ initially

$$U_\mathrm{R}^2 = 2\,\Psi_0 \int\limits_0^{+\infty} \left| e^{-(\ln 2)\cdot\left(\frac{f}{f_\mathrm{G}}\right)^2} \right|^2 \mathrm{d}f = 2\,\Psi_0 \int\limits_0^{+\infty} e^{-2(\ln 2)\cdot\left(\frac{f}{f_\mathrm{G}}\right)^2} \mathrm{d}f. \tag{2.97}$$

With the relationship (see Appendix F.2 or e.g. [11, 74])

$$\int\limits_0^\infty e^{-a^2 x^2}\,\mathrm{d}x = \frac{\sqrt{\pi}}{2\,a} \qquad \text{for} \qquad a > 0 \tag{2.98}$$

and

$$a = \frac{\sqrt{2\,\ln 2}}{f_\mathrm{G}} \tag{2.99}$$

results in

$$U_\mathrm{R}^2 = 2\,\Psi_0\frac{\sqrt{\pi}}{2}\frac{f_\mathrm{G}}{\sqrt{2\,\ln 2}} = \Psi_0\frac{\sqrt{\pi}f_\mathrm{G}}{\sqrt{2\,\ln 2}} \tag{2.100}$$

and finally

$$U_\mathrm{R}^2 = \Psi_0\,\frac{\sqrt{\pi}}{\sqrt{2\,\ln 2}}f_\mathrm{G}. \tag{2.101}$$

This means that even with the Gaussian low-pass filter as the receive filter, $U_\mathrm{R}^2 \sim f_\mathrm{G}$ applies, compared to the first-order RC low-pass filter, the noise power—at the same cutoff frequency—is however lower by a constant factor $(\sqrt{\pi/(2\,\ln 2)})/\pi \approx 0.48$. This results from the stronger band limitation of the Gaussian low-pass filter (stronger drop of the transfer function compared to that of the first-order RC low-pass filter with increasing frequency).

Approximation of a Gaussian Low-Pass Filter by a Rational Transfer Function
A Gaussian low-pass can be approximated by a rational[10] transfer function

$$G_\mathrm{e}(p) = \frac{1}{\left(1 + pT_\mathrm{g}\right)^n} \quad\bullet\!\!-\!\!\circ\quad g_\mathrm{e}(t) = \frac{e^{-\frac{t}{T_\mathrm{g}}}\cdot t^{n-1}}{T_\mathrm{g}^n\,(n-1)!} \quad \text{with } T_\mathrm{g} = \frac{1}{2\pi f_\mathrm{g}}. \tag{2.102}$$

---

[10] A *rational function* of nth degree is a function of the form $y(x) = a_n x^n + a_{n-1}x^{n-1} + \ldots + a_1 x + a_0$ with $n = 0, 1, 2, \ldots$ and the real coefficients a_n and $a_n \neq 0$. Functions that have a quotient form and in which rational functions are in the numerator and denominator are referred to as *general rational functions* [74].

with

$$f_G = f_g \cdot \sqrt{2^{\frac{2}{n}} - 1} \qquad \text{or} \qquad \omega_G = \frac{1}{T_g} \cdot \sqrt{2^{\frac{2}{n}} - 1}. \tag{2.103}$$

For $n \to \infty$, a Gaussian low-pass can thus be approximated by the transfer function of an RC low-pass of nth order (2.102). Even for $n \approx 10$, one obtains usable rational approximations for Gaussian low-passes, which are practically realizable in circuit technology.

Raised Cosine and Root Raised Cosine Filters

Description in Time and Frequency Domain In order for no intersymbol interference to occur at the detection point at the symbol clock sampling times, suitable filter functions or pulse shapes must be chosen. Often in communication systems for the cascade $G_s(f) \cdot G_e(f)$ of transmit and receive filters, a *cosine roll-off* transfer function with the roll-off factor r according to

$$G_N(f) = \begin{cases} 1 & \text{for } f \leq \frac{1-r}{2T_s} \\ \frac{1}{2}\left[1 + \sin\left(\frac{\pi\,(1-2f\,T_s)}{2r}\right)\right] & \text{for } \frac{1-r}{2T_s} \leq f \leq \frac{1+r}{2T_s} \\ 0 & \text{for } f > \frac{1+r}{2T_s} \end{cases} \tag{2.104}$$

with the weighting function

$$g_N(t) = \frac{1}{T_s}\,\mathrm{si}\left(\pi\,\frac{t}{T_s}\right)\frac{\cos\left(r\,\pi\,t/T_s\right)}{1 - 4(r\,t/T_s)^2} \tag{2.105}$$

is used. Such a filter is also referred to as a *Nyquist* filter or *Raised-Cosine* filter. It can be mathematically well described and handled and practically approximated. Figure 2.39 shows the weighting function and transfer function of a cosine roll-off or Nyquist filter

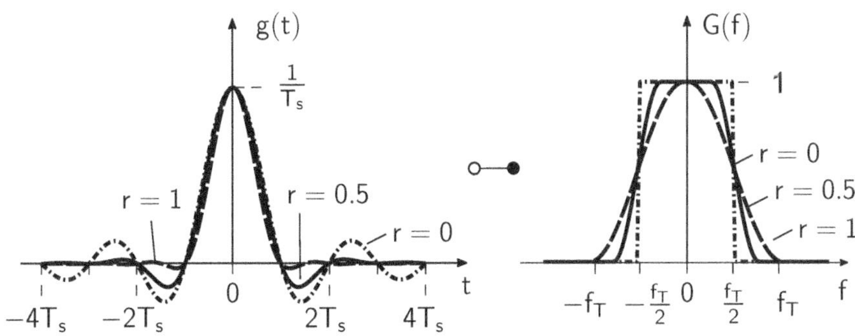

Fig. 2.39 Cosine roll-off or Nyquist filter in time and frequency domain

for exemplary roll-off factors r. With increasing roll-off factor r, the occupied bandwidth becomes larger and the overshoot in the time domain smaller. The bandwidth is

$$B = f_{\mathrm{T}} \cdot (1 + r) = \frac{1}{T_{\mathrm{s}}} \cdot (1 + r). \tag{2.106}$$

For $r = 0$, one obtains a rectangular spectrum of width $f_{\mathrm{T}} = 1/T_{\mathrm{s}}$.

Series Connection of Transmit and Receive Filters Conveniently, the Nyquist frequency response according to (2.104) is evenly distributed between transmit and receive filters, so that

$$G_{\mathrm{s}}(f) = G_{\mathrm{e}}(f) = \sqrt{G_{\mathrm{N}}(f)} \tag{2.107}$$

applies. Such a filter is also referred to as a *root* raised cosine or *root Nyquist*-filter. The transmit filter—as well as the receive filter—then each have the weighting function [3]

$$g_{\mathrm{s}}(t) = \begin{cases} \frac{1}{T_{\mathrm{s}}} \frac{(4rt/T_{\mathrm{s}})\cos(\pi(1+r)t/T_{\mathrm{s}})+\sin(\pi(1-r)t/T_{\mathrm{s}})}{(\pi t/T_{\mathrm{s}})(1-(4rt/T_{\mathrm{s}})^2)} & \text{for } t \neq 0, t \neq \pm\frac{T_{\mathrm{s}}}{4r} \\ \frac{1}{T_{\mathrm{s}}}\left(1 - r + \frac{4r}{\pi}\right) & \text{for } t = 0 \\ \frac{r}{\sqrt{2}T_{\mathrm{s}}}\left[\left(1 + \frac{2}{\pi}\right)\sin\left(\frac{\pi}{4r}\right) + \left(1 - \frac{2}{\pi}\right)\cos\left(\frac{\pi}{4r}\right)\right] & \text{for } t = \pm\frac{T_{\mathrm{s}}}{4r} \end{cases} . \tag{2.108}$$

If a Dirac impulse $U_{\mathrm{s}} T_{\mathrm{s}} \delta(t)$ is evaluated by the transmit filter $g_{\mathrm{s}}(t)$ according to (2.108), the result is the basic transmit impulse $x_{\mathrm{s}}(t) = U_{\mathrm{s}} T_{\mathrm{s}} g_{\mathrm{s}}(t)$ and after filtering with $G_{\mathrm{e}}(f)$, the basic receive impulse $x_{\mathrm{e}}(t) = U_{\mathrm{s}} T_{\mathrm{s}} g_{\mathrm{N}}(t)$ which has zeros at the symbol interval T_{s}. Therefore, no intersymbol interference occurs in such a transmission system.

Figure 2.40 shows the weighting function and transfer function of a root cosine roll-off or root Nyquist filter. It can be seen that the weighting function $g_{\mathrm{s}}(t)$ or the basic transmit impulse $x_{\mathrm{s}}(t)$ generally does not have zeros at the interval T_{s}. This is acceptable for transmission, as no absence of intersymbol interference is required at the output of the transmit filter—but only at the output of the receive filter.

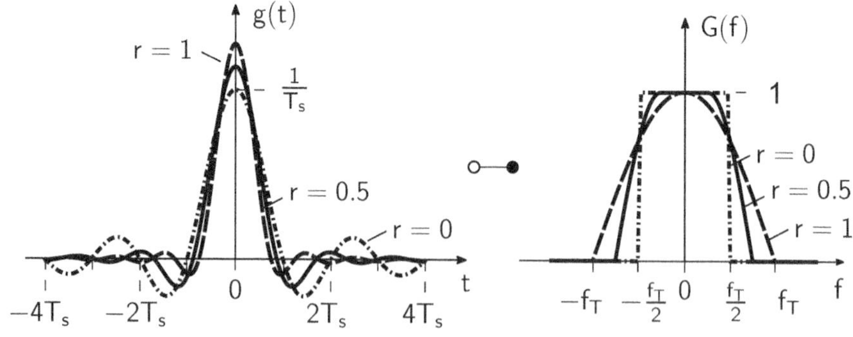

Fig. 2.40 Root raised cosine or root Nyquist filter in time and frequency domain

Filtering of White Noise with Root Raised Cosine Receive Low Pass Filter The calculation of the noise power at the output of a root cosine roll-off receive filter can be computationally advantageous for the roll-off factors $r = 0$ or $r = 1$. The transfer function $G_N(f)$ is point symmetric with respect to the points $(-f_T/2; 1/2)$ and $(f_T/2; 1/2)$ [52]: This means that the frequency response $G_N(f)$ (see Fig. 2.39) at the point $f = f_T/2$ or $f = -f_T/2$ for all roll-off factors r has the amplitude $G_N(f_T/2) = G_N(-f_T/2) = 1/2$. When the roll-off factor r changes, the area under $|G_e(f)|^2 = G_N(f)$ does not change.

The noise power at the detection point is for root cosine roll-off low-pass receive filters (calculation for $r = 0$):

$$U_R^2 = \Psi_0 \int_{-\infty}^{+\infty} |G_e(f)|^2 \, df = 2\,\Psi_0 \int_0^{\infty} |G_N(f)| \, df$$

$$= 2\,\Psi_0 \int_0^{f_T/2} 1^2 \, df = 2\,\Psi_0 \left[f \right]_0^{f_T/2} = 2\,\Psi_0 \left[\frac{f_T}{2} - 0 \right] = 2\,\Psi_0 \frac{f_T}{2}$$

and the result is

$$U_R^2 = \Psi_0 f_T = \frac{\Psi_0}{T_s}. \tag{2.109}$$

The noise power at the output of a root cosine roll-off low-pass receive filter when excited by white noise with the power spectral density Ψ_0 depends only on the clock frequency f_T or the symbol duration T_s, but not on the filter parameter roll-off factor r. Since the magnitude square of the receive filter transfer function $|G_e(f)|^2$ is relevant for the calculation of the noise power and for root cosine roll filters with the real transfer function $\sqrt{G_N(f)}$ the identity $|G_e(f)|^2 = \left| \sqrt{G_N(f)} \right|^2 = \left(\sqrt{|G_N(f)|} \right)^2 = |G_N(f)| = G_N(f)$ applies with the already explained point symmetry of the filter slopes, the noise power according to (2.109) is independent of r. The phase response of the receive filter $G_e(f)$ plays no role in this consideration.

2.2.3   Nyquist Criteria

In the preceding sections, it became clear that intersymbol interference generally occurs at the detection point, which affects the reliability and quality of the transmission. The influence of intersymbol interference can be captured in the eye diagram using the (half) vertical eye opening. It was also shown with the cosine roll-off pulse that, with the noise influence as disturbance on the transmission channel examined so far, special received pulse shapes exist with suitable transmit and receeive filtering that guarantee freedom from intersymbol interference at the detection point: Then the eye is fully open vertically and the half vertical eye opening is maximal. For this, it is necessary to sample

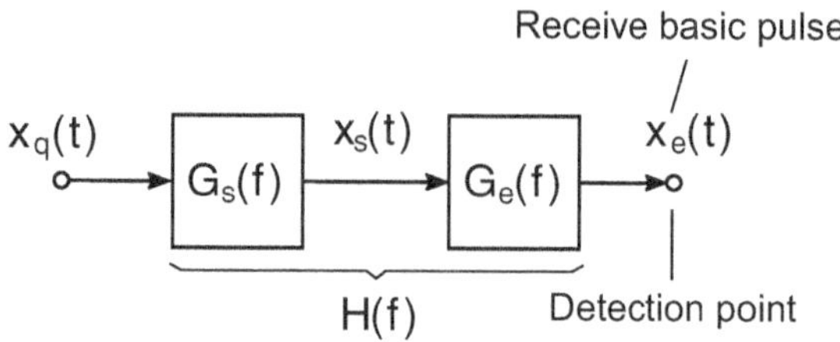

Fig. 2.41 On the Nyquist criteria: Block diagram for evaluating the useful signal

exactly at the optimal sampling time. If the optimal sampling time is not hit because the information about the symbol clock is not exactly available, this can—depending on the temporal duration of the eye opening—lead to a more or less strong reduction of the vertical eye opening and thus of the signal-to-noise ratio. A measure of the sensitivity of the signal quality with respect to the accuracy of the sampling time at the detection point is the horizontal eye opening: If the eye opening is large in time (horizontal), the tolerance to a deviation from the optimal sampling time is greater than with a very narrow eye opening.

In the following sections, the still open questions in this context will be examined:

- What should the optimal received pulse shape look like if the received signal at the detection point has to be free of intersymbol interference?
- How should the optimal received pulse shape be designed if the received signal at the detection point should have the greatest possible tolerance to deviations from the optimal sampling time?

The answer to these two questions leads to the so-called Nyquist criteria, which are to be introduced in the following.

For the considerations on the Nyquist criteria, the received basic pulse present at the detection point is examined according to the two questions raised, and properties are formulated that this received basic pulse or its amplitude spectral density must have so that no intersymbol interference occurs. Effects on this received basic pulse have the transmit filter $G_s(f)$ and the receive filter $G_e(f)$, see Fig. 2.41. Here, $x_q(t) = U_0 T_s \delta(t)$ denotes a single weighted Dirac impulse, which after evaluation by the transmit filter leads to the basic transmit pulse $x_s(t)$ and after subsequent receive filtering to the basic received pulse $x_e(t)$. The noise as disturbance plays no role here, as only the useful signal path is analyzed.

First Nyquist Criterion

Transmission free of intersymbol interference at the detection point at the optimal sampling time can be achieved if the basic receive pulse is limited to the duration $2\,T_s$, or if temporally extended basic receive pulses have zeros at the sampling points with the exception of $t = 0$ (non-causal representation) in the symbol clock.

The aim is to find a general rule for an intersymbol interference-free pulse shape $x_e(t)$ at the detection point. The requirement

$$x_e(t) = \begin{cases} 0 & \text{for } t = \nu\, T_s \\ x_e(0) & \text{for } t = 0 \end{cases} \tag{2.110}$$

for the basic receive pulse $x_e(t)$ can be formulated: It states that when sampling in the symbol clock $T_s = 1/f_T$, only the value $x_e(0)$ is different from zero. Sampling in the symbol clock corresponds to multiplication with a T_s periodic Dirac pulse sequence in the time domain and thus the requirement (2.110) can also be formulated as

$$x_e(t) \cdot C \sum_{\nu=-\infty}^{+\infty} \delta(t - \nu\, T_s) = C \sum_{\nu=-\infty}^{+\infty} x_e(\nu\, T_s) \cdot \delta(t - \nu\, T_s) = C \cdot x_e(0) \cdot \delta(t) \tag{2.111}$$

This means that after sampling in the symbol clock, only the value of the basic receive pulse $x_e(t = 0)$ at $t = 0$ is different from zero and all other symbol clock sampling values of the basic receive pulse disappear. A transformation of this requirement (2.111) into the frequency domain using the correspondence[11]

$$C \sum_{\nu=-\infty}^{+\infty} \delta(t - \nu\, T_s) \quad\bullet\!\!-\!\!\circ\quad C \cdot f_T \sum_{\mu=-\infty}^{+\infty} \delta(f - \mu\, f_T) \quad \text{with } f_T = \frac{1}{T_s} \tag{2.112}$$

provides with $U(f) * \delta(f - f_0) = U(f - f_0)$ (shift by convolution with pulse), see e.g. Appendix A, the corresponding formulation in the frequency domain

$$X_e(f) * C \cdot f_T \sum_{\mu=-\infty}^{+\infty} \delta(f - \mu f_T) = C \cdot f_T \sum_{\mu=-\infty}^{+\infty} X_e(f - \mu f_T). \tag{2.113}$$

A Dirac impulse in the time domain corresponds with a constant in the frequency domain and thus results with (2.111) and (2.113) in the condition for intersymbol interference freedom at the detection point:

$$\sum_{\mu=-\infty}^{+\infty} X_e(f - \mu f_T) = \text{const.} \tag{2.114}$$

The two multiplicative quantities C and T_s (or f_T) play no role; they are included in the constant.

---

[11] The constant C has the dimension of a time (unit s). It ensures that the time signal and amplitude spectral density have the introduced and so far used dimensions. For the examples, $C = 1$ s is set.

First Nyquist Criterion A signal transmission free of intersymbol interference is achieved when the amplitude spectral density of the basic receive pulse, repeated and summed up periodically with the clock frequency (f_T), results in a constant.

Insert: On the Mathematical Description of Sampling in Time and Frequency Domain

To mathematically describe the symbol clock sampling, the time function $x_\mathrm{e}(t)$ of the basic receive pulse is multiplied by the dimensionless function

$$x_\delta(t) = C \sum_{\nu=-\infty}^{+\infty} \delta(t - \nu T_\mathrm{s}) \tag{2.115}$$

Since a time pulse $\delta(t)$ has the dimension of an Frequency (with the unit 1/s), see e.g. Appendix A, C must have the dimension of a Time (with the unit s) so that $x_\delta(t)$ is dimensionless and thus the sampled basic receive pulse again has the dimension of an Amplitude (with the unit V). For calculations in the examples and tasks, $C = 1\,\mathrm{s}$ is chosen to ensure the correctness of the dimensions (or units) without obtaining additional amplitude scalings. Alternatively, for example, $C = T_\mathrm{s}$ can be chosen—which is also common.

The Fourier transformation of (2.115) results in (see e.g. Appendix A)

$$X_\delta(f) = \mathcal{F}\{x_\delta(t)\} = C \cdot f_\mathrm{T} \sum_{\mu=-\infty}^{+\infty} \delta(f - \mu f_\mathrm{T}) \tag{2.116}$$

and has the dimension of an time—as the dimension of a frequency pulse $\delta(f)$, see e.g. Appendix A.

The convolution of (2.116) with the amplitude spectral density $X_\mathrm{e}(f)$ of the basic receive pulse results in

$$X_\mathrm{e}(f) * X_\delta(f) = X_\mathrm{e}(f) * C \cdot f_\mathrm{T} \sum_{\mu=-\infty}^{+\infty} \delta(f - \mu f_\mathrm{T}) = C \cdot f_\mathrm{T} \sum_{\mu=-\infty}^{+\infty} X_\mathrm{e}(f - \mu f_\mathrm{T}),$$

$$\tag{2.117}$$

and is an amplitude spectral density with the dimension Amplitude $\cdot$ Time in the unit Vs or V/Hz.

The first Nyquist criterion was described and derived based on the basic receive pulse. The basic receive pulse is necessary for the metrological detection, e.g., in the eye diagram. For the description and derivation of intersymbol interference freedom, it is also sufficient to consider the total weight function $h(t)$ as a cascade of transmit and receive filters $h(t) = g_\mathrm{s}(t) * g_\mathrm{e}(t)$—or the corresponding total transfer function $H(f) = G_\mathrm{s}(f) \cdot G_\mathrm{e}(f)$—to derive and justify the first Nyquist criterion.

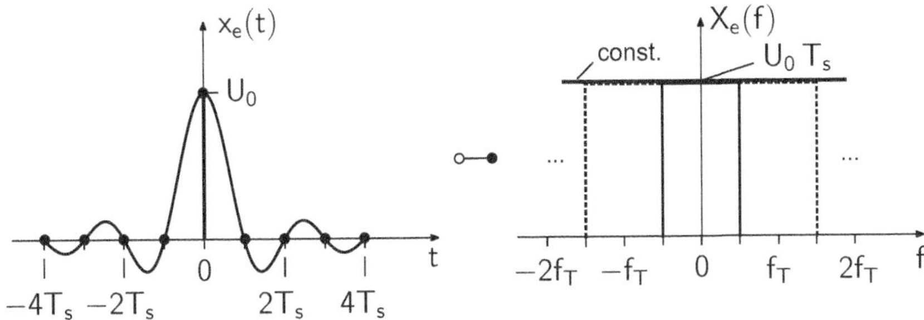

Fig. 2.42 On the first Nyquist criterion: Sinc pulse and rectangular spectrum

Examples for the First Nyquist Criterion

Example 2.2 (First Nyquist criterion with sinc pulse) In Fig. 2.42, the time characteristic and the spectrum of the sinc pulse $x_e(t) = U_0 \, \mathrm{si}(\pi f_T t)$ and the associated rectangular spectrum of width f_T are shown (solid lines): Sampling in the time domain at interval T_s results in the periodic repetition of the spectrum $X_e(f)$ with the clock frequency f_T (dashed spectra), so that the summation yields a constant (strong solid horizontal line). This corresponds in the time domain to a Dirac impulse at time $t = 0$: Only at $t = 0$ does the sampling of the receive pulse yield a contribution different from zero. The symbol clock sampling values of the received pulse $x_e(t)$ are also shown. Temporally adjacent receive pulses shifted by the symbol duration T_s—shown dashed in Fig. 2.43—do not influence the useful sampling value $x_e(t = 0)$.

This example corresponds to a raised cosine pulse and spectrum with the roll-off factor $r = 0$.

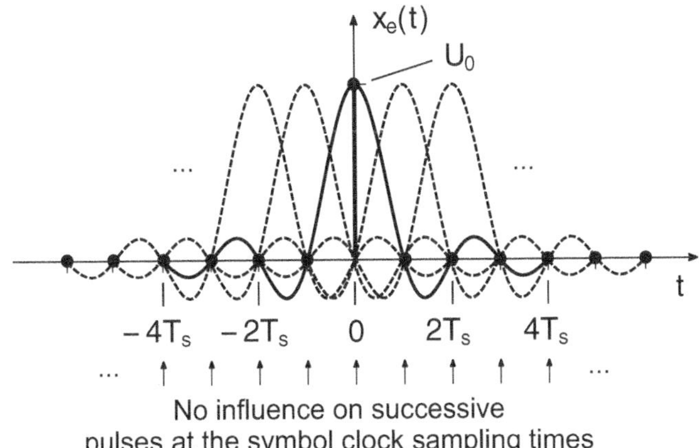

Fig. 2.43 On the first Nyquist criterion: Sinc pulse and temporally adjacent pulses

Example 2.3 (First Nyquist criterion with raised cosine pulse with $r = 1$) In another example, Fig. 2.44 shows the time characteristic and the spectrum for a cosine roll-off pulse and the associated spectrum for the roll-off factor $r = 1$. Here too, it becomes clear that the sum of the periodically repeated and summed spectra yields a constant and therefore the first Nyquist criterion is met. Temporally adjacent basic receive pulses shifted by the symbol duration T_s—shown dashed in Fig. 2.45—do not influence the useful sampling value $x_e(t = 0)$ again.   □

It can be said that raised cosine pulses for all roll-off factors r comply with the first Nyquist criterion and therefore lead to an intersymbol interference-free useful signal transmission, i.e., there is no influence of temporally successive pulses effective at the sampling time.

In the frequency domain, the filter slope, which protrudes into the adjacent spectrum and, when summed with the adjacent spectrum, results in a constant, is also called a *Nyquist* edge: The steeper this slope is, the stronger the overshoot of the pulse in the time domain, and the flatter this slope runs, the less the pulse overshoots in the time domain. All spectra with a Nyquist edge guarantee intersymbol interference freedom (e.g., cosine roll-off spectra, trapezoidal spectra) [48], as here, with all received basic pulses, the sam-

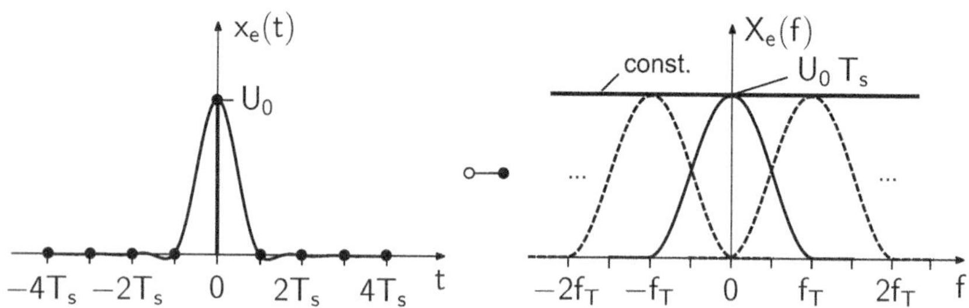

Fig. 2.44 On the first Nyquist criterion: Raised cosine pulse and spectrum for the roll-off factor $r = 1$

Fig. 2.45 On the first Nyquist criterion: Raised cosine pulse and temporally adjacent pulses for the roll-off factor $r = 1$

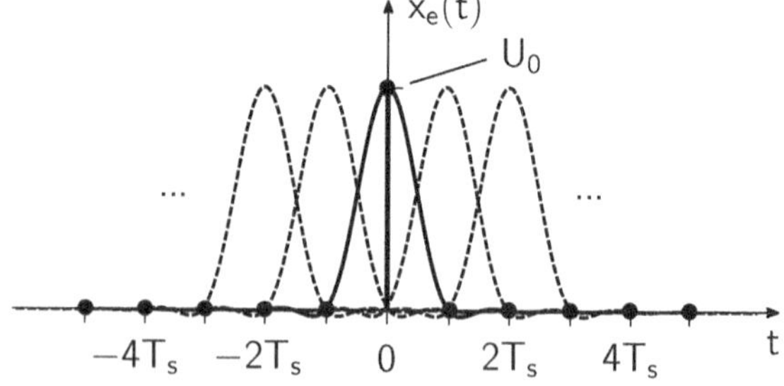

pling in the symbol clock T_s and the resulting periodicity of the spectra with the clock frequency f_T and their superposition leads to a constant.

Second Nyquist Criterion

The goal is to find a general rule for a pulse shape $x_e(t)$ at the detection point that is as insensitive as possible to inaccuracies in the sampling time. Then, the horizontal eye opening of the received useful signal at the detection point must be as large as possible. For this purpose, the requirements

1. Symmetry of the basic received pulse
2. No ISI contributions in the middle of the flank, i.e., zero crossings of the basic received pulse also in the middle between integer multiples of the symbol duration

are established and the former is expressed as

$$x_e\left(\frac{T_s}{2}\right) = x_e\left(\frac{T_s}{2} - T_s\right) = x_e\left(-\frac{T_s}{2}\right) \tag{2.118}$$

and the latter is formulated as

$$x_e\left(\frac{T_s}{2} + \nu T_s\right) = 0 \qquad \text{for} \qquad \nu \neq 0, -1 . \tag{2.119}$$

This means that the basic received pulse is a temporally even pulse (symmetrical to the vertical axis, first requirement) and that, starting from half the symbol duration $T_s/2$—except at $-T_s/2$ and $+T_s/2$—it must have zero crossings at multiples of the symbol duration T_s (second requirement), see Fig. 2.46. In summary, the requirements can be expressed as

$$x_e(t) = \begin{cases} x_e(0) & \text{for } t = 0 \\ \frac{x_e(0)}{2} & \text{for } t = \pm\frac{T_s}{2} \\ 0 & \text{for } t = \pm\nu\frac{T_s}{2} \quad \text{with} \quad \nu = 2, 3, 4, \dots \end{cases} \tag{2.120}$$

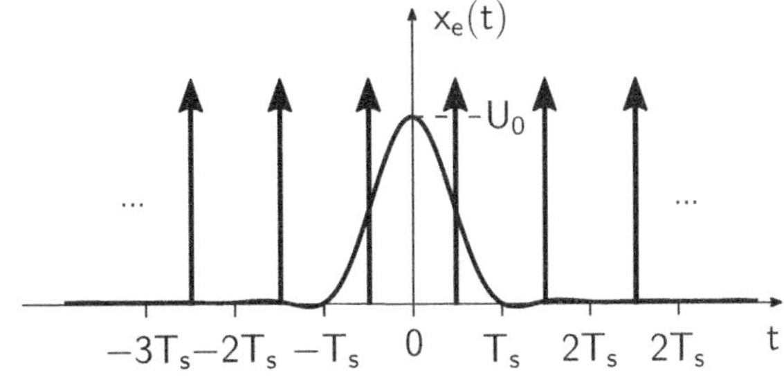

Fig. 2.46 For sampling in the symbol clock at the second Nyquist criterion using the example of the cosine roll-off pulse with $r = 1$

When multiplying with a Dirac pulse sequence periodic with T_s in the time domain (sampling in the symbol clock, now shifted by $T_\mathrm{s}/2$ compared to (2.111)

$$C \cdot x_\mathrm{e}(t) \cdot \sum_{\nu=-\infty}^{+\infty} \delta\left(t - \frac{T_\mathrm{s}}{2} - \nu\, T_\mathrm{s}\right) = C \cdot x_\mathrm{e}\left(\frac{T_\mathrm{s}}{2}\right) \cdot \left[\delta\left(t - \frac{T_\mathrm{s}}{2}\right) + \delta\left(t + \frac{T_\mathrm{s}}{2}\right)\right]$$

(2.121)

two Dirac pulses remain: Only the two pulses $\delta\left(t - \frac{T_\mathrm{s}}{2}\right)$ and $\delta\left(t + \frac{T_\mathrm{s}}{2}\right)$ coincide with the two non-zero values of the basic received pulse from the first requirement; all other pulses of the periodic pulse sequence fall into zero crossings of the basic received pulse $x_\mathrm{e}(t)$ and disappear after the multiplication of both. With the correspondences

$$\frac{1}{2}\left[\delta(t - t_0) + \delta(t + t_0)\right] \quad \circ\!\!-\!\!\bullet \quad \cos(2\pi\, t_0\, f) \tag{2.122}$$

and

$$C \sum_{\nu=-\infty}^{+\infty} \delta\left(t - \frac{T_\mathrm{s}}{2} - \nu\, T_\mathrm{s}\right) \quad \circ\!\!-\!\!\bullet \quad C \cdot f_\mathrm{T} \sum_{\mu=-\infty}^{+\infty} (-1)^\mu\, \delta(f - \mu\, f_\mathrm{T}) \tag{2.123}$$

the transformation into the frequency domain (here with $t_0 = T_\mathrm{s}/2$)

$$X_\mathrm{e}(f) * C \cdot f_\mathrm{T} \sum_{\mu=-\infty}^{+\infty} (-1)^\mu\, \delta(f - \mu f_\mathrm{T}) = C \cdot x_\mathrm{e}\left(\frac{T_\mathrm{s}}{2}\right) \cdot 2\cos(\pi\, T_\mathrm{s} f) \tag{2.124}$$

provides an equivalent formulation of (2.121) or the two original requirements, now in the frequency domain [65].

Insert: On the Derivation of the Correspondence of the Cosine Function in the Spectral Domain

With the help of the theorem for shifting in the time domain, the spectral functions for the Dirac pulses shifted by $\pm t_0$ are obtained

$$\delta(t + t_0) \quad \circ\!\!-\!\!\bullet \quad e^{\mathrm{j}2\pi\, f t_0}$$

$$\delta(t - t_0) \quad \circ\!\!-\!\!\bullet \quad e^{-\mathrm{j}2\pi\, f t_0}.$$

Comparing this with the relationship (see Appendix F.3 or e.g. [11, 74])

$$\cos(x) = \frac{1}{2}\left(e^{\mathrm{j}x} + e^{-\mathrm{j}x}\right), \tag{2.125}$$

it is recognizable that

$$\cos\left(2\pi f t_0\right) = \frac{1}{2}\left(e^{\mathrm{j}2\pi f t_0} + e^{-\mathrm{j}2\pi f t_0}\right) \tag{2.126}$$

applies and thus the correspondence

$$\frac{1}{2}\,[\delta(t - t_0) + \delta(t + t_0)] \quad \circ\!\!-\!\!\bullet \quad \cos(2\pi\, t_0\, f). \tag{2.127}$$

results. Therefore, two Dirac pulses symmetrically located to $t = 0$ in the time domain result in a cosine function in the frequency domain.

The condition for a pulse shape at the detection point that is as insensitive as possible to deviations from the optimal sampling time is therefore

$$\sum_{\mu=-\infty}^{+\infty} (-1)^{\mu}\, X_{\mathrm{e}}(f - \mu f_{\mathrm{T}}) = \mathrm{const.} \cdot \cos(\pi\, T_{\mathrm{s}} f). \tag{2.128}$$

The constant on the right side of (2.128) contains, for example, $x_{\mathrm{e}}(T_{\mathrm{s}}/2)$, C and f_{T}.

Second Nyquist Criterion If the sum of the amplitude density spectra, periodically repeated with the clock frequency and added with *alternating signs*, results in a cosine function, the horizontal eye opening is maximal.

Insert: Derivation of the Fourier Transform of the Shifted Periodic Pulse Train (Relation (2.123))
The $T_{\mathrm{s}}/2$ shifted with T_{s} periodic pulse train on the left side of (2.123) can be written as a convolution product

$$\sum_{\nu=-\infty}^{+\infty} \delta\!\left(t - \frac{T_{\mathrm{s}}}{2} - \nu\, T_{\mathrm{s}}\right) = \sum_{\nu=-\infty}^{+\infty} \delta(t - \nu\, T_{\mathrm{s}}) * \delta\!\left(t - \frac{T_{\mathrm{s}}}{2}\right) \tag{2.129}$$

The Fourier transformation yields

$$f_{\mathrm{T}} \sum_{\mu=-\infty}^{+\infty} \delta(f - \mu f_{\mathrm{T}}) \cdot \mathrm{e}^{-\mathrm{j}2\pi f \cdot \frac{T_{\mathrm{s}}}{2}} = f_{\mathrm{T}} \sum_{\mu=-\infty}^{+\infty} \mathrm{e}^{-\mathrm{j}\pi f\, T_{\mathrm{s}}}\, \delta(f - \mu f_{\mathrm{T}}) \tag{2.130}$$

and with the application of the selective property (only values at the frequencies $\mu f_{\mathrm{T}} = \mu/T_{\mathrm{s}}$ are preserved) further results

$$f_{\mathrm{T}} \sum_{\mu=-\infty}^{+\infty} \mathrm{e}^{-\mathrm{j}\pi \frac{\mu}{T_{\mathrm{s}}} T_{\mathrm{s}}}\, \delta(f - \mu f_{\mathrm{T}}) = f_{\mathrm{T}} \sum_{\mu=-\infty}^{+\infty} \mathrm{e}^{-\mathrm{j}\mu\pi}\, \delta(f - \mu f_{\mathrm{T}}). \tag{2.131}$$

Furthermore,

$$e^{-j\mu\pi} = \cos(\mu\pi) - j\sin(\mu\pi) \tag{2.132}$$

can be written, where for integer values μ

$$\sin(\mu\pi) = 0 \tag{2.133}$$

and

$$\cos(\mu\pi) = \begin{cases} +1 & \text{for } \ldots -6, -4, -2, 0, 2, 4, 6 \ldots \\ [1ex] -1 & \text{for } \ldots -5, -3, -1, 1, 3, 5, 7 \ldots \end{cases} \tag{2.134}$$

apply. This results in

$$e^{-j\mu\pi} = \cos(\mu\pi) = (-1)^{\mu} \tag{2.135}$$

and the formulation in the frequency domain reads

$$f_{\mathrm{T}} \sum_{\mu=-\infty}^{+\infty} (-1)^{\mu}\, \delta(f - \mu f_{\mathrm{T}}) \tag{2.136}$$

and thus the sought correspondence for (2.123) results

$$\sum_{\nu=-\infty}^{+\infty} \delta\left(t - \frac{T_{\mathrm{s}}}{2} - \nu T_{\mathrm{s}}\right) \quad \circ\!\!-\!\!\bullet \quad f_{\mathrm{T}} \sum_{\mu=-\infty}^{+\infty} (-1)^{\mu}\, \delta(f - \mu f_{\mathrm{T}}). \tag{2.137}$$

Example of the Second Nyquist Criterion

Example 2.4 (Second Nyquist criterion with root cosine roll-off pulse with $r = 1$) For a cosine roll-off receive pulse with the roll-off factor $r = 1$, in the time domain with (2.105)

$$x_{\mathrm{e}}(t) = U_0\, \mathrm{si}(\pi f_{\mathrm{T}} t)\, \frac{\cos(\pi f_{\mathrm{T}} t)}{1 - (2 f_{\mathrm{T}} t)^2} \tag{2.138}$$

and in the frequency domain, one obtains from (2.104) initially

$$X_{\mathrm{e}}(f) = \begin{cases} \frac{U_0 T_{\mathrm{s}}}{2}\left[1 + \sin\left(\frac{\pi}{2}(1 - 2f T_{\mathrm{s}})\right)\right] & \text{for } |f| \leq f_{\mathrm{T}} \\ 0 & \text{otherwise} \end{cases}. \tag{2.139}$$

A transformation yields

$$\sin\left(\frac{\pi}{2}(1 - 2f T_{\mathrm{s}})\right) = \sin\left(\frac{\pi}{2} - \pi f T_{\mathrm{s}}\right) = \sin\left(-\left(\pi f T_{\mathrm{s}} - \frac{\pi}{2}\right)\right) \tag{2.140}$$

and with the sine function as an odd function, i.e., $f(x) = -f(-x)$, one obtains

$$\sin\left(-\left(\pi f T_{\mathrm{s}} - \frac{\pi}{2}\right)\right) = -\sin\left(\pi f T_{\mathrm{s}} - \frac{\pi}{2}\right). \tag{2.141}$$

The non-zero part of the receive spectrum $X_{\mathrm{e}}(f)$ in the range $|f| \leq f_{\mathrm{T}}$ can thus be written as

$$X_{\mathrm{e}}(f) = \frac{U_0 T_{\mathrm{s}}}{2}\left[1 - \sin\left(\pi f T_{\mathrm{s}} - \frac{\pi}{2}\right)\right] \tag{2.142}$$

and with the trigonometric relationship $\sin(x - \pi/2) = -\cos(x)$ it results

$$X_{\mathrm{e}}(f) = \frac{U_0 T_{\mathrm{s}}}{2}\left[1 + \cos\left(\pi f T_{\mathrm{s}}\right)\right]. \tag{2.143}$$

Using the theorem

$$\cos^2(x) = \frac{1}{2}(1 + \cos(2x)) \tag{2.144}$$

finally results in

$$X_{\mathrm{e}}(f) = \begin{cases} U_0 T_{\mathrm{s}} \cos^2\left(\pi \frac{T_{\mathrm{s}}}{2}f\right) & \text{for} \quad |f| \leq f_{\mathrm{T}} \\ 0 & \text{otherwise} \end{cases}. \tag{2.145}$$

One obtains

$$\sum_{\mu=-\infty}^{+\infty} (-1)^{\mu} X_{\mathrm{e}}(f - \mu f_{\mathrm{T}}) = \sum_{\mu=-\infty}^{+\infty} (-1)^{\mu} U_0 T_{\mathrm{s}} \cos^2\left(\pi \frac{T_{\mathrm{s}}}{2}(f - \mu f_{\mathrm{T}})\right), \tag{2.146}$$

and the result is

$$\sum_{\mu=-\infty}^{+\infty} (-1)^{\mu} X_{\mathrm{e}}(f - \mu f_{\mathrm{T}}) = U_0 T_{\mathrm{s}} \cos\left(\frac{\pi f}{f_{\mathrm{T}}}\right). \tag{2.147}$$

In Fig. 2.47, a section of the spectral functions is graphically represented to illustrate the result (2.147): The original received spectrum $X_{\mathrm{e}}(f)$ (for $\mu = 0$) is plotted with a solid line, the receive spectra shifted by integer multiples of the clock frequency (for $\mu \neq 0$) are shown dashed. The cosine function resulting from the superposition of the periodically repeated with the clock frequency and added with alternating signs receive spectra is drawn with a bold line.

The second Nyquist criterion is met by cosine roll-off receive pulses with the roll-off factor $r = 1$; this does not apply to all other roll-off factors. $\square$

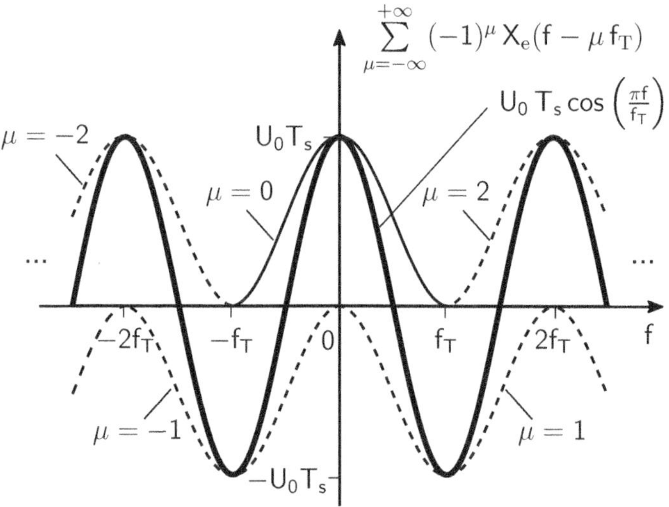

Fig. 2.47 On verifying the compliance with the second Nyquist criterion with a raised cosine receive pulse with the roll-off factor $r = 1$ in the spectral domain

Insights and Summary on the Nyquist Criteria

The first and second Nyquist criteria have established general rules that received pulse shapes must adhere to if they are to be optimal in terms of intersymbol interference freedom or robustness against deviations in clock information.

In particular, filters with synthetic band limitation[12] are suitable, which can be mathematically well described and therefore can be approximately realized with the help of digital signal processing. All filters with a rational transfer function do not lead to received pulse shapes that fulfill the first and second Nyquist criteria. In this case, only an approximation is possible.

Finally, the boundary conditions for compliance with the first and second Nyquist criteria in the frequency range should be illustrated again using an example. Figure 2.48 illustrates the spectrum of a Nyquist-1 basic received pulse $x_e(t)$. Characteristic here is the requirement that the frequency response at $f = f_T/2$ has fallen to half of its maximum value and that the Nyquist edge is point-symmetric to the points $(f_T/2;\ U_0 T_s/2)$ and $(-f_T/2;\ U_0 T_s/2)$. Consequently, in the example

$$X_e\left(f = \frac{f_T}{2}\right) = \frac{U_0 T_s}{2} \tag{2.148}$$

---

[12] Synthetic band limitation often refers to a type of filtering that is not originally based on rational transfer functions close to the circuit-technical realization, but originates from a mathematical consideration or optimization and usually requires a computer-based practical realization.

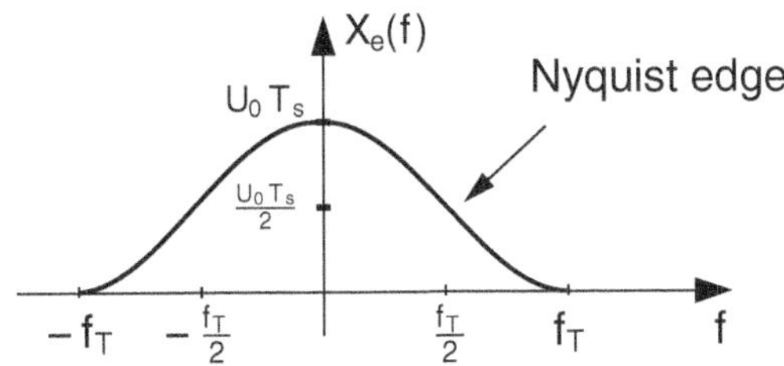

Fig. 2.48 Spectrum of a Nyquist-1 basic receive pulse $x_e(t)$

must apply. It should be noted again that instead of the basic received pulse $x_e(t)$, the corresponding weighting function $h(t)$ of a transmission system can be used to analyze compliance with the first Nyquist criterion. The first Nyquist criterion is met if, when sampling the basic received pulse $x_e(t)$ with the sampling period $T_a = T_s$, a constant results in the frequency range due to the periodicity of the received spectrum with the sampling frequency $f_a = 1/T_a = f_T$. Figure 2.49 illustrates the periodicity in the spectral range resulting from the sampling of the Nyquist-1 basic received pulse $x_e(t)$. Here, the condition (according to Fig. 2.49)

$$\sum_{\mu=-\infty}^{+\infty} X_e(f - \mu f_T) = U_0 T_s \tag{2.149}$$

must be met.

If, in addition, a maximization of the horizontal eye opening is sought, compliance with the second Nyquist criterion is necessary. In this case, the periodic continuation of the received spectrum with alternating signs must meet the condition

$$\sum_{\mu=-\infty}^{+\infty} (-1)^\mu X_e(f - \mu f_T) = U_0 T_s \cdot \cos(\pi T_s f). \tag{2.150}$$

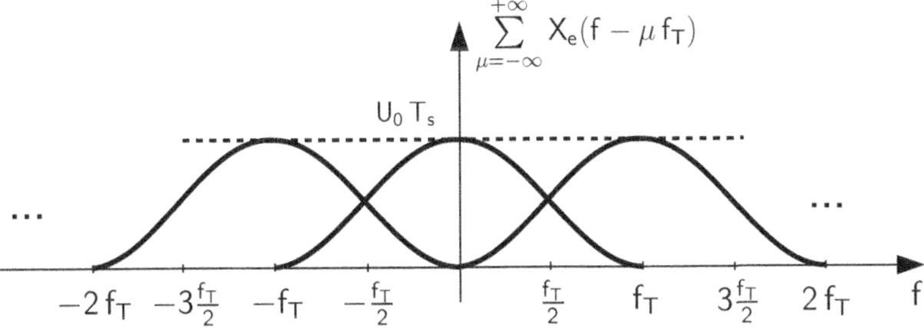

Fig. 2.49 On meeting the first Nyquist criterion in the spectral domain

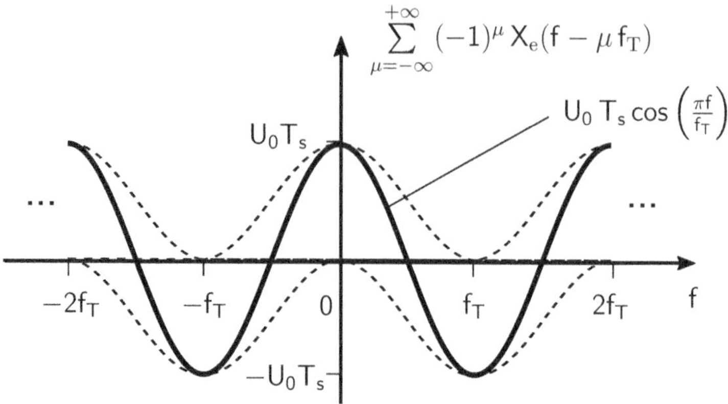

Fig. 2.50 To fulfill the second Nyquist criterion in the frequency range

Figure 2.50 illustrates compliance with the second Nyquist criterion in the frequency range.

2.2.4 Optimal Receive Filter: Matched Filter

After the previous considerations on the basic received pulse shape, it remains open how the optimal received pulse shape or the optimal receive filter must be designed if a maximum signal-to-noise ratio at the detection point with a known pulse shape at the receiver input and given noise interference is the optimization goal—if the optimal received pulse shape or the optimal receive filter is sought with respect to a maximum signal-to-noise ratio. The decision problem is first considered using the example of a single pulse ered.

A information signal is known in a receiver regarding to its form. It should be decided at the receiver output whether the received signal mixture, consisting of useful signal and disturbance, contains the useful message signal or not: A *decision problem* needs to be solved. Applications can be found in the receive of echo pulses in radar systems for location and in the frame synchronization of receivers, as well as in the digital transmission of messages, to which this type of filtering can be advantageously transferred and applied.

Figure 2.51 shows the arrangement. The problem statement is therefore more precise: The receive filter $G_e(f)$ •—o $g_e(t)$ should be chosen so that with minimal error probability it can be decided by sampling at the filter output whether a pulse of known form, disturbed by additive white noise with the power spectral density Ψ_0, is present at a certain time $t = T_A$ or not.

It is assumed that the useful signal $u_0(t)$ is known and is disturbed by white, Gaussian-distributed noise with the known power spectral density Ψ_0. The desired receive filter is a linear, time-invariant system that should have such a time or frequency characteristic that the error probability is minimized. Because of the monotonic relationship between

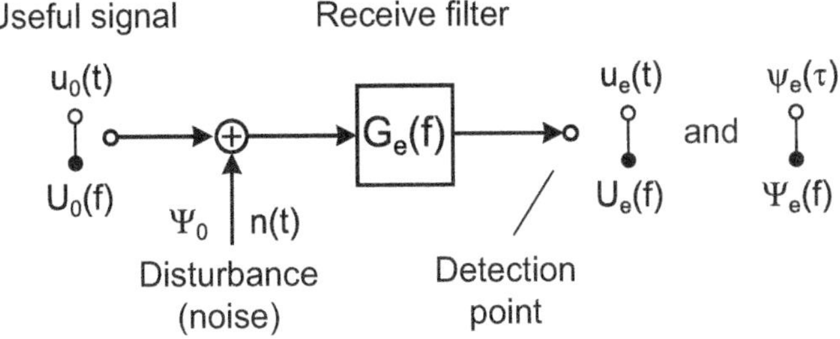

Fig. 2.51 Arrangement for determining the optimal receive filter

the error probability and the signal-to-noise ratio, the signal-to-noise ratio can be used as a quality criterion. For the derivation, a single rectangle pulse of amplitude U_0 and duration T_s is considered—in reference to the previously extensively described basic model of baseband transmission.

As signal power at the detection point, when considering a single pulse—in contrast to the previously considered transmission of a pulse *sequence*—the square of the total received signal amplitude $u_e(t = T_A)$ available at the sampling time T_A at the detection point. Since only a single pulse is considered, intersymbol interferences do not play a role. The noise power at the detection point is

$$U_R^2 = \psi_e(\tau = 0) = \Psi_0 \int_{-\infty}^{+\infty} |G_e(f)|^2 \, df. \tag{2.151}$$

The signal-to-noise ratio can thus be expressed as

$$\varrho = \frac{(u_e(T_A))^2}{U_R^2} = \frac{(u_e(T_A))^2}{\psi_e(0)} \ . \tag{2.152}$$

For a clear and concise calculation, the sampling time is chosen to be $T_A = 0$—this can generally be achieved, for example, by a suitable temporal shift of the pulse. The useful signal amplitude present at the detection point is thus $u_e(t = 0)$ and with the help of the relationship for the inverse Fourier transformation for $t = 0$ (see Appendix C), we get

$$u_e(0) = \int_{-\infty}^{+\infty} U_e(f) \, df = \int_{-\infty}^{+\infty} U_0(f) \cdot G_e(f) \, df. \tag{2.153}$$

With the results (2.151) and (2.153), the signal-to-noise ratio (2.152) is written as

$$\varrho = \frac{\left(\int_{-\infty}^{+\infty} U_0(f) \cdot G_e(f) \, df \right)^2}{\Psi_0 \int_{-\infty}^{+\infty} |G_e(f)|^2 \, df} \ . \tag{2.154}$$

$U_0(f)$ and Ψ_0 are known and thus given according to the conditions, the function $G_e(f)$ is sought and to be determined so that ϱ becomes maximum. This type of optimization is a task from the calculus of variations [11, 74]. For the solution, the Schwarz[13] inequality [11, 74]

$$\left(\int\limits_{-\infty}^{+\infty} U_0(f) \cdot G_e(f)\,df \right)^2 \leq \int\limits_{-\infty}^{+\infty} |U_0(f)|^2\,df \cdot \int\limits_{-\infty}^{+\infty} |G_e(f)|^2\,df \qquad (2.155)$$

can be used, which can be understood as a generalized triangle inequality. The equality sign applies for $U_0(f) = \lambda \cdot G_e(f)$. Here, λ is any arbitrary, positive, real factor with the dimension Time $\cdot$ Amplitude. For the signal-to-noise ratio, we get

$$\varrho \leq \frac{\int\limits_{-\infty}^{+\infty} |U_0(f)|^2\,df \cdot \int\limits_{-\infty}^{+\infty} |G_e(f)|^2\,df}{\Psi_0 \int\limits_{-\infty}^{+\infty} |G_e(f)|^2\,df} \qquad (2.156)$$

or after shortening

$$\varrho \leq \frac{1}{\Psi_0} \int\limits_{-\infty}^{+\infty} |U_0(f)|^2\,df. \qquad (2.157)$$

Using the PARSEVAL theorem, it is recognizable that

$$\int\limits_{-\infty}^{+\infty} |U_0(f)|^2\,df = \int\limits_{-\infty}^{+\infty} u_0^2(t)\,dt = E_0 \qquad (2.158)$$

is the energy of the transmission pulse and we get

$$\varrho \leq \frac{E_0}{\Psi_0} \qquad (2.159)$$

The upper limit for the signal-to-noise ratio thus depends only on the *energy of the input signal* and the *power spectral density of the noise*. The pulse shape does not matter.

The maximum signal-to-noise ratio

$$\varrho_{\mathrm{MF}} = \frac{E_0}{\Psi_0}, \qquad (2.160)$$

that can be achieved by matched filtering, is reached when the equality sign applies in the Schwarz inequality.

---

[13] Hermann Amandus Schwarz (1843–1921), German mathematician.

Thus, according to (2.154), one obtains for the transfer function of the optimal search filter

$$G_{\text{e opt}}(f) = \frac{1}{\lambda} \cdot U_0^*(f). \tag{2.161}$$

The optimal transfer function $G_{\text{e opt}}(f)$ of the receiving filter is, apart from a constant factor λ with the dimension Time $\cdot$ Amplitude (i.e., with the unit Vs or V/Hz) identical to the conjugate-complex amplitude density function $U_0^*(f)$ of the input or transmission pulse $u_0(t)$. The weighting function of the signal-adapted filter reads

$$g_{\text{e opt}}(t) = \frac{1}{\lambda} \cdot u_0(-t). \tag{2.162}$$

In Fig. 2.52, the transmission pulse and the weighting function of the signal-adapted filter are shown using the example of a temporally asymmetric pulse: It becomes clear that a temporal reflection of the transmission pulse occurs and that a scaling with the factor $1/\lambda$ is done to obtain the weighting function of the Matched Filter. In particular, it can be seen that the multiplication with the factor $1/\lambda$—where λ has the unit Vs—leads to the Matched-Filter-weighting function $g_{\text{e opt}}(t)$ having the unit V when a transmission pulse with the unit $1/s$ is used.

The optimal receive filter depends on the shape of the transmission pulse $u_0(t)$—hence the term *signal-adapted* or *matched* filter is used. However, the signal-to-noise ratio achievable at the output of such a filter is, as mentioned above, not dependent on the pulse shape—but only on the pulse energy E_0 and the spectral power density Ψ_0 of the disturbance.

The optimal search filter $G_{\text{e opt}}(f)$ amplifies the spectral components where the amplitude spectral density of the useful signal is large and the spectral power density of the noise is relatively small. Therefore, the signal-to-noise ratio at the sampling time becomes maximum.

Examples for the Matched Filter

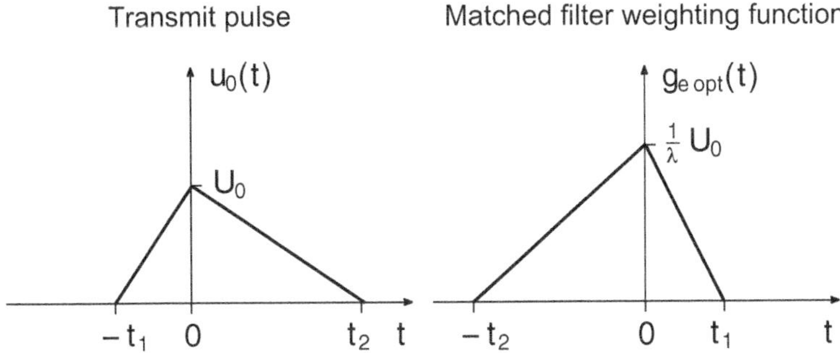

Fig. 2.52 Transmit pulse and weighting function of the signal-adapted receive filter (Matched Filter) using the example of a sawtooth-shaped transmission base pulse (non-causal representation)

Example 2.5 (Signal-to-noise ratio: Comparison of matched and suboptimal receive filter) In this example, an optimal, i.e., signal-adapted, and a non-optimal receiving filter are compared in terms of the achievable signal-to-noise ratio with a rectangular NRZ receiver input pulse. The weighting function of the transmit filter $g_s(t)$ and the receive filter $g_e(t)$ can be found in Fig. 2.53.

The resulting weighting function $h(t) = g_s(t) * g_e(t)$ is shown in Fig. 2.54.

The value $u_e(t = 0)$ that is decisive for signal detection in the case of single pulse consideration is determined by the maximum value $h(t = 0)$ and can be determined by evaluating the convolution integral. The maximum value within the convolution integral results from the maximum overlap of the two weighting functions $g_s(t)$ and $g_e(t)$ and can be calculated as

$$h(t = 0) = \int_0^{T_s} \frac{1}{T_s} \cdot \frac{1}{T_s}\, dt = \frac{1}{T_s} \tag{2.163}$$

If the transmit filter is now excited with a weighted Dirac pulse

$$u_q(t) = U_0\, T_s\, \delta(t) \tag{2.164}$$

it becomes apparent that at the optimal sampling time $t = 0$, the useful signal value

$$u_e(t = 0) = U_0\, T_s \cdot h(t = 0) = U_0\, T_s \cdot \frac{1}{T_s} = U_0 \tag{2.165}$$

is produced at the receive filter output.

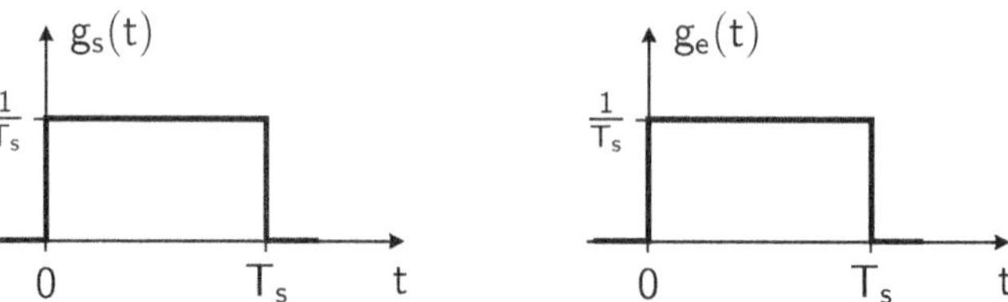

Fig. 2.53 Weighting function of transmit filter $g_s(t)$ and receive filter $g_e(t)$

Fig. 2.54 Weighting function of the cascade of transmit filter $g_s(t)$ and receive filter $g_e(t)$ according to the Matched-Filter principle (non-causal representation)

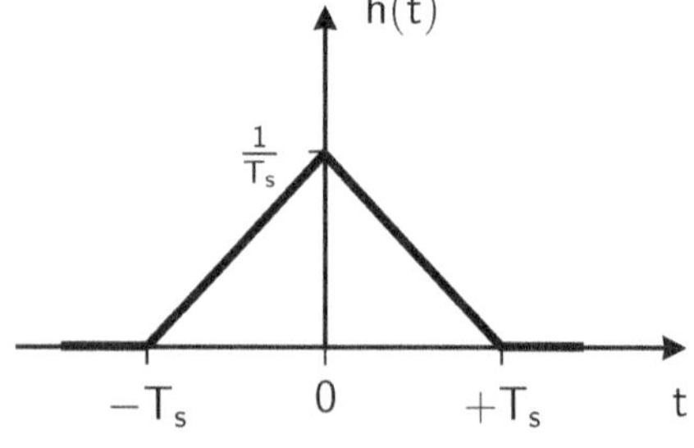

For the determination of the signal-to-noise ratio, in addition to $u_e(t = 0)$, the noise power U_R^2 needs to be calculated. In the considered case, this results in

$$U_R^2 = \Psi_0 \cdot \int_{-\infty}^{+\infty} |g_e(t)|^2 \, dt = \Psi_0 \cdot \int_0^{T_s} \left(\frac{1}{T_s}\right)^2 dt = \frac{\Psi_0}{T_s^2} T_s = \Psi_0 \frac{1}{T_s}. \qquad (2.166)$$

Thus, the signal-to-noise ratio at the output of the signal-matched filter follows

$$\varrho_{MF} = \frac{(u_e(t = 0))^2}{U_R^2} = \frac{U_0^2 T_s}{\Psi_0}. \qquad (2.167)$$

This result can also be obtained from (2.160) by dividing the energy $E_0 = U_0^2 T_s$ of the receiver input pulse by the spectral power density Ψ_0:

$$\varrho_{MF} = \frac{E_0}{\Psi_0} = \frac{U_0^2 T_s}{\Psi_0}. \qquad (2.168)$$

When the Matched-Filter condition is met, the signal-to-noise ratio at the sampling time takes the maximum possible value.

To illustrate this, a deviation from the Matched-Filter principle will now be examined. For this purpose, the form of the weighting function of the receive filter $g_e(t)$ shown in Fig. 2.55 is assumed and examined as an example—while the form of the receiver input pulse or the transmit filter remains the same.

The convolution of the weighting function

$$g_s(t) = \begin{cases} \frac{1}{T_s} & \text{for} \quad 0 < t < T_s \\ 0 & \text{otherwise} \end{cases} \qquad (2.169)$$

of the transmit filter with the weighting function

$$g_e(t) = \begin{cases} \frac{1}{T_s} \cdot \frac{t}{T_s} & \text{for} \quad 0 \leq t < T_s \\ 0 & \text{otherwise} \end{cases} \qquad (2.170)$$

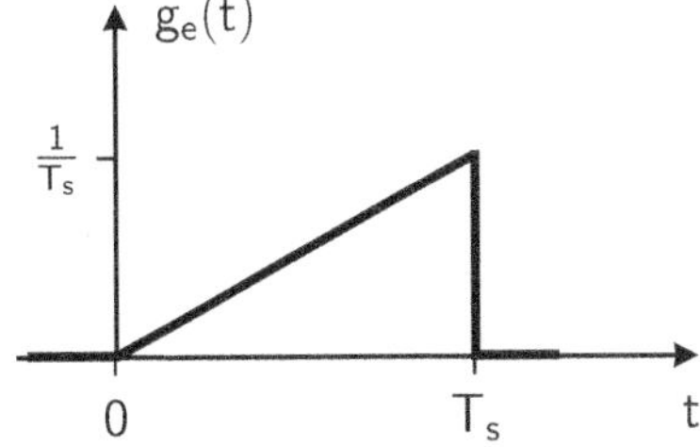

Fig. 2.55 Weighting function of the receive filter $g_e(t)$ in violation of the Matched-Filter principle

Fig. 2.56 Weighting function $h(t)$ in the example

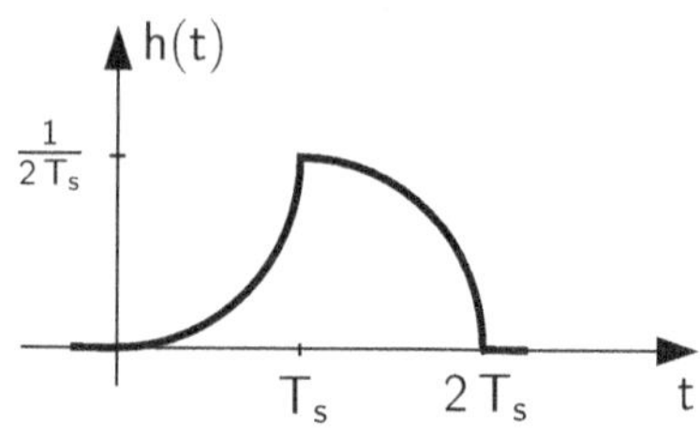

of the receive filter leads to the total weighting function

$$
h(t) = \begin{cases}
0 & \text{for } t < 0 \\
\frac{1}{2\,T_\mathrm{s}^3} \cdot t^2 & \text{for } 0 \le t \le T_\mathrm{s} \\
\frac{1}{2\,T_\mathrm{s}^3} \cdot (T_\mathrm{s}^2 - (t - T_\mathrm{s})^2) & \text{for } T_\mathrm{s} \le t \le 2\,T_\mathrm{s} \\
0 & \text{for } t > 2\,T_\mathrm{s}
\end{cases}
\tag{2.171}
$$

Figure 2.56 illustrates the resulting shape of the total weighting function $h(t)$.
The maximum value of $h(t) = g_\mathrm{s}(t) * g_\mathrm{e}(t)$ for the useful signal amplitude at the receive
filter output is now obtained as

$$
h(t = 0) = \int_0^{T_\mathrm{s}} \frac{1}{T_\mathrm{s}} \cdot \frac{1}{T_\mathrm{s}^2} \cdot t \, \mathrm{d}t = \frac{1}{2\,T_\mathrm{s}}.
\tag{2.172}
$$

To determine the available useful signal amplitude $u_\mathrm{e}(t = 0)$ at the receive filter output,
the transmit pulse shaper is excited with a weighted Dirac impulse

$$
u_\mathrm{q}(t) = U_0\, T_\mathrm{s}\, \delta(t) .
\tag{2.173}
$$

It turns out that at the optimal sampling time $t = 0$, the value

$$
u_\mathrm{e}(t = 0) = U_0\, T_\mathrm{s} \cdot h(t = 0) = U_0\, T_\mathrm{s} \cdot \frac{1}{2\,T_\mathrm{s}} = \frac{U_0}{2}
\tag{2.174}
$$

arises as the useful signal amplitude. The noise power results in

$$
U_\mathrm{R}^2 = \Psi_0 \cdot \int_{-\infty}^{+\infty} |g_\mathrm{e}(t)|^2 \, \mathrm{d}t = \Psi_0 \cdot \int_0^{T_\mathrm{s}} \left(\frac{t}{T_\mathrm{s}^2}\right)^2 \mathrm{d}t = \frac{\Psi_0}{T_\mathrm{s}^4}\frac{T_\mathrm{s}^3}{3} = \Psi_0 \frac{1}{3\,T_\mathrm{s}}.
\tag{2.175}
$$

This results in the signal-to-noise ratio

$$
\varrho_\text{no-MF-1} = \frac{(u_\mathrm{e}(t = 0))^2}{U_\mathrm{R}^2} = \frac{\left(\frac{U_0}{2}\right)^2}{\Psi_0 \frac{1}{3\,T_\mathrm{s}}} = \frac{\frac{U_0^2}{4}}{\Psi_0 \frac{1}{3\,T_\mathrm{s}}} = \frac{3}{4} \cdot \frac{U_0^2\, T_\mathrm{s}}{\Psi_0}.
\tag{2.176}
$$

As expected, it turns out that the achievable signal-to-noise ratio decreases due to the violation of the Matched-Filter principle compared to the optimal, signal-matched receive filtering. One obtains

$$\varrho_{\text{no-MF-1}} = \frac{3}{4} \, \varrho_{\text{MF}}. \tag{2.177}$$

The signal-to-noise ratio $\varrho_{\text{kein-MF-1}}$ is lower than in the example with signal-matched and thus optimal receive filtering with regard to a maximum signal-to-noise ratio. $\square$

Example 2.6 (Comparison of Matched Filter and RC first-order low-pass) A single rectangle impulse of amplitude U_0 and duration T_s is considered, which is transmitted over a channel disturbed by white, Gaussian noise of power spectral density Ψ_0. At the receiver, it should be decided by sampling whether the impulse was sent or not, i.e., whether it is present in the received signal mixture. A Matched filter and a RC first-order low-pass filter are used as receive filters alternatively: The achievable signal-to-noise ratios in both cases are to be calculated and compared with each other.

With smatched receive filtering with $g_e(t) = 1/(U_0 \, T_s) \cdot u_0(-t)$, the energy of the transmission impulse results in

$$E_0 = \int\limits_{-\infty}^{+\infty} u_0^2(t) \, dt = U_0^2 \cdot T_s \tag{2.178}$$

and the maximum achievable signal-to-noise ratio results in

$$\varrho_{\text{MF}} = \frac{E_0}{\Psi_0} = \frac{U_0^2 \, T_s}{\Psi_0}. \tag{2.179}$$

For the receive filtering with a RC first-order low-pass, the maximum useful signal sample value at time $t = T_A = T_s$ is obtained from the rectangular response of the received impulse (2.18) as

$$u_e(t = T_s) = U_0\left(1 - e^{-\frac{T_s}{T_g}}\right) = U_0\left(1 - e^{-2\pi f_g T_s}\right) \tag{2.180}$$

and the noise power at the detection point is according to (2.59)

$$U_R^2 = \Psi_0 \, \pi \, f_g, \tag{2.181}$$

so that the signal-to-noise ratio can be given as

$$\varrho_{RC} = \frac{U_0^2\left(1 - e^{-2\pi f_g T_s}\right)^2}{\Psi_0 \, \pi \, f_g}, \tag{2.182}$$

This signal-to-noise ratio again depends inversely on the signal and noise power of the cutoff frequency f_g of the receive filter, so that an optimization of the receive filter cutoff

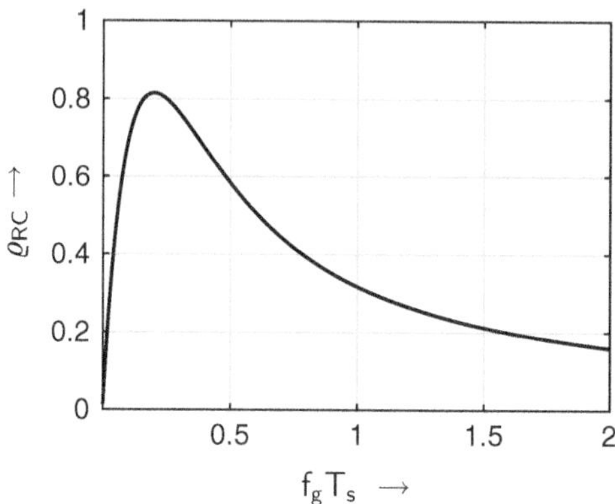

Fig. 2.57 Signal-to-noise ratio ϱ_{RC} as a function of the receive filter cutoff frequency f_g normalized to the symbol duration T_s

frequency is again necessary: It results for the case of the single impulse $f_g T_s \approx 0.2$ (see Fig. 2.57), so that the signal-to-noise ratio

$$\varrho_{RC} = \frac{U_0^2\left(1 - e^{-2\pi \cdot 0.2}\right)^2}{\Psi_0\, \pi \cdot 0.2/T_s} \tag{2.183}$$

only depends on the parameters of the transmit signal (U_0, T_s) and the disturbance Ψ_0. The optimization of the cutoff frequency results in the case of the single impulse in a smaller optimal value ($f_g T_s \approx 0.2$) than in the case of the transmission of a pulse sequence ($f_g T_s \approx 0.4$), since the eye remains open to smaller cutoff frequencies, at which it would still be closed due to the decision threshold U_E lying at $U_0/2$ in the case of transmission of a pulse sequence.

With exemplary given values $U_0 = 1\,\text{V}$, $T_s = 1\,\mu s$ and $\Psi_0 = 10^{-6}\,\text{V}^2/\text{Hz}$, the comparison results in

$$\varrho_{MF} = 1 \qquad \text{and} \qquad \varrho_{RC} = 0.8145 \tag{2.184}$$

It can be seen that even with a comparatively easy to implement RC low-pass filter, approximately 80 % of the performance of a signal-adapted filter can be achieved.

It should be noted again that these considerations refer to a single impulse and only the value at the optimal sampling time is evaluated for the useful signal amplitude. The intersymbol interferences known from information and data transmission with pulse sequences therefore play no role. $\qquad\qquad\square$

Example 2.7 (Signal-to-noise ratio when the matched filter condition is violated) If the receive filter is not adjusted to the transmitting filter characteristic, the achievable signal-to-noise ratio is reduced compared to that achievable with a matched receive filter. This example is intended to illustrate the signal-to-noise ratio that can be achieved when the transmitting and receiving filters have the weighting functions according to Fig. 2.58:

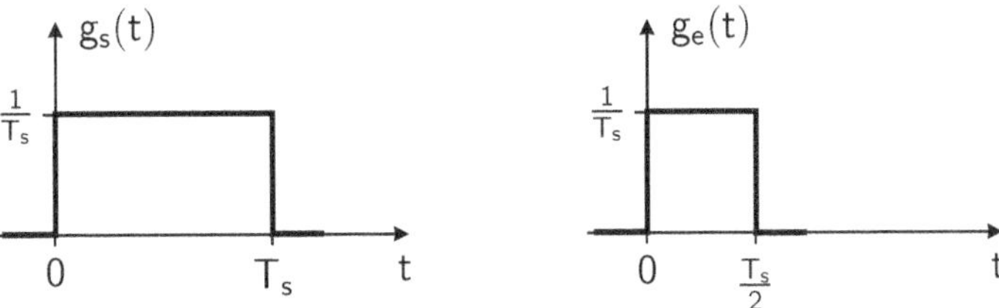

Fig. 2.58 Weighting functions of transmit filter $g_s(t)$ and receive filter $g_e(t)$ in an exemplary violation of the matched filter principle

So again, it is an NRZ transmit rectangular pulse of duration T_s, but the receive filter only has half the length $T_s/2$.

Figure 2.59 illustrates the resulting weighting function from the cascade of transmit and receive filters.

If the transmitting filter is excited with a weighted Dirac pulse

$$u_q(t) = U_0\, T_s\, \delta(t) \tag{2.185}$$

it is shown that at the optimal sampling time $t = 0$ the useful signal amplitude value

$$u_e(t = 0) = U_0\, T_s \cdot h(t = 0) = U_0\, T_s \cdot \frac{1}{2\, T_s} = \frac{U_0}{2} \tag{2.186}$$

is produced.

The noise power at the output of the receive filter is obtained by applying the Parseval theorem to

$$U_R^2 = \Psi_0 \cdot \int_{-\infty}^{+\infty} |g_e(t)|^2\, dt = \Psi_0 \cdot \int_{-T_s/4}^{T_s/4} \left(\frac{1}{T_s}\right)^2 dt = \frac{\Psi_0}{T_s^2}\frac{T_s}{2} = \Psi_0\, \frac{1}{2\, T_s}. \tag{2.187}$$

For the achievable signal-to-noise ratio, one obtains

$$\varrho_{\text{no-MF-2}} = \frac{(u_e(t = 0))^2}{U_R^2} = \frac{\left(\frac{U_0}{2}\right)^2}{\Psi_0 \frac{1}{2\, T_s}} = \frac{\frac{U_0^2}{4}}{\Psi_0 \frac{1}{2\, T_s}} = \frac{1}{2} \cdot \frac{U_0^2\, T_s}{\Psi_0}. \tag{2.188}$$

Fig. 2.59 Weighting function from cascade of transmit filter $g_s(t)$ and receive filter $g_e(t)$ in violation of the matched filter principle (non-causal representation)

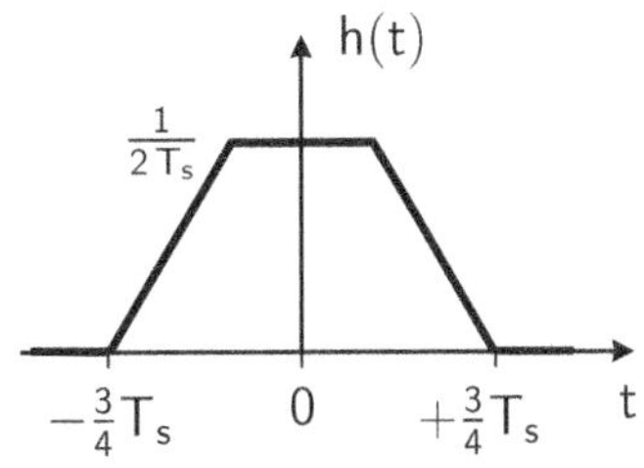

and it is smaller than the signal-to-noise ratio (2.167) in the case of a signal-adapted receive filter:

$$\varrho_{\text{no-MF-2}} = \frac{1}{2}\,\varrho_{\text{MF}}.$$ (2.189)

$\square$

The previous considerations on the detection of a single pulse within a received signal, which is disturbed by white, Gaussian noise, have shown that the signal-to-noise ratio becomes maximum when the matched filter principle is adhered to in the receive filtering.

It should be noted that these results can be directly transferred to the transmission and detection of NRZ pulse sequences, provided that the received pulses do not interfere with each other at the sampling time, i.e., there are no intersymbol interferences.

2.2.5 Aspects of Non-White Noise

The previous considerations on the basic model of baseband transmission focused on determining and maximizing the signal-to-noise ratio with white, Gaussian-distributed noise as disturbance. Under this boundary condition, it could be shown that a maximization of ϱ requires the use of a signal-adapted or matched filter, where it became clear that intersymbol interference causes a reduction of ϱ. Under the boundary condition of avoiding intersymbol interference, i.e., by adhering to the first Nyquist criterion, it could be shown that the maximum possible signal-to-noise ratio is achieved with the use of a matched filter.

If one wants to achieve the maximum possible signal-to-noise ratio ϱ under the influence of interference, the receiver concept must be modified. One can start from the following result: Although this relationship for a pulse sequence can only be maintained with simultaneous ISI-freedom of the received symbols, matched receive filtering has a fundamental importance in communications engineering. In the case of intersymbol interference present at the output of the receive filter, the following basic structures have proven to be useful:

Whitening Matched Filter
When using a whitening matched filter, the weighting function of the receive filter $g_e(t)$ is extended based on the matched-filter principle in such a way that the independence of the noise sample values at the output of the sampler can be guaranteed. The noise sample values are then white, hence the name of this filter concept. This is advantageous for receiver structures based on the MLSE (Maximum Likelihood Sequence Estimation) principle, such as the Viterbi detector [42].

Optimal Nyquist Filter

When using an optimal Nyquist filter, the weighting function of the receive filter $g_e(t)$ is extended based on the matched-filter principle in such a way that an ISI-free transmission is present at its output. The use of an optimal Nyquist filter thus leads to an interference-free useful signal transmission: There are no intersymbol interferences at the detection point. However, the sample values are generally disturbed by colored noise.

2.3   Multi-Level Baseband Transmission

2.3.1   Introduction and Motivation

After the basic model of information transmission for unipolar baseband transmission with rectangular NRZ transmit signals and RC first-order low-pass receive filter has been used to describe important relationships and the methodology for analyzing and optimizing transmission systems in detail, the consideration will now be extended to the case of multi-level transmission in the baseband. The alternative filter concepts discussed in the previous section will also be used.

Multi-level transmission is of interest in principle in two cases:

- With a constant bit rate or bit sequence frequency f_B, an increasing number of levels leads to a reduction in the symbol rate or symbol clock frequency f_T, and thus to a reduction in the bandwidth of the information-carrying signal. This can be advantageously used when the bandwidth of the transmit signal must be limited while maintaining the same bit rate requirements, either due to regulatory requirements, for example, when allocating frequency ranges, or due to the transmission medium itself limiting the usable bandwidth, for example, due to the increasing attenuation of a cable with increasing frequency. In these cases, it is necessary or advantageous to concentrate the transmission spectrum on a given frequency range or a frequency range with particularly low attenuation. With the knowledge from the previous sections, it can be determined that a reduction of the frequency range used for the signal requires spectrally narrower receive filters and reduces the noise power effective at the detection point. With limited average transmission power or a given amplitude range of the transmitter, the multi-level nature of the useful signal transmission means that the eye opening remaining at the detection point is reduced.
  The question therefore arises as to whether the decreasing vertical eye opening with increasing number of stages can be compensated for by the saving in bandwidth and the associated smaller noise power at the decision input.
- With a constant symbol rate or symbol clock frequency f_T and thus constant transmission bandwidth of the signal, an increase in the number of levels leads to a higher transferable bit rate or bit sequence frequency f_B—in a constant bandwidth. With the knowledge from the previous sections, it becomes clear that with the same used

frequency range, the noise power effective at the detection point remains the same. With limited average transmission power or a given modulation range, the multi-levels nature of the useful signal transmission again means that the usable vertical eye opening at the detection point is reduced.

In this case, the question arises as to how strongly the decreasing vertical eye opening with increasing number of levels affects the quality of the transmission, i.e. the signal-to-noise ratio. The noise power at the decision input remains unchanged, as the occupied bandwidth does not change.

These dependencies will be described and examined in detail in the following using the known system-theoretical and mathematical means.

2.3.2 Model of the Multi-Level Baseband Transmission System

The model for baseband transmission needs to be supplemented with regard to multi-level transmission: Fig. 2.60 shows the block diagram of the baseband transmission system. The component *Multi-level encoder* in the transmitter and the corresponding component *Multi-level decoder* in the receiver have been added. The multi-level encoder converts a group of bits into a symbol that originates from an *s*-level alphabet. The number of levels *s* of the transmission is often advantageously chosen as a power of two, as this practically allows a simple bit-symbol assignment of the binary bits to the *s*-valued symbols. The multi-level decoder reverses this assignment after the decision in the receiver and assigns a bit sequence to the received and decided symbol, so that a bit sequence can be handed over to the sink for use or possibly for further processing. The detector now has the task of deciding which of the *s* different symbols was sent.

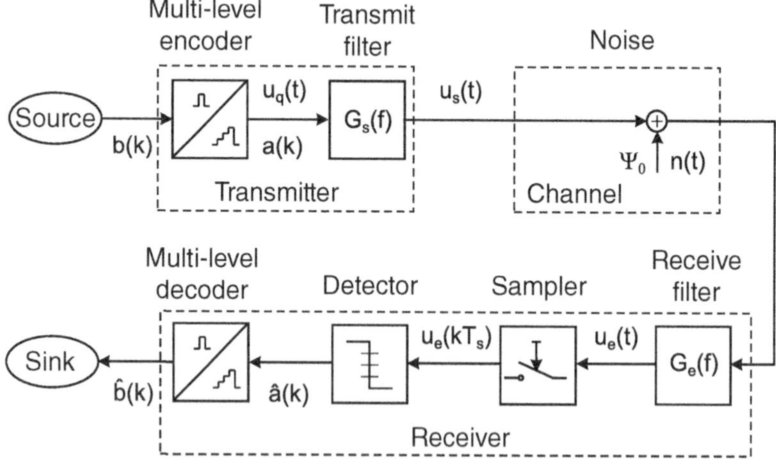

Fig. 2.60 Transmission system of multi-level information transmission in the baseband

Previously—in the two-level transmission—it only had to decide whether a '0' or a '1' was sent. All other components of the transmission system are already known from the previous sections.

Figure 2.60 shows the multi-level baseband transmission system: The digital message to be transmitted is generally initially in the form of a sequence $b(k)$ of binary signals (bits) with the elements $b[k] \in \{0, 1\}$, which is converted into an s-level source symbol sequence $a(k)$ by a mapping rule.[14] This source symbol sequence, assumed to be redundancy-free, is physically represented by the source signal

$$u_q(t) = U_s T_s \sum_{k=-\infty}^{+\infty} a[k] \cdot \delta(t - k T_s) \, . \tag{2.190}$$

The digital source emits weighted Dirac impulses $U_s T_s \delta(t)$ with dimensionless coefficients $a[k]$ at the symbol interval T_s; U_s denotes the half-level transmit amplitude. At the output of the transmit filter $G_s(f) = \mathcal{F}\{g_s(t)\}$, the transmit signal

$$u_s(t) = U_s T_s \sum_{k=-\infty}^{+\infty} a[k] \cdot g_s(t - k T_s) \tag{2.191}$$

suitable for transmission on the transmission channel in terms of spectrum and power is present. The signal arriving at the receiver input is composed additively of the useful signal at the transmit filter output and the Gaussian noise signal $n(t)$. The receive filter with the transfer function $G_e(f) = \mathcal{F}\{g_e(t)\}$ serves to shape the basic receive impulse and to limit the noise power. At the output of the overall channel

$$H(f) = G_s(f) \cdot G_e(f) \quad \bullet\!-\!\circ \quad h(t) = g_s(t) * g_e(t), \tag{2.192}$$

which includes all stages of the transmitter, the physical transmission channel, and the receiver that influence the original source signal spectrum. The received useful signal

$$u_e(t) = U_s T_s \sum_{k=-\infty}^{+\infty} a[k] \cdot h(t - k T_s) \tag{2.193}$$

is obtained. After sampling at the receive filter output at the symbol interval T_s, detection takes place. The sampling clock is determined by a receiver-side clock recovery from the received pulse sequence; in the following, the sampling clock is assumed to be known in the receiver (ideal clock synchronization). The estimated received symbol sequence $\hat{a}(k)$ is finally converted into the reconstructed received bit sequence $\hat{b}(k)$.

---

[14] For an even number of levels s, the (dimensionless) amplitude coefficients $a[k]$ come from the value set $\{\pm 1, \pm 3, \ldots, \pm(s - 1)\}$, while for an odd number of levels $a[k] \in \{0, \pm 2, \pm 4, \ldots, \pm(s - 1)\}$ applies.

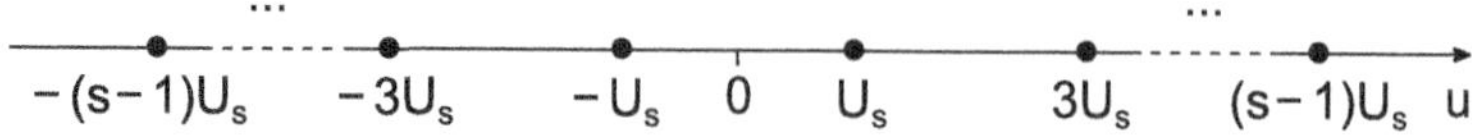

Fig. 2.61 Signal space representation of an s-level baseband constellation

With this type of representation and modeling of the signal, a bipolar[15] transmission with the amplitude levels $\pm U_s, \pm 3U_s, \ldots, \pm(s-1)U_s$, so that U_s is referred to as *half-level amplitude*. A suitable and common representation is the signal space diagram shown in Fig. 2.61: A signal space diagram is a representation in which the possible transmission amplitude levels—usually as points—are shown to get an overview of the number and arrangement of these amplitude stages. The individual elements $u_q[k]$ of the source signal according to (2.190) are obtained by multiplying the respective symbols $a[k]$ with the half-level amplitude U_s according to

$$u_q[k] = U_s \cdot a[k], \tag{2.194}$$

with $a[k] \in \{\pm 1, \pm 3, \ldots, \pm(s-1)\}$.

A bipolar transmission can have practical advantages over a unipolar transmission, e.g. in terms of the average transmission power to be expended or the absence of direct current. Figure 2.62 exemplarily shows signal sections for two- and four-level transmission with NRZ rectangular pulses, assuming the same modulation range, which ranges from $A_{min} = -(s-1)U_s$ to $A_{max} = (s-1)U_s$. It becomes clear that, assuming the same bit rate, the duration of the symbols is increased with a higher number of levels, but the distance between the individual amplitude levels, which is important for the reliability of the decision at the detection point, is reduced.

These qualitative insights are now to be formulated generally and mathematically: The clock frequency f_T, with which symbols are transmitted, depends on the number of levels s and is related to the bit sequence frequency f_B via

$$f_T = \frac{f_B}{\text{ld}(s)} . \tag{2.195}$$

Here, $\text{ld}(s)$ denotes the *dyadic* logarithm, i.e., the logarithm to the base 2: $\text{ld}(s) = \log_2(s)$. The duration of a symbol $T_s = 1/f_T$ and the bit duration $T_b = 1/f_B$ are therefore linked via

$$T_s = T_b \cdot \text{ld}(s) \tag{2.196}$$

---

[15] In a *bipolar* transmission, amplitude levels or voltage states with both polarities are used, i.e., the transmit signals are sent—depending on the symbol—with positive or negative voltage (in contrast to *unipolar* transmission). The amplitude levels are usually symmetric to the horizontal axis and thus to 0 V.

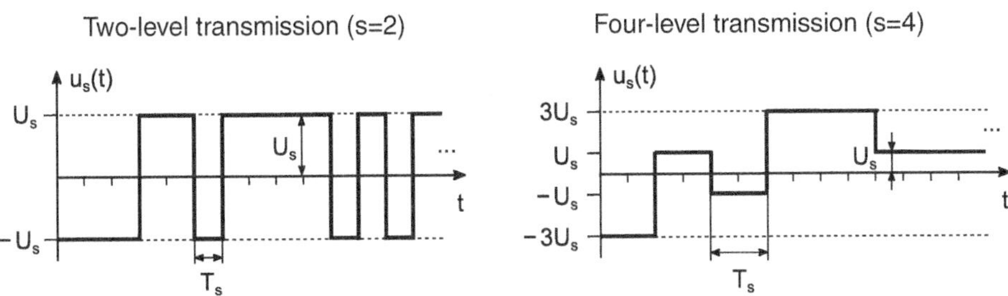

Fig. 2.62 Multi-level transmit signals in baseband transmission (examples: $s = 2$ and $s = 4$ with constant bit sequence frequency and predetermined maximum modulation range in the transmitter)

For the cases described at the beginning of the section on multi-level transmission, where multi-level transmission can be advantageously used, the conditions can now be specified more precisely:

- With a constant bit rate or bit sequence frequency f_B, the bit duration $T_b = 1/f_B$ is constant.
- With a constant symbol rate or clock frequency f_T, the symbol duration $T_s = 1/f_T$ is constant.

These quantities are interchangeable and can be converted into each other via (2.195).

2.3.3   Evaluation of the Useful Signal in Multi-Level Transmission

The useful signal is evaluated on its way to the detection point by the series connection of transmit and receive filter transfer functions according to

$$H(f) = G_s(f) \cdot G_e(f) \,. \tag{2.197}$$

If a Dirac impulse $x_q(t) = U_s\, T_s\, \delta(t) \circ\!\!-\!\!\bullet X_q(f) = U_s\, T_s$ passes through the transmit filter $g_s(t)$, the basic transmission impulse is obtained

$$x_s(t) = x_q(t) * g_s(t) = U_s\, T_s\, g_s(t) \quad \circ\!\!-\!\!\bullet \quad X_s(f) = U_s\, T_s\, G_s(f). \tag{2.198}$$

The basic receive impulse present at the detection point

$$x_e(t) = x_q(t) * h(t) = U_s\, T_s\, h(t) \quad \circ\!\!-\!\!\bullet \quad X_e(f) = U_s\, T_s\, H(f) \tag{2.199}$$

results from the evaluation of the Dirac pulse $x_q(t) = U_s\, T_s\, \delta(t)$ through the cascade of transmit and receive filter ($H(f) = G_s(f) \cdot G_e(f)$).

Fig. 2.63 Eye diagram
at the detection point with
intersymbol interference
($s = 2$)

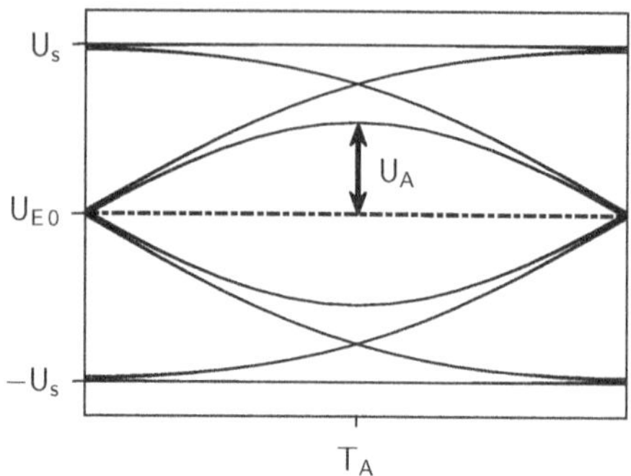

If a redundancy-free, bipolar source is assumed, an eye diagram is obtained that is symmetric to the horizontal axis—i.e., the time axis. If the basic receive impulses $x_e(t)$ are also temporally symmetric, an eye diagram is created that is also symmetric to the vertical axis—i.e., the amplitude axis. The optimal sampling time is then at $T_A = 0$. If s-level bipolar transmit signals with half the level amplitude U_s are used, an eye diagram with $(s-1)$ eyes(openings) and $(s-1)$ thresholds $U_{E\nu}$ (with $\nu = -(s/2 - 1), \ldots, (s/2 - 1)$, $\nu \in \mathbb{G}$) is obtained. The decision about the symbols is made with respect to the threshold $U_{E\nu}$, which lies in the middle between the two inner lines of the ν-th eye. As a measure of the quality of the useful signal at the detection point, the *half vertical eye opening U_A* is also used in multi-level transmission.

Figure 2.63 shows an eye diagram with $s = 2$ transmit amplitude levels with remaining intersymbol interference at the detection point, which occurs with NRZ rectangular pulse transmission and Gaussian receive filtering (example: $f_G = 0.45/T_s$). In contrast, Fig. 2.64 shows an eye diagram with $s = 4$ amplitude levels without intersymbol interference; the first Nyquist criterion is met: Such an eye diagram results, for example, from root cosine roll-off transmit and receive filtering (example: $r = 0.5$) and non-distorting transmission channel.

The inner boundary lines of an eye—especially when transmitting with remaining intersymbol interference, see Fig. 2.63—occur in the eye diagram when an unfavorable

Fig. 2.64 Eye diagram at
the detection point without
intersymbol interference
($s = 4$)

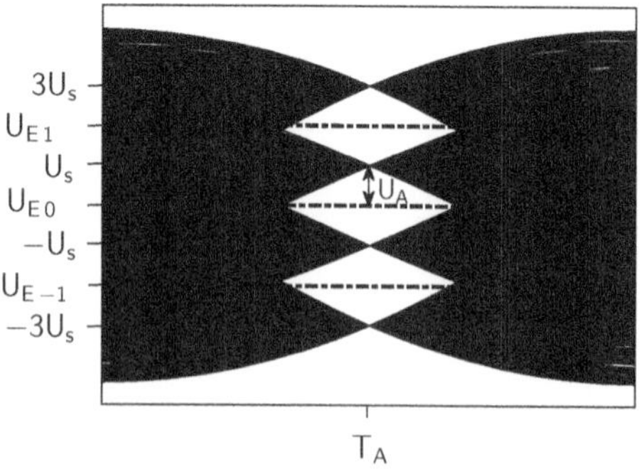

symbol sequence is transmitted, in which generally all precursors and successors of the neighboring pulses contribute to a reduction of the usable receive sampling value at the detection point at the optimal sampling time T_A [65]. To calculate the half vertical eye opening for the worst case, the middle eye with $U_{E0} = 0$ can be used, since the vertical eye opening for all eyes of a multi-level system is the same [44] and the determination of the half vertical eye opening U_A based on the middle eye often simplifies the calculation. The minimum half vertical eye opening occurs with received base pulses with monotonically progressing pulse edges by comparing a sampling value of the received signal $u_e(t)$ with the threshold U_{E0} when a pulse with the smallest positive amplitude $x_e(0)$ is to be detected between many consecutive pulses with the largest negative amplitude $-x_e(0) \cdot (s - 1)$ or vice versa a pulse with the smallest negative amplitude $-x_e(0)$ within a long sequence of pulses with the largest positive amplitude $x_e(0) \cdot (s - 1)$. In this case—with strong similarity to (2.28) and (2.31)—the half vertical eye opening results in bipolar, s-level transmission:

$$U_A = x_e(T_A) - \left[(s - 1) \cdot \sum_{\substack{k=-\infty \\ k \neq 0}}^{+\infty} |x_e(T_A - k\,T_s)| \right] - U_{E0}. \tag{2.200}$$

In bipolar transmission, the optimal threshold for the middle eye is at $U_{E0} = 0$ and (2.200) simplifies with temporally—i.e., with respect to the vertical amplitude axis—symmetrical received base pulse and resulting optimal sampling time $T_A = 0$ to [65]

$$U_A = x_e(0) - \left[2\,(s - 1) \cdot \sum_{k=1}^{\infty} |x_e(k\,T_s)| \right]. \tag{2.201}$$

With increasing number of levels s and specification of a maximum average transmission power or a maximum modulation amplitude of the transmitter, the vertical eye opening decreases: The useful signal at the detection point becomes more sensitive to interference. The vertical eye opening decreases with s-level transmission by division into $(s - 1)$ eyes [34]; the bandwidth required for signal transmission also decreases according to (2.195).

Example 2.8 (Calculation of the half vertical eye opening) A simple example for the calculation of the half vertical eye opening in bipolar, s-level transmission is considered: The received base pulse $x_e(t)$ is shown in Fig. 2.65. It is symmetric to the vertical axis. When sampled at the symbol rate, it has a main value of $x_e(0) = U_s$ at $t = 0$ and two contributions each with the amplitude $U_s/5$ at $\pm T_s$. All other symbol rate samples are zero. In two-level transmission, i.e. $s = 2$, the application of (2.201) results in

$$U_A = x_e(0) - \left[2\,(s - 1) \cdot \sum_{k=1}^{\infty} |x_e(k\,T_s)| \right] = x_e(0) - [2(2 - 1) \cdot x_e(T_s)] \tag{2.202}$$

Fig. 2.65 Triangular received base pulse for the example of calculating the half vertical eye opening

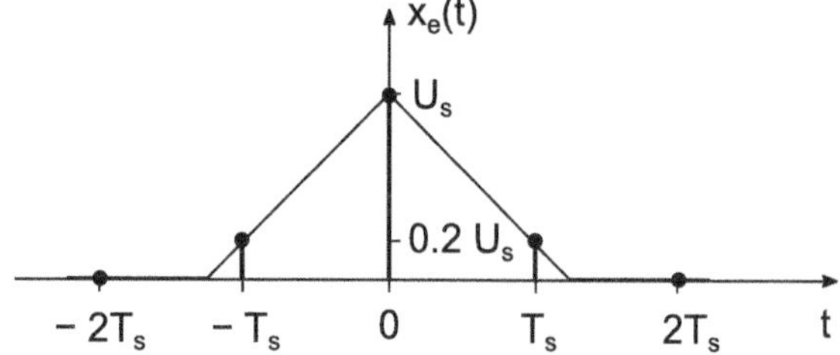

and we obtain

$$U_A = U_s - [2 \cdot 0.2\, U_s] = 0.6\, U_s \tag{2.203}$$

as the half vertical eye opening.

Figure 2.66 shows the practical verification of this calculation of the half vertical eye opening in two-level, bipolar transmission: In the worst case, the two adjacent pulses contribute to the reduction of the vertical eye opening, i.e., they occur at the positive considered pulse with maximum U_s at $t = 0$ each at $\pm T_s$ with negative amplitude $-U_s$. In this way, the amplitude of U_s at $t = 0$ is reduced by the amount $2 \cdot 0.2\, U_s$ and the previously calculated value of $U_A = 0.6\, U_s$ results as the half vertical eye opening.

If, on the other hand, a multi-level transmission is chosen, e.g. with $s = 4$ under otherwise identical boundary conditions, the same method results in

$$U_A = x_e(0) - \left[2\,(s-1) \cdot \sum_{k=1}^{\infty} |x_e(k\,T_s)| \right] = x_e(0) - [2(4-1) \cdot x_e(T_s)] \tag{2.204}$$

and we now obtain

$$U_A = U_s - [2 \cdot 3 \cdot 0.2\, U_s] = -0.2\, U_s \tag{2.205}$$

as the half vertical eye opening: This means that the eye is already closed at the detection point. For $s = 4$ (and larger level numbers), a transmission system with such a received base pulse is not suitable for reliable information transmission.

Fig. 2.66 On the creation of the half vertical eye opening for the worst case with triangular received base pulse ($s = 2$)

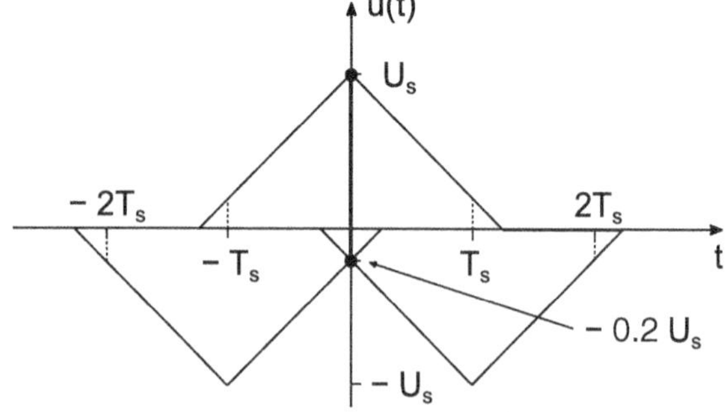

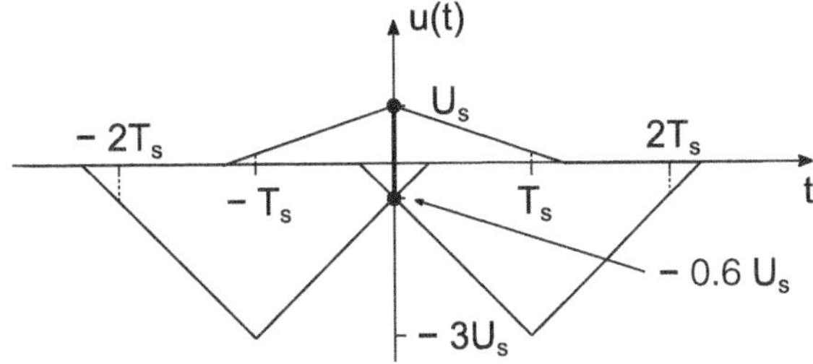

Fig. 2.67 On the creation of the half vertical eye opening for the worst case with a triangular receive base impulse ($s = 4$)

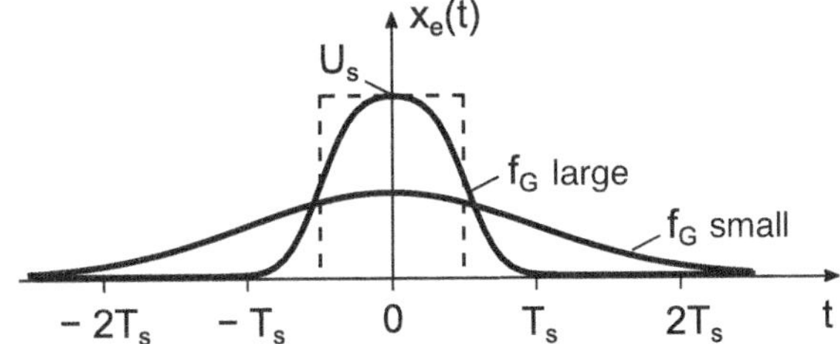

Fig. 2.68 Response to a rectangular impulse of a Gaussian low-pass filter (basic receive pulse) at high ($f_G = 1/T_s$) and low ($f_G = 0.2/T_s$) cutoff frequency

Figure 2.67 illustrates the creation of the already calculated half vertical eye opening in four-level, bipolar transmission: In the worst case, the two adjacent pulses contribute to the maximum reduction of the vertical eye opening, i.e., they occur at the positive considered pulse with maximum U_s at $t = 0$ each at $\pm T_s$ with the largest possible negative amplitude, i.e. with $-3\,U_s$. The amplitude of U_s at $t = 0$ is now reduced by the amount $2 \cdot 3 \cdot 0.2\,U_s$ and the already calculated value of $U_A = -0.2\,U_s$ results as the half vertical eye opening. ☐

The following discusses the two cases of multi-level baseband transmission

- with NRZ rectangular transmit impulses and receive filtering with Gaussian low-pass filter and
- with root cosine roll-off transmit impulses and receive filtering with root cosine roll-off low-pass filter

as examples (Fig. 2.68, 2.69).

Vertical Eye Opening for Multi-Level Transmission with NRZ Transmission Impulses and Gaussian Receive Filtering

Figure 2.70 shows eye diagrams with NRZ rectangle impulse transmit signal and Gaussian receive filtering for three exemplary level numbers and assumed same signal modulation range. It becomes clear that the half vertical eye opening decreases with increasing level number s. To capture the quality of the useful signal, the half vertical eye opening is calculated for s-level transmission.

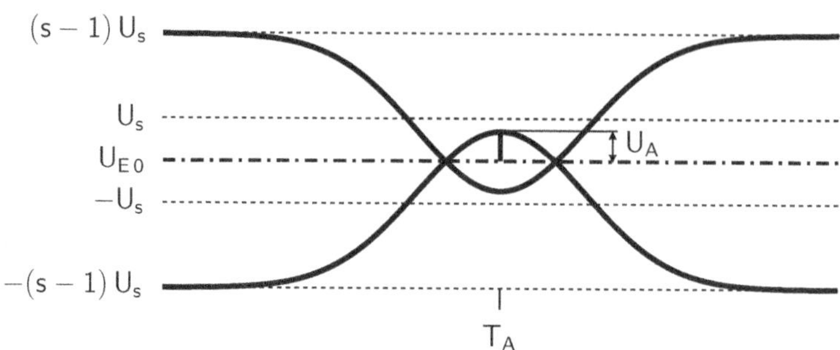

Fig. 2.69 On the calculation of the eye opening in multi-level transmission (Example: $s = 4$)

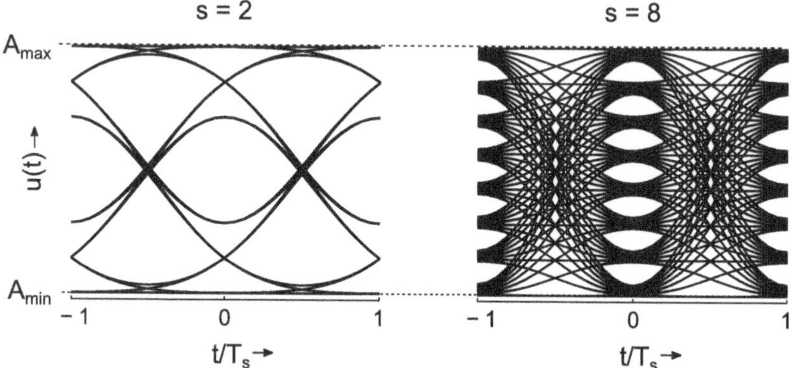

Fig. 2.70 Eye diagrams of multi-level signals in baseband transmission with remaining intersymbol interference (Examples: $s = 2$ and $s = 8$)

For the special case of NRZ pulse transmit and receive filtering with a Gaussian low-pass filter with the cutoff frequency f_G according to (2.85), the receive base impulse according to (2.89), now with the amplitude U_s, is effective at the detection point according to

$$x_e(t) = \frac{U_s}{2}\left[\mathrm{erf}\left(\frac{\pi f_G}{\sqrt{\ln(2)}}\left(t + \frac{T_s}{2}\right)\right) - \mathrm{erf}\left(\frac{\pi f_G}{\sqrt{\ln(2)}}\left(t - \frac{T_s}{2}\right)\right)\right]. \quad (2.206)$$

(see Fig. 2.68). The half vertical eye opening according to (2.200) simplifies for general s-level, bipolar transmission to

$$U_A = U_s\left[s \cdot \mathrm{erf}\left(\frac{\pi f_G}{\sqrt{\ln(2)}}\left(\frac{T_s}{2}\right)\right) - (s - 1)\right], \quad (2.207)$$

since in a long sequence of positive or negative transmission impulses—as the basis for calculating the half vertical eye opening for the worst case—the voltage at the detection point completely settles to $+(s - 1)U_s$ or $-(s - 1)U_s$.

Insert: Derivation of the Half Vertical Eye Opening in Multi-Level Transmission with NRZ Transmit Pulses and Gaussian Receive Filtering

If a Gaussian low-pass filter is fed with a bipolar, random, uniformly distributed NRZ rectangular pulse sequence with two amplitude levels $\pm U_s$, the eye diagram according to Fig. 2.63 is obtained at the filter output. The half vertical eye opening U_A is determined by the inner boundary lines of the eye diagram; therefore, for a calculation, the detection of a single positive pulse amidst a large number of negative pulses (or vice versa) must be considered. In the special case of the transmission of NRZ pulses, the filter output signal oscillates to a constant voltage ($+U_s$ or $-U_s$) during a long transmission pulse sequence of amplitude $+U_s$ or $-U_s$. In binary, bipolar transmission, the rectangular response of a Gaussian low-pass filter $x_e(t)$ can be written for the half vertical eye opening

$$U_A = 2\,x_e(0) - U_s \tag{2.208}$$

according to (2.206).[16] The function value $x_e(0)$ of the basic receive pulse (2.206)

$$x_e(0) = \frac{U_s}{2}\left[\mathrm{erf}\left(\frac{\pi f_G}{\sqrt{\ln(2)}}\left(0 + \frac{T_s}{2}\right)\right) - \mathrm{erf}\left(\frac{\pi f_G}{\sqrt{\ln(2)}}\left(0 - \frac{T_s}{2}\right)\right)\right]$$
$$x_e(0) = \frac{U_s}{2}\left[\mathrm{erf}\left(\frac{\pi f_G}{\sqrt{\ln(2)}}\left(\frac{T_s}{2}\right)\right) - \mathrm{erf}\left(\frac{\pi f_G}{\sqrt{\ln(2)}}\left(-\frac{T_s}{2}\right)\right)\right] \tag{2.209}$$

can, since $\mathrm{erf}(x)$ is an odd function and thus $\mathrm{erf}(x) = -\mathrm{erf}(-x)$ applies (see Appendix E and, for example, [66]), be simplified to

$$x_e(0) = \frac{U_s}{2}\left[2 \cdot \mathrm{erf}\left(\frac{\pi f_G}{\sqrt{\ln(2)}}\left(\frac{T_s}{2}\right)\right)\right]$$
$$x_e(0) = U_s\,\mathrm{erf}\left(\frac{\pi f_G}{\sqrt{\ln(2)}}\left(\frac{T_s}{2}\right)\right) \tag{2.210}$$

can be simplified. Thus, according to (2.208)

---

[16] The basic receive pulse (2.206) itself ranges in amplitude between 0 and U_s (see Fig. 2.68); however, the transmit signal in two-level, bipolar transmission ranges between $-U_s$ and $+U_s$ (see, for example, Fig. 2.62), thus with doubled amplitude. Therefore, in (2.208), the basic receive pulse must first be doubled so that the resulting pulse then ranges between 0 and $2\,U_s$. To calculate the half vertical eye opening U_A, the half level amplitude U_s must be subtracted from the value of the (doubled) basic receive pulse at the optimal sampling time, i.e., at $t = T_A = 0$, in (2.208). The vertical eye opening can take a maximum value of U_s according to (2.208)—at a very high receive filter cutoff frequency f_G. This approach to calculating the half vertical eye opening is possible because the useful signal—and thus the eye diagram—does not change in magnitude with a vertical shift.

$$U_A = 2\,U_s\,\mathrm{erf}\!\left(\frac{\pi f_G}{\sqrt{\ln(2)}}\left(\frac{T_s}{2}\right)\right) - U_s$$

$$U_A = U_s\left[2\cdot\mathrm{erf}\!\left(\frac{\pi f_G}{\sqrt{\ln(2)}}\left(\frac{T_s}{2}\right)\right) - 1\right]$$

$$(2.211)$$

the half vertical eye opening for bipolar binary transmission ($s = 2$) is obtained.

When extending to an s-level transmission under otherwise identical conditions, the middle eye (around the threshold $U_{E0} = 0\,\mathrm{V}$) is used for the calculation (Fig. 2.69).

The now valid approach

$$U_A = s\cdot x_e(0) - (s-1)\,U_s \tag{2.212}$$

leads with (2.210) to

$$U_A = U_s\left[s\cdot\mathrm{erf}\!\left(\frac{\pi f_G}{\sqrt{\ln(2)}}\left(\frac{T_s}{2}\right)\right) - (s-1)\right] \tag{2.213}$$

as the half vertical eye opening according to (2.207) for s-level baseband transmission.

In Fig. 2.70, eye diagrams of a baseband transmission with NRZ rectangular transmission pulses and Gaussian receive filtering are shown: At the detection point, intersymbol interferences remain, the eye is not fully open vertically in each case. It also becomes clear that an increasing number of levels s leads to a reduction of the eye opening. For the representation, an equal maximum amplitude range of $A_{\min} = -(s-1)U_s$ to $A_{\max} = (s-1)U_s$ was assumed for all numbers of levels.

Example 2.9 (Half vertical eye opening in the transmission of bipolar, s-level NRZ rectangular signals and receive filtering by a Gaussian low-pass filter) The half vertical eye opening is to be calculated according to the general relationship (2.201) and the result is to be compared with the compact formulation achieved by another method (2.207) for this case.

The basic receive impulse $x_e(t)$ according to (2.206) is non-negative and symmetric to the vertical axis. The optimal sampling time is $T_A = 0$ (see Fig. 2.68). According to (2.201), the result is

$$U_{\mathrm{A}} = x_{\mathrm{e}}(0) - \left[2\,(s-1)\cdot\sum_{k=1}^{\infty} |x_{\mathrm{e}}(k\,T_{\mathrm{s}})|\right] = x_{\mathrm{e}}(0) - \left[2\,(s-1)\cdot\sum_{k=1}^{\infty} x_{\mathrm{e}}(k\,T_{\mathrm{s}})\right],$$

$$(2.214)$$

under the assumption that the basic receive impulse considered here and especially its symbol clock sampling values only take positive values and thus $|x_{\mathrm{e}}(k\,T_{\mathrm{s}})| = x_{\mathrm{e}}(k\,T_{\mathrm{s}})$ applies (see Fig. 2.68). The first of the symbol clock sampling values are, using (2.206):

$$\begin{aligned}
x_{\mathrm{e}}(0) &= \frac{U_{\mathrm{s}}}{2}\left[\mathrm{erf}\left(\frac{\pi f_{\mathrm{G}}}{\sqrt{\ln(2)}}\left(0+\frac{T_{\mathrm{s}}}{2}\right)\right) - \mathrm{erf}\left(\frac{\pi f_{\mathrm{G}}}{\sqrt{\ln(2)}}\left(0-\frac{T_{\mathrm{s}}}{2}\right)\right)\right] \\
&= \frac{U_{\mathrm{s}}}{2}\left[\mathrm{erf}\left(\frac{\pi f_{\mathrm{G}}}{\sqrt{\ln(2)}}\left(\frac{T_{\mathrm{s}}}{2}\right)\right) - \mathrm{erf}\left(\frac{\pi f_{\mathrm{G}}}{\sqrt{\ln(2)}}\left(-\frac{T_{\mathrm{s}}}{2}\right)\right)\right] \\
&= \frac{U_{\mathrm{s}}}{2}\left[\mathrm{erf}\left(\frac{\pi f_{\mathrm{G}}}{\sqrt{\ln(2)}}\left(\frac{T_{\mathrm{s}}}{2}\right)\right) + \mathrm{erf}\left(\frac{\pi f_{\mathrm{G}}}{\sqrt{\ln(2)}}\left(\frac{T_{\mathrm{s}}}{2}\right)\right)\right] \\
&= U_{\mathrm{s}}\left[\mathrm{erf}\left(\frac{\pi f_{\mathrm{G}}}{\sqrt{\ln(2)}}\left(\frac{T_{\mathrm{s}}}{2}\right)\right)\right] \\
x_{\mathrm{e}}(T_{\mathrm{s}}) &= \frac{U_{\mathrm{s}}}{2}\left[\mathrm{erf}\left(\frac{\pi f_{\mathrm{G}}}{\sqrt{\ln(2)}}\left(T_{\mathrm{s}}+\frac{T_{\mathrm{s}}}{2}\right)\right) - \mathrm{erf}\left(\frac{\pi f_{\mathrm{G}}}{\sqrt{\ln(2)}}\left(T_{\mathrm{s}}-\frac{T_{\mathrm{s}}}{2}\right)\right)\right] \\
&= \frac{U_{\mathrm{s}}}{2}\left[\mathrm{erf}\left(\frac{\pi f_{\mathrm{G}}}{\sqrt{\ln(2)}}\left(\frac{3T_{\mathrm{s}}}{2}\right)\right) - \mathrm{erf}\left(\frac{\pi f_{\mathrm{G}}}{\sqrt{\ln(2)}}\left(\frac{T_{\mathrm{s}}}{2}\right)\right)\right] \\
x_{\mathrm{e}}(2T_{\mathrm{s}}) &= \frac{U_{\mathrm{s}}}{2}\left[\mathrm{erf}\left(\frac{\pi f_{\mathrm{G}}}{\sqrt{\ln(2)}}\left(2T_{\mathrm{s}}+\frac{T_{\mathrm{s}}}{2}\right)\right) - \mathrm{erf}\left(\frac{\pi f_{\mathrm{G}}}{\sqrt{\ln(2)}}\left(2T_{\mathrm{s}}-\frac{T_{\mathrm{s}}}{2}\right)\right)\right] \\
&= \frac{U_{\mathrm{s}}}{2}\left[\mathrm{erf}\left(\frac{\pi f_{\mathrm{G}}}{\sqrt{\ln(2)}}\left(\frac{5T_{\mathrm{s}}}{2}\right)\right) - \mathrm{erf}\left(\frac{\pi f_{\mathrm{G}}}{\sqrt{\ln(2)}}\left(\frac{3T_{\mathrm{s}}}{2}\right)\right)\right] \\
x_{\mathrm{e}}(3T_{\mathrm{s}}) &= \frac{U_{\mathrm{s}}}{2}\left[\mathrm{erf}\left(\frac{\pi f_{\mathrm{G}}}{\sqrt{\ln(2)}}\left(3T_{\mathrm{s}}+\frac{T_{\mathrm{s}}}{2}\right)\right) - \mathrm{erf}\left(\frac{\pi f_{\mathrm{G}}}{\sqrt{\ln(2)}}\left(3T_{\mathrm{s}}-\frac{T_{\mathrm{s}}}{2}\right)\right)\right] \\
&= \frac{U_{\mathrm{s}}}{2}\left[\mathrm{erf}\left(\frac{\pi f_{\mathrm{G}}}{\sqrt{\ln(2)}}\left(\frac{7T_{\mathrm{s}}}{2}\right)\right) - \mathrm{erf}\left(\frac{\pi f_{\mathrm{G}}}{\sqrt{\ln(2)}}\left(\frac{5T_{\mathrm{s}}}{2}\right)\right)\right]
\end{aligned}$$

$$\vdots$$

If these symbol clock sampling values are inserted into (2.214), the result for the half vertical eye opening is initially

$$U_\mathrm{A} = x_\mathrm{e}(0) - 2(s-1)(x_\mathrm{e}(T_\mathrm{s}) + x_\mathrm{e}(2T_\mathrm{s}) + x_\mathrm{e}(3T_\mathrm{s}) + \ldots)$$

$$U_\mathrm{A} = U_\mathrm{s}\left[\operatorname{erf}\left(\frac{\pi f_\mathrm{G}}{\sqrt{\ln(2)}}\left(\frac{T_\mathrm{s}}{2}\right)\right)\right] - \ldots$$

$$\ldots - 2(s-1)\frac{U_\mathrm{s}}{2}\left[\left(\operatorname{erf}\left(\frac{\pi f_\mathrm{G}}{\sqrt{\ln(2)}}\left(\frac{3T_\mathrm{s}}{2}\right)\right) - \operatorname{erf}\left(\frac{\pi f_\mathrm{G}}{\sqrt{\ln(2)}}\left(\frac{T_\mathrm{s}}{2}\right)\right)\right) + \ldots\right.$$

$$\ldots + \left(\operatorname{erf}\left(\frac{\pi f_\mathrm{G}}{\sqrt{\ln(2)}}\left(\frac{5T_\mathrm{s}}{2}\right)\right) - \operatorname{erf}\left(\frac{\pi f_\mathrm{G}}{\sqrt{\ln(2)}}\left(\frac{3T_\mathrm{s}}{2}\right)\right)\right) + \ldots$$

$$\left.\ldots + \left(\operatorname{erf}\left(\frac{\pi f_\mathrm{G}}{\sqrt{\ln(2)}}\left(\frac{7T_\mathrm{s}}{2}\right)\right) - \operatorname{erf}\left(\frac{\pi f_\mathrm{G}}{\sqrt{\ln(2)}}\left(\frac{5T_\mathrm{s}}{2}\right)\right)\right) + \ldots\right]$$

$$U_\mathrm{A} = U_\mathrm{s}\left[\operatorname{erf}\left(\frac{\pi f_\mathrm{G}}{\sqrt{\ln(2)}}\left(\frac{T_\mathrm{s}}{2}\right)\right)\right] - \ldots$$

$$\ldots - (s-1)U_\mathrm{s}\left[\operatorname{erf}\left(\frac{\pi f_\mathrm{G}}{\sqrt{\ln(2)}}\left(\frac{3T_\mathrm{s}}{2}\right)\right) - \operatorname{erf}\left(\frac{\pi f_\mathrm{G}}{\sqrt{\ln(2)}}\left(\frac{T_\mathrm{s}}{2}\right)\right) + \ldots\right.$$

$$\ldots + \operatorname{erf}\left(\frac{\pi f_\mathrm{G}}{\sqrt{\ln(2)}}\left(\frac{5T_\mathrm{s}}{2}\right)\right) - \operatorname{erf}\left(\frac{\pi f_\mathrm{G}}{\sqrt{\ln(2)}}\left(\frac{3T_\mathrm{s}}{2}\right)\right) + \ldots$$

$$\left.\ldots + \operatorname{erf}\left(\frac{\pi f_\mathrm{G}}{\sqrt{\ln(2)}}\left(\frac{7T_\mathrm{s}}{2}\right)\right) - \operatorname{erf}\left(\frac{\pi f_\mathrm{G}}{\sqrt{\ln(2)}}\left(\frac{5T_\mathrm{s}}{2}\right)\right) + \ldots\right].$$

It can be seen that consecutive summands each contain the same terms, which are associated with different signs, which almost all cancel each other out when the infinite sum is continued. Only the terms that contain the argument $T_\mathrm{s}/2$ remain, as well as another term that depends on where the infinite sum is broken off. For a useful result, a sufficient number of summands must be considered. For this last remaining term (here when the summation is terminated after $x_\mathrm{e}(3T_\mathrm{s})$: $\operatorname{erf}\left(\frac{\pi f_\mathrm{G}}{\sqrt{\ln(2)}}\left(\frac{7T_\mathrm{s}}{2}\right)\right)$), it should be noted that this tends towards 1 as k increases in (2.201), since $\lim_{x\to\infty}\operatorname{erf}(x) = 1$ applies (see Appendix E). It remains

$$U_\mathrm{A} = U_\mathrm{s}\left[\operatorname{erf}\left(\frac{\pi f_\mathrm{G}}{\sqrt{\ln(2)}}\left(\frac{T_\mathrm{s}}{2}\right)\right)\right] - \ldots$$

$$\ldots - (s-1)U_\mathrm{s}\left[-\operatorname{erf}\left(\frac{\pi f_\mathrm{G}}{\sqrt{\ln(2)}}\left(\frac{T_\mathrm{s}}{2}\right)\right) + \underbrace{\operatorname{erf}\left(\frac{\pi f_\mathrm{G}}{\sqrt{\ln(2)}}\left(\frac{7T_\mathrm{s}}{2}\right)\right)}_{\approx 1}\right]$$

$$U_\mathrm{A} = U_\mathrm{s}\left[\operatorname{erf}\left(\frac{\pi f_\mathrm{G}}{\sqrt{\ln(2)}}\left(\frac{T_\mathrm{s}}{2}\right)\right)\right] - (s-1)U_\mathrm{s}\left[-\operatorname{erf}\left(\frac{\pi f_\mathrm{G}}{\sqrt{\ln(2)}}\left(\frac{T_\mathrm{s}}{2}\right)\right) + 1\right],$$

so that

$$U_\mathrm{A} = U_\mathrm{s}\left[\operatorname{erf}\left(\frac{\pi f_\mathrm{G}}{\sqrt{\ln(2)}}\left(\frac{T_\mathrm{s}}{2}\right)\right) + (s-1)\operatorname{erf}\left(\frac{\pi f_\mathrm{G}}{\sqrt{\ln(2)}}\left(\frac{T_\mathrm{s}}{2}\right)\right) - (s-1)\right]$$

$$(2.215)$$

or

$$U_A = U_s \left[s \cdot \mathrm{erf}\left(\frac{\pi f_G}{\sqrt{\ln(2)}} \left(\frac{T_s}{2} \right) \right) - (s - 1) \right] \tag{2.216}$$

results. If this result is compared with the result achieved by another method (2.207) for the half vertical eye opening, it becomes clear that both methods lead to identical formulations. $\square$

Vertical Eye Opening for Root Raised Cosine Transmit Pulses and Root Raised Cosine Receive Filtering

When a s-level, bipolar baseband transmission system with root raised cosine transmit pulses and root raised cosine receive filtering is used, a total useful signal transmission is created from the transmit filter input to the receive filter output, which meets the first Nyquist criterion: Therefore, there are no intersymbol interferences at the sampling points and it applies to the half vertical eye opening

$$U_A = U_s. \tag{2.217}$$

The entire half-level transmit amplitude U_s is available at the detection point as a half vertical eye opening U_A for decision making.

In Fig. 2.71, eye diagrams of a baseband transmission with root raised cosine transmit pulses and identical receive filtering are shown (each with the roll-off factor $r = 0.5$): At the detection point, no intersymbol interferences are effective at the optimal sampling point, the eye is fully open vertically. An increasing number of levels s also leads to a reduction in the vertical eye opening. Compared to Fig. 2.70, the difference can be seen that the amplitude range, in which the receiver must operate linearly, is larger than the range from $-(s-1)U_s$ to $(s-1)U_s$, since due to the cosine-roll-off band limitation in transmit and receive filter, an overshoot of the pulses occurs in the time domain.

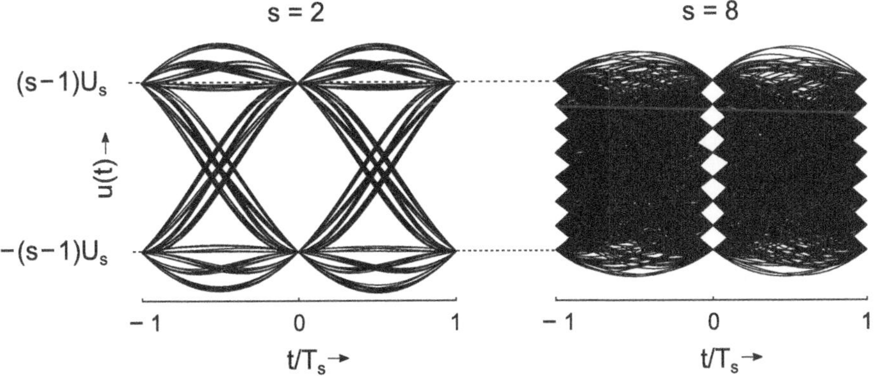

Fig. 2.71 Eye diagrams of multi-level signals in baseband transmission without intersymbol interferences (Examples: $s = 2$ and $s = 8$)

Examples of the Transmit Signal in Multi-Level Transmission

Example 2.10 (Transmit signal and signal space constellation of a rectangular, two-level, and bipolar baseband signal) A transmitter sends a rectangular, two-stage, and bipolar baseband signal $u_s(t)$ with the amplitude levels $\pm U_s$ and the pulse width T_s. According to the structure of the baseband transmitter unit shown in Fig. 2.72, the symbol set of the source signal applies: $u_q[k] \in \{-U_s, +U_s\}$.

Let's now transmit the data sequence $\dots 1\,0\,1\,\dots$ and sketch the characteristic of the baseband signal.

The resulting transmit signal for a rectangular pulse shaping (see also Fig. 2.73), i.e.

$$g_s(t) = \begin{cases} 1/T_s & |t| < T_s/2 \\ 0 & \text{otherwise} \end{cases}, \qquad (2.218)$$

is shown for a signal space constellation according to Fig. 2.74 in Fig. 2.75. The individual signal points generally correspond to the product of half-level amplitude and symbol to be transmitted: $U_s \cdot a[k]$.

For example, if a symbol duration of $T_s = 1\,\text{ms}$ is assumed, a symbol rate of $1000\,\text{Symbole/s}$ or a symbol clock frequency of $f_T = 1\,\text{kHz}$ results. Since each symbol in two-stage transmission ($s = 2$) transmits exactly one bit, a bit rate of $1000\,\text{bit/s}$ or a bit sequence frequency of $f_B = f_T \cdot \text{ld}\,s = 1\,\text{kHz} \cdot \text{ld}\,2 = 1\,\text{kHz}$ is obtained. $\square$

Example 2.11 (Transmit signal and signal space constellation of a rectangular, four-level, and bipolar baseband signal) Instead of a two-level input alphabet with

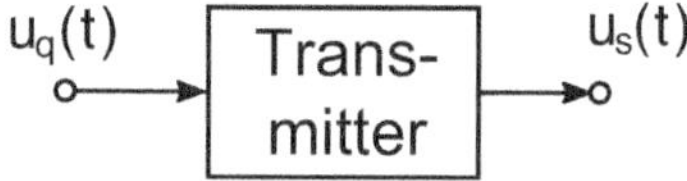

Fig. 2.72 Baseband transmission system: Generation of the transmit signal

Fig. 2.73 Weighting function of the transmit filter $g_s(t)$ (non-causal representation)

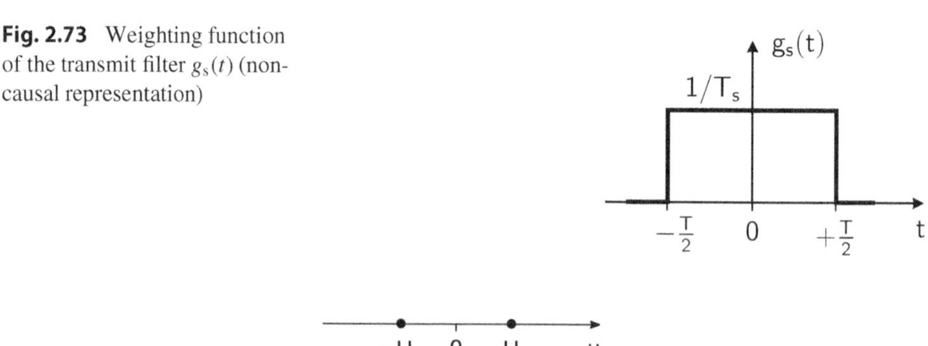

Fig. 2.74 Signal space constellation for two-level, bipolar baseband transmission ($s = 2$)

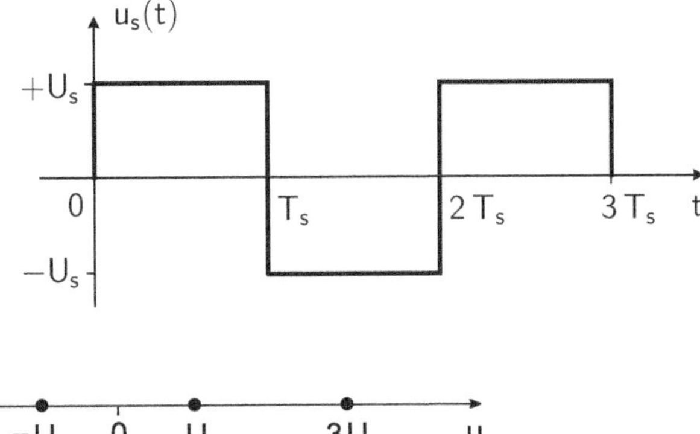

Fig. 2.75 Two-level transmit signal $u_s(t)$ with rectangular pulse shaping

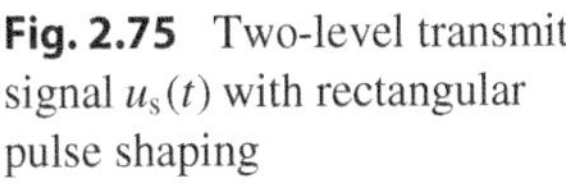

Fig. 2.76 Signal space constellation for four-level, bipolar baseband transmission ($s = 4$)

$u_q[k] \in \{-U_s, +U_s\}$, the four-level input alphabet $u_q[k] \in \{-3U_s, -U_s, +U_s, +3U_s\}$ outlined in Fig. 2.76 is used as a basis. The symbol duration is again $T_s = 1$ ms.

The data transmission rate (bit rate) of the system is now to be analyzed again. Comparing Figs. 2.74 with 2.76 shows that the number of points in the signal space has now doubled. Thus, each symbol now transmits 2 bit. Since the symbol duration has not changed, 1000 Symbol/s are also transmitted in this case ($f_T = 1$ kHz). However, the bit rate now increases to the value 2000 bit/s (bit sequence frequency of $f_B = f_T \cdot \mathrm{ld}\,s = 1$ kHz $\cdot$ ld4 $= 2$ kHz) since each symbol can now transmit two bits.

Due to the fact that the symbol duration has not changed, both baseband signals occupy the same bandwidth. The advantage of a higher bit rate is offset by a higher susceptibility to disturbance, as the distances between the signal points now decrease, i.e., U_s becomes smaller for example with the same amplitude range or the same transmission power (see, for example, Figs. 2.70 or 2.71). $\qquad\Box$

Average Transmit Power of a Multi-Level Baseband Signal

An important characteristic of a transmit signal is its average transmit power. It depends on the baseband signal constellation, i.e., the arrangement and number of amplitude levels, as well as the transmit pulse shape and thus the transmit filtering.

First, the influence of the s-level, random and uniformly distributed transmit signal amplitude levels on the transmission power is calculated for a redundancy-free source. An s-level, bipolar baseband constellation with the half level amplitude U_s is considered according to Fig. 2.77.

The average transmit power of this constellation can be formulated as

$$P_s = 2 \cdot \frac{1}{s}\left[(U_s)^2 + (3\,U_s)^2 + (5\,U_s)^2 + \ldots + ((s-1)\,U_s)^2\right]$$

$$P_s = 2 \cdot \frac{U_s^2}{s}\left[1^2 + 3^2 + 5^2 + \ldots + (s-1)^2\right]$$

$$\tag{2.219}$$

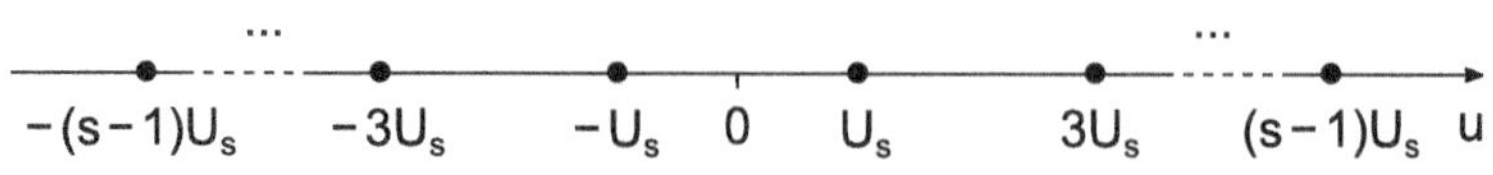

$$-(s-1)U_s \qquad -3U_s \qquad -U_s \quad 0 \quad U_s \qquad 3U_s \qquad (s-1)U_s \quad u$$

Fig. 2.77 Signal space diagram for calculating the average transmit power of a baseband constellation

Here, only the signal points at positive amplitude values were explicitly used; the negative signal points can be taken into account due to the symmetry of the arrangement by doubling the value calculated in this way. The relationship (2.219) can be expressed as

$$P_s = \frac{2\,U_s^2}{s} \sum_{v=1}^{s/2} (2v-1)^2 \qquad \text{for } s \text{ even} \tag{2.220}$$

It contains the sum of the sequence of odd natural numbers with the general solution [11, 74]

$$\sum_{k=1}^{n} (2k-1)^2 = \frac{n(4n^2-1)}{3}. \tag{2.221}$$

With the upper limit of the sum valid here

$$n = \frac{s}{2} \tag{2.222}$$

the solution required in (2.220) is obtained

$$\sum_{v=1}^{s/2} (2v-1)^2 = \frac{\frac{s}{2}\left(4\left(\frac{s}{2}\right)^2 - 1\right)}{3} = \frac{4}{3}\left(\frac{s}{2}\right)^3 - \frac{1}{3}\frac{s}{2} = \frac{1}{6}\left(s^3 - s\right). \tag{2.223}$$

This results in the average transmission power of the s-level baseband constellation as

$$P_s = \frac{2\,U_s^2}{s} \cdot \frac{1}{6}\left(s^3 - s\right) = \frac{2\,U_s^2}{s} \cdot \frac{s}{6}\left(s^2 - 1\right)$$
$$P_s = \frac{U_s^2}{3}\left(s^2 - 1\right). \tag{2.224}$$

If the transmit filtering $G_s(f)$ is now also taken into account, the actual (system-theoretical) average transmit power is obtained with the power spectral density of the transmit signal [26, 65]

$$\Psi_s(f) = \frac{U_s^2 T_s}{3}\left(s^2 - 1\right) |G_s(f)|^2. \tag{2.225}$$

to [65]

$$P_{\rm s} = \int\limits_{-\infty}^{+\infty} \Psi_{\rm s}(f)\,\mathrm{d}f = \frac{U_{\rm s}^2 T_{\rm s}}{3}\left(s^2 - 1\right) \int\limits_{-\infty}^{+\infty} |G_{\rm s}(f)|^2\,\mathrm{d}f. \tag{2.226}$$

For the average transmit power of transmit signals at the output of the transmit filter $G_{\rm s}(f)$, the portion $\int\limits_{-\infty}^{+\infty} |G_{\rm s}(f)|^2\,\mathrm{d}f$ added multiplicatively by the pulse shaping is still needed according to (2.226). This is for NRZ rectangular pulse shaping or root raised cosine filtering

$$\int\limits_{-\infty}^{+\infty} |G_{\rm s}(f)|^2\,\mathrm{d}f = \frac{1}{T_{\rm s}}, \tag{2.227}$$

so that the transmit power according to (2.224)

$$P_{\rm s} = \frac{U_{\rm s}^2 T_{\rm s}}{3}\left(s^2 - 1\right) \cdot \frac{1}{T_{\rm s}} = \frac{U_{\rm s}^2}{3}\left(s^2 - 1\right) \tag{2.228}$$

results.

For an NRZ transmit signal, this can be demonstrated comparatively easily in the time domain using the Parseval's theorem (e.g., [48]) via

$$\int\limits_{-\infty}^{+\infty} |G_{\rm s}(f)|^2\,\mathrm{d}f = \int\limits_{-\infty}^{+\infty} |g_{\rm s}(t)|^2\,\mathrm{d}t = \int\limits_{-T_{\rm s}/2}^{+T_{\rm s}/2} \left|\frac{1}{T_{\rm s}}\right|^2\,\mathrm{d}t = \left(\frac{1}{T_{\rm s}}\right)^2 \cdot T_{\rm s} = \frac{1}{T_{\rm s}} \tag{2.229}$$

while it can be shown for root raised cosine transmit pulses in the same way as for the filtering of white noise with root raised cosine receive low-pass filter, which leads to the result (2.109) (here in the calculation for $r = 0$):[17]

$$\int\limits_{-\infty}^{+\infty} |G_{\rm s}(f)|^2\,\mathrm{d}f = 2 \int\limits_{0}^{\infty} |G_{\rm N}(f)|\,\mathrm{d}f$$

$$= 2 \int\limits_{0}^{f_{\rm T}/2} 1^2\,\mathrm{d}f = 2\Big[f\Big]_{0}^{f_{\rm T}/2} = 2\left[\frac{f_{\rm T}}{2} - 0\right] = 2\frac{f_{\rm T}}{2} = f_{\rm T} = \frac{1}{T_{\rm s}}.$$

---

[17] The calculation is shown here specifically for the roll-off factor $r = 0$, but is valid for all roll-off factors $r = 0\ldots 1$: For the calculation, the magnitude square of the transmist filter transfer function $|G_{\rm s}(f)|^2$ is relevant and for root raised cosine filter with the real transfer function $\sqrt{G_{\rm N}(f)}$, $|G_{\rm s}(f)|^2 = \left|\sqrt{G_{\rm N}(f)}\right|^2 = \left(\sqrt{|G_{\rm N}(f)|}\right)^2 = |G_{\rm N}(f)| = G_{\rm N}(f)$ applies: Due to the previously explained point symmetry of the filter flanks of $G_{\rm N}(f)$, the result of (2.227) and thus the transmission power is independent of the roll-off factor r.

As a result, the average transmit power of transmit signals with NRZ rectangular pulse shaping or with root raised cosinefiltering corresponds to the average transmit power of the s-level constellation calculated according to (2.224). The average transmit power ultimately reads

$$P_s = \frac{U_s^2}{3}\left(s^2 - 1\right). \tag{2.230}$$

Insert: Relationship Between Average Transmit Power and Transmitter Amplitude Range

Now that the transmit power of multi-level baseband signals has been worked out, the relationship to the amplitude range of the transmitter, ranging from $A_{\min} = -(s-1)U_s$ to $A_{\max} = (s-1)U_s$, which was referred to, for example, in the representation of the eye diagram in Fig. 2.70, should be established. Strictly speaking, this amplitude range applies to transmit pulses without overshoot, e.g. for NRZ rectangular pulses; for transmit pulses with overshooting parts—such as root raised cosine pulses—this overshoot must be taken into account, and it also depends on the roll-off factor r. Such overshooting transmit pulses are not analyzed in this introductory consideration regarding the maximum amplitude range. For the calculation given here, transmit pulses without overshoot are assumed.

First, the positive maximum value $A_{\max} = (s-1)U_s$ is calculated, this is rearranged according to the half level amplitude $U_s = A_{\max}/(s-1)$ and inserted into (2.230). One obtains

$$P_s = \frac{A_{\max}^2}{3(s-1)^2}\left(s^2 - 1\right). \tag{2.231}$$

Using the binomial formula $a^2 - b^2 = (a+b)(a-b)$, $s^2 - 1 = (s+1)(s-1)$ can be written and it results

$$P_s = \frac{A_{\max}^2}{3(s-1)(s-1)}(s+1)(s-1) = \frac{A_{\max}^2}{3(s-1)}(s+1). \tag{2.232}$$

The transmit power P_s can therefore be expressed as a function of the maximum amplitude $A_{\max}$ as

$$P_s = \frac{A_{\max}^2}{3}\frac{(s+1)}{(s-1)} \tag{2.233}$$

and since $A_{\max}$ and $A_{\min}$ lie symmetrically to the horizontal axis in bipolar transmission, i.e. $A_{\min} = -A_{\max}$ applies, also

$$P_s = \frac{A_{\min}^2}{3}\frac{(s+1)}{(s-1)} \tag{2.234}$$

can be specified. Conversely, maximum and minimum amplitude can be written as a function of the transmit power:

$$A_{\max} = +\sqrt{3P_s \frac{(s-1)}{(s+1)}} \quad \text{or} \quad A_{\min} = -\sqrt{3P_s \frac{(s-1)}{(s+1)}}. \quad (2.235)$$

2.3.4   Evaluation of the Disturbance in Multi-Level Transmission

The noise signal $n(t)$ is evaluated on the way to the detection point, unlike the useful signal, only by the receive filter $G_e(f)$ (see Fig. 2.60) and also leads in multi-level transmission to the square of the effective value U_R of the noise signal as noise power according to (2.53).

Due to the dependence of the symbol clock frequency f_T on the number of levels s according to (2.195), with a constant bit rate or bit sequence frequency f_B and increasing number of levels s, a decreasing clock frequency f_T is obtained, so that the receive filter bandwidth decreases—and thus ultimately the noise power U_R^2 effective at the detection point decreases.

2.3.5   Quality of Transmission: Signal-to-Noise Ratio

The quality of transmission is again described by the signal-to-noise ratio according to (2.60). The discussed dependencies of the useful signal and the disturbance on the number of levels s are captured by the calculation of the eye opening and the noise power, so that the definition of the signal-to-noise ratio can be maintained and is also valid for the case of multi-level transmission.

In the considerations regarding the dependence of the quality of transmission on the receive filter cutoff frequency f_g, a counteracting dependence of the transmission quality on this cutoff frequency became clear: If f_g increases, the half vertical eye opening U_A of the useful signal at the detection point and also the noise power U_R^2 increase—these dependencies cause the counteracting dependence of the transmission quality on eye opening and noise power from the considered parameter f_g. Assuming, for example, an equal bit sequence frequency f_B, the dependence of the transmission quality on the number of levels s is similar: Here, with an increasing number of levels s, the half vertical eye opening U_A decreases on the one hand and the noise power U_R^2 on the other hand—and again a counteracting dependence of the transmission quality on eye opening and noise power from the now considered parameter s results.

These dependencies are each captured and represented by the signal-to-noise ratio ϱ according to (2.60).

The error probability resulting from the signal-to-noise ratio in multi-level transmission will be calculated in the following section.

2.3.6   Error Probability in Multi-Level Baseband Transmission

After the quality criteria signal-to-noise ratio and error probability for a two-level—i.e., binary—transmission have been introduced and thoroughly analyzed, this section will extend, in particular, the quality criterion of error probability to multi-level methods (see, for example, [26, 52]). The signal-to-noise ratio ϱ is, as discussed in the previous section, also suitable for multi-level transmission.

To determine the error probability in multi-level baseband transmission, only the range of positive amplitudes is used for calculation for reasons of symmetry. For example, one obtains for $s = 4$

$$P_{\mathrm{f}} = 2\,\frac{1}{s}\left[\int\limits_{0}^{\infty} p(u + U_{\mathrm{s}})\,\mathrm{d}u + \int\limits_{-\infty}^{2U_{\mathrm{s}}} p(u - 3U_{\mathrm{s}})\,\mathrm{d}u + \int\limits_{2U_{\mathrm{s}}}^{\infty} p(u - U_{\mathrm{s}})\,\mathrm{d}u\right] \qquad (2.236)$$

and generally

$$P_{\mathrm{f}} = 2\,\frac{1}{s}\left[\int\limits_{0}^{\infty} p(u + U_{\mathrm{s}})\,\mathrm{d}u + \int\limits_{-\infty}^{2U_{\mathrm{s}}} p(u - 3U_{\mathrm{s}})\,\mathrm{d}u + \dots \right.$$
$$\left. \dots + \int\limits_{(s-2)U_{\mathrm{s}}}^{\infty} p(u - (s - 3)U_{\mathrm{s}})\,\mathrm{d}u\right], \qquad (2.237)$$

if the highest amplitude level is $(s - 1)U_{\mathrm{s}}$ (Fig. 2.78). It is assumed that only signals of immediately adjacent amplitude levels lead to incorrect decisions; the influence of more

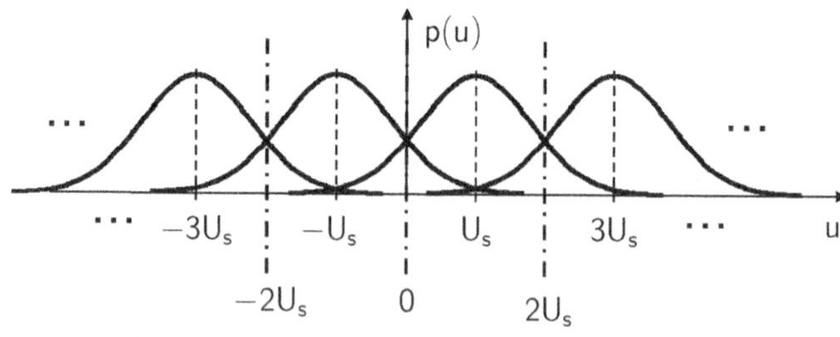

Fig. 2.78 For the calculation of the error probability of multi-level baseband transmission (example: $s = 4$)

distant amplitude levels is neglected. This is justified by the fact that in practically usable transmission systems, the noise power—and thus the width of the probability density functions of the noise signal—must be sufficiently small in relation to the distance of the amplitude levels in order to result in a usable signal-to-noise ratio. Due to the symmetry of all Gaussian probability density functions, i.e., with equal noise power for all amplitude levels, one obtains with

$$\int_{U_s}^{\infty} p(u)\,du = \frac{1}{2}\,\mathrm{erfc}\left(\sqrt{\frac{\varrho}{2}}\right) \tag{2.238}$$

—taking into account the fact that $(s-1)$ of these integrals occur (see, for example, (2.236))—the symbol error probability symbol error probability s-level baseband transmission as

$$P_f = 2 \cdot \frac{1}{s} \cdot (s-1) \cdot \frac{1}{2}\,\mathrm{erfc}\left(\sqrt{\frac{\varrho}{2}}\right)$$
$$P_f = \frac{s-1}{s}\,\mathrm{erfc}\left(\sqrt{\frac{\varrho}{2}}\right). \tag{2.239}$$

The relationship between symbol and bit error probability is largely determined by the type of bit-symbol assignment. If, for example, a Gray coding [6] is assumed, adjacent transmission symbols differ by one bit, so that the most likely symbol error—the misinterpretation of a received symbol as an adjacent transmission symbol—leads to only a bit error [5].

When using Gray coding[18], adjacent transmission symbols differ only in one bit position. Figure 2.79 exemplifies the resulting signal space diagram for the case $s = 8$.

The bit error probability is then for sufficiently large signal-to-noise ratios ϱ, which lead to one bit error per occurring symbol error, $\mathrm{ld}(s)$ lower than the symbol error probability, and it results

$$P_b = \frac{P_f}{\mathrm{ld}(s)} = \frac{s-1}{s\,\mathrm{ld}(s)}\left[1 - \mathrm{erf}\left(\sqrt{\frac{\varrho}{2}}\right)\right] = \frac{s-1}{s\,\mathrm{ld}(s)}\,\mathrm{erfc}\left(\sqrt{\frac{\varrho}{2}}\right). \tag{2.240}$$

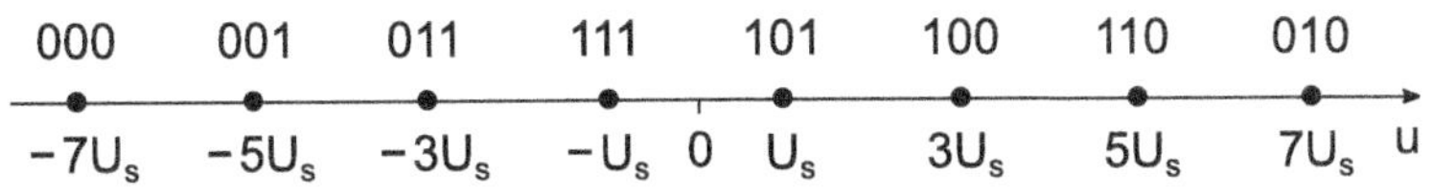

000   001   011   111   101   100   110   010

$-7U_s$ $-5U_s$ $-3U_s$ $-U_s$ 0 U_s $3U_s$ $5U_s$ $7U_s$ u

Fig. 2.79   Baseband constellation (example: $s = 8$) with Gray coding

---

[18] Frank Gray (1887–1969), American physicist.

2.3.7 Various Definitions of Signal-to-Noise Ratios and Their Relationship to Error Probability

Introduction and Basics

Previous considerations have shown that with an unchanged structure of the transmitter unit, the receive filtering leads to different sizes regarding ϱ and P_{f} or P_{b}. This circumstance is particularly interesting in multi-level transmission and critical for assessing the performance of transmission methods. The achievable transmission quality depends on many parameters, which in many cases also influence each other. The aim of an analysis or optimization is to make statements about how a transmission system should be designed so that it is as efficient as possible under given boundary conditions. For this, a fair and meaningful comparison of transmission methods and systems is indispensable. The following parameters should be considered:

- *Transmit power:* The transmit power directly influences the half vertical eye opening. Therefore, a higher transmit power is helpful for improved transmission quality, as in this case the distance between the signal points increases and the susceptibility to disturbances decreases under the influence of noise.
- *Symbol and bit sequence frequency:* The symbol sequence or clock frequency f_{T} significantly determines the bandwidth required for transmission together with the transmit filter used. The bit sequence frequency f_{B} results from the symbol sequence frequency and the level size s of the method.
- *Noise power density of the noise source:* The intensity of the disturbance should also be taken into account when comparing different transmission methods. It is usually given as the noise power density Ψ_0 of the noise source.

Definition of Comparison Criteria

Quality criteria are used to describe the performance of digital transmission systems. Different definitions are used. In addition to the signal-to-noise ratio at the detection point or at the receiver *output* defined in Eq. (2.60), the quotients at the receiver *input*

$$\varrho_{\mathrm{s}} = \frac{\text{symbol energy}}{\text{noise power spectral density}} = \frac{E_{\mathrm{s}}}{\Psi_0} \tag{2.241}$$

and

$$\varrho_{\mathrm{b}} = \frac{\text{bit energy}}{\text{noise power spectral density}} = \frac{E_{\mathrm{b}}}{\Psi_0} \tag{2.242}$$

have proven to be useful criteria for comparing digital transmission methods. They each represent a ratio of a useful signal energy (symbol or bit energy) and a noise power density and are therefore dimensionless quantities. They occur at the input of the receiver, e.g., before the receive filter. They each relate the signal energy (bit or symbol energy) present at the receiver input to the noise power density and do not directly provide information about the expected quality of the transmission. However, they can be advantageously used for the evaluation of digital transmission methods—e.g., given properties of the transmitter such as transmit power and constant bit sequence or clock frequency and the disturbance in the form of noise power spectral density. Thus, given the signal energy, the received signal may have experienced strong or weak signal distortions on the transmission path; only the quality of the subsequent signal processing (type of receive filtering and possibly the subsequent discrete signal processing) determine the quality of the transmission.

On the other hand, the signal-to-noise ratio ϱ in the definition according to (2.60) describes the expected quality of the transmission—it is also referred to as the detection signal-to-noise ratio, as it is effective at the detection point. It appears, for example, in the argument of the complementary error function for calculating the symbol and bit error probability, e.g., according to (2.80) and (2.240)—and it can be used directly to describe the quality of the transmission.

Figure 2.80 shows where in the transmission system the individual signal-to-noise ratios are each defined.

If one wants to achieve the highest possible output-side signal-to-noise ratio under the given conditions, the receive filter must be designed according to the matched filter principle in the case of disturbance by additive, white Gaussian noise. It is also of interest whether the resulting overall impulse response of the transmission system fulfills the first Nyquist criterion.

Example 2.12 (Signal-to-noise ratio definitions in two-level transmission) In the case of an intersymbol interference-free received signal at the detection point, i.e., at the receive filter output, the following applies

$$U_A = U_s, \tag{2.243}$$

i.e., the half vertical eye opening U_A corresponds to half-level transmit amplitude U_s. Together with the noise power U_R^2 effective after receive filtering at the sampling time according to (2.53), the signal-to-noise ratio results according to (2.60) with (2.243) to

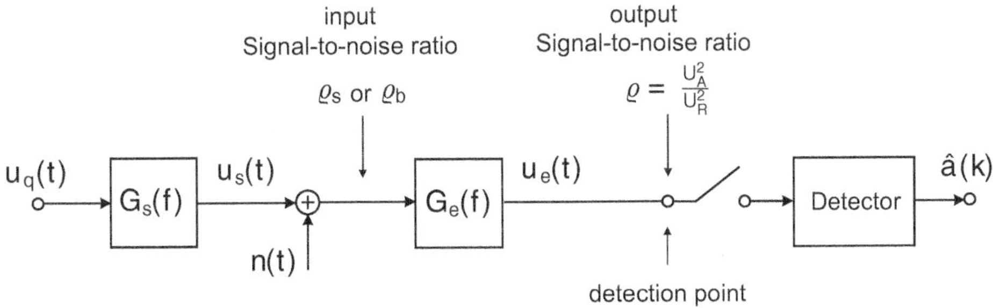

Fig. 2.80 For the definition of signal-to-noise ratios

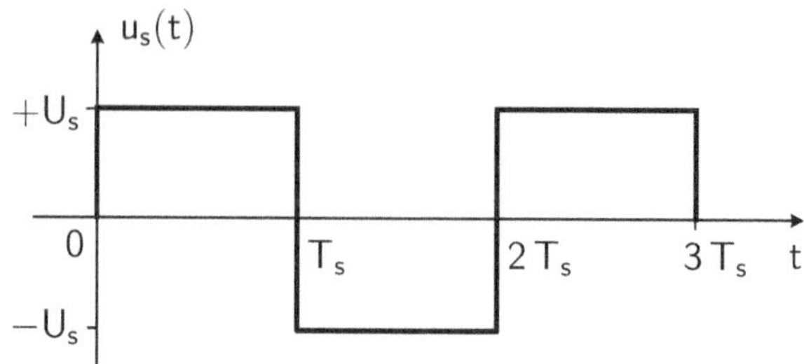

Fig. 2.81 Transmit signal $u_s(t)$ with rectangular pulse shaping

$$\varrho = \frac{U_A^2}{U_R^2} = \frac{U_s^2 \, T_s}{\Psi_0} = \frac{E_s}{\Psi_0}, \tag{2.244}$$

which is linked with the symbol error probability P_f for a redundancy-free source for s-level baseband transmission over

$$P_f = \frac{s-1}{s}\left[1 - \mathrm{erf}\left(\sqrt{\frac{\varrho}{2}}\right)\right] \tag{2.245}$$

[65].

In this case, the numerator in (2.244) in the form $U_s^2 \, T_s$ with reference to Fig. 2.81 corresponds to the symbol energy $E_s = P_s \, T_s$, which in the case of a two-level transmission coincides with the bit energy (i.e., $E_s = E_b$ for $s = 2$).

The signal-to-noise ratio at the receive filter output (i.e., the detection signal-to-noise power ratio) then corresponds in magnitude to the signal-to-noise ratio at the receive filter input. In this case, the entire symbol energy present at the receiver input can be used for decision-making at the detection point. $\square$

Comparison of Multi-Level Transmission Methods

In the following, digital transmission methods with different numbers of levels will be compared. It will be discussed which of the defined input-side signal-to-noise ratios is suitable under which conditions.

Comparison for Constant Symbol Clock Frequency If digital transmission methods are compared at the same clock frequency f_T and thus at a fixed ϱ_s, then according to (2.241)

$$\varrho_s = \frac{E_s}{\Psi_0} = \frac{P_s \cdot T_s}{\Psi_0}. \tag{2.246}$$

For all methods, a fixed (constant) transmit power P_s, a constant symbol duration T_s, and a constant noise power density Ψ_0 are then assumed, i.e., the properties of the transmitter unit and the noise source are constant. The use of ϱ_s is thus tied to a constant symbol rate or clock frequency f_T.

Example 2.13 (Comparison of transmission methods at the same symbol clock frequency f_T) Two transmission methods with different constellation sizes are to be compared and it is to be examined how the use of ϱ_s according to (2.241) and ϱ_b according to (2.242) affect this.

Method I: $T_s = 1\,\text{ms}$, $s = 2$, $\Longrightarrow T_{b1} = T_s/\text{ld}\,2 = T_s = 1\,\text{ms}$, thus resulting in a bit sequence frequency of $f_{B1} = 1/T_{b1} = 1\,\text{kHz}$

Method II: $T_s = 1\,\text{ms}$, $s = 4$, $\Longrightarrow T_{b2} = T_s/\text{ld}\,4 = T_s/2 = 0.5\,\text{ms}$, thus resulting in a bit sequence frequency of $f_{B2} = 1/T_{b2} = 2\,\text{kHz}$

The input-side signal-to-noise ratios are now for Method I

$$\varrho_{s1} = \frac{E_{s1}}{\Psi_0} = \frac{P_{s1} \cdot T_s}{\Psi_0} \tag{2.247}$$

and for Method II

$$\varrho_{s2} = \frac{E_{s2}}{\Psi_0} = \frac{P_{s2} \cdot T_s}{\Psi_0}. \tag{2.248}$$

If the two methods are to be compared at the same E_s/Ψ_0—which can be useful, for example, when determining the symbol or bit error probability—it becomes clear that the transmission power must be identical in both cases and it applies $P_{s1} = P_{s2}$: This is a sensible condition for a fair comparison. However, it should be noted that different bit sequence frequencies (bit rates) f_{B1} and f_{B2} are transmitted.

If the comparison of both methods were based on the size ϱ_b, the results for Method I

$$\varrho_{b1} = \frac{E_{b1}}{\Psi_0} = \frac{P_{s1} \cdot T_{b1}}{\Psi_0} \tag{2.249}$$

and for Method II

$$\varrho_{b2} = \frac{E_{b2}}{\Psi_0} = \frac{P_{s2} \cdot T_{b2}}{\Psi_0}. \tag{2.250}$$

would then have to apply for equality of ϱ_{b1} and ϱ_{b2}

$$P_{s1} \cdot T_{b1} = P_{s2} \cdot T_{b2} = P_{s2} \cdot \frac{T_{b1}}{2} \qquad \Longrightarrow \qquad P_{s2} = 2\,P_{s1}. \tag{2.251}$$

In a comparison based on $\varrho_b = E_b/\Psi_0$ at a constant clock frequency f_T, Method II would contain twice the transmission power of Method I: Therefore, a fair comparison is not possible on this basis. $\qquad\qquad\square$

Comparison for Constant Bit Sequence Frequency If, on the other hand, one wants to compare methods on the basis of a constant bit rate f_B, the quotient

$$\varrho_b = \frac{E_b}{\Psi_0} = \frac{P_s \cdot T_b}{\Psi_0} \tag{2.252}$$

according to (2.242) is to be used. This assumes for all methods a fixed (constant) transmission power P_s, a constant bit duration T_b and a constant noise power density Ψ_0, i.e. the properties of the transmission unit and the noise source are constant. The use of ϱ_b is thus tied to a constant bit rate or bit sequence frequency f_B.

Example 2.14 (Comparison of transmission methods at the same bit sequence frequency f_B) Two transmission methods with different constellation sizes are to be compared and it is to be examined how the use of ϱ_s and ϱ_b affect this.

Method I: $T_b = 1\,\text{ms}$, $s = 2$, $\Longrightarrow T_{s1} = T_b \cdot \text{ld}\,2 = T_b = 1\,\text{ms}$, thus resulting in a symbol sequence frequency of $f_{T1} = 1/T_{s1} = 1\,\text{kHz}$

Method II: $T_b = 1\,\text{ms}$, $s = 4$, $\Longrightarrow T_{s2} = T_b \cdot \text{ld}\,4 = T_b \cdot 2 = 2\,\text{ms}$, thus resulting in a symbol sequence frequency of $f_{T2} = 1/T_{s2} = 0.5\,\text{kHz}$

The input-side signal-to-noise ratios are now for Method I

$$\varrho_{b1} = \frac{E_{b1}}{\Psi_0} = \frac{P_{s1} \cdot T_b}{\Psi_0} = P_{s1} \cdot \frac{T_b}{\Psi_0} \tag{2.253}$$

and for Method II

$$\varrho_{b2} = \frac{E_{b2}}{\Psi_0} = \frac{P_{s2} \cdot T_b}{\Psi_0} = P_{s2} \cdot \frac{T_b}{\Psi_0}. \tag{2.254}$$

If the two methods are to be compared at the same E_b/Ψ_0—which can be useful, for example, when determining the symbol or bit error probability—it becomes clear that the transmission powers must be identical in both cases and it applies $P_{s1} = P_{s2}$: This is a sensible condition for a fair comparison. However, it should be noted that different symbol sequence frequencies (symbol rates) f_{T1} and f_{T2} are transmitted. This can be an important criterion, for example, given the bandwidth of the transmission channel.

If the comparison of both methods were carried out on the basis of the size ϱ_s, the results for Method I

$$\varrho_{s1} = \frac{E_{s1}}{\Psi_0} = \frac{P_{s1} \cdot T_{s1}}{\Psi_0} \tag{2.255}$$

and for Method II

$$\varrho_{s2} = \frac{E_{s2}}{\Psi_0} = \frac{P_{s2} \cdot T_{s2}}{\Psi_0}. \tag{2.256}$$

would then have to apply for equality of ϱ_{s1} and ϱ_{s2}

$$P_{s1} \cdot T_{s1} = P_{s2} \cdot T_{s2} = P_{s2} \cdot 2\,T_{s1} \qquad \Longrightarrow \qquad P_{s1} = 2\,P_{s2}. \tag{2.257}$$

In a comparison based on $\varrho_s = E_s/\Psi_0$ at constant bit sequence frequency f_B, Method I would contain twice the transmission power of Method II: Therefore, a fair comparison is not possible on this basis. $\qquad\square$

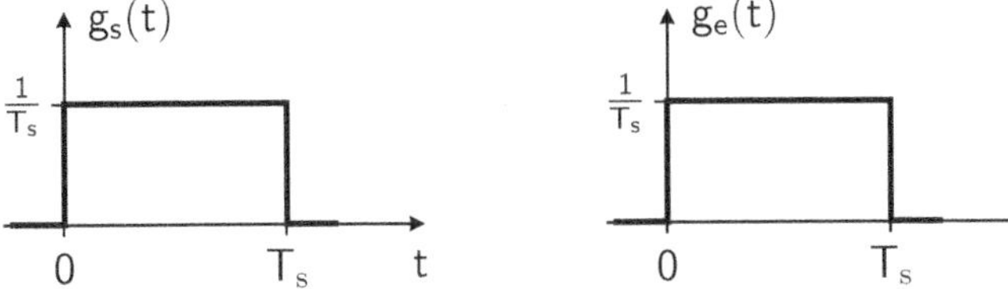

Fig. 2.82 Weighting functions of transmit filter $g_s(t)$ and receive filter $g_e(t)$

Various Definitions of Signal-to-Noise Ratios and Their Relationship to Error Probability

Ratio of Average Symbol Energy to Noise Power Spectral Density It is assumed that transmission methods with the same properties regarding the transmit filtering, the clock frequency $f_T = 1/T_s$, and the average transmission power P_s are used. The characteristics of transmit and receive filters can be found in Fig. 2.82.

Due to the same clock frequency $f_T = 1/T_s$ used for all methods, the same noise power

$$U_R^2 = \Psi_0 \frac{1}{T_s}. \tag{2.258}$$

is obtained for all methods after the receive filtering. An increase in the constellation size s thus leads to a reduced half vertical eye opening U_A at the detection point and thus to a reduction of the signal-to-noise ratio available for decision making (see Fig. 2.86). The bit rate to be transmitted with an increase in the step number in a given channel bandwidth—at a fixed clock frequency f_T and thus also fixed symbol duration T_s—results in

$$f_B = f_T \cdot \mathrm{ld}(s). \tag{2.259}$$

Due to the constant symbol duration T_s and transmission power P_s for all examined methods, the input-side signal-to-noise ratio ϱ_s can be used according to (2.241) (see Fig. 2.83). This signal-to-noise ratio is a quantity that is effective at the *input* of the receive filter according to Fig. 2.83 and is the ratio of usable energy of the useful signal symbol E_s and the noise power density Ψ_0. It should not be confused with the signal-to-noise ratio ϱ that is effective at the detection point and thus at the *output* of the receive filter (Figs. 2.84, 8.85 and 8.86).

Ratio of Average Bit Energy to Noise Power Spectral Density If, on the other hand, one wants to analyze the performance of digital transmission methods at a constant bit rate f_B, the clock frequency decreases via the approach

$$f_B = f_T \cdot \mathrm{ld}(s) \tag{2.260}$$

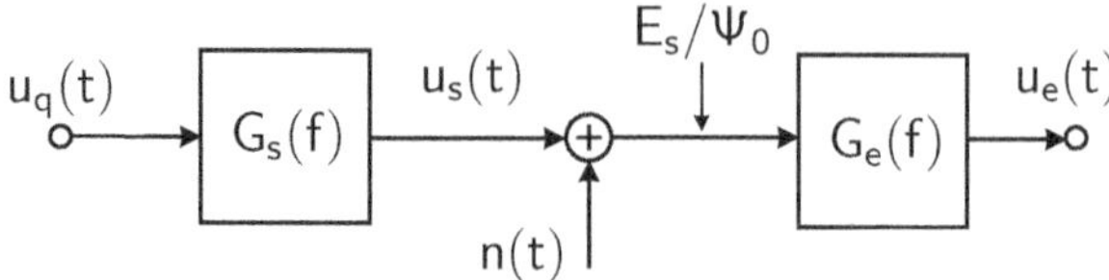

Fig. 2.83 On the definition of the ratio of average symbol energy to noise power density E_s/Ψ_0 at the receiver input when comparing different baseband transmission methods based on a constant symbol sequence frequency or clock frequency f_T

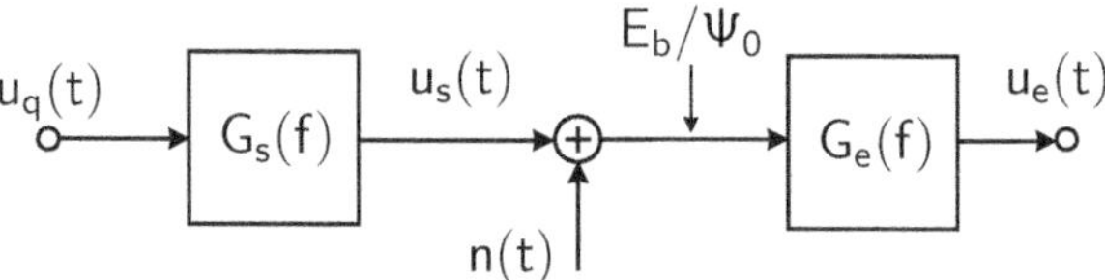

Fig. 2.84 On the definition of the ratio of average bit energy to noise power density E_b/Ψ_0 at the receiver input when comparing different baseband transmission methods based on a constant bit sequence frequency f_B

Fig. 2.85 Bit error probability in two-level, bipolar transmission in the baseband

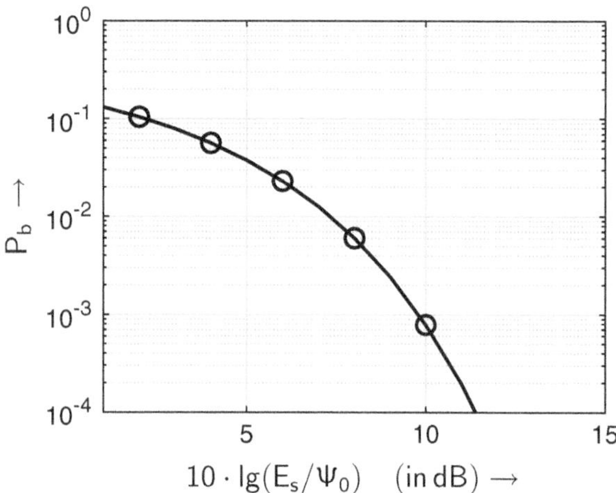

with increasing number of levels s

$$f_T = \frac{1}{T_s} = \frac{f_B}{\mathrm{ld}(s)}.$$ (2.261)

Due to the associated increase in symbol duration T_s, the noise power decreases over

$$U_R^2 = \Psi_0 \frac{1}{T_s} = \Psi_0 f_T = \Psi_0 \frac{f_B}{\mathrm{ld}(s)}$$ (2.262)

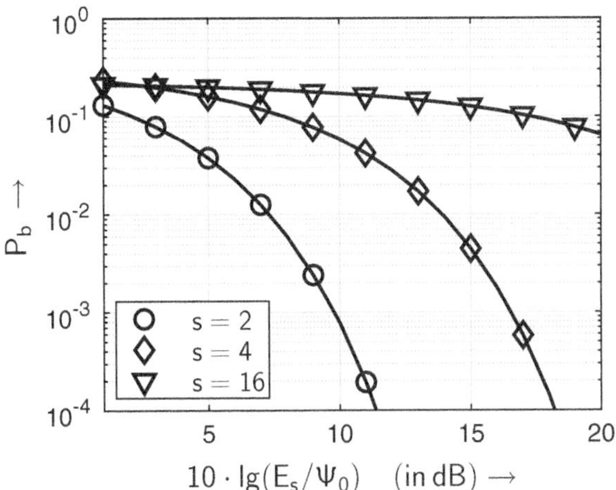

Fig. 2.86 Bit error probability P_b as a function of E_s/Ψ_0 in multi-level baseband transmission with constant symbol clock frequency f_T

with increasing number of levels s. It remains to be investigated at this point whether a reduction in the available eye opening for decision—due to an increase in the number of levels s—can be compensated by saving bandwidth, i.e., reducing the effective noise power at the detection point. The general relationships between bit and symbol energy on the one hand

$$E_b = P_s\, T_b \tag{2.263}$$

$$E_s = P_s\, T_s = \mathrm{ld}(s)\, E_b, \tag{2.264}$$

and between bit and symbol duration on the other hand

$$T_s = \mathrm{ld}(s)\, T_b. \tag{2.265}$$

In this case, the input signal-to-noise ratio

$$\frac{E_s}{\Psi_0} = \frac{P_s\, T_s}{\Psi_0} = \frac{P_s\, \mathrm{ld}(s)\, T_b}{\Psi_0} \tag{2.266}$$

shows that this depends on the number of levels s at a constant bit rate $f_B = 1/T_b$. Consequently, for the comparison of baseband transmission methods at a constant bit rate or bit sequence frequency f_B, due to the different symbol durations for the individual methods, not the ratio $\varrho_s = E_s/\Psi_0$ but the ratio $\varrho_b = E_b/\Psi_0$ of *bit* energy to noise power density must be used according to (2.242) (see Fig. 2.84)

Signal-to-Noise Ratios and Error Probability For the special case of a Nyquist-1 system, the average bit error probability and the bit error probability for the worst case coincide. A transmission system is then referred to as a Nyquist-1 system when the first

Nyquist criterion is adhered to from the input of the transmit filter to the output of the receive filter, so that no intersymbol interference is effective at the detection point at the optimal sampling time.

With the relationship

$$\varrho = \frac{U_A^2}{U_R^2} = \frac{U_s^2\, T_s}{\Psi_0} = \frac{E_s}{\Psi_0} \tag{2.267}$$

valid for Nyquist-1 systems assuming matched filtering and constant symbol clock frequency $f_T = 1/T_s$, the bit error probability of an s-level baseband transmission is obtained

$$P_b = \frac{s-1}{s\,\mathrm{ld}(s)}\left[1 - \mathrm{erf}\left(\sqrt{\frac{\varrho}{2}}\right)\right] = \frac{s-1}{s\,\mathrm{ld}(s)}\,\mathrm{erfc}\left(\sqrt{\frac{E_s}{2\,\Psi_0}}\right) \tag{2.268}$$

Figure 2.85 shows the course of the bit error probability P_b as a function of the input-side signal-to-noise ratio $\varrho_s = E_s/\Psi_0$ for $s = 2$.

Figure 2.86 shows the course of the bit error probability P_b according to (2.268) as a function of the input-side signal-to-noise ratio $\varrho_s = E_s/\Psi_0$ when using higher-level baseband constellations and the same symbol clock frequency f_T.

Figure 2.87 shows the course of the bit error probability as a function of the input-side signal-to-noise ratio $\varrho_b = E_b/\Psi_0$ when using higher-level baseband constellations and the same bit sequence frequency f_B. The reduction in the eye opening available for decision-making associated with the increase in the number of levels s is not compensated by the bandwidth savings and the associated reduction in noise power at the decision point associated with the transition to higher level numbers s (see Fig. 2.87).

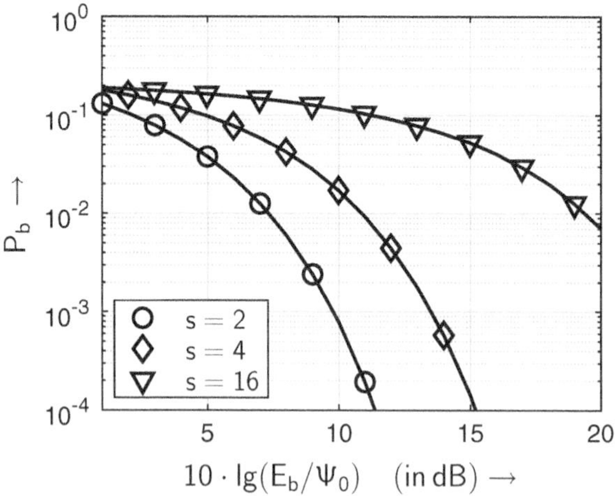

Fig. 2.87 Bit error probability P_b as a function of E_b/Ψ_0 in multi-level baseband transmission with constant bit sequence frequency f_B

2.3.8   Signal-Matched Receive Filtering in Multi-Level Baseband Transmission

The receive filters (Matched Filter) previously classified as optimal in terms of the achievable signal-to-noise ratio for the detection of single pulses can also be advantageously used for the transmission of data or pulse *sequences* typical for digital information transmission, in order to achieve the largest possible signal-to-noise ratio, as the conditions at the detection point are very similar to those assumed in the derivation of the Matched Filter for single pulses, especially in the case of intersymbol interference-free useful signal transmission and white noise interference.

Therefore, examples are revisited and considered for the transmission of pulse sequences with intersymbol interference-free receive pulse shape, which were previously discussed in the signal-matched filtering of single pulses.

Example 2.15 (Eye diagram and vertical eye opening with signal-adapted receive filter for bipolar NRZ rectangular pulse sequences) For the construction of the eye diagram, all precursors and successors of the neighboring pulses that have an influence on the observation interval (typically within the symbol duration T_s) must be taken into account. Table 2.2 shows the precursors and successors of the neighboring pulses to be considered when selecting the observation interval for two-level, bipolar transmission.

The partial sequences resulting from the different polarities of the transmission pulse sequences shown in Table 2.2 in the observation interval are shown in Figs. 2.88 and 2.89.

Figure 2.90 illustrates the resulting eye diagram, which results from the signal sequences shown in Figs. 2.88 and 2.89.

When choosing an unfavorable sequence of symbols, the inner boundary lines of the eye are formed. In the chosen example, sequences 3 and 6 lead to these inner boundary lines of the eye.

Table 2.2 Polarities for transmission pulse sequences to be considered in the construction of the eye diagram

Sequence	$a[k-1]$	$a[k]$	$a[k+1]$
1	$+1$	$+1$	$+1$
2	$+1$	$+1$	-1
3	$+1$	-1	$+1$
4	$+1$	-1	-1
5	-1	$+1$	$+1$
6	-1	$+1$	-1
7	-1	-1	$+1$
8	-1	-1	-1

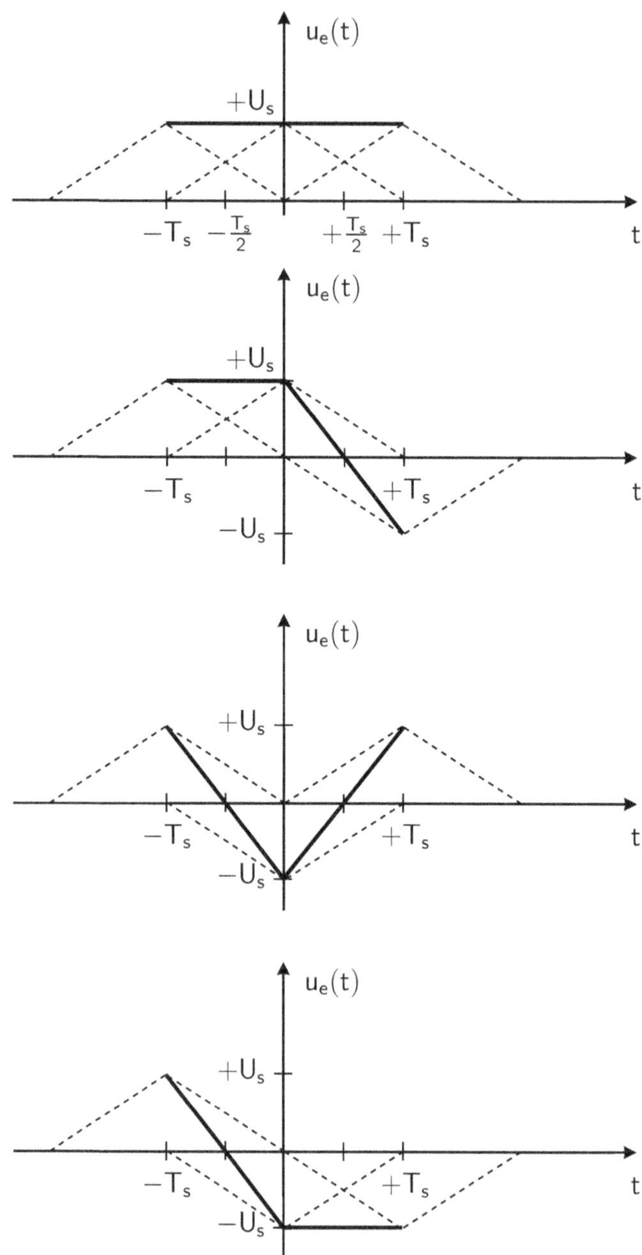

Fig. 2.88 On the formation of an eye diagram using sequence numbers 1-4 (see Table 2.2)

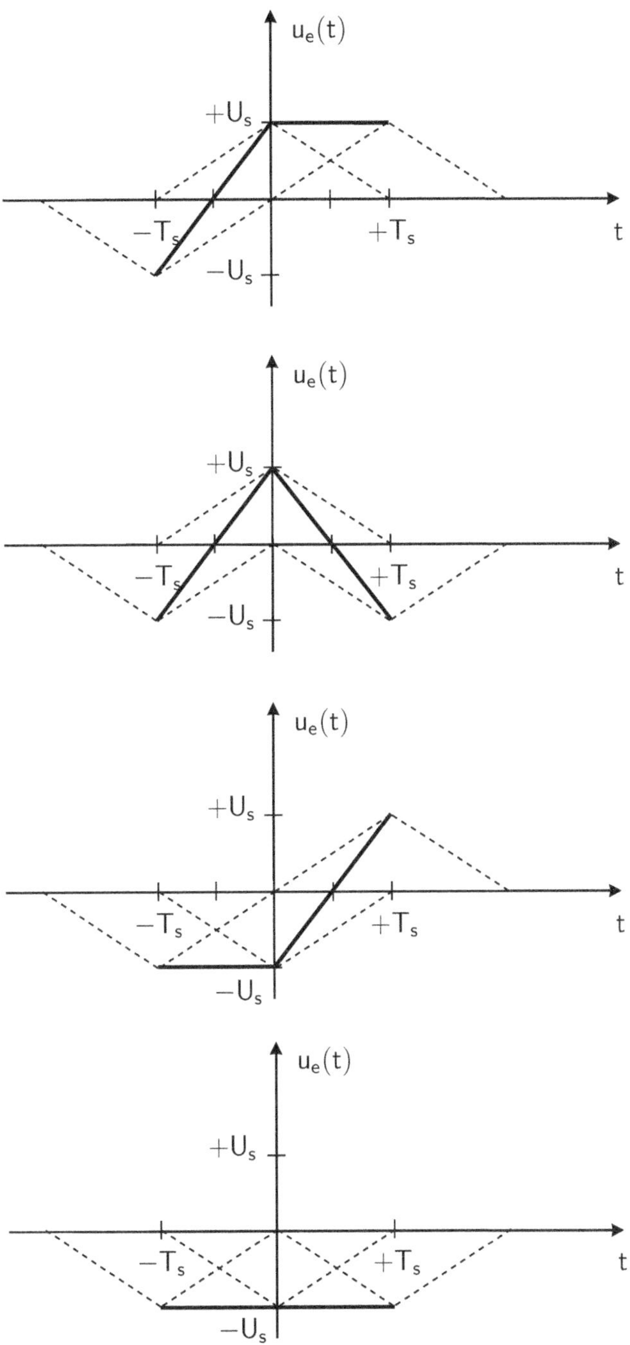

Fig. 2.89 On the formation of an eye diagram using sequence numbers 5-8 (see Table 2.2)

Regarding the first Nyquist criterion, the following statement can be made based on the eye diagram shown in Fig. 2.90: The first Nyquist criterion is met, as an ISI-free transmission is possible, i.e., consecutively transmitted symbols do not interfere with each other at the sampling time (i.e., at $T_A = 0$).   □

Example 2.16 (Signal-to-noise ratio: Comparison of signal-adapted and suboptimal receive filter with bipolar pulse sequences) In this example, an optimal, i.e., signal-adapted, and a non-optimal receive filter are compared in the bipolar transmission of a data or pulse sequence with respect to the achievable signal-to-noise ratio with rectangular basic receive pulse.

The weighting function of transmit filter $g_s(t)$ and receive filter $g_e(t)$ in signal-adapted receive filtering can be found in Fig. 2.91.

The resulting weighting function $h(t) = g_s(t) * g_e(t)$ is shown in Fig. 2.92.

It shows that, as expected, intersymbol interferences can be avoided, since the temporal extension of $h(t)$ does not exceed the value $2 T_s$. Consequently, there can be no overlay of preceding and subsequent pulses at the sampling time: The half vertical eye opening is determined solely by the maximum value of $h(t)$. The determination of the maximum value of $h(t)$ can be done by evaluating the convolution integral. The maximum value within the convolution integral results from the maximum overlap of the two weighting functions $g_s(t)$ and $g_e(t)$ and can be calculated as

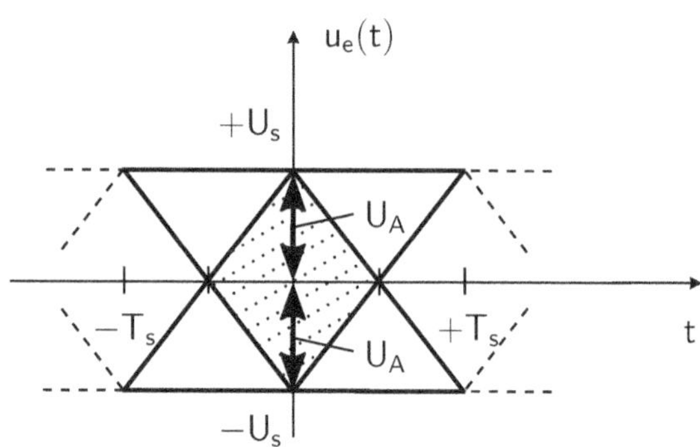

Fig. 2.90 Constructed eye diagram (see also Figs. 2.88 and 2.89) with NRZ rectangular pulse shaping and signal-adapted receive filtering

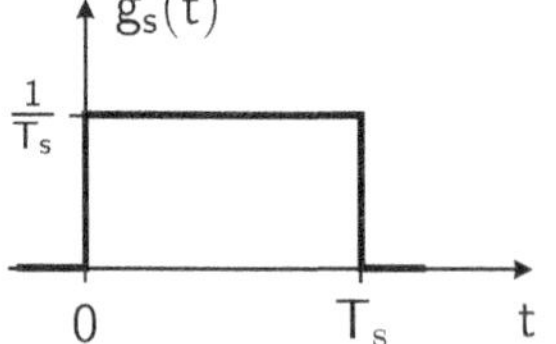

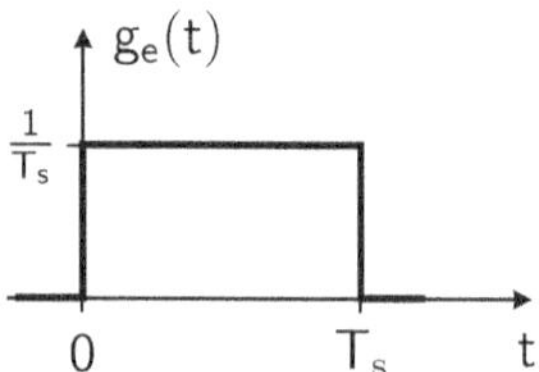

Fig. 2.91 Weighting function of transmit filter $g_s(t)$ and receive filter $g_e(t)$

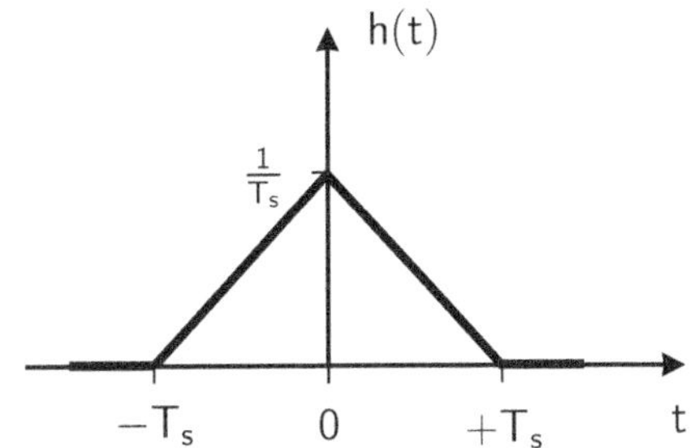

Fig. 2.92 Weighting function of the cascade of transmit filter $g_s(t)$ and receive filter $g_e(t)$ according to the matched filter principle (non-causal representation)

$$h_A = \int_0^{T_s} \frac{1}{T_s} \cdot \frac{1}{T_s} \, dt = \frac{1}{T_s}. \tag{2.269}$$

Since no intersymbol disturbances need to be considered, it is sufficient to excite the transmit filter with a weighted Dirac pulse. If the transmit filter is now excited with a weighted Dirac pulse

$$u_q(t) = U_s\, T_s\, \delta(t) \tag{2.270}$$

it shows that at the optimal sampling time $t = 0$ (non-causal representation) the value

$$U_A = U_s\, T_s \cdot \frac{1}{T_s} = U_s \tag{2.271}$$

arises, which is independent of preceding and subsequent transmitted pulses due to the assumed ISI-free transmission.

For the determination of the signal-to-noise ratio at the detection point, in addition to the determination of U_A, the calculation of the noise power U_R^2 is required. In the considered case, this results in

$$U_R^2 = \Psi_0 \cdot \int_{-\infty}^{+\infty} |g_e(t)|^2 \, dt = \Psi_0 \cdot \int_0^{T_s} \left(\frac{1}{T_s}\right)^2 dt = \frac{\Psi_0}{T_s^2}\, T_s = \Psi_0\, \frac{1}{T_s}. \tag{2.272}$$

Thus, in this case, the signal-to-noise ratio follows

$$\varrho_{MF} = \frac{U_s^2\, T_s}{\Psi_0}. \tag{2.273}$$

Under the conditions outlined at the beginning, the signal-to-noise ratio takes the maximum possible value.

To illustrate this, a deviation from the matched filter principle will now be examined. For this purpose, the form of the weighting function of the receive filter $g_e(t)$ shown in

Fig. 2.93 is assumed and examined, while the form of the receiver input pulse or the transmit filter remains the same.

Since in this case too, the resulting impulse response does not exceed the value $2\,T_\mathrm{s}$, intersymbol interference at the output of the receive filter can be avoided in principle. Consequently, preceding and subsequent pulses do not need to be considered for the calculation of the half vertical eye opening. The maximum value of $h(t) = g_\mathrm{s}(t) * g_\mathrm{e}(t)$ results in

$$h_\mathrm{A} = \int_0^{T_\mathrm{s}} \frac{1}{T_\mathrm{s}} \cdot \frac{1}{T_\mathrm{s}^2} \cdot t \, \mathrm{d}t = \frac{1}{2\,T_\mathrm{s}}. \tag{2.274}$$

Since no intersymbol interference needs to be considered, it is again sufficient to excite the transmit filter with a weighted Dirac pulse

$$u_\mathrm{q}(t) = U_\mathrm{s}\,T_\mathrm{s}\,\delta(t) \tag{2.275}$$

and it turns out that at the optimal sampling time $t = 0$, the value

$$U_\mathrm{A} = U_\mathrm{s}\,T_\mathrm{s} \cdot h_\mathrm{A} = U_\mathrm{s}\,T_\mathrm{s} \cdot \frac{1}{2\,T_\mathrm{s}} = \frac{U_\mathrm{s}}{2} \tag{2.276}$$

arises. The noise power U_R^2 results in

$$U_\mathrm{R}^2 = \Psi_0 \cdot \int_{-\infty}^{+\infty} |g_\mathrm{e}(t)|^2 \, \mathrm{d}t = \Psi_0 \cdot \int_0^{T_\mathrm{s}} \left(\frac{t}{T_\mathrm{s}^2}\right)^2 \mathrm{d}t = \frac{\Psi_0}{T_\mathrm{s}^4}\frac{T_\mathrm{s}^3}{3} = \Psi_0\,\frac{1}{3\,T_\mathrm{s}}. \tag{2.277}$$

Thus, in this case, the signal-to-noise ratio follows

$$\varrho_\text{no-MF-1} = \frac{U_\mathrm{A}^2}{U_\mathrm{R}^2} = \frac{\left(\frac{U_\mathrm{s}}{2}\right)^2}{\Psi_0\,\frac{1}{3\,T_\mathrm{s}}} = \frac{\frac{U_\mathrm{s}^2}{4}}{\Psi_0\,\frac{1}{3\,T_\mathrm{s}}} = \frac{3}{4} \cdot \frac{U_\mathrm{s}^2\,T_\mathrm{s}}{\Psi_0}. \tag{2.278}$$

As expected, it turns out that the signal-to-noise ratio decreases due to the violation of the matched filter principle compared to optimal receive filtering. One obtains

$$\varrho_\text{no-MF-1} = \frac{3}{4}\,\varrho_\text{MF}. \tag{2.279}$$

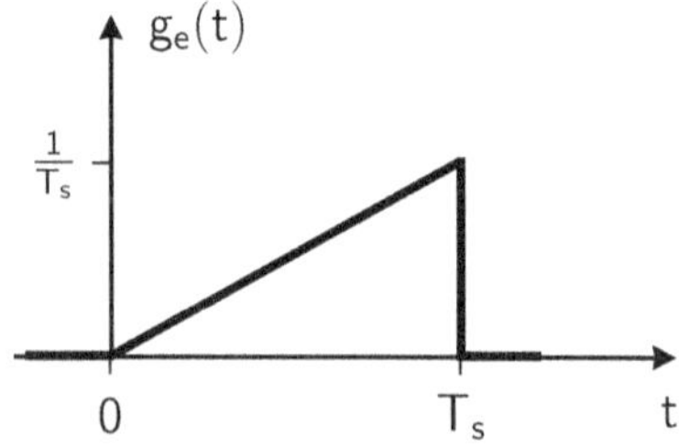

Fig. 2.93 Weighting function of the receive filter $g_\mathrm{e}(t)$ in case of violation of the matched filter principle

The signal-to-noise ratio $\varrho_{\text{no-MF-1}}$ is lower than with signal-adapted and thus optimal recepive filtering with respect to a maximum signal-to-noise ratio. $\square$

Example 2.17 (Signal-to-noise ratio in case of violation of the matched filter condition with bipolar pulse sequences) If the receive filter is not matched to the transmit filter characteristic (see Fig. 2.94), intersymbol interference can be avoided, for example, since the temporal extension of the resulting impulse response from transmit and receive filter, i.e., $h(t)$, does not exceed the value $2\,T_s$. However, the achievable signal-to-noise ratio decreases due to the lack of adaptation of the receive filter characteristic to the transmit filter characteristic.

Figure 2.95 illustrates the resulting weighting function from the cascade of transmit pulse shaper and receive filter.

If the transmit filter is excited with a weighted Dirac pulse

$$u_q(t) = U_s\, T_s\, \delta(t) \tag{2.280}$$

it turns out that at the optimal sampling time for two-stage bipolar transmission, the value

$$U_A = \frac{U_s}{2} \tag{2.281}$$

arises, which is independent of preceding and subsequent pulses due to the temporal extension of $h(t)$ (smaller than $2\,T_s$).

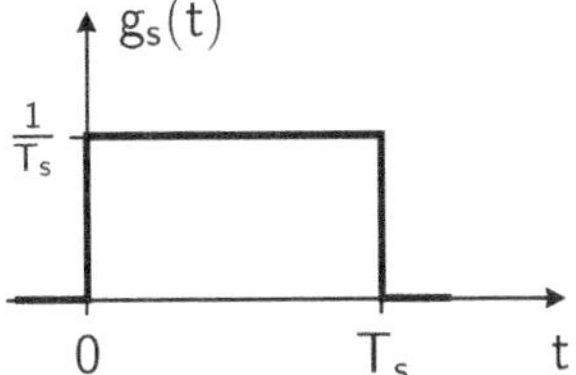
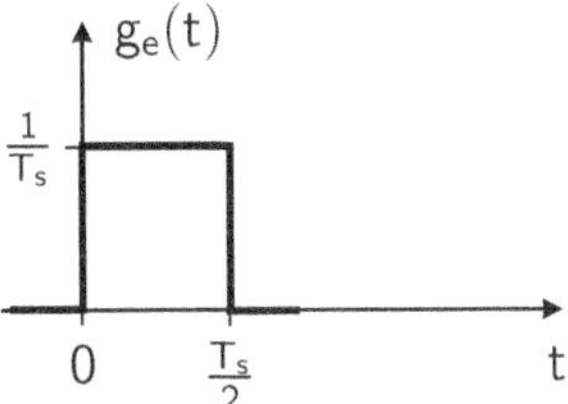

Fig. 2.94 Weighting function of transmit filter $g_s(t)$ and receive filter $g_e(t)$ in the case of exemplary violation of the matched filter principle

Fig. 2.95 Weighting function from cascade of transmit filter $g_s(t)$ and receive filter $g_e(t)$ in case of violation of the matched filter principle (non-causal representation)

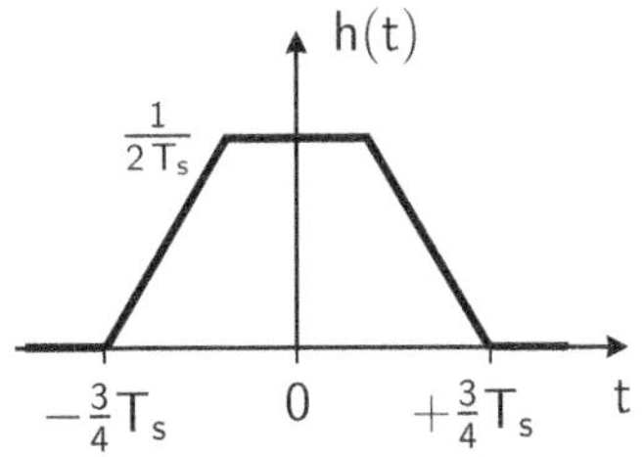

The noise power at the output of the receive filter results from the application of the PARSEVAL theorem to

$$U_R^2 = \Psi_0 \cdot \int_{-\infty}^{+\infty} |g_e(t)|^2 \, dt = \Psi_0 \cdot \int_{-T_s/4}^{T_s/4} \left(\frac{1}{T_s}\right)^2 dt = \frac{\Psi_0}{T_s^2} \frac{T_s}{2} = \Psi_0 \frac{1}{2\,T_s}. \tag{2.282}$$

The resulting signal-to-noise ratio is

$$\varrho_{\text{no-MF-2}} = \frac{U_A^2}{U_R^2} = \frac{\left(\frac{U_s}{2}\right)^2}{\Psi_0 \frac{1}{2T_s}} = \frac{\frac{U_s^2}{4}}{\Psi_0 \frac{1}{2T_s}} = \frac{1}{2} \cdot \frac{U_s^2\,T_s}{\Psi_0}. \tag{2.283}$$

and is half as large as with signal-adapted receive filtering:

$$\varrho_{\text{no-MF-2}} = \frac{1}{2}\,\varrho_{\text{MF}}. \tag{2.284}$$

$\square$

2.3.9   Comparison of Different Multi-Level Baseband Transmission Methods

Transmission System and System Description
Multi-level methods can be used to either reduce the occupied bandwidth or symbol clock frequency f_T at the same bit rate f_B with increasing number of levels s, or to increase the bit rate with increasing number of levels s at the same symbol clock frequency f_T-for example, with a fixed available transmission bandwidth.

The starting point is the transmission system shown in Fig. 2.96: The transmit filter $g_s(t)$ and the receive filter $g_e(t)$ have the weighting functions shown in Fig. 2.96: The resulting transmit signal $u_s(t)$ is again an NRZ rectangular signal and the receive filter $g_e(t)$ is a filter matched to the transmit filter $g_s(t)$, i.e., a matched filter. The resulting weighting function is obtained via

$$h(t) = g_s(t) * g_e(t), \tag{2.285}$$

it is also shown in Fig. 2.96.
It is assumed that the transmit power is fixed according to (2.230), so that a half-level transmit amplitude

$$U_s = \sqrt{\frac{3P_s}{(s^2 - 1)}} \tag{2.286}$$

results. The half vertical eye opening is $U_A = U_s$, as the first Nyquist criterion is met. The noise power at the detection point is

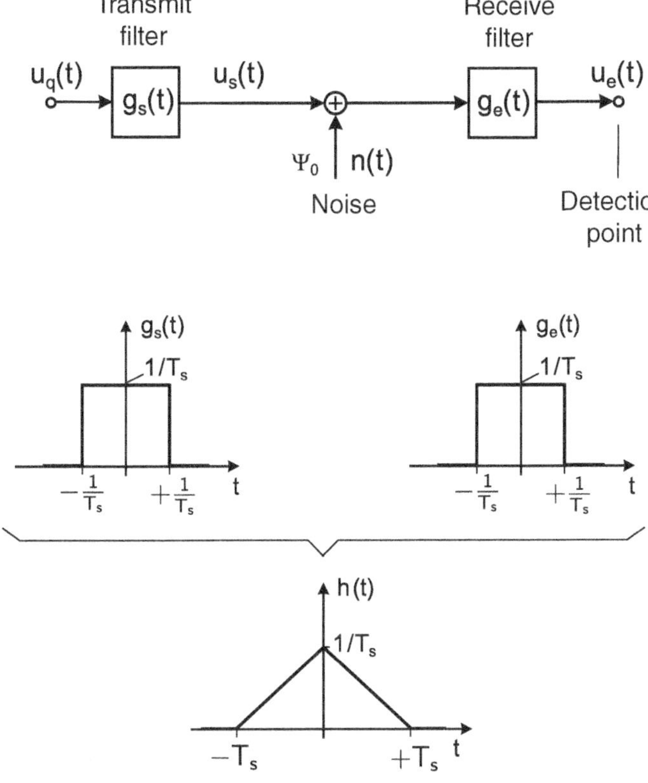

Fig. 2.96 For comparison of multi-level baseband methods: Transmission system and filter characteristics

$$U_R^2 = \Psi_0 \int_{-\infty}^{+\infty} |G_e(f)|^2 \, df = \Psi_0 \int_{-\infty}^{+\infty} |g_e(t)|^2 \, dt = \Psi_0 \int_{-T_s/2}^{+T_s/2} \left(\frac{1}{T_s}\right)^2 dt = \frac{\Psi_0}{T_s^2} T_s = \frac{\Psi_0}{T_s}$$

$$(2.287)$$

and finally, the signal-to-noise ratio

$$\varrho = \frac{U_A^2}{U_R^2} = \frac{\left(\sqrt{\frac{3P_s}{(s^2-1)}}\right)^2}{\frac{\Psi_0}{T_s}} = \frac{\frac{3P_s}{(s^2-1)}}{\frac{\Psi_0}{T_s}} = \frac{3P_s T_s}{\Psi_0 (s^2 - 1)} \tag{2.288}$$

is obtained as a function of the number of levels s. Here, $E_s = P_s T_s$ is the energy of a symbol. Furthermore, the relationships (2.195) and (2.196) between bit rate and clock frequency $f_T = f_B/\mathrm{ld}(s)$ or between bit duration and symbol duration $T_s = T_b \cdot \mathrm{ld}(s)$, which are each linked via the number of levels s, apply.

Multi-Level Transmission at a Constant Bit Rate

At a constant bit rate, the symbol duration can be written as

$$T_\mathrm{s} = \frac{1}{f_\mathrm{T}} \quad \text{and} \quad f_\mathrm{T} = \frac{f_\mathrm{B}}{\mathrm{ld}(s)} \tag{2.289}$$

as

$$T_\mathrm{s} = \frac{\mathrm{ld}(s)}{f_\mathrm{B}} \tag{2.290}$$

This results in the signal-to-noise ratio

$$\varrho = \frac{3P_\mathrm{s}\,\mathrm{ld}(s)}{\Psi_0 f_\mathrm{B}\left(s^2 - 1\right)}, \tag{2.291}$$

with (2.288), which depends only on the given values for the transmit power P_s, for the noise power density Ψ_0, and for the bit rate f_B—and on the number of levels s, which is variable.

How a change in the number of levels s affects the signal-to-noise ratio is examined in the following numerical example.

Example 2.18 (Multi-level transmission at constant bit rate f_B) The following quantities are given

- Bit rate or bit sequence frequency: $f_\mathrm{B} = 1\,\mathrm{kHz}$ (bit duration: $T_\mathrm{b} = 1/f_\mathrm{B} = 1\,\mathrm{ms}$)
- Average transmit power: $P_\mathrm{s} = 1\,\mathrm{V}^2$
- Noise power spectral density: $\Psi_0 = 10^{-3}\,\mathrm{V}^2/\mathrm{Hz}$

In Fig. 2.97, the function of the signal-to-noise ratio according to (2.291) for the given quantities is shown as a function of the number of levels s. It can be seen that the signal-to-noise ratio decreases with increasing number of levels. $\qquad\Box$

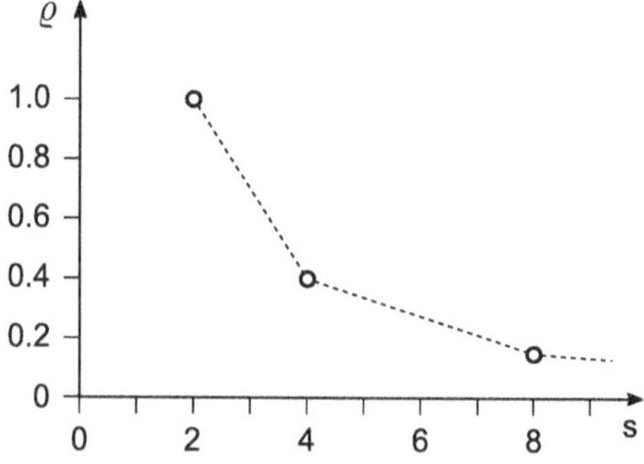

Fig. 2.97 Signal-to-noise ratio ϱ as a function of the number of levels s at constant bit rate f_B

The numerical example shows that as the number of levels s increases, the signal-to-noise ratio in the AWGN channel at constant bit rate f_B decreases: As the number of levels s increases, the half vertical eye opening U_A becomes smaller and the symbol clock frequency f_T and thus the bandwidth of the receive filter also decrease. The reduced receive filter bandwidth results in a decreasing noise power U_R^2 as the number of levels s increases. The conclusion is that the decrease in noise power is not sufficient to compensate for the shrinking eye opening: Thus, the signal-to-noise ratio decreases with increasing number of levels and constant transmit power. In this case, a small number of levels s should be chosen for transmission if the bit error probability is to be minimized. If, on the other hand, the used bandwidth is to be as small as possible, a large number of levels s should be used, which, however, has limiting effects on the achievable transmission quality—as the considerations have shown.

Multi-Level Transmission at a Constant Symbol Clock Frequency
At constant clock frequency or symbol duration, the symbol duration

$$T_s = \frac{1}{f_T} \tag{2.292}$$

remains constant and with (2.288) the signal-to-noise ratio

$$\varrho = \frac{3P_s}{\Psi_0 f_T \left(s^2 - 1\right)}, \tag{2.293}$$

is obtained directly, which only depends on the given values for the transmit power P_s, for the noise power density Ψ_0 and for the clock frequency f_T—and on the number of levels s, which is again variable.

How a change in the number of levels s affects the achievable signal-to-noise ratio is analyzed again in the following numerical example.

Example 2.19 (Multi-level transmission at constant symbol clock frequency f_T) The following quantities are given

- Symbol clock frequency or symbol rate: $f_T = 1\,\text{kHz}$ (Symbol duration: $T_s = 1/f_T = 1\,\text{ms}$)
- Average transmit power: $P_s = 1\,\text{V}^2$
- Noise power spectral density: $\Psi_0 = 10^{-3}\,\text{V}^2/\text{Hz}$

In Fig. 2.98, the function of the signal-to-noise ratio according to (2.291) for the given quantities is shown as a function of the number of levels s. It can be seen again that the signal-to-noise ratio decreases with increasing number of levels, but now somewhat more strongly than in Fig. 2.97. $\square$

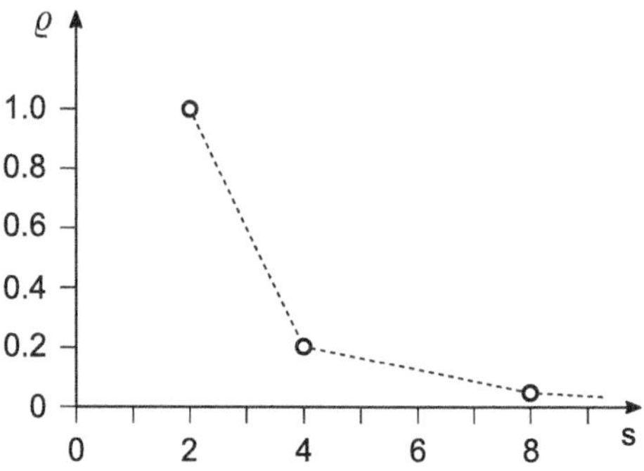

Fig. 2.98 Signal-to-noise ratio ϱ as a function of the number of levels s at constant symbol clock frequency or symbol rate f_T

It can be seen from the numerical example that as the number of levels s increases, the signal-to-noise ratio in the AWGN channel decreases at a constant clock frequency f_T: As the number of levels s increases, the half vertical eye opening U_A becomes smaller and the symbol clock frequency f_T remains constant—and thus the bandwidth of the receive filter. The reduced vertical eye opening causes a decrease in the signal-to-noise ratio—at constant transmit power—as the number of levels s increases and the noise power U_R^2 remains constant. However, as the number of levels s increases, the bit rate is increased by a factor of $ld(s)$, i.e., at $s = 4$, the bit rate doubles compared to $s = 2$, and at $s = 8$, it triples.

This example illustrates an important basic law of communications engineering: If the bit rate or bit sequence frequency f_B is increased at a constant signal bandwidth—here given by the symbol clock frequency f_T—the achievable transmission quality decreases—here represented by the signal-to-noise ratio ϱ. The transmission quality can also be equivalently stated as the probability of error. This is briefly referred to as the interchangeability of bit rate and error probability.

2.3.10 Numerical Example: Multi-Level Baseband Transmission with NRZ Transmit Signal and Gaussian Receive Filter

Introduction and Task Definition

Using a numerical example, the procedure described in the previous sections for describing and optimizing a multi-level baseband transmission system, which is operated over a distortion-free channel with pure noise disturbance, is now again illustrated using a specific task. The transmission system under consideration is shown in Fig. 2.99.

A baseband system is now considered in which improvements in transmission quality can be achieved compared to the basic model, but intersymbol interference remains at the detection point in the optimal sampling time.

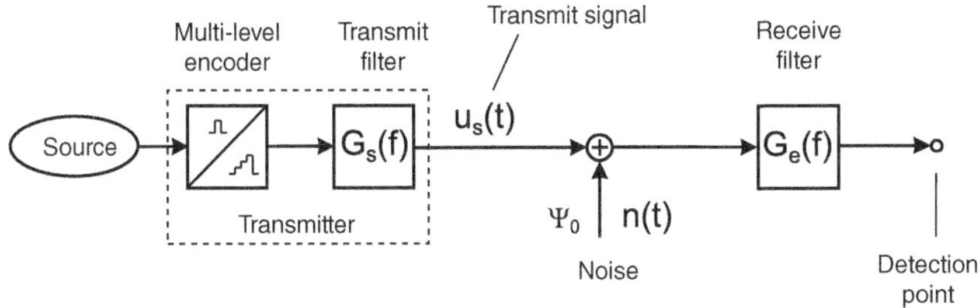

Fig. 2.99 Transmission system for the numerical example with bipolar, multi-level baseband transmission

The transmitter of a baseband transmission system transmits a bipolar, s-level random sequence of rectangular NRZ pulses with the half-level amplitude U_s and the pulse width T_s. The average transmission power P_s is given; from this, the half-level transmit amplitude U_s is determined. The influence of the channel consists solely of the additive addition of a noise disturbance. This disturbance is modeled as white, Gaussian noise with the power spectral density Ψ_0. The receive filter is a Gaussian low-pass filter with the cutoff frequency f_G.

The achievable quality of transmission, i.e., the signal-to-noise ratio and the bit error probability, is to be determined. To do this, the individual necessary quantities (U_A, U_R^2) must be calculated and possibly optimized (f_G). The two-level transmission ($s = 2$) is first examined to allow a comparison with the numerical example of unipolar transmission. Subsequently, selected exemplary level numbers s are considered to illustrate the influence of the level number.

Given Values

The following numerical values are given:

- Signal: Transmit power $P_s = 1\,\mathrm{V}^2$, bit rate $f_B = 1\,\mathrm{MHz}$, bipolar transmission,
- Disturbance: Power spectral density $\Psi_0 = 2 \cdot 10^{-8}\,\frac{\mathrm{V}^2}{\mathrm{Hz}}$,
- Number of levels: Initially $s = 2$, later variable number of levels s.

In the example, a two-level, bipolar transmission with Gaussian receive low-pass filter is initially chosen to clearly demonstrate the changes compared to the numerical example with the two-level, unipolar transmission with RC first-order low-pass filter according to the basic model. Subsequently, the achievable signal-to-noise ratio with variable number of levels is examined to illustrate the change in transmission quality with variable number of stages and constant, predetermined bit rate f_B.

Procedure and Results

From the given average transmission power (2.230), the half-level transmit amplitude

$$P_s = \frac{U_s}{3}\left(s^2 - 1\right) \qquad \Longrightarrow \qquad U_s = \sqrt{\frac{3\,P_s}{s^2 - 1}} \qquad (2.294)$$

is firstly calculated. For $P_s = 1\,\text{V}^2$ and $s = 2$, the half-level transmit amplitude $U_s = 1\,\text{V}$ is obtained.

The receive filter is predefined in its structure as a Gaussian low-pass filter, but its cutoff frequency f_G is freely selectable and thus optimizable. Therefore, the half vertical eye opening according to (2.207) and the noise power according to (2.101) are determined as a function of the receive filter cutoff frequency. In Fig. 2.100, the half vertical eye opening and in Fig. 2.101 the noise power at the detection point, each as a function of the receive filter cutoff frequency f_G are shown. The typical curves are recognizable: The half vertical eye opening is—with a closed eye—initially zero and tends towards a maximum of $U_s = 1\,\text{V}$ given by the transmit amplitude as the receive filter cutoff fre-

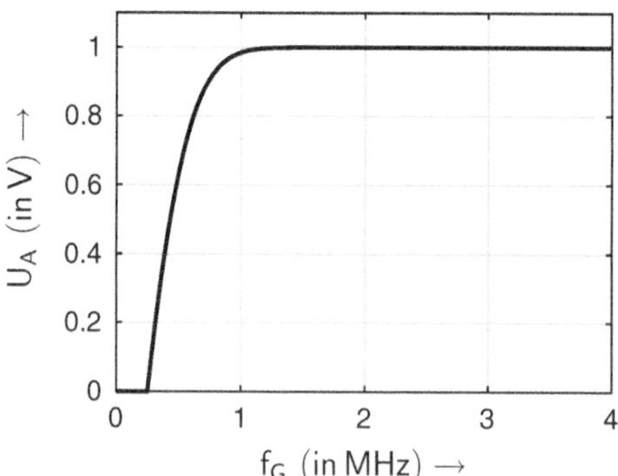

Fig. 2.100 Half vertical eye opening U_A as a function of the receive filter cutoff frequency f_G

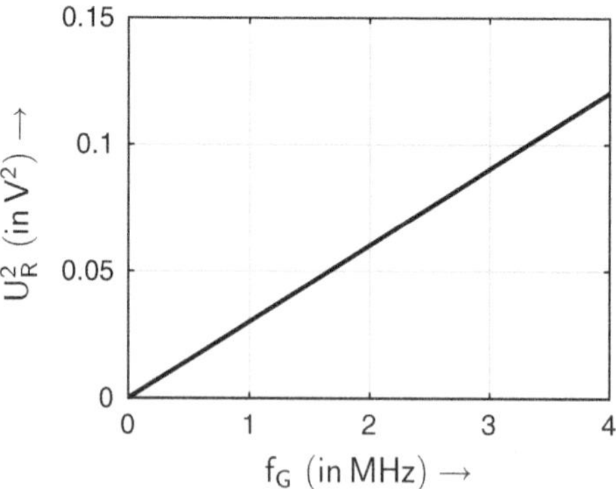

Fig. 2.101 Noise power U_R^2 as a function of the receive filter cutoff frequency f_G

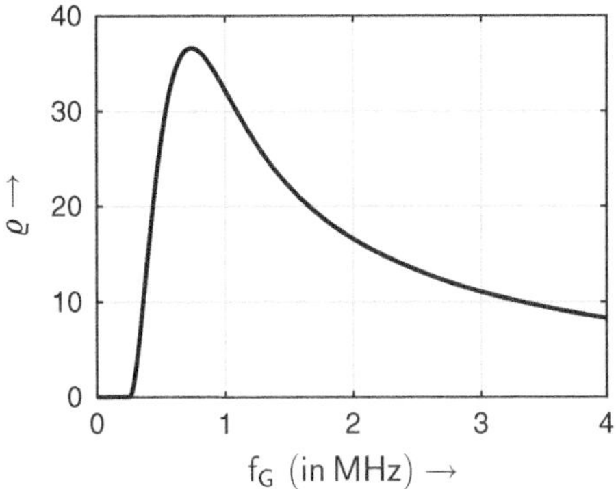

Fig. 2.102 Signal-to-noise ratio ϱ as a function of the receive filter cutoff frequency f_G

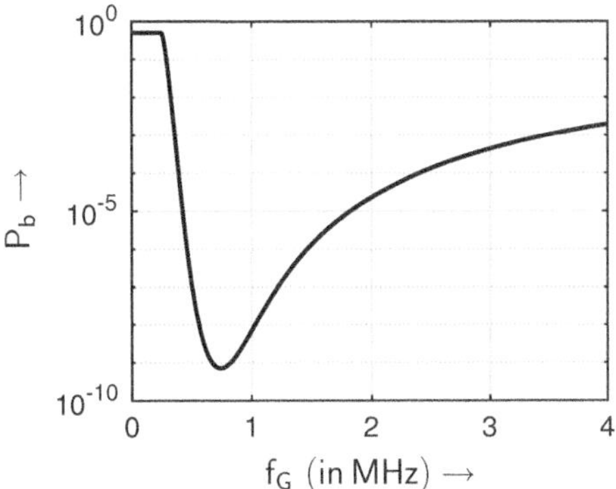

Fig. 2.103 Bit error probability P_f as a function of the receive filter cutoff frequency f_G

quency increases, while the noise power linearly increases without limit with increasing receive filter cutoff frequency f_G.

If the signal-to-noise ratio according to (2.60) and the bit error probability according to (2.240) are now calculated as a function of the receive filter cutoff frequency f_G and shown in Figs. 2.102 and 2.103, it is possible to determine the optimal cutoff frequency $f_{\mathrm{G\,opt}}$ of the receive filter. It is read from one of the diagrams where the signal-to-noise ratio becomes maximum—or the bit error probability minimum—and $f_{\mathrm{G\,opt}} = 740\,\mathrm{kHz}$ is obtained.

Numerical Results for Optimized Transmission System

Once the optimal cutoff frequency $f_{\mathrm{G\,opt}}$ of the receive filter has been determined, all other values for signal and disturbance at the detection point can be determined and the quality of transmission for an optimized transmission system can be specified. The individual relationships, which have been formulated as functions of the cutoff frequency f_G of the receive filter, are evaluated at the point $f_\mathrm{G} = f_{\mathrm{G\,opt}}$. The numerical results are compiled in Table 2.3.

Table 2.3 Preset values and numerical results for a two-level baseband system with NRZ transmission pulses at optimal receive filter cutoff frequency $f_{G\,opt}$ of the Gaussian low-pass filter

Size	Numerical Value
Average transmit power	$P_s = 1\,\text{V}^2$
Number of levels	$s = 2$
Half-level transmit amplitude	$U_s = 1\,\text{V}$
Noise power spectral density	$\Psi_0 = 2 \cdot 10^{-8}\,\text{V}^2/\text{Hz}$
Pulse duration (symbol duration)	$T_s = 1\,\mu\text{s}$
Optimal receive filter cutoff frequency	$f_{G\,opt} = 740\,\text{kHz}$
Half vertical eye opening	$U_A = 0.903\,\text{V}$
Noise power	$U_R^2 = 0.022\,\text{V}^2$
Signal-to-noise ratio	$\varrho_{max} = 36{,}63$
Bit error probability	$P_{b\,min} = 7.15 \cdot 10^{-10}$

The maximum achievable signal-to-noise ratio is $\varrho_{max} = 36.63$ or $\varrho_{max}^* = 10\lg\varrho_{max} = 15.64\,\text{dB}$; it leads to a bit error probability of $P_{b\,min} = 7.15 \cdot 10^{-10}$.

The optimal receive filter cutoff frequency $f_{G\,opt}$ is below the half clock or symbol sequence frequency $f_T = 1/T_s$, which corresponds to the bit sequence frequency or bit rate f_B in two-level transmission. As a result, the half vertical eye opening U_A has a value that is below the maximum half vertical eye opening U_s: This maximum half vertical eye opening would be reached at a (very) large receive filter cutoff frequency f_G (see, for example, Fig. 2.100). This would in turn cause a (strong) increase in noise power compared to $f_{G\,opt}$ (see Fig. 2.101). All these factors together lead to the optimal cutoff frequency in the size that results in its optimization with respect to a maximum signal-to-noise ratio or a minimum bit error probability.

It can be seen that with Gaussian low-pass filtering at the receiver, an improvement in transmission quality can be achieved compared to the system with first-order *RC* receive low-pass filter: The signal-to-noise ratio increases significantly or the error probability becomes correspondingly smaller. There are still intersymbol interferences at the optimal receive filter cutoff frequency $f_{G\,opt}$ at the detection point, so that the half vertical eye opening U_A is still not maximum—but significantly larger than in the unipolar numerical example—is; the noise power U_R^2 remains in a similar order of magnitude. These improvements are due to the type of receive filtering and the bipolar transmission method, which makes more efficient use of the given transmit power.

Some Diagrams for Optimized Transmission System

This section is intended to illustrate with graphical representations how some time and spectral functions look in an optimized transmission system.

Figure 2.104 shows the amplitude frequency response of the receive filter $G_e(f)$ and Fig. 2.105 the eye diagram of the useful signal at the detection point—each for the case of the optimized cutoff frequency of the receive low-pass filter ($f_{G\,opt} = 740\,\text{kHz}$). It can

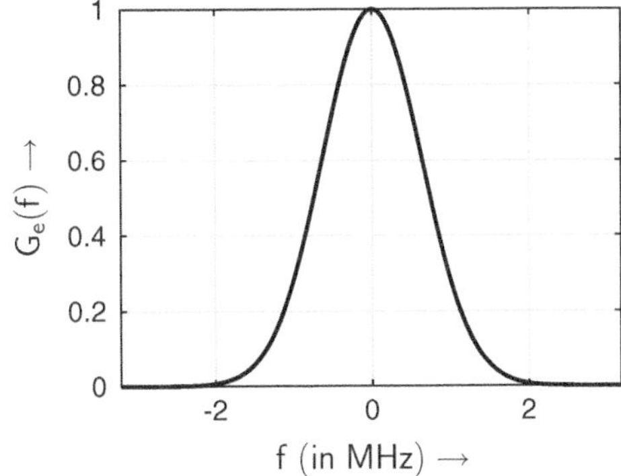

Fig. 2.104 Amplitude response of the receive filter at optimal cutoff frequency $f_{\mathrm{G\,opt}}$

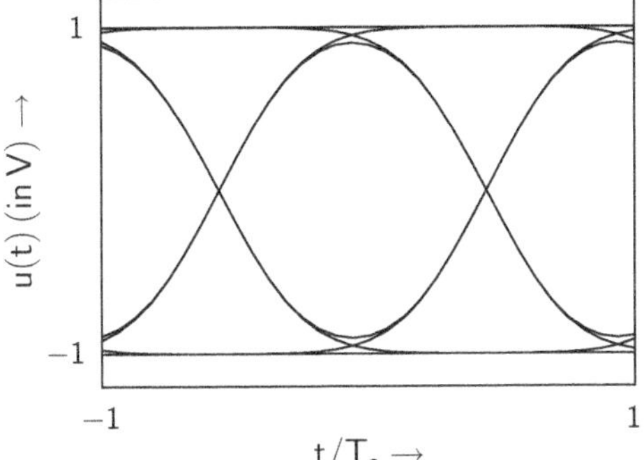

Fig. 2.105 Eye diagram of the useful signal at the detection point with optimal receive filter cutoff frequency $f_{\mathrm{G\,opt}}$

be seen that the eye is not fully open vertically, as the band limitation of the receivefilter starts relatively early and thus in the optimized transmission system the useful signal is deformed, but above all the noise can be relatively well limited in its power on the receiver side.

Influence of the Number of Signalling Levels

If the maximum amplitude range or the average transmit power is limited and thus predetermined in the transmitter, the half-level transmit amplitude U_s decreases with increasing number of levels s: One obtains $(s-1)$ eye openings, which become smaller with increasing number of levels s (see, for example, Fig. 2.70).

To illustrate the influence of increasing number of levels s, the numerical example is now extended in such a way that, with otherwise identical specifications regarding signal and interference, different numbers of levels are allowed and considered. With the same predetermined bit rate $f_{\mathrm{B}} = 1\,\mathrm{MHz}$ and average transmission power $P_s = 1\,\mathrm{V}^2$, the respective half-level transmit amplitudes U_s according to (2.294) and clock frequencies

f_T according to (2.195) are to be calculated and thus the previously described calculations for eye opening U_A, noise power U_R^2, signal-to-noise ratio ϱ and possibly the bit error probability P_b are to be carried out, with an optimization of the receive filter cutoff frequency f_G to be carried out for each number of levels.

Some results for the numerical example with s-level baseband transmission are shown in Table 2.4.

In Fig. 2.106, the maximum achievable signal-to-noise ratio ϱ_{max} at optimal receive filter cutoff frequency $f_{G\,opt}$ is shown as a function of the number of levels s to demonstrate the achievable transmission quality. It becomes clear that with increasing number of levels s, the transmission quality of the baseband transmission system decreases—here represented by the maximum achievable signal-to-noise ratio ϱ_{max} at optimal receive filter cutoff frequency.

Table 2.4 Some results for the s-level numerical example with NRZ transmission pulses and Gaussian receive filter at constant bit sequence frequency $f_B = 1\,\mathrm{MHz}$

s	2	4	8	16	32
U_s (in V)	1,000	0.447	0.218	0.108	0.054
f_T (in MHz)	1,000	0.500	0.333	0.250	0.200
$f_{G\,opt}$ (in kHz)	740.0	445.0	333.3	277.5	240.0
U_A (in V)	0.903	0.416	0.205	0.103	0.052
U_R^2 (in V^2)	0.022	0.013	0.010	0.008	0.007
ϱ_{max}	36.63	12.90	4.184	1.274	0.371
$P_{b\,min}$	$7.15 \cdot 10^{-10}$	$1.22 \cdot 10^{-4}$	$1.19 \cdot 10^{-2}$	0.061	0.105

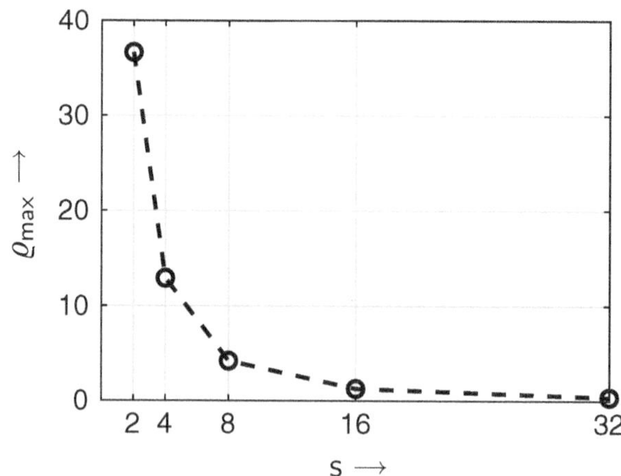

Fig. 2.106 Signal-to-noise ratio ϱ_{max} as a function of the number of levels s in bipolar, multi-level baseband transmission with NRZ transmit pulses and Gaussian receive filtering at a predetermined bit sequence frequency $f_B = 1\,\mathrm{MHz}$

Due to the reduced clock frequency f_T according to (2.195) with increasing number of levels at constant bit sequence frequency f_B, the receive filter can operate at a lower optimal cutoff frequency—and the noise power at the detection point decreases with increasing number of levels s. However, the effect of the reduced vertical eye opening due to multi-staging predominates, so that the signal-to-noise ratio ϱ decreases or the bit error probability P_b increases when the number of levels s is increased in a distortion-free channel with pure noise disturbance—provided the bandwidth necessary for the transmission of the signal is available and is not limited by the channel. Only in real channels with bandwidth limitation, e.g. when transmitting over copper cables, it can be advantageous to work with a higher number of levels (than $s = 2$) to achieve improved transmission quality (see Chap. 4).

2.3.11 Numerical Example: Multi-level Baseband Transmission with Root Raised Cosine Transmit and Receive Filters

Introduction and Task Definition

Using a numerical example, the procedure described in the previous sections for describing and optimizing a multi-level baseband transmission system, which is operated over a distortion-free channel with pure noise disturbance, is again illustrated using a specific task. The transmission system under consideration is shown in Fig. 2.99. We now consider a baseband system in which, by suitable choice of the transmit and receive filters, no intersymbol interference is effective at the detection point in the optimal sampling time, i.e., the first Nyquist criterion is met.

The transmitter of a baseband transmission system transmits a bipolar, s-level random sequence, which is filtered in the transmitter with a root-cosine roll-off filter with the roll-off factor r, with half-level transmit amplitude U_s and the symbol clock frequency f_T. The influence of the channel consists solely of the additive addition of a noise disturbance. This disturbance is modeled as white, Gaussian noise with the spectral power density Ψ_0. The receive filter is a root raised cosine low-pass filter, also with the roll-off factor r.

The achievable quality of the transmission, i.e., the signal-to-noise ratio and the bit error probability, is to be determined. To do this, the individual necessary quantities must be calculated. First, a two-level baseband transmission ($s = 2$) is analyzed to allow a comparison with the numerical example of unipolar transmission and with the detailed example of bipolar transmission with Gaussian receive low-pass filter. Subsequently, the transmission with selected exemplary level numbers s is considered to clarify the influence of the level number on the transmission quality.

Given Values

The following numerical values are given:

- Signal: Transmit power $P_s = 1\,\text{V}^2$, bit rate $f_B = 1\,\text{MHz}$, bipolar transmission,
- Disturbance: Power spectral density $\Psi_0 = 2 \cdot 10^{-8}\,\frac{\text{V}^2}{\text{Hz}}$,
- Number of levels: Initially $s = 2$, later variable level number s.

In the example, a two-level, bipolar transmission with root raised cosine transmit and receive filters is first examined to highlight the changes compared to the numerical example with the two-level, unipolar transmission with RC first-order low-pass filter according to the basic model and the immediately previously treated example with NRZ transmit pulse shaping and Gaussian receive low-pass in bipolar transmission. Subsequently, the achievable signal-to-noise ratio with variable level number is again analyzed to make visible the change in transmission quality with variable level number and constant, predetermined bit rate f_B.

Procedure and Results

From the given average transmit power, the half-level transmit amplitude

$$P_s = \frac{U_s}{3}\left(s^2 - 1\right) \qquad \Longrightarrow \qquad U_s = \sqrt{\frac{3\,P_s}{s^2 - 1}} \tag{2.295}$$

is first calculated. The result for $P_s = 1\,\text{V}^2$ and $s = 2$ is the half-level transmit amplitude $U_s = 1\,\text{V}$.

The transmit and receive filters are predefined in their structure as root-cosine roll-off low-pass filters. The received useful signal is therefore free of intersymbol interference at the optimal sampling time. The half vertical eye opening after (2.217) therefore corresponds to the half-level transmit amplitude, i.e. $U_A = U_s$, and the noise power is determined according to (2.109): Both the half vertical eye opening U_A and the noise power U_R^2 are independent of the roll-off factor r.

If the signal-to-noise ratio is now calculated according to (2.60) and the bit error probability according to (2.240), a single numerical value is obtained in each case: Both quantities are also independent of the roll-off factor. There is no dependence on a quantity comparable to the receive filter cutoff frequency here—and therefore no corresponding optimization is possible or necessary.

Numerical Results for Optimized Transmission System

Numerical results for some important parameters of this transmission system are compiled in Table 2.5.

The maximum achievable signal-to-noise ratio is $\varrho = 50.0$ or $\varrho^* = 10\lg\varrho = 16.99\,\text{dB}$; it leads to a bit error probability of $P_b = 7.69 \cdot 10^{-13}$.

Table 2.5 Given values and numerical results for two-level baseband system with root raised cosine transmit and receive filters

Parameter	Numerical Value
Average Transmit Power	$P_{\mathrm{s}} = 1\,\mathrm{V}^2$
Number of Levels	$s = 2$
Half-level Transmit Amplitude	$U_{\mathrm{s}} = 1\,\mathrm{V}$
Noise Power Spectral Density	$\Psi_0 = 2 \cdot 10^{-8}\,\mathrm{V}^2/\mathrm{Hz}$
Pulse Duration (Symbol Duration)	$T_{\mathrm{s}} = 1\,\mu\mathrm{s}$
Half Vertical Eye Opening	$U_{\mathrm{A}} = 1.00\,\mathrm{V}$
Noise Power	$U_{\mathrm{R}}^2 = 0.02\,\mathrm{V}^2$
Signal-to-Noise Ratio	$\varrho = 50{,}0$
Bit Error Probability	$P_{\mathrm{b}} = 7.69 \cdot 10^{-13}$

Because in the baseband transmission system with root-cosine roll-off transmit and receive filters, no intersymbol interference remains at the detection point, and also the band limitation of the receive filter causes a beneficial limitation of the noise power, there is a further improvement in transmission quality compared to the previously considered example system with NRZ transmit pulses and Gaussian receive low-pass filter: The signal-to-noise ratio becomes larger or the bit error probability smaller. In particular, the now larger half vertical eye opening due to the absence of ISI contributes significantly to the improvement of the transmission quality; the noise power is only slightly reduced.

Some Diagrams of the Transmission System
This section is intended to illustrate, through graphical representations, how some time and spectral functions look like in the transmission system.

Figure 2.107 shows the amplitude frequency response of the receive filter $G_{\mathrm{e}}(f)$ with a roll-off factor of $r = 0.5$ and Fig. 2.108 shows the eye diagram of the useful signal at the detection point, each with the number of levels $s = 2$. It can be seen that the eye is fully open vertically at the optimal sampling time.

Fig. 2.107 Amplitude response of the receive filter with root raised cosine transmit and receive filtering (Example: $r = 0.5$)

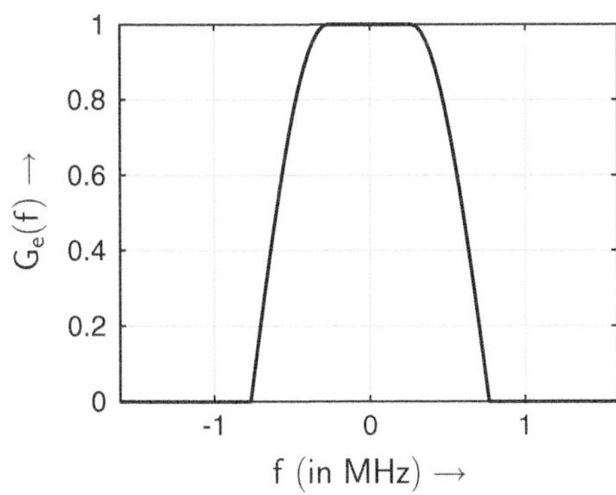

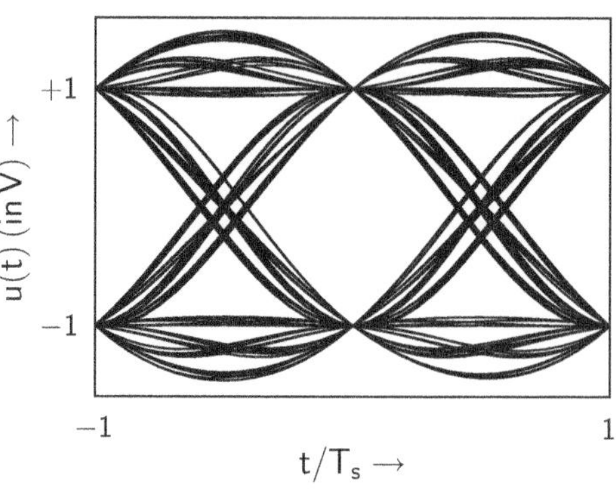

Fig. 2.108 Eye diagram of the useful signal at the detection point with root raised cosine transmit and receive filtering (Example: $r = 0.5$)

Influence of the Number of Signalling Levels

If the maximum modulation range or the average transmit power is limited and thus predetermined in the transmitter, the half-level transmit amplitude U_s decreases with increasing number of levels s: One obtains $(s - 1)$ eye openings, which become smaller with increasing number of levels s (see, for example, Fig. 2.71).

To illustrate the influence of the number of levels s on the performance of the transmission system, the numerical example is expanded again under otherwise identical conditions regarding signal and noise, so that the number of steps is variable while the bit rate remains constant.

Given a predetermined bit rate $f_B = 1\,\text{MHz}$ and average transmist power $P_s = 1\,\text{V}^2$, the respective symbol clock frequencies f_T according to (2.195) and half-level transmit amplitudes U_s according to (2.295) are to be determined and thus the previously described calculations for eye opening U_A, noise power U_R^2, signal-to-noise ratio ϱ, and possibly the bit error probability P_b are to be carried out.

Some results for the numerical example with s-level baseband transmission are shown in Table 2.6.

In Fig. 2.109, the achievable signal-to-noise ratio ϱ as a function of the number of steps s is shown to indicate the achievable transmission quality.

Table 2.6 Some results for the s-level numerical example with root raised cosine roll-off transmit and receive filters at constant bit sequence frequency $f_B = 1\,\text{MHz}$

s	2	4	8	16	32
U_s (in V)	1.000	0.447	0.218	0.108	0.054
f_T (in MHz)	1.000	0.500	0.333	0.250	0.200
U_A (in V)	1.000	0.447	0.218	0.108	0.054
U_R^2 (in V^2)	0.020	0.010	0.007	0.005	0.004
ϱ	50.00	20.00	7.143	2.353	0.733
P_b	$7.69 \cdot 10^{-13}$	$2.90 \cdot 10^{-6}$	$2.20 \cdot 10^{-3}$	0.029	0.076

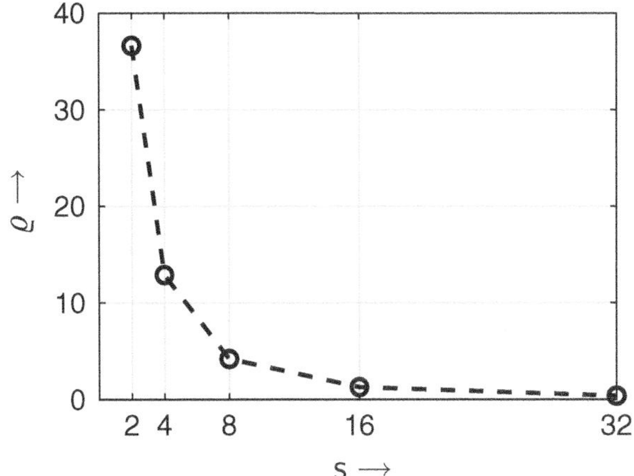

Fig. 2.109 Signal-to-noise ratio ϱ as a function of the number of levels s in bipolar, multi-level baseband transmission with root-cosine roll-off transmit and receive filtering at a given bit sequence frequency $f_B = 1\,\text{MHz}$

It becomes clear again that the transmission quality of the baseband transmission system decreases with increasing number of levels s.

Due to the reduced clock frequency with increasing number of levels according to (2.195), the receive filter can operate at a smaller bandwidth—and the noise power at the detection point decreases with increasing number of levels s. The effect of the reduced vertical eye opening due to multi-level operation predominates again, so that the signal-to-noise ratio decreases or the bit error probability increases when the number of levels s is increased in a distortion-free channel with pure noise disturbance. Only in real channels with bandwidth limitation, e.g., when transmitting over copper cable, it may be advantageous to work with a higher number of levels (than $s = 2$) to achieve improved transmission quality (see Chap. 4).

2.3.12 Discussion on Quality Criteria

When a digital transmission system is to be examined or optimized, a measure for the quality or goodness of data transmission must be found or defined, which is accessible to calculation or measurement [1]. A quality or goodness criterion forms the bridge between a general formulation of the goal of an investigation or optimization and the mathematical treatment of the resulting analysis or optimization problem [53]. The quality criterion chosen for the analysis or optimization of a transmission system determines to what extent the results obtained with a model of the transmission system correspond to the physical-technical reality and thus whether and to what extent they are practically usable [32].

As mentioned before, the most meaningful criterion for assessing the quality (or goodness) of a digital information transmission is the average bit error probability: It provides a statement about how many bits are incorrectly detected or interpreted by the receiver on average in relation to a number of transmitted bits. Often this error probability is difficult to calculate, as it depends on many characteristics of the specific transmission method (e.g., type of bit-symbol assignment, channel coding) and the transmission channel (e.g., time and frequency dispersion). Therefore, surrogate quality criteria are often used for the optimization of transmission systems—such as the signal-to-noise ratio—which must be in a monotonic relationship with the error probability in order to provide meaningful and practically usable results [1].

Different tasks in the analysis, design, and optimization of digital transmission systems may require different quality criteria: If general transmission possibilities on a given medium are to be captured, often many parameters are variable. In this case, quality criteria are useful that allow a clear investigation and optimization based on an analytical calculation. Above all, ratios of influencing variables must be correctly reproduced. If, on the other hand, a transmission system is to be evaluated or dimensioned in terms of its absolute performance, quality criteria are required that provide a statement about the actual average transmission quality. Then parameters of the transmission method and the characteristics of the involved systems are given or limited to a meaningful, usually narrow range, and the average error probability must be determined via a calculation or simulation [1].

2.4   Summary and Bibliographic Notes

The aim of this chapter was to introduce and describe basic principles of information transmission with digital signals over disturbed transmission channels. To this end, a physically intuitive basic model of baseband transmission over a non-distorting channel with pure noise disturbance was first used, with the help of which important characteristics of the useful signal and interference and their evaluation on the transmission path through the components of the transmission system were described and analyzed. An important insight is that every information transmission is disturbed and thus arrangements to limit the effects of the disturbances occurring on the transmission channel must be applied.

From the basic model, a number of starting points for improving the transmission quality of the baseband transmission system emerge. Therefore, in the further course of the chapter, criteria for evaluating the transmission quality were defined and derived. The considerations were extended towards optimality criteria for intersymbol interference freedom, robust pulse shapes with respect to deviations from the optimal sampling point, and for achieving the maximum signal-to-noise ratio at the detection point. Finally, an extension was made with regard to multi-level transmission in the baseband, which can be used to increase the throughput or the transferable bit rate or to improve the transmission quality given bandwidth limitation. Detailed examples of the aspects of message transmission introduced and discussed in this chapter round off the considerations.

Representations of the basics of transmission engineering, which are suitable for a deeper study of the contents dealt with in this chapter, can be found, for example, in [22, 26, 29, 42, 48, 55, 65, 72] (in German) and in English, for example, in [3, 5, 6, 52, 62].

2.5 Problems

Problem 2.1 (Baseband transmission with NRZ pulses and *RC* first-order low-pass)

The transmitter of a baseband transmission system sends a unipolar, rectangular NRZ binary signal $u_0(t)$ (NRZ: Non Return To Zero) with the amplitude U_0 and the pulse width T_s. The useful signal is disturbed on the transmission path by white, Gaussian noise with the constant noise power spectral density Ψ_0 (in V^2/Hz). The receive filter is a first-order *RC* low-pass filter (a real pole at the 3-dB cutoff frequency ω_g or f_g).

To be determined are:

a) the basic receive pulse $x_e(t)$ at the detection point and the resulting function of the received useful signal $u_e(t)$,

b) the half vertical eye opening U_A with redundancy-free coding as a function of the receive filter cutoff frequency f_g,

c) the function of the effective value square U_R^2 of the noise at the detection point as a function of the receive filter cutoff frequency f_g,

d) the signal-to-noise ratio ϱ at the detection point, also as a function of the receive filter cutoff frequency f_g,

e) the signal-to-noise ratio ϱ at the optimal receive filter cutoff frequency $f_{g\,opt}$ with the given values $U_0 = 5\,V$, $T_s = 1\,\mu s$ and $\Psi_0 = 10^{-6}\,V^2/Hz$ and

f) the bit error probability P_b.

Problem 2.2 (Nyquist Criteria)

A source sends Dirac pulses $u_q(t) = U_0\,T_s\,\delta(t)$. The spectrum of a basic receive pulse $x_e(t)$ is known with $X_e(f) = U_0\,T_s\,si^2(\pi f T_s)$. $T_s = 1\,\mu s$ and $U_0 = 2\,V$ are given. It is to be checked whether this basic receive pulse

a) meets the first Nyquist criterion and/or

b) meets the second Nyquist criterion.

If the two Nyquist criteria (or one of the two) are met, this must be demonstrated.

The spectrum of the received pulse $X_e(f)$ should—taking into account the spectrum of the source signal—be evenly distributed on the transmit filter $G_s(f)$ and receive filter $G_e(f)$ assuming an ideal or distortion-free channel.

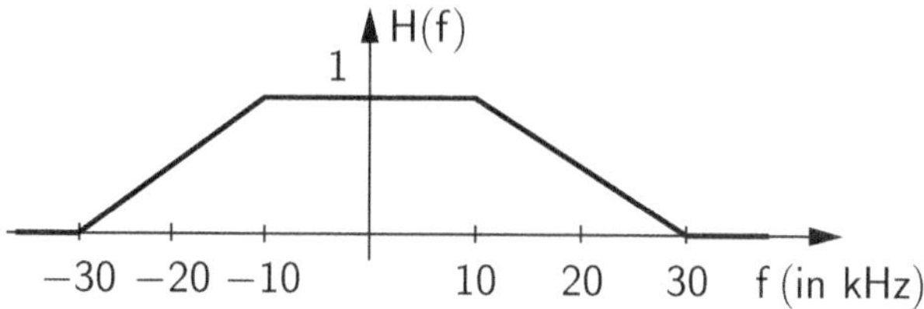

Fig. 2.110 Target total transfer function $H(f) = G_s(f) \cdot G_e(f)$ of the data transmission system to be designed

c) What magnitude do the transfer functions of the transmit and receive filters have?

d) The signal-to-noise ratio at the detection point for the detection of a single pulse is to be calculated when the useful signal is disturbed by white, Gaussian noise with the constant spectral noise power density $\Psi_0 = 10^{-6} \text{ V}^2/\text{Hz}$.

Problem 2.3 (Nyquist Criteria and Symbol Clock Frequency)

A data transmission system is to be designed. The desired total transfer function from the transmit and receive filters can be found in Fig. 2.110.

With which symbol rate or clock frequency can this system transmit without intersymbol interference if the channel is distortion-free?

Problem 2.4 (Matched Filter)

An information transmitter transmits a rectangle impulse $u_0(t)$ with the amplitude U_0 and the pulse width T_s. The useful signal is disturbed by white, Gaussian noise with the constant spectral noise power density Ψ_0.

The signal-to-noise ratio present at the detection point is to be calculated when the following are used as receive filters alternatively

a) a first-order RC low-pass filter,
b) a Gaussian low-pass filter,
c) a matched filter

are applied.

2.6   Solutions of the Problems

Solution of Problem 2.1 This task can be advantageously solved using the extensively derived relationships and results for the basic model of baseband transmission of NRZ rectangular pulses and receive filtering with RC low-pass in Chap. 2 and especially in reference to the section *Numerical example: Baseband transmission over a channel with pure noise disturbance*.

a) The time function $x_e(t)$ of the basic receive pulse at the detection point is derived from the extensively derived result (2.18) to

$$x_{\mathrm{e}}(t) = U_0 \left[\left(1 - e^{-\frac{t}{T_{\mathrm{g}}}} \right) \cdot 1(t) - \left(1 - e^{-\frac{t-T_{\mathrm{s}}}{T_{\mathrm{g}}}} \right) \cdot 1(t - T_{\mathrm{s}}) \right]. \tag{2.296}$$

It is shown in Fig. 2.6. An exemplary resulting received useful signal $u_{\mathrm{e}}(t)$ is shown in Fig. 2.9.

b) The half vertical eye opening is derived from (2.35), there with derivation, as a function of the receive filter cutoff frequency f_{g} to

$$U_{\mathrm{A}} = U_0 \left(\frac{1}{2} - e^{-2\pi f_{\mathrm{g}} T_{\mathrm{s}}} \right) \qquad \text{for} \qquad f_{\mathrm{g}} \geq \frac{\ln 2}{2\pi T_{\mathrm{s}}}. \tag{2.297}$$

c) The function of the noise effective square value U_{R}^2 at the detection point as a function of the receive filter cutoff frequency f_{g} is according to (2.59), there with derivation,

$$U_{\mathrm{R}}^2 = \Psi_0 \, \pi \, f_{\mathrm{g}}. \tag{2.298}$$

d) For the signal-to-noise ratio ϱ at the detection point, also as a function of the receive filter cutoff frequency f_{g}, one obtains after (2.60) with the results for the half vertical eye opening U_{A} and the noise power U_{R}^2

$$\varrho = \frac{U_{\mathrm{A}}^2}{U_{\mathrm{R}}^2} = \frac{U_0^2 \left(\frac{1}{2} - e^{-2\pi f_{\mathrm{g}} T_{\mathrm{s}}} \right)^2}{\Psi_0 \, \pi \, f_{\mathrm{g}}}, \tag{2.299}$$

where practically usable results are only to be expected for vertically opened eye and thus for $f_{\mathrm{g}} \geq \ln 2/(2\pi T_{\mathrm{s}})$.

With the given numerical values, the numerical results are calculated below.

Optimal receive filter cutoff frequency (via representation of the signal-to-noise ratio ϱ as a function of f_{g} and reading off the optimal cutoff frequency $f_{\mathrm{g\,opt}}$ there, where the signal-to-noise ratio is maximum): $f_{\mathrm{g\,opt}} = 394\,\mathrm{kHz}$ (or as an approximation $f_{\mathrm{g\,opt}} \cdot T_{\mathrm{s}} \approx 0.4 \longrightarrow f_{\mathrm{g\,opt}} \approx 400\,\mathrm{kHz}$).

If one sets as the receive filter cutoff frequency f_{g} the optimal cutoff frequency $f_{\mathrm{g\,opt}} = 394\,\mathrm{kHz}$ into the obtained equations, one obtains the further numerical results:
$U_{\mathrm{A}} = 2.079\,\mathrm{V}, U_{\mathrm{R}}^2 = 1.238\,\mathrm{V}^2, \varrho = 3.493 \, \hat{=} \, 5.432\,\mathrm{dB}$

e) The error probability for binary transmission is derived from (2.80), there with derivation, to

$$P_{\mathrm{f}} = \frac{1}{2} \left[1 - \mathrm{erf}\left(\sqrt{\frac{\varrho}{2}} \right) \right] = \frac{1}{2} \, \mathrm{erfc}\left(\sqrt{\frac{\varrho}{2}} \right). \tag{2.300}$$

In binary or two-level transmission, this probability P_{f} that a received symbol is decided incorrectly with respect to the decision threshold, coincides with the bit error probability P_{b}.

As a numerical result, one obtains (e.g., when using a computer algebra or spreadsheet program) $P_{\mathrm{f}} = 3.08 \cdot 10^{-2}$.

If one uses the table of the erfc function in Appendix E, the argument $x = \sqrt{\varrho/2} = 1.3216$ must first be calculated. As a (rough) value, one obtains by reading off for $x = 1.3 \longrightarrow \text{erfc}(x) = 0.0659$ and finally by division by 2 the approximate value $P_f \approx 3.3 \cdot 10^{-2}$ for the bit error probability. An improvement in accuracy when using the table of the erfc function can be achieved if the more accurate tables available in reference works, worked out to several decimal places, are used.

Solution of Problem 2.2 First, the given received signal is described in the time and frequency domain using the relationship

$$x_e(t) = U_0 \begin{cases} 1 - \frac{|t|}{T_s} & \text{for} \quad |t| \le T_s \\ 0 & \text{for} \quad |t| > T_s \end{cases} \quad \circ\!\!-\!\!\bullet \quad X_e(f) = U_0\, T_s\, \text{si}^2\left(\pi f T_s\right) \quad (2.301)$$

and is shown in Fig. 2.111.

a) Compliance with the first Nyquist criterion: The basic receive pulse $x_e(t)$ is sampled at the symbol rate, i.e., multiplied by a periodic Dirac pulse sequence (see Appendix A), and one obtains

$$x_e(t) \cdot C \sum_{k=-\infty}^{+\infty} \delta(t - k\, T_s) = C \sum_{k=-\infty}^{+\infty} x_e(k\, T_s) \cdot \delta(t - k\, T_s). \quad (2.302)$$

According to Fig. 2.112, all Dirac pulses outside the range $-T_s < t < T_s$ are multiplied by zero; they therefore disappear in the result. Only the Dirac pulse evaluated with $x_e(0) \cdot C$ at $t = 0$ remains, i.e.,

$$x_e(0) \cdot C \cdot \delta(t), \quad (2.303)$$

whose spectrum is a constant

$$C \cdot f_T \sum_{\mu=-\infty}^{+\infty} X_e(f - \mu f_T) = C \cdot x_e(0) = \text{const.} \quad (2.304)$$

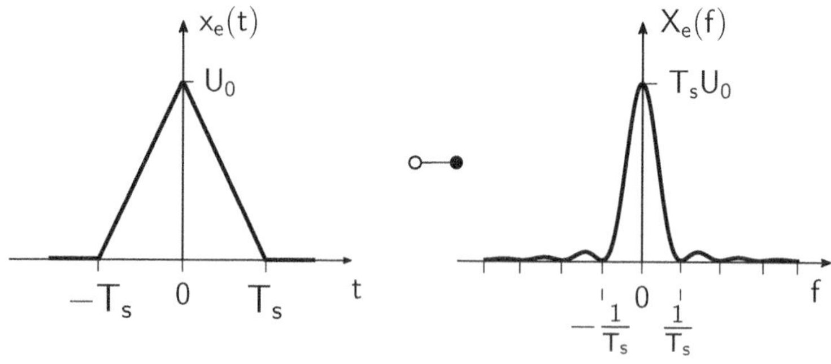

Fig. 2.111 Triangle basic receive pulse in time and frequency domain

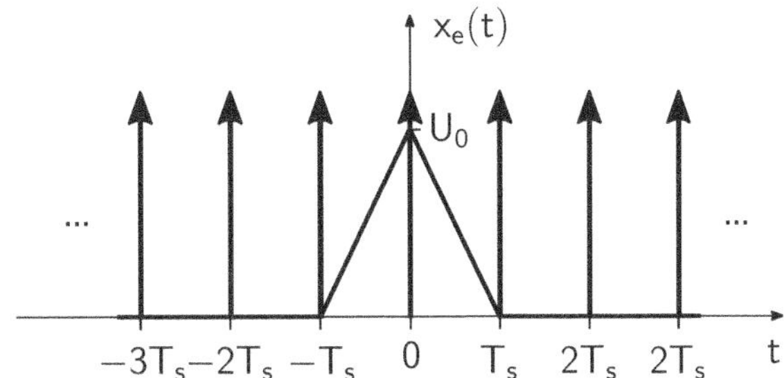

Fig. 2.112 Illustration of the triangular basic receive pulse with sampling at the symbol rate to check for compliance with the first Nyquist criterion

$\Longrightarrow$ The first Nyquist criterion is therefore met.

b) Compliance with the second Nyquist criterion: The basic receive pulse $x_\mathrm{e}(t)$ is again sampled at the symbol rate, i.e., multiplied by a periodic Dirac pulse sequence (see Appendix A), but now shifted by $T_\mathrm{s}/2$, and one obtains

$$x_\mathrm{e}(t) \cdot C \sum_{k=-\infty}^{+\infty} \delta\left(t - \frac{T_\mathrm{s}}{2} - k\,T_\mathrm{s}\right). \tag{2.305}$$

According to Fig. 2.113, all Dirac pulses outside the range $-T_\mathrm{s} < t < T_\mathrm{s}$ are again multiplied by zero; they therefore disappear in the result. What remains is

$$x_\mathrm{e}\left(\frac{T_\mathrm{s}}{2}\right) \cdot C \cdot \delta\left(t - \frac{T_\mathrm{s}}{2}\right) + x_\mathrm{e}\left(-\frac{T_\mathrm{s}}{2}\right) \cdot C \cdot \delta\left(t + \frac{T_\mathrm{s}}{2}\right) \tag{2.306}$$

or

$$x_\mathrm{e}\left(\frac{T_\mathrm{s}}{2}\right) \cdot C \cdot \left[\delta\left(t - \frac{T_\mathrm{s}}{2}\right) + \delta\left(t + \frac{T_\mathrm{s}}{2}\right)\right]. \tag{2.307}$$

With the correspondences

$$\delta(t + T) \;\circ\!\!-\!\!\bullet\; e^{j2\pi fT}$$
$$\delta(t - T) \;\circ\!\!-\!\!\bullet\; e^{-j2\pi fT}$$

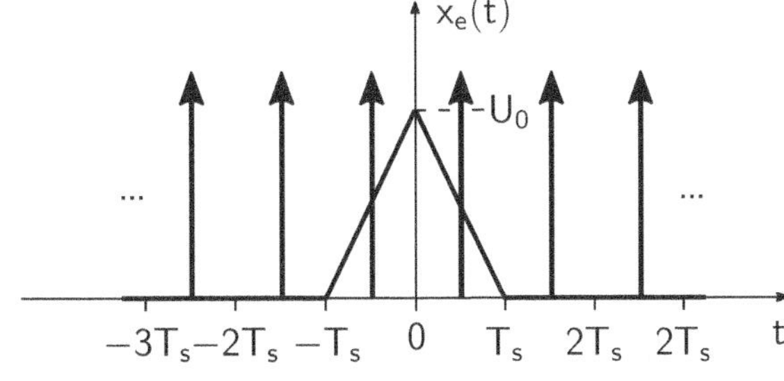

Fig. 2.113 Illustration of the triangular basic receive pulse with sampling at the symbol rate to check for compliance with the second Nyquist criterion

and the relationship

$$\cos(x) = \frac{1}{2}\left(e^{jx} + e^{-jx}\right) \tag{2.308}$$

one can write

$$\sum_{\mu=-\infty}^{+\infty} (-1)^{\mu} X_{\mathrm{e}}(f - \mu f_{\mathrm{T}}) = \left[e^{j2\pi f \frac{T_s}{2}} + e^{-j2\pi f \frac{T_s}{2}}\right] \cdot C \cdot x_{\mathrm{e}}\left(\frac{T_s}{2}\right)$$

$$= x_{\mathrm{e}}\left(\frac{T_s}{2}\right) \cdot C \cdot 2\cos(\pi f T_s)$$

$$= \text{const.} \cdot \cos(\pi f T_s)$$

and a cosine-shaped characteristic is recognizable.

$\Longrightarrow$ The second Nyquist criterion is therefore also met.

c) The source sends Dirac pulses with the area $U_0 T_s$ as the source signal and the received spectrum according to the task is $X_{\mathrm{e}}(f) = U_0 T_s \, \mathrm{si}^2(\pi f T_s)$. This leaves for the series connection of transmit and receive filter the portion

$$G_{\mathrm{s}}(f) \cdot G_{\mathrm{e}}(f) = \mathrm{si}^2(\pi f T_s). \tag{2.309}$$

With equal distribution on the transmit and receive side, the transmit and receive filter transfer function results in

$$G_{\mathrm{s}}(f) = G_{\mathrm{e}}(f) = \mathrm{si}(\pi f T_s). \tag{2.310}$$

Transmit and receive filters are thus root-Nyquist filters, since the series connection of both meets the first—and here also the second—Nyquist criterion. In addition, in this special case, the transmit and receive filters are Nyquist systems themselves, as they each meet the first Nyquist criterion. However, this does not always have to be the case—and it is not necessary for reliable signal transmission—to maintain the first Nyquist criterion via the cascade of transmit and receive filters.

d) The receive filter $G_{\mathrm{e}}(f)$ is a filter matched to the transmit filter $G_{\mathrm{s}}(f)$ or to the transmit spectrum $X_{\mathrm{s}}(f) = U_0 T_s G_{\mathrm{s}}(f)$ (Matched Filter). Since the channel does not cause distortions, the signal-to-noise ratio can therefore be advantageously calculated using the relationship valid for signal-matched filtering (2.160):

$$\varrho = \frac{E_0}{\Psi_0}. \tag{2.311}$$

The energy E_0 of the transmit pulse is obtained via

$$E_0 = \int_{-\infty}^{+\infty} u_0^2(t)\, dt = \int_{-\frac{T_s}{2}}^{+\frac{T_s}{2}} U_0^2\, dt = U_0^2\left[\frac{T_s}{2} - \left(-\frac{T_s}{2}\right)\right] = U_0^2 T_s, \tag{2.312}$$

and with the given noise power spectral density Ψ_0, the desired signal-to-noise ratio results in

$$\varrho = \frac{E_0}{\Psi_0} = \frac{U_0^2 \, T_s}{\Psi_0} = \frac{2^2 \, \text{V}^2 \cdot 10^{-6} \, \text{s}}{10^{-6} \, \text{V}^2/\text{Hz}} = 4. \tag{2.313}$$

Solution of Problem 2.3 The task is to dimension a data transmission system in such a way that, given a total transmission function $H(f) = G_s(f) \cdot G_e(f)$, an intersymbol interference-free transmission with maximum symbol rate is realized (see also Fig. 2.114).

An intersymbol interference-free transmission can be achieved after sampling in the symbol clock with the sampling frequency $f_a = 1/T_s$ if a constant results due to the periodicity that occurs in the frequency range. As shown in Fig. 2.115, the sampling frequency must assume the value

$$f_a = \frac{1}{T_s} = 40 \text{ kHz}$$

in order for a constant to result in the frequency range. Since the value $f_T = 1/T_s$ represents the symbol sequence or clock frequency, in this case, a symbol sequence frequency of $f_T = 40\,\text{kHz}$ can be transmitted interference-free over the given system. A statement about the possible bit sequence frequency cannot be derived, as no information is given about the constellation level of the transmission.

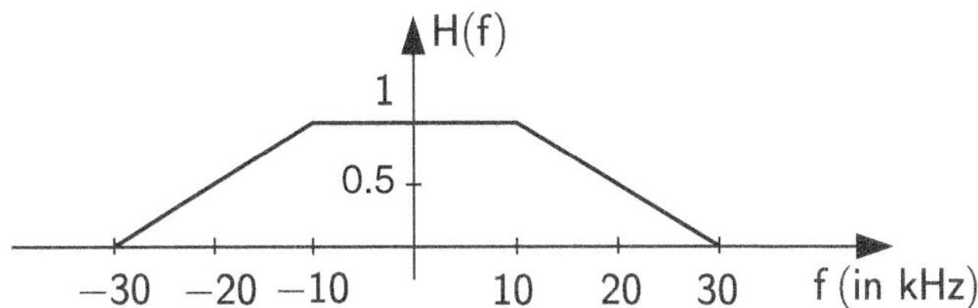

Fig. 2.114 Total transfer function $H(f) = G_s(f) \cdot G_e(f)$ of a given data transmission system

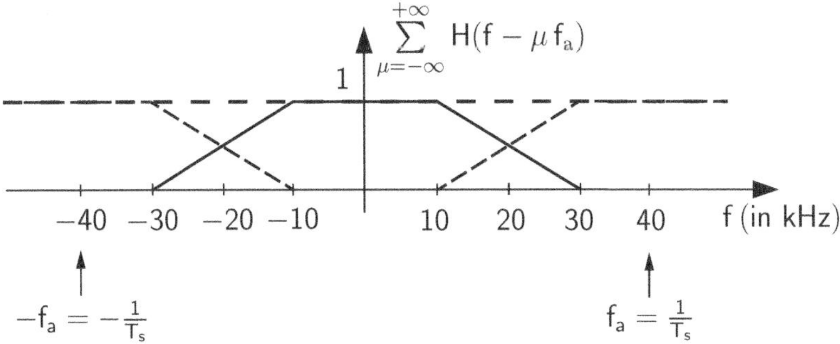

Fig. 2.115 Illustration of sampling in the frequency range

Solution of Problem 2.4

a) The signal-to-noise ratio achievable with a first-order *RC* low-pass filter is calculated
   in the example *Comparison of Matched Filter and First-Order RC Low-Pass filter*.
   This result is used here.

   For the receive filtering with a first-order *RC* low-pass filter, the maximum useful signal sample value at the time $t = T_A = T_s$ is obtained from the rectangular response of
   the received pulse (2.18) to

$$u_e(t = T_s) = U_0\left(1 - e^{-\frac{T_s}{T_g}}\right) = U_0\left(1 - e^{-2\pi f_g T_s}\right) \tag{2.314}$$

   and the noise power at the detection point is according to (2.59)

$$U_R^2 = \Psi_0\,\pi f_g, \tag{2.315}$$

   so that the signal-to-noise ratio

$$\varrho_{RC} = \frac{U_0^2\left(1 - e^{-2\pi f_g T_s}\right)^2}{\Psi_0\,\pi f_g} \tag{2.316}$$

   results. The optimization of the cutoff frequency of the receive filter f_g results in the
   case of the single pulse $f_g T_s \approx 0.2$ (see Fig. 2.57), so that the signal-to-noise ratio

$$\varrho_{RC} = \frac{U_0^2\left(1 - e^{-2\pi \cdot 0.2}\right)^2}{\Psi_0\,\pi \cdot 0.2/T_s} \tag{2.317}$$

   is obtained. With the given example numerical values, the result is

$$\varrho_{RC} = 0.8145. \tag{2.318}$$

b) For the receive filtering with a Gaussian low-pass filter, the maximum useful signal
   sample value at the time $t = T_A = 0$ is obtained from the rectangular response of the
   received pulse (2.89) to

$$u_e(t = 0) = U_0\left[\mathrm{erf}\left(\frac{\pi f_G}{\sqrt{\ln(2)}}\left(\frac{T_s}{2}\right)\right)\right] \tag{2.319}$$

   and the noise power at the detection point is according to (2.101)

$$U_R^2 = \Psi_0\,\frac{\sqrt{\pi}}{\sqrt{2\ln 2}}f_G, \tag{2.320}$$

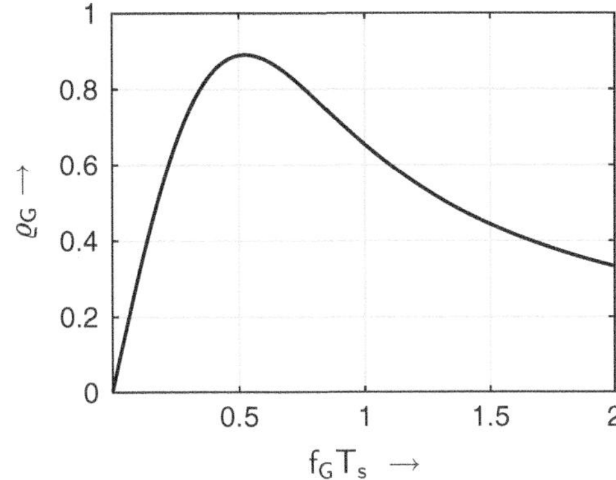

Fig. 2.116 Signal-to-noise ratio ϱ_G as a function of the receive filter cutoff frequency f_G normalized to the symbol duration T_s

so that the signal-to-noise ratio

$$\varrho_G = \frac{U_0^2 \left[\operatorname{erf}\left(\frac{\pi f_G}{\sqrt{\ln(2)}}\left(\frac{T_s}{2}\right)\right)\right]^2}{\Psi_0 \frac{\sqrt{\pi}}{\sqrt{2\ln 2}} f_G} \tag{2.321}$$

results. The optimization of the cutoff frequency of the receive filter f_G results in the case of the single pulse $f_G T_s \approx 0.525$ (see Fig. 2.116), so that with the given example numerical values, the signal-to-noise ratio

$$\varrho_G = 0.8901 \tag{2.322}$$

is obtained.

c) For the matched receive filtering, the energy of the transmitted pulse results in

$$E_0 = \int_{-\infty}^{+\infty} u_0^2(t)\,\mathrm{d}t = U_0^2 \cdot T_s \tag{2.323}$$

and thus the signal-to-noise ratio achievable according to (2.160) is

$$\varrho_{MF} = \frac{E_0}{\Psi_0} = \frac{U_0^2 T_s}{\Psi_0}. \tag{2.324}$$

With the given example numerical values, the result is

$$\varrho_{MF} = 1. \tag{2.325}$$

Summary of the results:

In Table 2.7, the results for the maximum achievable signal-to-noise ratios for the three considered types of receive filters are compiled. The maximum value is achieved

Table 2.7 Results for the signal-to-noise ratios with different types of receive filtering using the example of single pulse transmission

RC first-order low-pass	Gaussian low-pass	Matched Filter
$\varrho_{RC} = 0.8145$	$\varrho_{G} = 0.8901$	$\varrho_{MF} = 1$

with a signal-adapted filter (Matched Filter); however, such a filter is not always easy to implement. The approximately practically realizable Gaussian low-pass filter allows about 89 % of the performance of a Matched Filter and the first-order RC low-pass filter still allows about 81 % with comparatively low implementation effort.

Transmission Channel Copper Cable

3

Abstract

Communication transmission channels are the heart of transmission systems: They establish a connection between a transmitter and a receiver, which allows information to be transmitted between the two endpoints via physical-technical means. Therefore, they are often at the center of communications engineering considerations or they form the starting point for the design of transmission systems, with the help of which information is to be transmitted over a given channel. After a brief general introduction to transmission channels and their classification, the focus of this chapter is on the description of copper lines as a channel for communication transmission. Starting from the electrical properties of copper two-wire lines, a system-theoretical description of the transmission channel copper cable is developed via the transmission line theory, which results in a cable transfer function and which is suitable and appropriate for the analysis and optimization of transmission systems that use such channels for information transmission. Finally, special types of couplings in communication cables are examined and also described system-theoretically, which result from the constructive structure of such cables and which are referred to as crosstalk.

3.1 Introduction to Communication Channels

A communication channel in a communication system represents the physical connection between a transmitter and a receiver, over which information-bearing signals can be transmitted. It is to be assumed as a component of a communication system. Its influence on the transmitted signal is determined by the interaction of properties of the transmitted

© The Editor(s) (if applicable) and The Author(s), under exclusive license to Springer Fachmedien Wiesbaden GmbH, part of Springer Nature 2025
C. Lange and A. Ahrens, *Digital Transmission Engineering*,
https://doi.org/10.1007/978-3-658-46789-0_3

signal (such as its bandwidth) with the channel properties (e.g., frequency dependence of the channel).

For wired digital information transmission with high data rates in long-distance traffic, telecommunications networks provide sufficient and scalable capacities with fiber-optic-based optical transmission engineering—with reference to the transmission medium and the transmission engineering. The requirements for data transmission speed are constantly increasing and often the distance from the access node (or the former exchange of the telephone networks) to the end customer or subscriber is a bottleneck in terms of data transmission speed. This last distance between the access node of the telecommunications network and the end customer or subscriber is, depending on geographical conditions, typically a few hundred meters to a few kilometers long. It is usually bridged or realized by the wired access networks, which—until a switch to fiber optic technology—are largely made up of multi-pair, symmetric copper cables, the local connection cables. These copper-based access networks were originally installed as part of the telephone networks for analog telephony voice transmission and they have also been used for digital information transmission for several decades, as they are almost universally available [71]. An important example of copper-based data signal transmission is modern DSL technology (digital subscriber line), through which today's broadband services are provided over these copper lines as modulated signals. Wired access networks are gradually being converted to optical transmission engineering. Due to the high financial and time expenditure required for this, this conversion towards an access network with fiber optic lines into every building or apartment will probably take a longer time. Thus, the symmetric pair of conductors within multi-pair cables is a very important transmission medium for supplying end customers or subscribers with wired broadband services. It will also retain its importance on short transmission routes in the medium term, e.g. within buildings and for special applications, as this allows the already installed communication network infrastructure to continue to be used and operated.

3.1.1 Basic Transmission Principles for Communication Channels

In order to design efficient wired communication systems, the properties of the symmetric pair of conductors within multi-pair cables as a transmission channel need to be suitably described in terms of pulse transmission and disturbance effects need to be captured and analyzed. The influence of the transmission channel will be modeled in the following by a transfer function $G_k(f)$ and an additive, Gaussian-distributed noise signal $n(t)$ with the power spectral density Ψ_0 [65] (Fig. 3.1). The importance of the superposition of white, zero-mean, Gaussian noise lies in the fact that in practice a multitude of independent disturbance or noise processes disturb the transmission. The superposition of these individual processes then leads to a Gaussian distribution according to the central limit theorem [48, 52], which means that almost every real channel has an AWGN component.

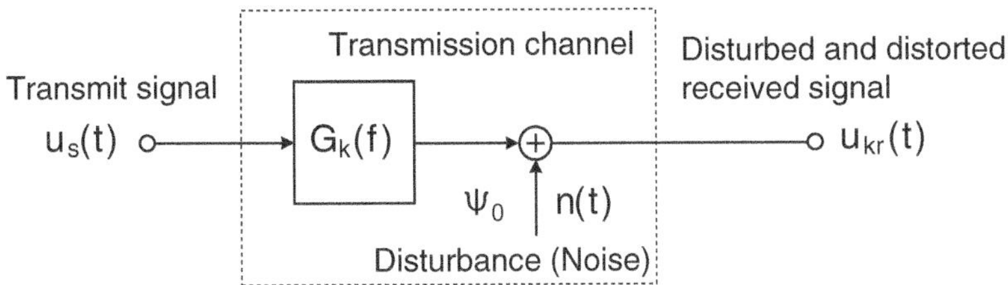

Fig. 3.1 Model of the cable transmission channel

Figure 3.1 shows a communication channel that maps the input signal (transmit signal) $u_s(t)$ to the generally distorted and noise-disturbed output signal $u_{kr}(t)$. The transmission channel is assumed to be time-invariant, i.e., the properties of the transmission channel do not change over time.[1] It is described by the weighting function $g_k(t)$, as the inverse Fourier transform of the transfer function $G_k(f)$. It would be desirable if the condition

$$u_{kr}(t) = g_0\, u_s(t - \tau_0) + n(t) \tag{3.1}$$

could be fulfilled. In this case, a constant delay τ_0 and attenuation g_0 would result for the transmit signal $u_s(t)$. In addition, the transmit signal is affected by disturbances, modeled by white, Gaussian-distributed noise.

Assuming a constant delay and attenuation, the weighting function of a communication channel becomes

$$g_k(t) = g_0\delta(t - \tau_0). \tag{3.2}$$

The Fourier transformation of the weighting function leads to the transfer function

$$G_k(f) = g_0 \cdot e^{-j\,2\pi f\,\tau_0}. \tag{3.3}$$

Figure 3.2 illustrates the resulting characteristics of the ideal channel in the time and frequency domain. In this case, distortion-free transmission can be ensured regardless of the data transmission speed. It should be noted that this can also be realized if the condition of a constant magnitude and delay characteristic only applies to the frequency range actually occupied by a specific input signal. However, the demand for a constant magnitude and delay characteristic for a frequency interval occupied by the input signal can only be met to a limited extent in practice.

---

[1] This assumption of time invariance of the channel can only be approximately fulfilled in practice, as even wired transmission channels, among other things, have a temperature dependence, so that their properties are strictly speaking time-variant. However, the influence of this temperature dependency and the resulting temporal dependence of the channel characteristics is usually significantly less than the relatively rapid changes in transmission properties, for example, of radio channels.

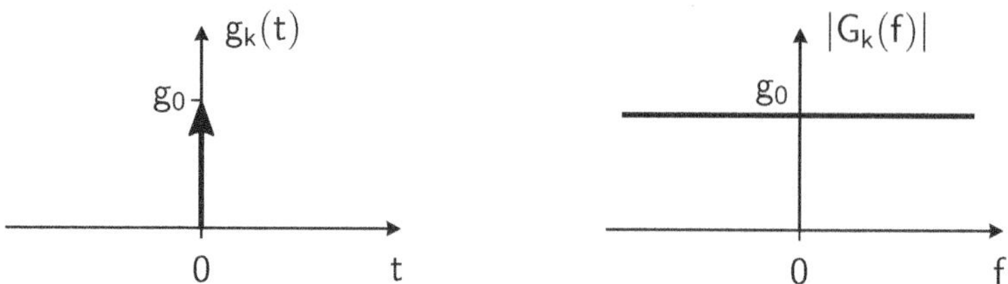

Fig. 3.2 Characteristics of the ideal channel in the time and frequency domain

Real transmission channels will always exhibit more or less strong frequency dependence due to frequency-dependent components and elements. Figure 3.3 illustrates the time and frequency characteristics of an exemplary real channel with non-ideal properties.

The properties of the transmit signal—here the frequency range occupied by the transmit signal—determine whether distortion-free transmission can be achieved given the channel characteristics. This becomes increasingly difficult with rising demands on the data rate, as a higher data rate increases the frequency range occupied by the transmit signal. In this case, the individual components of the transmit signal experience different amplitude and phase distortions, making distortion-free transmission unachievable. Channels with this property belong to the class of *time-dispersive* channels and exhibit *frequency-selective* behavior [42].

While the frequency-dependent characteristic of the channel can largely be disregarded at comparatively low bandwidth of the transmit signal and thus low data transmission speed (see Fig. 3.4), a greater distortion of the transmit signal arises at higher data transmission speed and thus greater bandwidth of the transmit signal (see Fig. 3.5): *Signal distortions* occur due to the influence of the transmission channel, which can become visible at the detection point as intersymbol interference.

3.1.2 Classification of Communication Channels

Communication channels can be found in various forms. In addition to copper cables (symmetrical, twisted pairs or coaxial cables), fiber optics or the radio channel play a

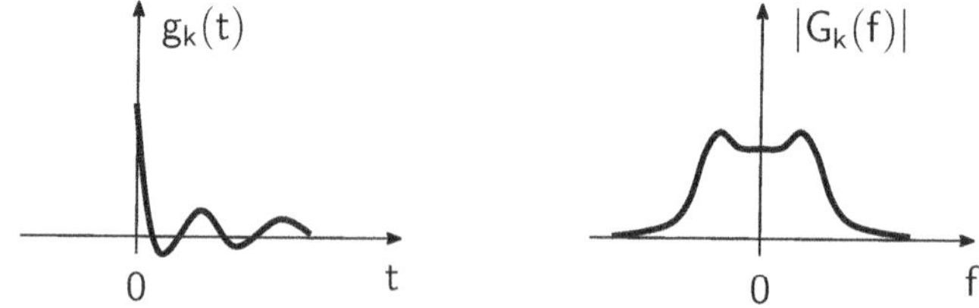

Fig. 3.3 Characteristics of a real channel in the time and frequency domain

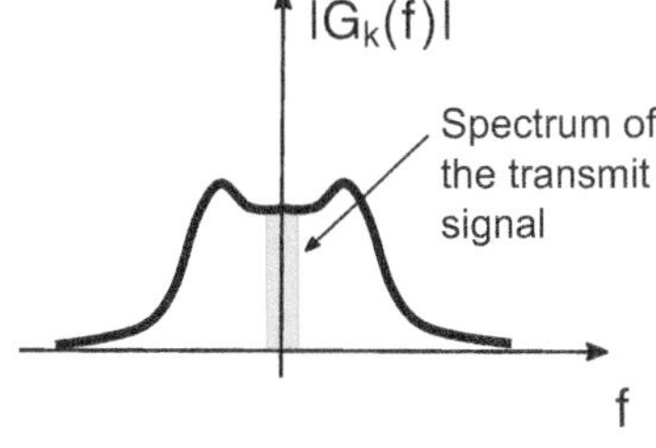

Fig. 3.4 Transmit signal and transmission channel (in the frequency range) at slow-speed transmission

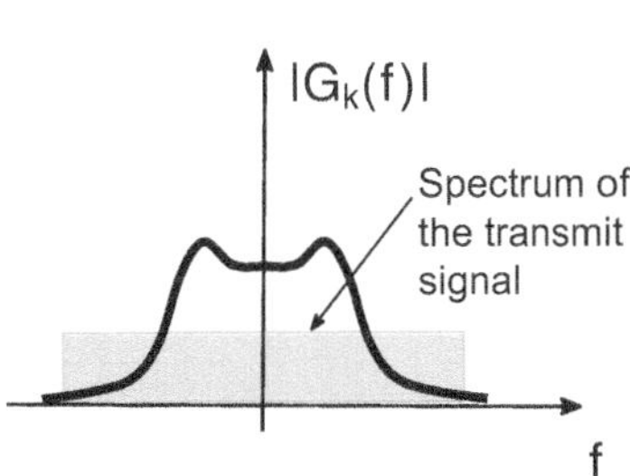

Fig. 3.5 Transmit signal and transmission channel (in the frequency range) at high-speed transmission

dominant role in information exchange and thus for the high-rate transmission of digital signals.

Such communication channels can be described by typical characteristics, such as the effect of the channel depending on the properties of the transmit signal. While at low data transmission speed all frequency components of the transmit signal are affected in the same way, for example by a frequency-independent attenuation, and therefore the transmission channel can also be described by a frequency-independent factor, at faster data transmission the individual frequency components of the transmit signal receive different evaluations. This behavior can be captured by a frequency-dependent transfer function. In Fig. 3.6, a classification of the transmission channels is given, which results depending on the properties of the transmit signal.

3.2   Introduction to Copper Lines

3.2.1   Introductory Remarks

In the previous section, some fundamentals of the description of communication channels were introduced. These will now be applied to the modeling of copper lines or cables as transmission channels.

In the transmission of information-bearing signals over a physical transmission channel, *signal distortions* generally occur and the transmission is additionally influenced by

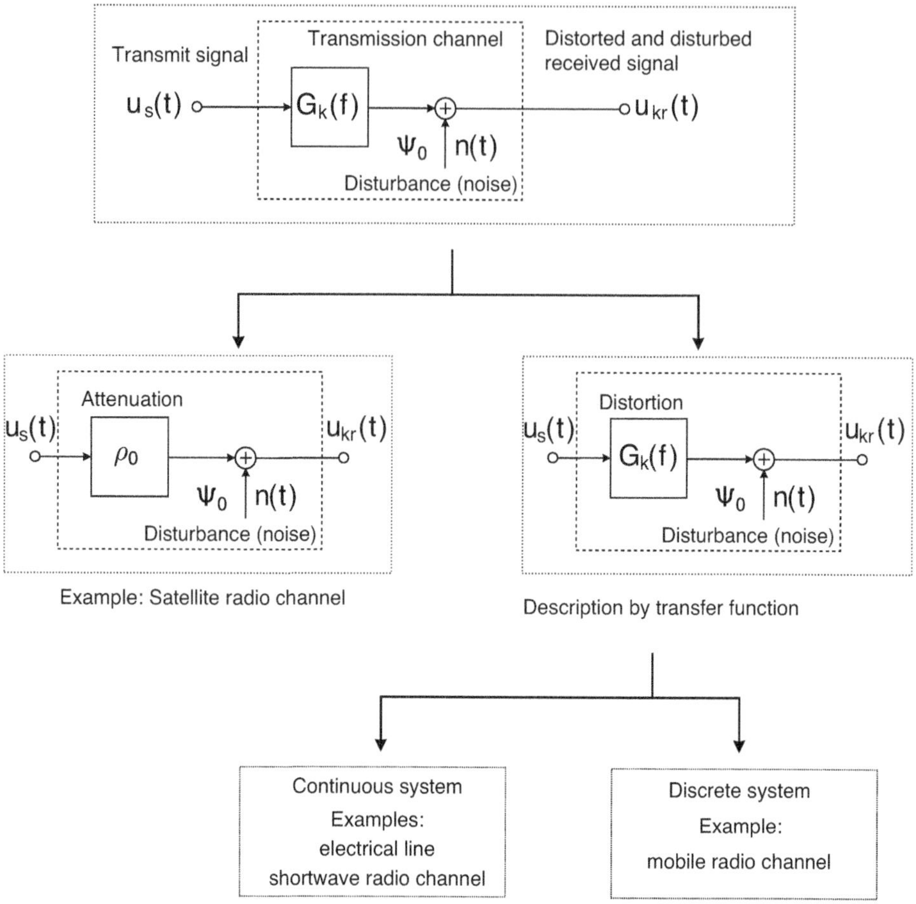

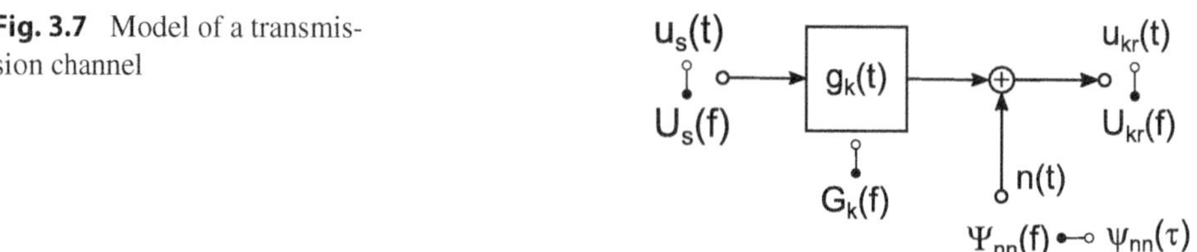

Fig. 3.6 Classification of communication channels according to system properties

noise disturbances. A general system-theoretical model of the transmission channel is shown in Fig. 3.7.

Linear signal distortions can advantageously be modeled using system theory by a channel transfer function $G_k(f)$ or by a weighting function $g_k(t)$ of the channel. Examples are signal distortions caused by frequency dependencies in the transmission path

Fig. 3.7 Model of a transmission channel

such as a frequency-dependent line attenuation in the transmission over copper cables or by filters in modules of the transmitter or receiver. Transmission channels can be time-variant and thus change their transmission properties over time. A prominent example of time-variant channels is the mobile wireless channel, whose transmission characteristics can change significantly in very short time intervals. Electrical lines or cables can also change their transmission properties over time, but typically on different time scales: For example, electrical conductors change their characteristics daily or seasonally with changing temperature. Similarly, cables can change their transmission properties due to aging, for example, if the insulation between the conductors changes due to aging processes. Since this time variability of the transmission properties of copper lines is comparatively slow, e.g., compared to the data transmission speed, it will not be further considered and modeled in the course of the book.

Noise disturbances, as already shown in the general channel description, are often modeled as white, Gaussian noise, i.e., as a stochastic process with a constant power density Ψ_0, whose amplitudes have a Gaussian or normal distribution. The theoretical basis is the central limit theorem [48, 52].

3.2.2 Electrical Lines and Cables

First, the terms *line* and *cable* should be defined more precisely: A *line* is generally understood to be a—usually technical—facility that can transmit a substance or matter, a physical state or energy or information over a distance. For example, there are lines for gas, water, and district heating. The lines used for the transport of electrical energy and for the transmission of information based on electromagnetic waves are accordingly referred to as *electrical lines or electrical wires*. These lines are the subject of the *transmission line theory* discussed in this chapter. In *cables*, for practical and economic reasons, several or a multitude of electrical lines or wires are often combined to protect them from environmental influences, for example. In practice, the terms *line, wire* and *cable* are often not very sharply distinguished, the meaning usually emerges from the respective context. For example, the high or medium voltage lines used for the transport of electrical energy are often overhead lines and the lines or wires combined in communication cables are often referred to as *cables*, even if perhaps only a single pair of wires consisting of two individual copper wires is meant.

Metallic conductors or electrical cables and wires are used in different forms in various areas of electrical engineering: Such wires and cables are used both for the supply of electrical energy as lines at various voltage levels and for the transmission of information as communication cables. With the help of the transmission line theory, the properties and behavior of electrical lines or wires for these very different fields of application in electrical power engineering and in communications engineering can be described universally and uniformly. A typical difference is that power supply lines are usually operated at *one* single frequency, while frequency bands are always transmitted on communication cables.

With the help of the transmission line theory, the transmission behavior of metallic wires is examined in this chapter first universally and then for harmonic excitation, i.e. for sinusoidal or cosine-shaped input variables in the steady state. Subsequently, practically important special cases are considered depending on the operating frequency. The focus is on information transmission and the particularly important case of the lossy transmission line (RC line) for the transmission of digital signals is analyzed and modeled in more detail. Important line parameters are derived and specified in each case.

In order to be able to advantageously describe and model the influence and effect of electrical wires on the transmission over communication cables on the signals with the means of signal and system theory, a transfer function for the significant case of the RC line in digital signal transmission is subsequently determined and a rational approximation is determined for this.

3.3   Transmission Line Theory

Electrical lines or wires are systems with distributed parameters. They serve for the transmission of electrical energy or power or the transmission of information through electrical signals.

Electrical energy propagates along metallic wires in the form of electromagnetic waves. The electrical processes occurring on wires are thus wave processes, i.e., the currents and voltages that occur are functions of time t and position x.

Figure 3.8 shows an example of a two-wire line consisting of two parallel conductors. The line has the length ℓ and the metallic conductors running at a distance d each have the radius r_0. The length ℓ of the line is defined as the one-way distance from the transmitter to the receiver or from the generator to the consumer.

The considerations on the theory of wires in this chapter largely follow the presentations in [46, 50, 56, 57, 67].

3.3.1   Electrical Line Parameters

An electrical line according to Fig. 3.8 is an electrical circuit consisting of two electrical conductors—a forward and a return conductor—which generally have an ohmic resistance R, a mutual inductance L, a mutual capacitance C, and a conductance G [58]: The resistance R provides information about the losses in metallic conductors, the conductance G describes the insulation losses and the dielectric losses in the insulation between the conductors, while the capacitance C reflects the capacitance existing between the conductors and the inductance L expresses the external and internal inductance of the line.

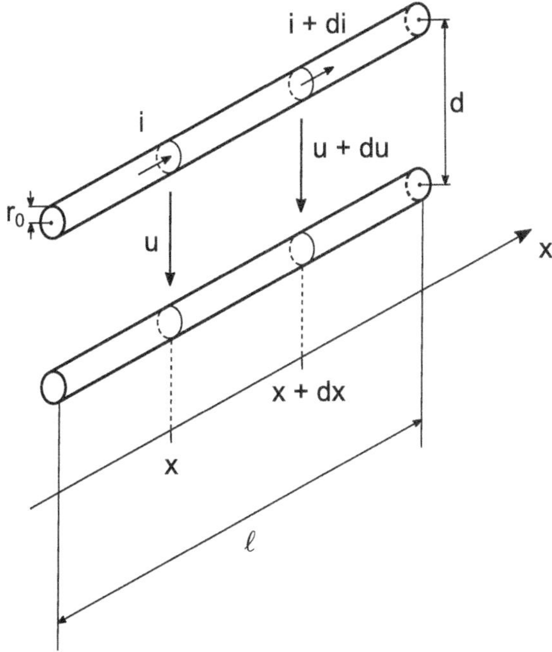

Fig. 3.8 Schematic representation of a two-wire line

An electrical line with its electrical properties—and thus its transmission behavior—can therefore in principle be understood and described as a passive electrical network consisting of the electrical elements *resistance R, conductance G, capacitance C* and *inductance L*, if the spatial extent of the line is also taken into account and recorded.

The electrical circuit elements describing the behavior of a line can be obtained from the material properties of the conductor and insulation material in conjunction with the constructive-geometric arrangement of the line [43].

Electrical Switching Elements for Describing Electrical Lines
The following considers possible approximate calculation options for the electrical switching elements of a two-wire line consisting of two parallel conductors of length ℓ.

The resistance R of a conductor can be specified with the help of the length 2ℓ of the conductor, its cross-sectional area $r_0^2\pi$ (circular area) and the specific resistance ρ or the specific conductivity of the conductor material as

$$R = \frac{2\ell\rho}{r_0^2\pi} \tag{3.4}$$

[43, 56]. It captures the losses in the metallic conductors. The length 2ℓ of the conductor is the loop length that is effective in the event of a short circuit at the end of the line.

The insulation conductance

$$G = \ell \, \frac{1}{R'_{\text{insulation}}} \tag{3.5}$$

describes the insulation losses and the dielectric losses in the insulation between the conductors. Here, $R'_{\text{insulation}}$ is an insulation resistance related to a unit of length (e.g. 1 km) that depends on the material of the insulation.

The capacitance

$$C = \ell \, \frac{\varepsilon_0 \varepsilon_{\text{r}} \pi}{\ln \frac{d}{r_0}} \tag{3.6}$$

describes the capacitance existing between the conductors. It depends on the line geometry (r_0 and d) and on the dielectric constant ($\varepsilon_0 \varepsilon_{\text{r}}$) of the insulation. Here, $\varepsilon_0 = 8.855 \cdot 10^{-12}$ As/Vm is the influence constant (the dielectric constant of the vacuum) and the relative dielectric constant or the relative permittivity $\varepsilon_{\text{r}} > 1$ is a material constant [21, 43, 50]. The relationship (3.6) applies to $d \gg r_0$ [50] and thus e.g. for parallel wire overhead lines, but strictly speaking not exactly for two-wire lines in cables; however, it also provides practically useful approximate values for this case.

The inductance

$$L = \ell \, \frac{\mu_0 \mu_{\text{r}}}{\pi} \ln \frac{d}{r_0} \tag{3.7}$$

captures on the one hand the external inductance occurring outside the electrical conductor material, which is determined by the geometry and magnetic material properties of the line, and on the other hand the internal inductance acting within the electrical conductors, which is caused by magnetic fields in the conductors. It depends on the line geometry (r_0 and d) and on the permeability ($\mu_0 \mu_{\text{r}}$), where $\mu_0 = 1.256 \cdot 10^{-6}$ Vs/Am is the permeability of the vacuum and μ_{r} as relative permeability is a material constant [43, 50]. The relationship (3.7) also applies to $d \gg r_0$ [50] and thus e.g. again for parallel wire overhead lines, but strictly speaking not exactly for two-wire lines in cables; however, it also provides practically useful approximate values for this case.

If the circuit elements obtained in this way are related to the length $\ell = 1$ km, the *kilometric line parameters* or *primary line parameters* or *transmission line parameters per unit length,* which are common in the description of electrical lines of different lengths, are obtained and they are used in the following.

Transmission Line Parameters per Unit Length of Electrical Lines
In order to describe lines independently of the length ℓ of the line under consideration, quantities related to the unit of length, the *transmission line parameters per unit length* or *primary line parameters* are introduced. They are usually related to 1 km line length and are referred to as

- resistance per unit length R' in the unit Ω/km,
- conductance per unit length G' in the unit S/km,
- capacitance per unit length C' in the unit F/km,
- inductance per unit length L' in the unit H/km.

These quantities thus represent line constants and allow a uniform description of lines of variable length. For the line parameters related to the length ℓ, the switching elements given below are obtained.

The *resistance per unit length*

$$R' = \frac{\rho}{r_0^2 \pi} \tag{3.8}$$

acts in the direction of transmission and is generally frequency-dependent due to the skin effect. Approximate values for resistances per unit length of communication cables are in the range of $R' = 10 \ldots 100\,\Omega/\text{km}$ and for overhead lines in the order of magnitude of $R' = 1 \ldots 10\,\Omega/\text{km}$ [57].

The *conductance per unit length*

$$G' = \frac{1}{R'_{\text{insulation}}} \tag{3.9}$$

acts perpendicular to the direction of transmission. It can also be specified using the loss factor $\tan\delta$:

$$G' = \omega C' \cdot \tan\delta. \tag{3.10}$$

Typical values for the kilometric conductances of cables (and overhead lines) are in the range of $G' = 0.1 \ldots 1\,\mu\text{S}/\text{km}$. The loss factor $\tan\delta$ depends on the insulating material used, the structural design, and often on frequency and temperature. It should be as small and constant as possible for all types of lines; typical values are $\tan\delta \approx 10^{-4} \ldots 10^{-3}\,(\ldots 10^{-1})$ [46, 57].

The *capacitance per unit length*

$$C' = \frac{\varepsilon_0 \varepsilon_r \pi}{\ln \frac{d}{r_0}} \tag{3.11}$$

is effective between the two conductors of a pair of conductors and is almost frequency-independent. Approximate values for capacitance per unit length of communication cables are in the range of $C' = 20 \ldots 50\,\text{nF}/\text{km}$ and for overhead lines in the order of magnitude of $C' = 5 \ldots 7\,\text{nF}/\text{km}$ [46, 57].

The *inductance per unit length*

$$L' = \frac{\mu_0 \mu_r}{\pi} \ln \frac{d}{r_0} \tag{3.12}$$

is associated with the magnetic field of the current in the direction of transmission. The external inductance is almost frequency-independent, the internal inductance, however, depends on the frequency. Typical values for kilometric inductances of communication cables are in the range of $L' = 0.3 \ldots 1\,\text{mH/km}$ and for overhead lines in the order of magnitude of $L' = 1 \ldots 2.5\,\text{mH/km}$ [46, 57].

3.3.2 Homogeneous Line

The transmission line parameters per unit length R', G', C' and L' are generally functions of time t and position x. However, for practically important lines, they can be assumed to be time-independent, i.e., they do not change with time. In addition, for mathematical treatment, it can be assumed that the line has constant electrical properties over its entire length, as, for example, the material, conductor diameter, insulation, and geometric-constructive design of cables can be considered constant. Such a line is referred to as a *homogeneous line*.

The homogeneous line is thus a model that does not exactly describe reality, but represents a sufficiently accurate approximation for many practically significant applications. Furthermore, a linear behavior of the line should be assumed, i.e. the electrical and magnetic parameters of the line should not depend on the size of the current and the voltage on the line.

For the analysis of the electrical behavior of lines, one assumes a line with constant electrical and magnetic properties and decomposes the line of length ℓ into short line elements of length dx and sums up the effects of all these infinitesimally short line elements: Therefore, an infinitesimal line element according to Fig. 3.9 is considered.

The line element of length dx includes the resistance $R'dx$, the conductance $G'dx$, the capacitance $C'dx$, and the inductance $L'dx$.

The ohmic resistance causes a voltage drop when current flows. The current is coupled with a magnetic alternating field, which induces a voltage: Therefore, the voltage along the line is not constant. The displacement current in the dielectric and the conduction current due to the finite conductivity of the dielectric (e.g., due to imperfect insulation, discharge currents, etc.) cannot be neglected, especially at high voltages or frequencies: The current also changes along the line.

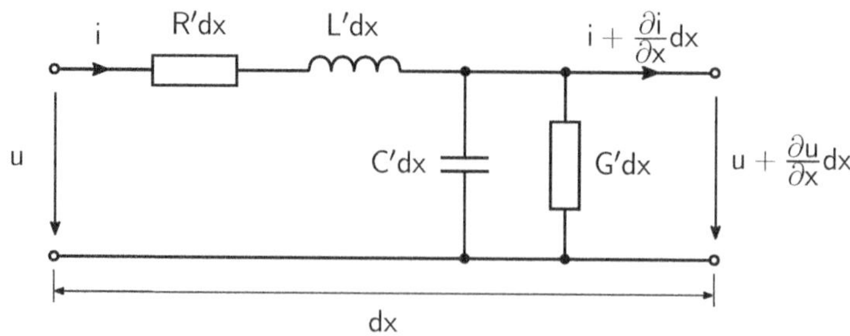

Fig. 3.9 Infinitesimal line element

If one assumes a homogeneous line, such a line can be approximated by a series connection of infinitely many infinitesimal line elements (see Fig. 3.10). Line elements of finite length are thus an approximation.

From the equivalent circuit diagram (Fig. 3.9) of a line element, the necessary mathematical relationships for the voltage and current are determined [50] using the application of Kirchhoff's laws (mesh and node rule, e.g., [43]). The instantaneous values of the voltage and the current at the input of a line element are denoted by u and i respectively, their instantaneous values at its output are then $u + (\partial u/\partial x)\, dx$ and $i + (\partial i/\partial x)\, dx$.

Insert: Explanation of Voltage and Current Change

The absolute change, for example, of the voltage at the point $x + dx$ compared to the voltage at the point x, when the time changes by dt, can be written as

$$du = \frac{\partial u}{\partial x} dx + \frac{\partial u}{\partial t} dt \qquad (3.13)$$

(differential calculus for functions with several variables [11]). If the same time moment is considered at the points x and $x + dx$, $dt = 0$ and one gets

$$du = \frac{\partial u}{\partial x} dx \qquad (3.14)$$

for the voltage change along the path element dx. This depends only on the step length dx and the tendency $\partial u/\partial x$ with which the voltage changes locally [57].

A similar consideration can be made for the current change along the path element dx.

If the network is passive, both du and di are negative.

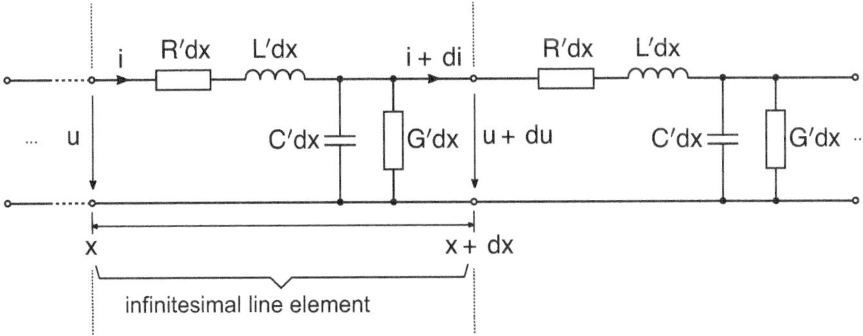

Fig. 3.10 Equivalent circuit for a homogeneous line (approximation)

The voltage change or the voltage drop at the line element of length dx is (according to the mesh rule $\sum U = 0$)

$$\underbrace{u}_{\text{voltage at line input}} - \underbrace{\left(u + \frac{\partial u}{\partial x}\mathrm{d}x\right)}_{\text{voltage at line output}} = \underbrace{i \cdot R'\mathrm{d}x}_{\text{voltage drop at resistance}} + \underbrace{\frac{\partial i}{\partial t}L'\mathrm{d}x}_{\text{voltage drop at inductance}} \,. \tag{3.15}$$

The current change at the line element of length dx can (according to the node rule $\sum I = 0$) be expressed as

$$\underbrace{i}_{\text{current at line input}} = \underbrace{\left(i + \frac{\partial i}{\partial x}\mathrm{d}x\right)}_{\text{current at line output}} + \underbrace{\left(u + \frac{\partial u}{\partial x}\mathrm{d}x\right)G'\mathrm{d}x}_{\text{insulation current}} + \underbrace{C'\mathrm{d}x\frac{\partial}{\partial t}\left(u + \frac{\partial u}{\partial x}\mathrm{d}x\right)}_{\text{current through capacitance}} \,.$$

$$\tag{3.16}$$

Equations (3.15) and (3.16) are first-order partial differential equations.

Insert: Derivation of the Transmission Line Equations

The transformation of the relationship (3.15) for the voltage characteristic by resolving the brackets and division by dx results in

$$u - \left(u + \frac{\partial u}{\partial x}\mathrm{d}x\right) = i \cdot R'\mathrm{d}x + \frac{\partial i}{\partial t}L'\mathrm{d}x$$

$$-\frac{\partial u}{\partial x}\mathrm{d}x = i \cdot R'\mathrm{d}x + \frac{\partial i}{\partial t}L'\mathrm{d}x \qquad\qquad \Big| : \mathrm{d}x$$

$$-\frac{\partial u}{\partial x} = R' \cdot i + L'\frac{\partial i}{\partial t}.$$

If one also transforms the relationship (3.16) for the current characteristic by resolving the brackets, summarizing, and dividing by dx, one obtains

$$i = \left(i + \frac{\partial i}{\partial x}\mathrm{d}x\right) + \left(u + \frac{\partial u}{\partial x}\mathrm{d}x\right)G'\mathrm{d}x + C'\mathrm{d}x\frac{\partial}{\partial t}\left(u + \frac{\partial u}{\partial x}\mathrm{d}x\right)$$

$$i = i + \frac{\partial i}{\partial x}\mathrm{d}x + uG'\mathrm{d}x + \frac{\partial u}{\partial x}G'\mathrm{d}x^2 + C'\mathrm{d}x\frac{\partial u}{\partial t} + C'\mathrm{d}x^2\frac{\partial}{\partial t}\left(\frac{\partial u}{\partial x}\right)$$

$$i = i + \frac{\partial i}{\partial x}\mathrm{d}x + uG'\mathrm{d}x + C'\mathrm{d}x\frac{\partial u}{\partial t} \qquad \Big| -i \qquad \Big| -\frac{\partial i}{\partial x}\mathrm{d}x$$

$$-\frac{\partial i}{\partial x}\mathrm{d}x = uG'\mathrm{d}x + C'\mathrm{d}x\frac{\partial u}{\partial t} \qquad\qquad \Big| : \mathrm{d}x$$

$$-\frac{\partial i}{\partial x} = G' \cdot u + C'\frac{\partial u}{\partial t}.$$

In the transition from the second to the third line, the two higher-order terms containing the term dx^2 were neglected, as they have no significant influence: The distance dx is very—infinitesimally—small, so that the terms of the relationships affected by dx^2 are much smaller than those containing dx, and they can therefore be neglected and omitted.

If one resolves the brackets in (3.15) and (3.16), neglects higher-order terms in the transformation of (3.16) and divides by dx, one obtains the *basic equations of transmission line theory*, also known as *transmission line equations*:

$$-\frac{\partial u}{\partial x} = R' \cdot i + L'\frac{\partial i}{\partial t} \tag{3.17}$$

$$-\frac{\partial i}{\partial x} = G' \cdot u + C'\frac{\partial u}{\partial t}. \tag{3.18}$$

If one differentiates (3.17) with respect to x and (3.18) with respect to t, the relationships

$$-\frac{\partial^2 u}{\partial x^2} = \frac{\partial}{\partial x}\left(R' \cdot i + L'\frac{\partial i}{\partial t}\right) = R'\frac{\partial i}{\partial x} + L'\frac{\partial}{\partial x}\left(\frac{\partial i}{\partial t}\right) \tag{3.19}$$

and

$$-\frac{\partial}{\partial t}\left(\frac{\partial i}{\partial x}\right) = \frac{\partial}{\partial t}\left(G' \cdot u + C'\frac{\partial u}{\partial t}\right) = G'\frac{\partial u}{\partial t} + C'\frac{\partial^2 u}{\partial t^2}. \tag{3.20}$$

are obtained.

Insert: Derivation of the Telegraph Equations
From (3.20) one obtains by swapping the differentiation with respect to x and t—which is allowed for continuous functions –

$$\frac{\partial}{\partial t}\left(\frac{\partial i}{\partial x}\right) = \frac{\partial}{\partial x}\left(\frac{\partial i}{\partial t}\right) = -G'\frac{\partial u}{\partial t} - C'\frac{\partial^2 u}{\partial t^2} \tag{3.21}$$

and from (3.18) one obtains by rearranging

$$\frac{\partial i}{\partial x} = -G' \cdot u - C'\frac{\partial u}{\partial t}. \tag{3.22}$$

If one substitutes (3.21) and (3.22) into (3.19), the result is

$$-\frac{\partial^2 u}{\partial x^2} = R'\left(-G'\cdot u - C'\frac{\partial u}{\partial t}\right) + L'\left(-G'\frac{\partial u}{\partial t} - C'\frac{\partial^2 u}{\partial t^2}\right)$$

$$-\frac{\partial^2 u}{\partial x^2} = -R'G'u - R'C'\frac{\partial u}{\partial t} - L'G'\frac{\partial u}{\partial t} - L'C'\frac{\partial^2 u}{\partial t^2}$$

and one obtains

$$\frac{\partial^2 u}{\partial x^2} = R'G'u + \left(R'C' + L'G'\right)\frac{\partial u}{\partial t} + L'C'\frac{\partial^2 u}{\partial t^2}. \tag{3.23}$$

From (3.17) to (3.20) it follows (see insert)

$$\frac{\partial^2 u}{\partial x^2} = R'G'u + \left(R'C' + L'G'\right)\frac{\partial u}{\partial t} + L'C'\frac{\partial^2 u}{\partial t^2}. \tag{3.24}$$

In a similar way, one obtains from (3.17), (3.19) and (3.20)

$$\frac{\partial^2 i}{\partial x^2} = R'G'i + \left(R'C' + L'G'\right)\frac{\partial i}{\partial t} + L'C'\frac{\partial^2 i}{\partial t^2}. \tag{3.25}$$

These equations (3.24) and (3.25) are called wave equations, they are also referred to as *telegraph* or *telegrapher's equations*.[2] The name comes from the fact that these equations were used by Oliver Heaviside[3] to study propagation processes on electrical telegraph lines in theoretical telegraphy.

These are partial differential equations of the second order. They are valid for any temporal course of voltage and current on the line, i.e., they also apply, for example, to compensation processes.

3.3.3 Transmission Line Equations with Harmonic Excitation

The general solution of the telegraph equations for the spatial-temporal course of the voltage and current functions, taking into account the initial and boundary conditions, is very complicated. Therefore, the special case of the line with (time-) harmonic excitation, i.e., with temporally sinusoidal or cosine-shaped time processes, for the steady state is now examined, which is important for electrical and communication engineering practice.

---

[2] The formulations $\partial^2 u/\partial x^2$ or $\partial^2 u/\partial t^2$ denote the second derivations of a quantity u with respect to the position x or time t.

[3] Oliver Heaviside (1850–1925): British mathematician and physicist.

A harmonic progression of the voltage u_x and the current i_x at the point x is assumed according to

$$u_x = U_x\, e^{j\omega t}$$

$$i_x = I_x\, e^{j\omega t}$$

The partial derivatives with respect to t and x are

$$\frac{\partial u_x}{\partial t} = U_x \cdot j\omega \cdot e^{j\omega t} \qquad \text{and} \qquad \frac{\partial i_x}{\partial t} = I_x \cdot j\omega \cdot e^{j\omega t}$$

$$\frac{\partial u_x}{\partial x} = \frac{\partial U_x}{\partial x} \cdot e^{j\omega t} \qquad \text{and} \qquad \frac{\partial i_x}{\partial x} = \frac{\partial I_x}{\partial x} \cdot e^{j\omega t}.$$

Temporal derivatives $\partial/\partial t$—as in complex alternating current calculation, see e.g. [50]—were replaced by multiplication with $j\omega$. The quantities U_x and I_x are complex amplitudes that are not time-variant. By substituting into (3.17) and (3.18), we first obtain

$$-\frac{\partial U_x}{\partial x} \cdot e^{j\omega t} = R' \cdot I_x\, e^{j\omega t} + L' \cdot I_x \cdot j\omega \cdot e^{j\omega t}$$

$$-\frac{\partial I_x}{\partial x} \cdot e^{j\omega t} = G' \cdot U_x\, e^{j\omega t} + C' \cdot U_x \cdot j\omega \cdot e^{j\omega t}$$

and finally the transmission line equations for harmonic excitation become

$$-\frac{\partial U_x}{\partial x} = \left(R' + j\omega L' \right) \cdot I_x \tag{3.26}$$

and

$$-\frac{\partial I_x}{\partial x} = \left(G' + j\omega C' \right) \cdot U_x . \tag{3.27}$$

With the abbreviated notations for the complex longitudinal resistance

$$Z' = R' + j\omega L' \tag{3.28}$$

and the complex transversal conductance

$$Y' = G' + j\omega C' \tag{3.29}$$

(3.26) and (3.27) can be written more compactly as

$$-\frac{\partial U_x}{\partial x} = Z' I_x \tag{3.30}$$

and

$$-\frac{\partial I_x}{\partial x} = Y' U_x . \tag{3.31}$$

The partial differential equations (3.24) and (3.25) are transformed into ordinary differential equations (3.26) and (3.27) or (3.30) and (3.31) by analyzing the propagation processes on lines under harmonic excitation.

By further differentiating (3.26) and (3.27) with respect to x, we obtain

$$-\frac{\partial^2 U_x}{\partial x^2} = \left(R' + j\omega L'\right) \cdot \frac{\partial I_x}{\partial x} = Z' \cdot \frac{\partial I_x}{\partial x}$$

$$-\frac{\partial^2 I_x}{\partial x^2} = \left(G' + j\omega C'\right) \cdot \frac{\partial U_x}{\partial x} = Y' \cdot \frac{\partial U_x}{\partial x}$$

and by substituting (3.26) and (3.27) into these relationships, the basic line equations or the telegraph equations for harmonic excitation take the form

$$\frac{\partial^2 U_x}{\partial x^2} = \left(R' + j\omega L'\right) \cdot \left(G' + j\omega C'\right) \cdot U_x = Z' \cdot Y' \cdot U_x \tag{3.32}$$

and

$$\frac{\partial^2 I_x}{\partial x^2} = \left(R' + j\omega L'\right) \cdot \left(G' + j\omega C'\right) \cdot I_x = Z' \cdot Y' \cdot I_x \; . \tag{3.33}$$

These are ordinary second-order differential equations with constant coefficients. Voltage and current quantities no longer appear mixed in the equations, but separately: Equation (3.32) describes the voltage profile and (3.33) is valid for the current profile. The complex amplitudes U_x and I_x are now only functions of the spatial coordinate x.

The propagation constant

$$\gamma = \sqrt{(R' + j\omega L')(G' + j\omega C')} \tag{3.34}$$

is defined.

With the propagation constant γ, the wave or telegraph equations for harmonic excitation can also be written more compactly as

$$\frac{\partial^2 U_x}{\partial x^2} = \gamma^2 \cdot U_x \tag{3.35}$$

and

$$\frac{\partial^2 I_x}{\partial x^2} = \gamma^2 \cdot I_x. \tag{3.36}$$

The aim is to calculate the voltage profile U_x and the current profile I_x as a function of the spatial coordinate x. As an approach to solving the present homogeneous linear differential equation with constant coefficients, we use

$$U_x = U \cdot e^{cx}. \tag{3.37}$$

This results in the voltage profile

$$\frac{\partial^2 U_x}{\partial x^2} = U \cdot c^2 \cdot e^{cx} = c^2 \cdot U_x. \tag{3.38}$$

By comparing (3.38) with (3.35), it becomes clear that $c^2 = \gamma^2$ applies and thus $c = \pm\gamma$.

With the integration constants U_1 and U_2, we thus obtain the solution

$$U_x = U_1 \cdot e^{-\gamma x} + U_2 \cdot e^{+\gamma x}. \tag{3.39}$$

For the current profile, with (3.26) we get

$$I_x = -\frac{1}{R' + j\omega L'} \cdot \frac{\partial U_x}{\partial x} = -\frac{1}{Z'} \cdot \frac{\partial U_x}{\partial x} \tag{3.40}$$

and by performing the differentiation contained in (3.40) from (3.39)

$$I_x = \frac{\gamma}{Z'} \cdot U_1 \cdot e^{-\gamma x} - \frac{\gamma}{Z'} \cdot U_2 \cdot e^{+\gamma x}. \tag{3.41}$$

The quantity

$$Z_{\mathrm{W}} = \frac{Z'}{\gamma} = \frac{R' + j\omega L'}{\sqrt{(R' + j\omega L')(G' + j\omega C')}} = \sqrt{\frac{R' + j\omega L'}{G' + j\omega C'}} \tag{3.42}$$

is introduced and referred to as *characteristic impedance*.

Finally, using (3.42), the result for voltage and current at the point x are the relationships

$$U_x = U_1 \cdot e^{-\gamma x} + U_2 \cdot e^{+\gamma x} \tag{3.43}$$

and

$$I_x = \frac{U_1}{Z_{\mathrm{W}}} \cdot e^{-\gamma x} - \frac{U_2}{Z_{\mathrm{W}}} \cdot e^{+\gamma x}. \tag{3.44}$$

The integration constants U_1 and U_2 can be determined from initial conditions. For this, the conditions at the beginning of the line ($x = 0$) are considered. There, with (3.43) and (3.44)

$$U_x = U_0 = U_1 + U_2 \tag{3.45}$$

and

$$I_x Z_{\mathrm{W}} = I_0 Z_{\mathrm{W}} = U_1 - U_2, \tag{3.46}$$

from which, by solving this system of equations for U_1 and U_2

$$U_1 = \frac{U_0 + I_0 Z_{\mathrm{W}}}{2} \tag{3.47}$$

and

$$U_2 = \frac{U_0 - I_0 Z_{\mathrm{W}}}{2} \tag{3.48}$$

appear. If these intermediate results (3.47) and (3.48) are inserted into (3.43) and (3.44), the results are

$$U_x = \frac{U_0 + I_0 Z_W}{2} \cdot e^{-\gamma x} + \frac{U_0 - I_0 Z_W}{2} \cdot e^{+\gamma x} \tag{3.49}$$

$$U_x = U_0 \cdot \frac{e^{+\gamma x} + e^{-\gamma x}}{2} - I_0 Z_W \cdot \frac{e^{+\gamma x} - e^{-\gamma x}}{2} \tag{3.50}$$

and

$$I_x = \frac{U_0 + I_0 Z_W}{2} \cdot \frac{1}{Z_W} \cdot e^{-\gamma x} - \frac{U_0 - I_0 Z_W}{2} \cdot \frac{1}{Z_W} \cdot e^{+\gamma x} \tag{3.51}$$

$$I_x = I_0 \cdot \frac{e^{+\gamma x} + e^{-\gamma x}}{2} - \frac{U_0}{Z_W} \cdot \frac{e^{+\gamma x} - e^{-\gamma x}}{2} \tag{3.52}$$

or when using the hyperbolic functions—see Appendix F.1 or e.g. [11, 74]—finally

$$U_x = U_0 \cosh \gamma x - I_0 Z_W \sinh \gamma x \tag{3.53}$$

and

$$I_x = I_0 \cosh \gamma x - \frac{U_0}{Z_W} \sinh \gamma x. \tag{3.54}$$

These equations describe the quantities voltage U_x and current I_x at the position x of the line under consideration, depending on the input quantities U_0 and I_0.

If you solve the equations (3.53) and (3.54) for the input quantities U_0 and I_0, one gets with

$$U_0 = U_x \cosh \gamma x + I_x Z_W \sinh \gamma x \tag{3.55}$$

and

$$I_0 = I_x \cosh \gamma x + \frac{U_x}{Z_W} \sinh \gamma x \tag{3.56}$$

a relationship between the voltage U_0 or the current I_0 at the input of the line and the voltage U_x or the current I_x at the point x—i.e., at a distance x from the beginning of the line.

Insert: Solving the System of Equations for Voltage U_x and Current I_x at Position x for the Input Quantities U_0 and I_0

Starting from the system of equations consisting of (3.53) and (3.54)

$$U_x = U_0 \cosh \gamma x - I_0 Z_W \sinh \gamma x \tag{3.57}$$

and

$$I_x = I_0 \cosh \gamma x - \frac{U_0}{Z_\mathrm{W}} \sinh \gamma x \qquad (3.58)$$

first (3.57) is rearranged for U_0 and (3.58) for I_0 one gets

$$U_0 = \frac{U_x + I_0 Z_\mathrm{W} \sinh \gamma x}{\cosh \gamma x} \qquad (3.59)$$

and

$$I_0 = \frac{I_x + \frac{U_0}{Z_\mathrm{W}} \sinh \gamma x}{\cosh \gamma x}. \qquad (3.60)$$

If (3.60) is inserted into (3.59), the result is

$$U_0 = \frac{1}{\cosh \gamma x}\left(U_x + \frac{1}{\cosh \gamma x}\left(I_x + \frac{U_0}{Z_\mathrm{W}} \sinh \gamma x\right) Z_\mathrm{W} \sinh \gamma x\right)$$

$$U_0 = \frac{U_x}{\cosh \gamma x} + \frac{I_x \cdot Z_\mathrm{W} \sinh \gamma x}{\cosh^2 \gamma x} + \frac{U_0 \cdot Z_\mathrm{W} \sinh^2 \gamma x}{Z_\mathrm{W} \cosh^2 \gamma x}.$$

Further rearrangement yields

$$U_0 - U_0 \frac{\sinh^2 \gamma x}{\cosh^2 \gamma x} = \frac{U_x}{\cosh \gamma x} + \frac{I_x \cdot Z_\mathrm{W} \sinh \gamma x}{\cosh^2 \gamma x}$$

$$U_0 \left(\frac{\cosh^2 \gamma x - \sinh^2 \gamma x}{\cosh^2 \gamma x}\right) = \frac{U_x}{\cosh \gamma x} + \frac{I_x \cdot Z_\mathrm{W} \sinh \gamma x}{\cosh^2 \gamma x},$$

and with the identity (see Appendix F.1 or e.g. [11, 74])

$$\cosh^2 \gamma x - \sinh^2 \gamma x = 1 \qquad (3.61)$$

via

$$U_0 \left(\frac{1}{\cosh^2 \gamma x}\right) = \frac{U_x}{\cosh \gamma x} + \frac{I_x \cdot Z_\mathrm{W} \sinh \gamma x}{\cosh^2 \gamma x}$$

$$U_0 = \frac{U_x \cdot \cosh^2 \gamma x}{\cosh \gamma x} + \frac{I_x \cdot Z_\mathrm{W} \sinh \gamma x \cdot \cosh^2 \gamma x}{\cosh^2 \gamma x}$$

finally the relationship (3.55)

$$U_0 = U_x \cosh \gamma x + I_x Z_\mathrm{W} \sinh \gamma x \qquad (3.62)$$

for the voltage at the input of the line is obtained. If you consider that voltage and current at any point x of the line are linked by the characteristic impedance Z_W—and in particular also $I_0 = U_0/Z_W$, $I_x = U_x/Z_W$ and $U_x = I_x Z_W$ apply –, a division of (3.62) by the characteristic impedance Z_W gives the equation (3.56)

$$I_0 = I_x \cosh \gamma x + \frac{U_x}{Z_W} \sinh \gamma x. \tag{3.63}$$

for the current at the input of the line [50].

3.3.4 Line Parameters

In the previous section on the behavior of electrical lines under harmonic excitation, the line parameters propagation constant γ and characteristic impedance Z_W were introduced: They are also referred to as *secondary line parameters*. These and some other related characteristics of electrical lines are discussed in this section.

Propagation, Attenuation, and Phase Constant
The propagation constant γ according to (3.34) is in general complex, so it can be represented in the form

$$\gamma = \alpha + \mathrm{j}\beta \tag{3.64}$$

The attenuation constant

$$\alpha = \mathrm{Re}\{\gamma\} = \sqrt{\frac{1}{2}\left(R'G' - \omega^2 L'C'\right) + \frac{1}{2}\sqrt{\left(R'^2 + \omega^2 L'^2\right)\left(G'^2 + \omega^2 C'^2\right)}} \tag{3.65}$$

describes the energy decrease along the line due to losses and the phase constant

$$\beta = \mathrm{Im}\{\gamma\} = \sqrt{-\frac{1}{2}\left(R'G' - \omega^2 L'C'\right) + \frac{1}{2}\sqrt{\left(R'^2 + \omega^2 L'^2\right)\left(G'^2 + \omega^2 C'^2\right)}} \tag{3.66}$$

describes the change in phase angle between output and input quantity when adjusting the termination resistance Z_a to the characteristic impedance Z_W of the line.

Insert: Derivation of the Attenuation and Phase Constant
Starting from the representation (3.64) of the propagation constant using the attenuation and the phase constant as $\gamma = \alpha + \mathrm{j}\beta$, we obtain with their definition (3.34)

$$\gamma^2 = \alpha^2 + \mathrm{j}2\alpha\beta - \beta^2 = (R' + \mathrm{j}\omega L')(G' + \mathrm{j}\omega C')$$
$$\gamma^2 = R'G' + \mathrm{j}\omega\left(L'G' + R'C'\right) - \omega^2 L'C'.$$

For the real and imaginary part of the quantity γ^2, it applies

$$\mathrm{Re}\{\gamma^2\} = \alpha^2 - \beta^2 = R'G' - \omega^2 L'C' \tag{3.67}$$

and

$$\mathrm{Im}\{\gamma^2\} = 2\alpha\beta = \omega(L'G' + R'C') \tag{3.68}$$

and also

$$|\gamma|^2 = \sqrt{\left(R'^2 + \omega^2 L'^2\right)\left(G'^2 + \omega^2 C'^2\right)} = \alpha^2 + \beta^2. \tag{3.69}$$

Based on these relationships, the attenuation constant α and the phase constant β can be calculated. To calculate the attenuation constant α we use

$$\left(\alpha^2 - \beta^2\right) + \left(\alpha^2 + \beta^2\right) = 2\alpha^2 \tag{3.70}$$

from which

$$\alpha = \sqrt{\frac{1}{2}\left(\alpha^2 - \beta^2\right) + \frac{1}{2}\left(\alpha^2 + \beta^2\right)} \tag{3.71}$$

results and finally, by inserting (3.67) and (3.69) into (3.71), the relationship (3.65)

$$\alpha = \sqrt{\frac{1}{2}\left(R'G' - \omega^2 L'C'\right) + \frac{1}{2}\sqrt{\left(R'^2 + \omega^2 L'^2\right)\left(G'^2 + \omega^2 C'^2\right)}} \tag{3.72}$$

is obtained.

The relationship (3.66) for the phase constant is obtained in a very similar way: We set

$$\left(\alpha^2 + \beta^2\right) - \left(\alpha^2 - \beta^2\right) = 2\beta^2 \tag{3.73}$$

from which

$$\beta = \sqrt{-\frac{1}{2}\left(\alpha^2 - \beta^2\right) + \frac{1}{2}\left(\alpha^2 + \beta^2\right)} \tag{3.74}$$

is obtained and by inserting (3.67) and (3.69) now into (3.74), the relationship (3.66)

$$\beta = \sqrt{-\frac{1}{2}\left(R'G' - \omega^2 L'C'\right) + \frac{1}{2}\sqrt{\left(R'^2 + \omega^2 L'^2\right)\left(G'^2 + \omega^2 C'^2\right)}} \tag{3.75}$$

results.

For the entire line length ℓ, the amplitude is attenuated or damped by the *attenuation measure*

$$a = \alpha \cdot \ell. \tag{3.76}$$

The total phase rotation at the output of the line compared to the line input is

$$b = \beta \cdot \ell \tag{3.77}$$

and is referred to as *phase-* or *angle measure*. This results in the complex *transmission-* or *propagation measure* of the line

$$g = \gamma \cdot \ell = a + jb. \tag{3.78}$$

The quantities denoted with the word part *measure* refer to the entire line length ℓ.

Characteristic Impedance

The characteristic impedance Z_W according to (3.42) is generally also complex and can be written as

$$Z_W = R_W + jX_W . \tag{3.79}$$

It is the ratio of the voltage wave propagating in one direction and the current wave propagating in the same direction. It is a characteristic constant of the line and depends only on the line parameters R', G', C' and L' (see Eq. (3.42)), but not on the termination of the line.

Reflection Coefficient

We will now consider the conditions at the end of the line. The line is terminated at its output with a resistance Z_a (see Fig. 3.11). To describe the conditions at the end of the line and their effects on, for example, the line input, the reflection coefficient is defined by the ratio of the reflected wave to the forward wave, and it results in

$$n = \frac{Z_a - Z_W}{Z_a + Z_W}. \tag{3.80}$$

The reflection coefficient n therefore depends on the termination resistance Z_a and the characteristic impedance Z_W of the line.

The derivation of the relationship (3.80) is given in a subsequent insert following [57], or it can be taken directly from the specialized literature, e.g. [57].

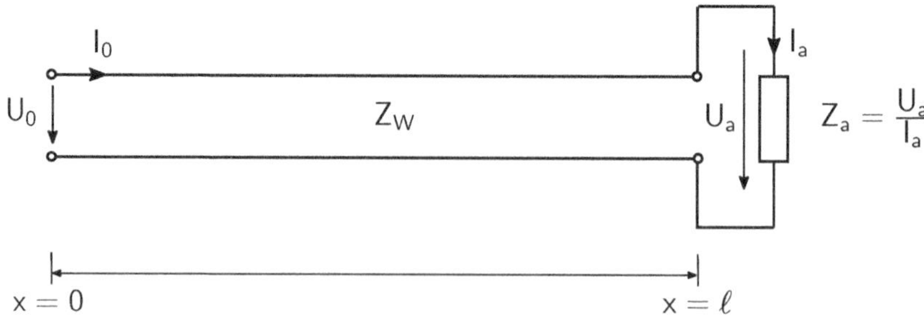

Fig. 3.11 Conditions at the line output

Insert: Derivation of the Reflection Coefficient

If we denote the voltage wave arriving at the end of the line by U_a' and the reflected voltage wave by U_a'', and the corresponding forward and backward current waves by I_a' and I_a'', then with (3.43) and (3.44)

$$U_{x=\ell} = U_a = U_1 \cdot e^{-\gamma \ell} + U_2 \cdot e^{+\gamma \ell} = U_a' + U_a'' \tag{3.81}$$

and

$$I_{x=\ell} = I_a = I_1 \cdot e^{-\gamma \ell} - I_2 \cdot e^{+\gamma \ell} = I_a' + I_a''. \tag{3.82}$$

It applies at the end of the line

$$Z_a = \frac{U_a}{I_a} \tag{3.83}$$

as well as

$$U_a' = U_a - U_a''$$
$$I_a' = I_a - I_a''.$$

If the last equation is multiplied on both sides by the characteristic impedance Z_W, the result for this equation is initially

$$I_a' \cdot Z_W = I_a \cdot Z_W - I_a'' \cdot Z_W, \tag{3.84}$$

and, taking into account $Z_W = U_a'/I_a' = -U_a''/I_a''$ (due to the counter-directionality of forward and backward wave [57]), the equation system

$$U_a' = U_a - U_a''$$
$$U_a' = I_a Z_W + U_a''.$$

If this equation system is solved for U_a' and U_a'', the result is

$$U_a' = \frac{U_a + I_a Z_W}{2}$$
$$U_a'' = \frac{U_a - I_a Z_W}{2}.$$

From this, we obtain the reflection coefficient

$$n = \frac{U_a''}{U_a'} = \frac{2(U_a - I_a Z_W)}{2(U_a + I_a Z_W)} = \frac{\frac{U_a}{I_a} - Z_W}{\frac{U_a}{I_a} + Z_W}. \tag{3.85}$$

and finally as a result (3.80)

$$n = \frac{Z_a - Z_W}{Z_a + Z_W}. \tag{3.86}$$

In particular, three typical operating states of a line are distinguished:

- Short circuit: The line is short-circuited at its output.
- Open circuit: The line has an open output.
- Matching: The line is terminated at its end with the characteristic impedance.

Characteristic values of the reflection coefficient n result for the typical operating states of the line:

$$\begin{aligned}
Z_\mathrm{a} &= 0 \quad &\text{(short circuit)} \quad &: \quad n = -1 \\
Z_\mathrm{a} &= \infty \quad &\text{(open)} \quad &: \quad n = +1 \\
Z_\mathrm{a} &= Z_\mathrm{W} \quad &\text{(matching)} \quad &: \quad n = 0
\end{aligned}$$

In the case of matching, no reflected wave is created due to $n = 0$, i.e., the energy reaching the end of the line is completely converted in the termination resistance Z_a. This corresponds to the behavior of an *electrically long* line from the beginning of the line: In this case, too, there is no reflected wave, since no energy arrives at the end of the line and therefore no wave can be reflected there. For a line operated in a matched manner, the ratio of complex voltage U_x to complex current I_x is constant at every point x of the line and equal to the characteristic impedance Z_W.

Input Impedance

In contrast to the characteristic impedance Z_W, the input impedance Z_0 of a line is the quotient of total voltage and total current at the line input. These can each be composed of several waves (forward-travelling and reflected wave). The input impedance of a line is therefore dependent on the line termination.

In Fig. 3.12, the conditions on the line are considered exemplarily for the case when only the line parameters R' and G' are taken into account : This is valid for comparatively low frequencies, when the inductive resistance and capacitive conductance components are approximately ineffective due to $\omega L' \ll R'$ and $\omega C' \ll G'$, and can therefore be neglected. If we further assume that the termination resistance Z_a at the end of the line is not identical to the characteristic impedance Z_W, i.e., $Z_\mathrm{a} \neq Z_\mathrm{W}$ applies, then a part of the electromagnetic wave running from the input to the output of the line is reflected and there exist forward and reflected waves on the line. Then the voltage U_x at the point x on the line is composed of the two parts of the forward and the reflected wave.

Voltage and current drop according to (3.43) and (3.44) each following an exponential function with the distance x: This means that per distance $\mathrm{d}x$, the voltage or current drop occurs by the same factor.

For the input impedance, characteristic values are obtained for the three typical operating states of the line:

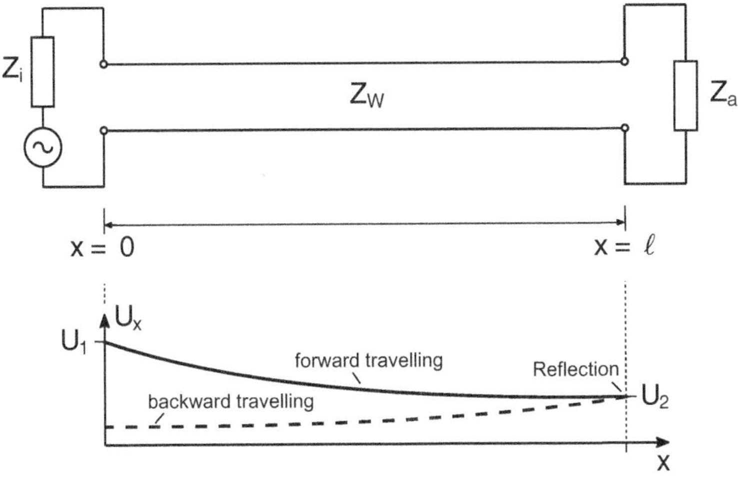

Fig. 3.12 On the interpretation of characteristic impedance and input impedance of a line

$$
\begin{aligned}
Z_a = 0 \quad &\text{(short circuit)} \quad : \quad n = -1 \quad \Rightarrow \quad Z_{0K} = Z_W \tanh \gamma\ell \\
Z_a = \infty \quad &\text{(open)} \quad : \quad n = +1 \quad \Rightarrow \quad Z_{0L} = Z_W \coth \gamma\ell \\
Z_a = Z_W \quad &\text{(matching)} \quad : \quad n = 0 \quad \Rightarrow \quad Z_{0A} = Z_W
\end{aligned}
$$

From these relationships for the input impedance of the line in characteristic operating states, in particular with

$$
Z_W = \sqrt{Z_{0L} \cdot Z_{0K}} \tag{3.87}
$$

a possibility to practically measure the characteristic impedance Z_W, as well as through the quotient

$$
\frac{Z_{0K}}{Z_{0L}} = \tanh^2 \gamma\ell \tag{3.88}
$$

an option to determine the propagation constant γ with known line length ℓ appears.

The derivation of the relationships for the input impedances of the three characteristic operating states of a line is shown in a subsequent insert, following [57], and can be taken from the specialized literature, e.g., [57].

Insert: Derivation of the Input Impedance

The input impdance of a line is defined as

$$
Z_0 = \frac{\text{overall voltage at the line input}}{\text{overall current at the line input}} = \frac{U_0}{I_0} . \tag{3.89}
$$

The total voltage U_0 at the entrance of the line consists of a forward-travelling component U_0' and a backward-travelling component U_0''. Similarly, the total input current consists of forward-travelling and backward-travelling waves I_0' and I_0''. Thus, the input impedance can be written as

$$Z_0 = \frac{U_0' + U_0''}{I_0' + I_0''}.$$

(3.90)

Furthermore, U_a' and U_a'' denote the forward- and backward-travelling voltage components at the line output, and I_a' and I_a'' the corresponding forward- and backward-travelling current components there.

This results in

$$U_0'' = U_a'' \cdot e^{-\gamma \ell} = n\, U_a' \cdot e^{-\gamma \ell} = n\, U_0' \cdot e^{-\gamma \ell} \cdot e^{-\gamma \ell} = n\, U_0' \cdot e^{-2\gamma \ell}$$

(3.91)

and

$$I_0'' = I_a'' \cdot e^{-\gamma \ell} = -n\, I_a' \cdot e^{-\gamma \ell} = -n\, I_0' \cdot e^{-\gamma \ell} \cdot e^{-\gamma \ell} = -n\, I_0' \cdot e^{-2\gamma \ell}$$

(3.92)

and it follows

$$Z_0 = \frac{U_0' + n\, U_0' \cdot e^{-2\gamma \ell}}{I_0' - n\, I_0' \cdot e^{-2\gamma \ell}} = \frac{U_0'}{I_0'} \cdot \frac{1 + n\, e^{-2\gamma \ell}}{1 - n\, e^{-2\gamma \ell}} = Z_W \cdot \frac{1 + n\, e^{-2\gamma \ell}}{1 - n\, e^{-2\gamma \ell}}.$$

(3.93)

The longer the line, the less the influence of the line termination on the input impedance due to attenuation.

An extension with $e^{\gamma \ell}$ yields

$$Z_0 = Z_W \cdot \frac{e^{\gamma \ell} + n\, e^{-\gamma \ell}}{e^{\gamma \ell} - n\, e^{-\gamma \ell}}.$$

(3.94)

In the following, the input impedances of the line for the three characteristic operating states short circuit, open circuit, and matching are determined.

For a short circuit at the end of the line, $Z_a = 0$ applies and thus $n = -1$. The input impedance is

$$Z_{0K} = Z_W \cdot \frac{e^{\gamma \ell} - e^{-\gamma \ell}}{e^{\gamma \ell} + e^{-\gamma \ell}} \cdot \frac{\frac{1}{2}}{\frac{1}{2}} = Z_W \cdot \frac{\sinh \gamma \ell}{\cosh \gamma \ell} = Z_W \tanh \gamma \ell.$$

(3.95)

For an open circuit at the end of the line, $Z_a = \infty$ applies (open line end) and thus $n = +1$ The input impedance is

$$Z_{0L} = Z_W \cdot \frac{e^{\gamma \ell} + e^{-\gamma \ell}}{e^{\gamma \ell} - e^{-\gamma \ell}} \cdot \frac{\frac{1}{2}}{\frac{1}{2}} = Z_W \cdot \frac{\cosh \gamma \ell}{\sinh \gamma \ell} = Z_W \coth \gamma \ell.$$

(3.96)

If the line is terminated by the characteristic impedance, $Z_a = Z_W$ applies and for the reflection coefficient $n = 0$ The input impedance is then identical to the characteristic impedance of the line:

$$Z_{0A} = Z_W. \tag{3.97}$$

Open circuit and short circuit at the end of the line are operating states of the line that can be relatively easily established for measurement purposes. Then the input impedances Z_{0K} and Z_{0L} of the line can be measured and the product

$$Z_{0K} \cdot Z_{0L} = Z_W^2 \tanh \gamma \ell \cdot \coth \gamma \ell = Z_W^2 \tag{3.98}$$

provides a way to experimentally determine the characteristic impedance of a line by

$$Z_W = \sqrt{Z_{0L} \cdot Z_{0K}} \,. \tag{3.99}$$

Also, the propagation constant γ can be measured and calculated for a known line length ℓ using the quotient

$$\frac{Z_{0K}}{Z_{0L}} = \tanh^2 \gamma \ell \tag{3.100}$$

and

$$\tanh \gamma \ell = \sqrt{\frac{Z_{0K}}{Z_{0L}}} \qquad \Longrightarrow \qquad \gamma = \frac{1}{\ell} \cdot \operatorname{arctanh} \sqrt{\frac{Z_{0K}}{Z_{0L}}} \tag{3.101}$$

Once the characteristic impedance Z_W and propagation constant γ are determined, the primary line parameters (or transmission line parameters per unit length) can be calculated from them

$$R' = \operatorname{Re}\{\gamma \cdot Z_W\} \qquad ; \qquad L' = \frac{1}{\omega} \cdot \operatorname{Im}\{\gamma \cdot Z_W\} \tag{3.102}$$

and

$$G' = \operatorname{Re}\left\{\frac{\gamma}{Z_W}\right\} \qquad ; \qquad C' = \frac{1}{\omega} \cdot \left\{\frac{\gamma}{Z_W}\right\} \tag{3.103}$$

and thus their generally frequency-dependent characteristic can be calculated [51].

Wavelength

With harmonic excitation, a sinusoidal or cosine-shaped voltage or current wave that decreases according to an exponential function can be observed on the line. The wavelength λ of this wave is the distance between two adjacent points, between which there is a phase difference of 2π. The phase angle β valid for the distance of 1 km (measured in $°$/km or rad/km) thus relates to this distance of 1 km like 2π to the wavelength λ. This results in:

$$\frac{\beta}{1} = \frac{2\pi}{\lambda} \qquad \Longrightarrow \qquad \lambda = \frac{2\pi}{\beta} \quad \text{(in km)}. \tag{3.104}$$

Phase Velocity and Time

Within the time period $T = 1/f$, the voltage or current wave advances by the wavelength λ. With the speed $v = s/t$, the phase velocity follows

$$v_\mathrm{p} = \frac{\lambda}{T} = \frac{2\pi}{\beta} \cdot \frac{1}{T} = \frac{2\pi f}{\beta} \qquad \Longrightarrow \qquad v_\mathrm{p} = \frac{\omega}{\beta}. \tag{3.105}$$

If we denote the kilometric phase time τ_p as the time a voltage or current component of the wave needs to cover the distance 1 km, we get with $v_\mathrm{p} = (1\,\mathrm{km})/\tau_\mathrm{p}$

$$\tau_\mathrm{p} = \frac{1}{v_\mathrm{P}} \qquad \Longrightarrow \qquad \tau_\mathrm{p} = \frac{\beta}{\omega}. \tag{3.106}$$

For the transit time, which a phase state of a wave needs to travel the distance ℓ (in km), the following applies accordingly

$$t_\mathrm{p} = \tau_\mathrm{p} \cdot \ell = \frac{\ell}{v_\mathrm{P}} = \frac{\beta \cdot \ell}{\omega} = \frac{b}{\omega} \qquad \Longrightarrow \qquad t_\mathrm{p} = \frac{\beta \cdot \ell}{\omega}. \tag{3.107}$$

Illustratively: If we were to mark a single sine or cosine oscillation at the line input at one point—e.g., at a zero point or at a maximum - with a color spot, it would take the time τ_p until this color spot has covered the distance 1 km or the time t_P until it has covered the distance ℓ and has arrived at the line output.

The points of equal phase of an electromagnetic wave spread out on a line with the phase velocity v_p.

Group Delay and Group Velocity

The phase velocity is generally frequency-dependent according to (3.105), i.e., the phase velocity is different for different frequencies. In communications engineering, frequency bands or frequency groups are often transmitted, whose frequency components thus have different phase velocities: Therefore, it is important to define parameters that take this situation and application into account.

The kilometric group delay τ_g, which describes the time a frequency group needs to cover a distance of 1 km on the line, is defined as

$$\tau_\mathrm{g} = \frac{\text{phase difference}}{\text{frequency difference}} = \frac{\mathrm{d}\beta}{\mathrm{d}\omega} \qquad \Longrightarrow \qquad \tau_\mathrm{g} = \frac{\mathrm{d}\beta}{\mathrm{d}\omega}. \tag{3.108}$$

For the group delay t_g of a line of length ℓ, we get

$$t_\mathrm{g} = \frac{\mathrm{d}\beta \cdot \ell}{\mathrm{d}\omega} = \frac{\mathrm{d}b}{\mathrm{d}\omega}. \tag{3.109}$$

The group velocity results in

$$v_g = \frac{\ell}{t_g} = \frac{d\omega \cdot \ell}{db} = \frac{d\omega \cdot \ell}{d\beta \cdot \ell} \qquad \Longrightarrow \qquad v_g = \frac{d\omega}{d\beta}. \tag{3.110}$$

The group velocity and the group delay are quantities that describe the propagation of frequency groups or frequency bands over a line of length ℓ (or the length 1 km in the case of the kilometric group delay). The transmission of frequency groups or frequency bands is a very typical application in digital communications engineering and therefore of great importance. For a distortion-free or low-distortion transmission, it is desirable that the group delay is constant, so that all frequency components of a frequency group to be transmitted experience the same propagation speed and no distortions or distortions of the signal occur because different frequency components of the same signal propagate at different speeds on the line.

Influence of the Attenuation and Phase Constant on Signal Distortions
In the transmission of signals on lines in communications engineering, it is an important task to transmit message signals as undistorted as possible electromagnetically along the lines. These message signals often consist of a multitude of frequency components, which can be broken down into harmonic oscillations (frequency components), for which the relationships for lines with harmonic excitation are valid and applicable. In order to be able to assemble an undistorted image of the transmitted signal from the partial components at the end of the line, the partial components must be evenly attenuated on the line and the same phase velocity must be effective for all partial components. If the line deviates from the former condition (same attenuation for all frequency components), amplitude distortions occur, if it does not meet the latter requirement (same phase velocity for all frequency components), this causes phase distortions.

A distortion-free transmission results when

- the attenuation is constant and thus not frequency-dependent, i.e., it applies

$$a(\omega) = \text{const.}, \tag{3.111}$$

- the phase measure depends linearly on the frequency and thus the temporal position of the frequency components to each other remains the same, i.e., it applies

$$b(\omega) = \text{const.} \cdot \omega. \tag{3.112}$$

With a linear dependence of the phase course on the frequency, the group delay

$$t_g = \frac{db(\omega)}{d\omega} = \text{const.} \qquad \text{or} \qquad \tau_g = \frac{d\beta(\omega)}{d\omega} = \text{const.} \tag{3.113}$$

is constant and the temporal position of the individual frequency components of the signal to each other is preserved.

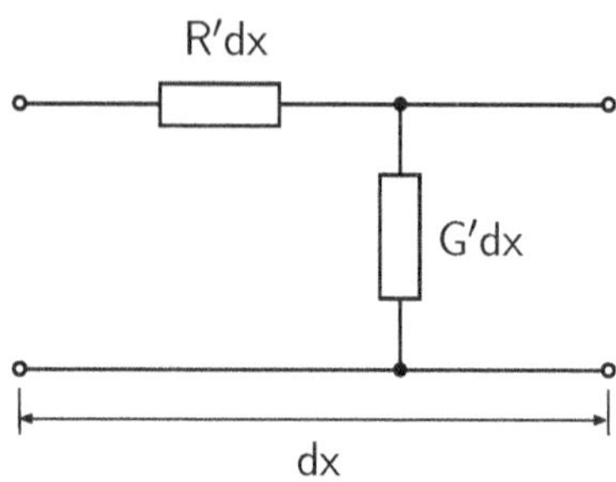

Fig. 3.13 Equivalent circuit of the line element for low frequency ($\omega \approx 0$)

3.3.5 Special Characteristics of Electrical Lines Depending on Frequency

The frequency range used plays a significant role in the transmission of digital signals. It is therefore important to understand how lines behave at different frequencies. The following will therefore examine typical special cases of lines depending on the frequency, which have a special significance in practice. In particular, the so-called *highly attenuated line* or *lossy line*, which is particularly important for the transmission of digital signals, will be described in more detail. The consideration begins with the direct current or direct voltage properties of electrical lines and continues in the direction of increasing frequency. The frequency division is to be understood in such a way that it depends on the line's element characteristics per unit length, i.e., a low, slightly increased or very high frequency in this categorization occurs when the respective conditions for the effectiveness of the individual line components occur. The frequency ranges are therefore to be understood relative to the line characteristic and therefore represent different absolute frequency ranges for different lines.

Direct Signal Properties of Lines

The properties of lines at low frequencies are important for direct signal transmission (direct current or direct voltage transmission), for low-frequency signals, and for the transmission of electrical energy at grid frequency (50 Hz). Since for the frequency $\omega \approx 0$, the resistance and conductance components of the line element result

$$R' \gg \omega L'$$
$$G' \gg \omega C'.$$

The equivalent circuit of the line element under these conditions is shown in Fig. 3.13.

The characteristic impedance results with $\omega \approx 0$ from (3.42)

$$Z_{\mathrm{W}} = \sqrt{\frac{R' + j\omega L'}{G' + j\omega C'}} \quad \Longrightarrow \quad Z_{\mathrm{W}} = \sqrt{\frac{R'}{G'}}. \tag{3.114}$$

The characteristic impedance in this frequency range is real and depends only on the resistance (R') and conductance per unit length (G') of the line.

After (3.65) for very low frequencies ($\omega \approx 0$) the attenuation constant is obtained

$$\alpha = \sqrt{R'G'} \tag{3.115}$$

and with (3.66) the phase constant is

$$\beta = 0. \tag{3.116}$$

The attenuation is constant in the considered frequency range according to (3.115) and the phase changes linearly with the frequency according to (3.116) (with the proportionality factor 0): There is no deformation of the signal, the transmission is distortion-free. The signal is—merely—attenuated on the transmission path. Intersymbol interferences on the transmission path can be avoided.

Lossy Line (*RC* Line)

A frequency range that is significant for the transmission of analog and digital signals over electrical lines is located at frequencies that are slightly above the considered low-frequency range. If the frequency ω is slightly higher, the situation for the resistance and conductance components of the line element applies as follows

$$R' \gg \omega L'$$
$$G' \ll \omega C'.$$

The equivalent circuit of the line element valid under these conditions is shown in Fig. 3.14. Since the ohmic resistance in the longitudinal branch and the capacitive conductance in the transverse branch must be taken into account, this operating frequency range of a line is also called the *RC* range.

The characteristic impedance of the *RC* line results from (3.42) under the mentioned conditions of the dominant resistance and capacitance to

$$Z_{\mathrm{W}} = \sqrt{\frac{R' + \mathrm{j}\omega L'}{G' + \mathrm{j}\omega C'}} \quad \Longrightarrow \quad Z_{\mathrm{W}} = \sqrt{\frac{R'}{\mathrm{j}\omega C'}}. \tag{3.117}$$

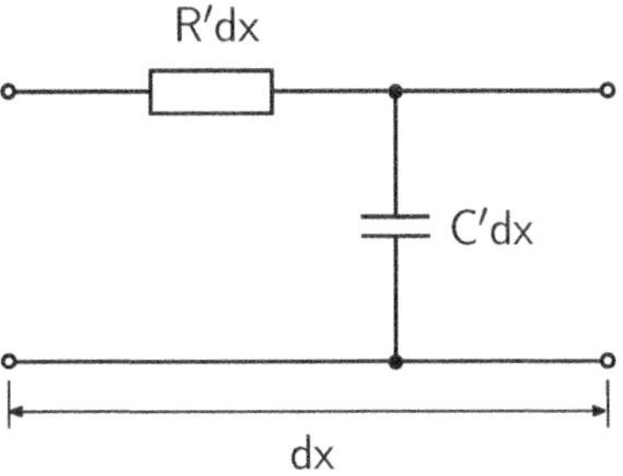

Fig. 3.14 Equivalent circuit of the line element for slightly higher frequency

A transformation results in

$$Z_W = \sqrt{\frac{R'}{j\omega C'}} = \frac{1}{\sqrt{j}}\sqrt{\frac{R'}{\omega C'}} = \sqrt{\frac{R'}{\omega C'}} \cdot e^{-j\frac{\pi}{4}}$$

$$= \sqrt{\frac{R'}{\omega C'}} \cdot \left(\cos\frac{\pi}{4} - j\sin\frac{\pi}{4}\right) = \sqrt{\frac{R'}{\omega C'}} \cdot \left(\frac{1}{\sqrt{2}} - j\frac{1}{\sqrt{2}}\right)$$

$$Z_W = \sqrt{\frac{R'}{2\omega C'}} \cdot (1-j)$$

and it becomes apparent that the characteristic impedance of a line in the *RC* range is complex and acts capacitively. Especially at low frequencies, it has strongly capacitive properties, while it becomes almost real at high frequencies.

The propagation constant (3.34) simplifies in the *RC* range to

$$\gamma = \sqrt{R' \cdot j\omega C'} = \sqrt{j\omega R'C'}. \tag{3.118}$$

To enable a decomposition into attenuation and phase constant, (3.118) is transformed according to

$$\gamma = \sqrt{j\omega R'C'} = \sqrt{j} \cdot \sqrt{\omega R'C'} = \sqrt{\omega R'C'} \cdot e^{j\frac{\pi}{4}}$$

$$= \sqrt{\omega R'C'} \cdot \left(\cos\frac{\pi}{4} + j\sin\frac{\pi}{4}\right) = \sqrt{\omega R'C'} \cdot \left(\frac{1}{\sqrt{2}} + j\frac{1}{\sqrt{2}}\right)$$

$$\gamma = \sqrt{\frac{\omega R'C'}{2}} \cdot (1+j) \,.$$

This gives the attenuation constant

$$\alpha = \text{Re}\{\gamma\} = \sqrt{\frac{\omega R'C'}{2}} \sim \sqrt{\omega} \tag{3.119}$$

and the phase constant

$$\beta = \text{Im}\{\gamma\} = \sqrt{\frac{\omega R'C'}{2}} \sim \sqrt{\omega} \,. \tag{3.120}$$

Both α and β are frequency-dependent with $\sqrt{\omega}$: This is referred to as a *root characteristic* of the line in the *RC* range. Due to the frequency dependence of the attenuation and phase constant, signals in amplitude and phase are distorted when transmitted over an *RC* line. Generally, under these conditions, intersymbol interference is caused by the transmission channel; only with very narrow-band signals a nearly ISI-free transmission can be achieved.

For the per-kilometer group delay of the *RC* line

$$\tau_g = \frac{d\beta}{d\omega} = \frac{d}{d\omega}\left(\sqrt{\frac{R'C'}{2}}\,\omega^{(1/2)}\right) = \frac{1}{2}\sqrt{\frac{R'C'}{2}}\,\omega^{(-1/2)} \tag{3.121}$$

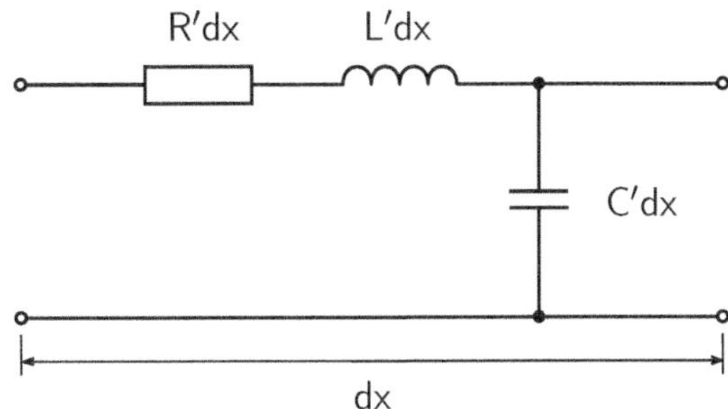

Fig. 3.15 Equivalent circuit of the line element for medium to high frequency

applies, leading to the result

$$\tau_g = \frac{1}{2}\sqrt{\frac{R'C'}{2\omega}}.$$ (3.122)

This per-kilometer group delay τ_g is therefore not constant, but frequency-dependent.

Low-Loss Line

At further increased frequency, a frequency range is reached where the conditions

$$R' \approx \omega L'$$

$$G' \ll \omega C'$$

apply. It is where the transmission line parameters per unit length R', L' and C' are effective. In this range, attenuation losses occur, which are smaller compared to the RC line: Therefore, the line in this frequency range is referred to as *low-loss* line.

In Fig. 3.15, the equivalent circuit of the line element is shown, which is valid under these boundary conditions.

The characteristic impedance of the low-loss line results from (3.42) under the mentioned conditions to

$$Z_W = \sqrt{\frac{R' + j\omega L'}{G' + j\omega C'}} \implies Z_W = \sqrt{\frac{R' + j\omega L'}{j\omega C'}}.$$ (3.123)

This relationship can be rewritten as

$$Z_W = \sqrt{\frac{R' + j\omega L'}{j\omega C'}} = \sqrt{\frac{j\omega L'\left(\frac{R'}{j\omega L'} + 1\right)}{j\omega C'}} \approx \sqrt{\frac{L'}{C'}}\sqrt{\frac{R'}{j\omega L'}} = \sqrt{\frac{L'}{C'}}e^{-j\frac{R'}{2\omega L'}}.$$ (3.124)

If we first consider all transmission line parameters per unit length, we get for the propagation constant (3.34) of the low-loss line

$$\gamma = j\omega\sqrt{L'C'}\left(1 + \frac{1}{j\omega}\left(\frac{R'}{2L'} + \frac{G'}{2C'}\right)\right). \tag{3.125}$$

This results in the attenuation constant

$$\alpha = \mathrm{Re}\{\gamma\} = \frac{R'}{2}\sqrt{\frac{C'}{L'}} + \frac{G'}{2}\sqrt{\frac{L'}{C'}} \tag{3.126}$$

and the phase constant

$$\beta = \mathrm{Im}\{\gamma\} = \omega\sqrt{L'C'}. \tag{3.127}$$

Since the conductance per unit length G' can often be neglected—according to the mentioned prerequisites for the low-loss line—the attenuation constant (3.126) can be simplified to

$$\alpha \approx \frac{R'}{2}\sqrt{\frac{C'}{L'}} \tag{3.128}$$

The attenuation constant α is not dependent on the frequency ω and is therefore constant and the phase constant β is linearly dependent on the frequency ω: Therefore, a signal transmitted over such a low-loss line is not distorted. This results in the—further—designation for a line in this frequency range, *distortion-free line*. Since the conductance G' according to the assumed boundary conditions in the considered frequency range is small, the second summand $G'/2 \cdot \sqrt{L'/C'}$ in (3.126) can often be neglected in practical calculations of the attenuation constant α.

For the kilometric group delay of the low-loss line applies

$$\tau_g = \frac{\mathrm{d}\beta}{\mathrm{d}\omega} = \frac{\mathrm{d}}{\mathrm{d}\omega}\left(\omega\sqrt{L'C'}\right) = \sqrt{L'C'}. \tag{3.129}$$

This group delay is therefore constant and not frequency-dependent.

Insert: Derivation of the Propagation Constant of the Low-Loss Line
The starting point for deriving the propagation constant of the low-loss line is the relationship (3.34), which is transformed:

$$\gamma = \sqrt{(R' + j\omega L')(G' + j\omega C')} = \sqrt{j\omega L'\left(1 + \frac{R'}{j\omega L'}\right)\cdot j\omega C'\left(1 + \frac{G'}{j\omega C'}\right)}$$

$$\gamma = j\omega\sqrt{L'C'}\cdot\sqrt{1 + \frac{R'}{j\omega L'}}\cdot\sqrt{1 + \frac{G'}{j\omega C'}}.$$

If we use the approximation $\sqrt{1+x} \approx 1 + x/2$ [74] under the condition $R'/\omega L' < 1$ and $G'/\omega C' < 1$, the result is

$$\gamma \approx j\omega\sqrt{L'C'}\left(1+\frac{R'}{j2\omega L'}\right)\left(1+\frac{G'}{j2\omega C'}\right). \tag{3.130}$$

Multiplying out gives

$$\gamma \approx j\omega\sqrt{L'C'}\left(1+\frac{R'}{j2\omega L'}+\frac{G'}{j2\omega C'}+\frac{R'G'}{(j2\omega)^2 L'C'}\right), \tag{3.131}$$

in which the product of the small sizes in the last summand in the bracket can be neglected, so that

$$\gamma \approx j\omega\sqrt{L'C'}\left(1+\frac{1}{j\omega}\left(\frac{R'}{2L'}+\frac{G'}{2C'}\right)\right). \tag{3.132}$$

and thus the relationship (3.125) is obtained.

From this result for the propagation constant γ of the distortion-free line, the attenuation constant α can be found as real part thereof according to

$$\alpha = \mathrm{Re}\{\gamma\} = \mathrm{Re}\left\{j\omega\sqrt{L'C'}\left(1+\frac{1}{j\omega}\left(\frac{R'}{2L'}+\frac{G'}{2C'}\right)\right)\right\}$$

$$= \mathrm{Re}\left\{j\omega\sqrt{L'C'}+\frac{j\omega\sqrt{L'C'}}{j\omega}\left(\frac{R'}{2L'}+\frac{G'}{2C'}\right)\right\}$$

$$= \mathrm{Re}\left\{j\omega\sqrt{L'C'}+\sqrt{L'C'}\left(\frac{R'}{2L'}+\frac{G'}{2C'}\right)\right\}$$

$$= \sqrt{L'C'}\left(\frac{R'}{2L'}+\frac{G'}{2C'}\right) = \frac{R'\cdot\sqrt{L'C'}}{2L'}+\frac{G'\cdot\sqrt{L'C'}}{2C'}$$

$$\alpha = \frac{R'}{2}\sqrt{\frac{C'}{L'}}+\frac{G'}{2}\sqrt{\frac{L'}{C'}}$$

as well as the phase constant β as imaginary part

$$\beta = \mathrm{Im}\{\gamma\} = \mathrm{Im}\left\{j\omega\sqrt{L'C'}\left(1+\frac{1}{j\omega}\left(\frac{R'}{2L'}+\frac{G'}{2C'}\right)\right)\right\}$$

$$= \mathrm{Im}\left\{j\omega\sqrt{L'C'}+\frac{j\omega\sqrt{L'C'}}{j\omega}\left(\frac{R'}{2L'}+\frac{G'}{2C'}\right)\right\}$$

$$= \mathrm{Im}\left\{j\omega\sqrt{L'C'}+\sqrt{L'C'}\left(\frac{R'}{2L'}+\frac{G'}{2C'}\right)\right\}$$

$$\beta = \omega\sqrt{L'C'}$$

and thus the relationships (3.126) and (3.127) can be derived.

A closer look at the attenuation constant α (3.126) shows that it has a minimum depending on L': It is given by the condition

$$\frac{\mathrm{d}\alpha}{\mathrm{d}L'} = \frac{G'}{4\sqrt{C'}}L'^{(-1/2)} - \frac{R'\sqrt{C'}}{4}L'^{(-3/2)} = 0 \tag{3.133}$$

and one obtains

$$\frac{G'}{\sqrt{C'}} - \frac{R'\sqrt{C'}}{L'} = 0$$

$$G'L' - R'C' = 0.$$

For a minimum of the attenuation constant, this results in the condition

$$R'C' = G'L'. \tag{3.134}$$

In practice, $R'C' > G'L'$ often applies, as R' is usually large and G' is small. Therefore, to maintain the condition (3.134), L' needs to be increased.

This has practical significance and wide application in analog transmission engineering, especially in telegraphy and telephony, with the *Krarup* and *Pupin*-lines:

- In Krarup lines[4], the inductance is increased by wrapping the line with thin iron wire with high permeability. The Krarup line is a homogeneous line. The wrapping with iron wire causes additional hysteresis and eddy current losses, which lead to an increase in the effective resistance of the line. Krarup lines were used as submarine cables and as land cables in overhead lines (as intermediate or end cables). They have not been used in communications engineering for many decades.
- In Pupin lines[5], an increase in inductance is achieved by inserting coils (so-called Pupin coils) at regular intervals in lines (typically e.g. every 1.7 km). As a result, the Pupin line is no longer a homogeneous line. An important application of this technique was telephony. Pupin lines are also no longer used in modern communication networks. They may still be found in rare individual cases in existing and not yet modernized special networks.

If the inductance is increased in one of the ways described above, this achieves the minimum attenuation $\sqrt{R'G'}$ according to (3.115) in an extended low-frequency range up to a cut-off frequency determined by the additionally increased inductance, above which the attenuation increases very sharply. Since this behavior of lines is very disadvantageous and therefore unsuitable for broadband digital transmission, these methods are no longer used today—and have not been used for some time.

---

[4] Carl Emil Krarup (1872–1909): Danish civil engineer.
[5] Mihajlo Idvorski Pupin (1854–1935): Serbian-American physicist and writer.

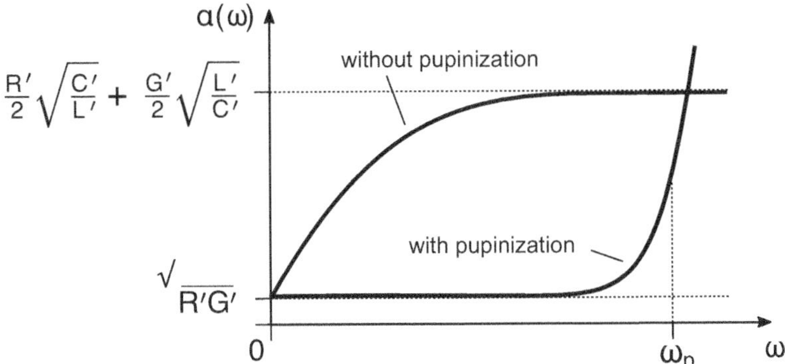

Fig. 3.16 Principle characteristic of the attenuation constant of a line without and with an increase of the inductance layer L' as a function of angular frequency

The effect that the increase of L' by e.g. Pupinization has on the frequency-dependent course of the attenuation constant α in principle, is shown in Fig. 3.16: The line attenuation α is initially kept at the minimal value $\alpha_{min} = \sqrt{R'G'}$ valid for $\omega \approx 0$ by the increased attenuation layer L' (see (3.115)) and rises steeply above a cut-off frequency ω_p

Further information on Pupin and Krarup lines can be found in e.g. [56, 57].

Lossless Line

At very high frequencies, i.e. for $\omega \to \infty$, the conditions

$$R' \ll \omega L'$$
$$G' \ll \omega C',$$

and resistance and conductance per unit length can be neglected in the considerations, i.e. $R' = 0$ and $G' = 0$ are assumed. These assumptions are mostly applicable in high and ultra-high frequency technology -- thus, at radio frequencies (RF). Therefore, the term *RF-line* is also used for the lossless line.

In Fig. 3.17, the equivalent circuit of the line element is shown, which applies to the lossless line. This results in another name for the lossless line: *LC*-line, since the induc-

Fig. 3.17 Equivalent circuit of the line element for very high frequency

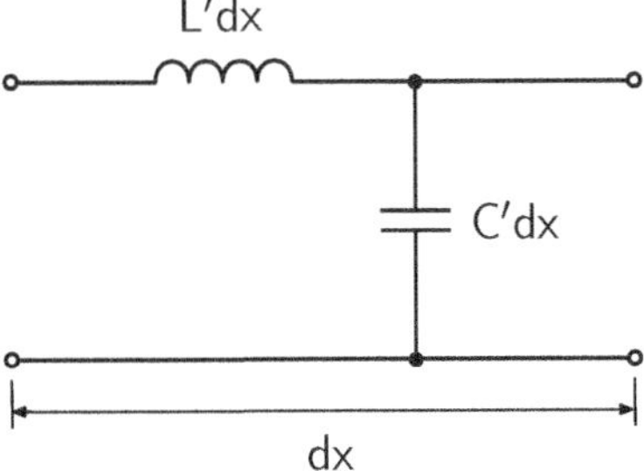

tive resistance in the longitudinal branch and the capacitive conductance in the transverse branch must be considered.

The characteristic impedance of the lossless line results from (3.42) under the assumed conditions to

$$Z_W = \sqrt{\frac{R' + j\omega L'}{G' + j\omega C'}} \quad \Longrightarrow \quad Z_W = \sqrt{\frac{L'}{C'}}. \tag{3.135}$$

The characteristic impedance of the lossless line is frequency-independent and real.

With (3.65) we obtain for the lossless line the attenuation constant under the given conditions

$$\alpha = 0 \tag{3.136}$$

and with (3.66) the phase constant

$$\beta = \omega\sqrt{L'C'}. \tag{3.137}$$

The propagation constant becomes

$$\gamma = j\beta = j\omega\sqrt{L'C'}. \tag{3.138}$$

Thus, the lossless line does not cause any attenuation and the phase is linearly dependent on the frequency ω: Signals transmitted over a lossless line are not altered in their form and thus no distortions occur.

For the kilometric group delay of the lossless line, the following applies

$$\tau_g = \frac{d\beta}{d\omega} = \frac{d}{d\omega}\left(\omega\sqrt{L'C'}\right) = \sqrt{L'C'}. \tag{3.139}$$

The group delay of the lossless line is constant and not frequency-dependent.

Summary of Special Cases

The special cases of lines and their transmission properties in harmonic processes, particularly with regard to the occurring signal distortions in amplitude and phase, form an important basis for the system-theoretical modeling of the transmission characteristics of lines and communication cables.

This modeling forms the bridge that is built from the line or telegraph equations of theoretical electrical engineering using the tools of system theory towards application in communications engineering.

In Tab. 3.1, important parameters of the line in harmonic processes are summarized.

The principle characteristics of the attenuation constant α and the phase constant β as a function of frequency are shown in Fig. 3.18.

Tab. 3.1 Overview of line properties for harmonic excitation of lines

Direct signal line	RC line	Low-loss line	Lossless line
($\omega \approx 0$)	(ω slightly higher)	(ω further increased)	(ω very high)
$R' \gg \omega L'$	$R' \gg \omega L'$	$R' \approx \omega L'$	$R' \ll \omega L'$
$G' \gg \omega C'$	$G' \ll \omega C'$	$G' \ll \omega C'$	$G' \ll \omega C'$
$\alpha = \sqrt{R'G'}$	$\alpha = \sqrt{\frac{\omega R' C'}{2}} \sim \sqrt{\omega}$	$\alpha \approx \frac{R'}{2}\sqrt{\frac{C'}{L'}} = \text{const.}$	$\alpha \approx 0$
$\beta = 0$	$\beta = \sqrt{\frac{\omega R' C'}{2}} \sim \sqrt{\omega}$	$\beta = \omega\sqrt{L'C'} \sim \omega$	$\beta = \omega\sqrt{L'C'} \sim \omega$
$Z_\mathrm{W} = \sqrt{\frac{R'}{G'}}$	$Z_\mathrm{W} = \sqrt{\frac{R'}{j\omega C'}}$	$Z_\mathrm{W} = \sqrt{\frac{R'+j\omega L'}{j\omega C'}}$	$Z_\mathrm{W} = \sqrt{\frac{L'}{C'}}$

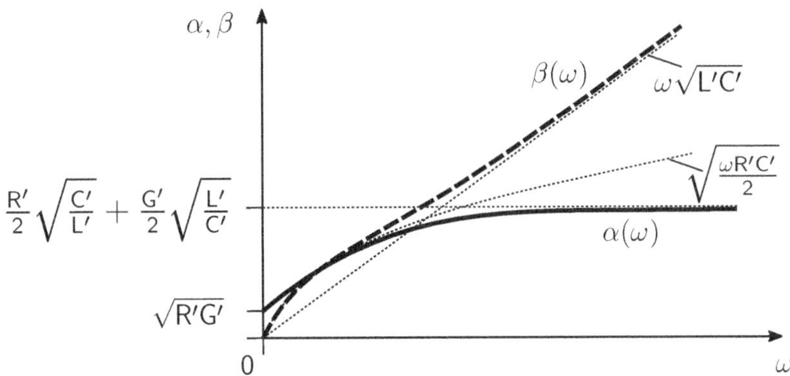

Fig. 3.18 Function graph of the damping and phase constant of a line as a function of frequency

3.4   Cable Transfer Function

3.4.1   Transfer Function of a Line

In order to apply the knowledge gained about the distorting properties of a line in communication theory and practice, it is useful to describe the properties of a line or a cable as a transfer function.

The starting point is the relationship

$$U_x = U_1 \cdot e^{-\gamma x} + U_2 \cdot e^{+\gamma x} \tag{3.140}$$

for the voltage on the line at the position x. This consists of forward and backward waves.

If a a line is terminated by its characteristic impedance at its output, there is no backward-travelling wave and the output voltage at the end of the line is, i.e. at $x = \ell$:

$$\underbrace{U_{x=\ell}}_{\text{voltage at the line output}} = \underbrace{U_1}_{\text{voltage at the line input}} \cdot e^{-\gamma\ell}. \tag{3.141}$$

The transfer function thus becomes

$$G_{\mathrm{k}}(\omega) = \frac{U_{x=\ell}}{U_1} \quad\Longrightarrow\quad G_{\mathrm{k}}(\omega) = e^{-\gamma\ell}. \tag{3.142}$$

It is a function of the frequency ω, as the propagation constant γ is frequency-dependent.

3.4.2 Transfer Function of the *RC* Line

The lossy line (*RC* line) is of particular importance for the wired transmission of digital signals.

If the propagation constant valid for the *RC* range is inserted according to (3.118)

$$\gamma = \sqrt{j\omega R'C'} \tag{3.143}$$

into (3.142) one obtains the transfer function of the *RC* line

$$G_{\mathrm{k}}(\omega) = e^{-\gamma\ell} = e^{-\ell\sqrt{j\omega R'C'}} = e^{-\ell\sqrt{j\frac{\omega}{\omega_0}}} \tag{3.144}$$

or with the frequency variable $f = \omega/(2\pi)$

$$G_{\mathrm{k}}(f) = e^{-\ell\sqrt{j\frac{f}{f_0}}}. \tag{3.145}$$

The cable characteristic frequency ω_0 or f_0 is a cable constant:

$$\omega_0 = 2\pi f_0 = \frac{1}{R'C'}. \tag{3.146}$$

It depends on the kilometric line parameters R' and C'—and thus on cable properties, such as the conductor material and diameter, the constructive design of the cable, and the type and material of the insulation.

If one takes into account the signal propagation time, i.e. $\tau_0 \cdot \ell$, over a line of length ℓ, one finally obtains the transfer function

$$G_{\mathrm{k}}(\omega) = e^{-\ell\sqrt{j\frac{\omega}{\omega_0}}} \cdot e^{-j\omega\tau_0\ell} \tag{3.147}$$

and as a weighting function

$$g_{\mathrm{k}}(t) = \frac{1}{\sqrt{\pi}} \frac{2\omega_0}{\ell^2} \left[\frac{\ell^2}{2\omega_0(t-\tau_0\ell)} \right]^{\frac{3}{2}} e^{-\frac{\ell^2}{2\omega_0(t-\tau_0\ell)}} \cdot 1(t). \tag{3.148}$$

The system-theoretical description of the properties can thus be done in the frequency and time domain with the relationships

Tab. 3.2   Characteristic frequencies of some cable types

$f_0 = \frac{\omega_0}{2\pi}$ (in MHz $\cdot$ km^2)	Cable type (Designation)	Conductor diameter (in mm)
0.178	Copper two-wire line	0.6
0.259	Copper two-wire line	0.8
0.827	Micro coaxial cable	0.8/2.7
2.686	Small coaxial cable	1.2/4.4
13.66	Standard coaxial cable	2.6/9.5

$$G_{\mathrm{k}}(\omega) = \mathrm{e}^{-\ell\sqrt{\mathrm{j}\frac{\omega}{\omega_0}}} \cdot \mathrm{e}^{-\mathrm{j}\omega\tau_0\ell} \quad\bullet\!-\!\!\circ\quad g_{\mathrm{k}}(t) = \frac{1}{\sqrt{\pi}}\frac{2\omega_0}{\ell^2}\left[\frac{\ell^2}{2\omega_0(t-\tau_0\ell)}\right]^{\frac{3}{2}} \mathrm{e}^{-\frac{\ell^2}{2\omega_0(t-\tau_0\ell)}} \cdot 1(t).$$

The size ω_0 or f_0 is a cut-off frequency related to one kilometer; it results in a cut-off frequency in conjunction with the line length ℓ, so that the line acts as a band-limiting low-pass filter: This causes signal distortions. The size τ_0 causes a delay, i.e., a propagation time, of the signal through the transmission over the line.

The transfer function determined here is valid for two-wire lines in the RC range and also for coaxial cables.

Some cable characteristic frequencies are compiled in Tab. 3.2.

For the optimization of point-to-point transmission systems, the absolute signal propagation time is often of secondary importance, as it does not influence the frequency-selective behavior of the copper line transmission channel: Therefore, the part of the transfer function caused by the propagation time can be neglected in such cases. However, there are also applications where this propagation time becomes significant, e.g., in ring network structures, where it must be taken into account.

In Fig. 3.19, an exemplary amplitude response of the transfer function of a copper two-wire line is shown, and Fig. 3.20 shows the weighting function of this line.

For the system-theoretical description and analysis of lines and cables, as well as for equalization, i.e., the compensation of signal distortions caused by the line, and for the related optimization of transmission systems, a rational transfer function is often advantageous and desirable. Therefore, a rational approximation of the cable transfer function of the RC line is now determined.

3.4.3   Rational Approximation of the Transfer Function of the RC Line

The transfer function

$$G_{\mathrm{k}}(\omega) = \mathrm{e}^{-\ell\sqrt{\mathrm{j}\frac{\omega}{\omega_0}}} \tag{3.149}$$

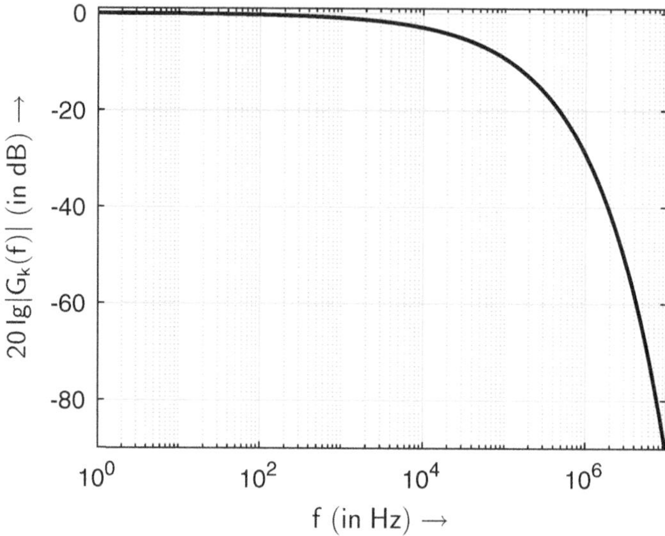

Fig. 3.19 Amplitude response of the transfer function of a copper two-wire line (Parameters: characteristic frequency $f_0 = 0.178\,\text{MHz} \cdot \text{km}^2$, length $\ell = 2\,\text{km}$)

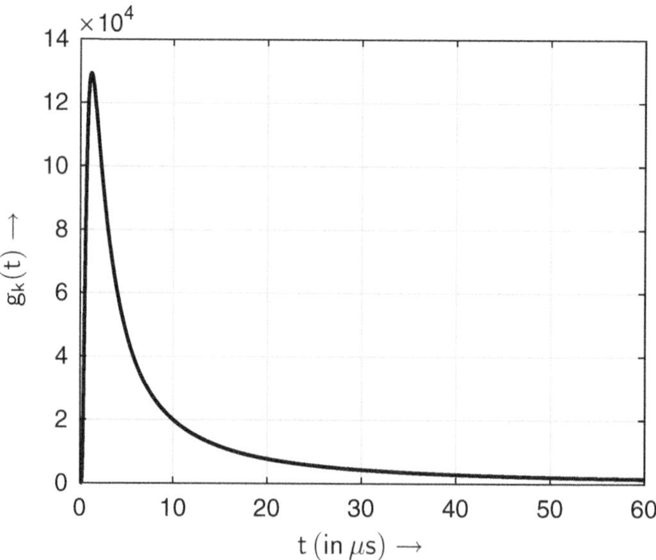

Fig. 3.20 Weighting function of a copper two-wire line (Parameters: characteristic frequency $f_0 = 0.178\,\text{MHz} \cdot \text{km}^2$, length $\ell = 2\,\text{km}$)

which is crucial for signal distortions, is of the form [29]

$$e^{-x} \qquad \text{with} \qquad x = \ell \sqrt{j \frac{\omega}{\omega_0}} . \tag{3.150}$$

It holds

$$\cosh x = \frac{e^x + e^{-x}}{2} \qquad \text{and therefore} \qquad e^x + e^{-x} = 2 \cosh x. \tag{3.151}$$

For $x \to \infty$, the term e^{-x} approaches zero and we initially obtain

$$e^x \approx 2 \cosh x \tag{3.152}$$

and finally as an approximation for large x

$$e^{-x} \approx \frac{1}{2 \cosh x} . \tag{3.153}$$

The function $\cosh x$ can be represented by

$$\cosh x = \prod_{\nu=1}^{\infty} \left(1 + \frac{1}{(2\nu - 1)^2} \frac{4x^2}{\pi^2} \right) \tag{3.154}$$

or, written out

$$\cosh x = \left(1 + \frac{4x^2}{\pi^2} \right) \cdot \left(1 + \frac{4x^2}{9\pi^2} \right) \cdot \left(1 + \frac{4x^2}{25\pi^2} \right) \cdots \tag{3.155}$$

as an infinite product. If this is inserted into the cable transfer function, it results in

$$G_k(\omega) \approx \frac{1}{2} \frac{1}{\displaystyle\prod_{\nu=1}^{\infty} \left(1 + j\frac{\omega}{\omega_\nu} \right)} \qquad \text{with} \qquad \omega_\nu = \frac{\pi^2 (2\nu - 1)^2}{4\ell^2} \omega_0 \tag{3.156}$$

or with the frequency variable f

$$G_k(f) \approx \frac{1}{2} \frac{1}{\displaystyle\prod_{\nu=1}^{\infty} \left(1 + j\frac{f}{f_\nu} \right)} \qquad \text{with} \qquad f_\nu = \frac{\pi^2 (2\nu - 1)^2}{4\ell^2} f_0 \tag{3.157}$$

or

$$G_k(p) \approx \frac{1}{2} \frac{1}{\displaystyle\prod_{\nu=1}^{\infty} (1 + pT_\nu)} \qquad \text{with} \qquad T_\nu = \frac{1}{\omega_\nu} = \frac{4\ell^2}{\pi^2 (2\nu - 1)^2 \omega_0} . \tag{3.158}$$

The cable transfer function is thus approximated as a low-pass filter by a rational function with an infinite number of simple real poles with the pole frequencies ω_ν or f_ν [29]. The position of the pole frequencies, relative to the frequency of the first

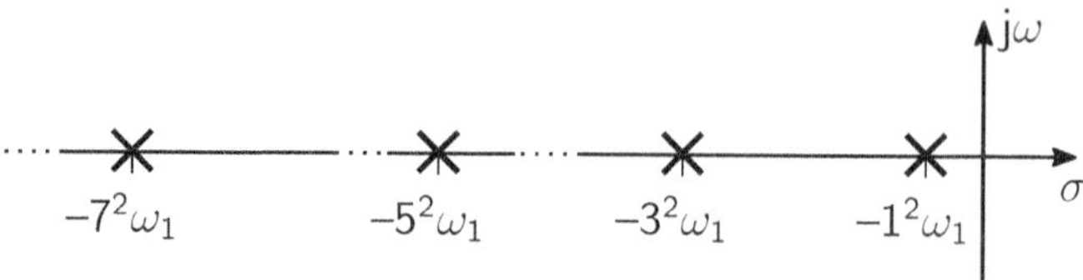

Fig. 3.21 Pole-zero plot of the rational approximation of a copper two-wire line

cable pole f_1, corresponds to the sequence of the squares of the odd natural numbers $\{1, 3^2, 5^2, 7^2, 9^2, \ldots\}$. The strength of the low-pass effect depends on the cable type (ω_0) and the cable length (ℓ). Practically useful approximations are obtained when the infinite product is truncated at a pole frequency of finite order, taking into account the signal bandwidth. This model describes the distorting properties of the cable, i.e., the deformation or distortion of the pulse shape; the signal's travel time from the transmitter to the receiver is not captured.

In Fig. 3.21, the rational approximation of the transfer function of an exemplary cable is shown in the pole-zero plot, which allows additional insights into the position of the poles in the complex plane.

The approximation of the cable transfer function according to (3.156) is inaccurate for low frequencies according to (3.153) (maximum attenuation deviation of 6 dB at $f = 0$ compared to (3.144), see Fig. 3.22). However, the attenuation of 6 dB at low frequencies correctly describes that if a line is terminetad by a resistance matched to the characteristic line impedance, a voltage U applied at the cable input appears at the cable output at low frequency with half the transmission amplitude $U/2$.

The quality of approximation in approximating the cable transfer function can be improved by supplementing the cable transfer function (3.157) with an additional real pole-zero pair in the ratio 1:2 (according to [29]). The cable transfer function is then approximated by

$$G_k(f) \approx \frac{1}{2} \frac{\left(1 + j\frac{f}{f_{01}}\right)}{\left(1 + j\frac{f}{0.5 f_{01}}\right) \prod_{\nu=1}^{\infty} \left(1 + j\frac{f}{f_\nu}\right)} \tag{3.159}$$

with $f_{01} = 0.231 f_0/\ell^2$ [29]. An exemplary cable transfer function (3.144) and its pole-zero approximations (3.157) and (3.159) are shown in Fig. 3.22.

Example 3.1 (Rational cable transfer function with two poles). Given is a copper two-wire line in a cable of length $\ell = 2\,\text{km}$ with a wire diameter of $d_0 = 0.6\,\text{mm}$. A rational transfer function in p and the weighting function are to be determined, if the approximation can be truncated after the second pole.

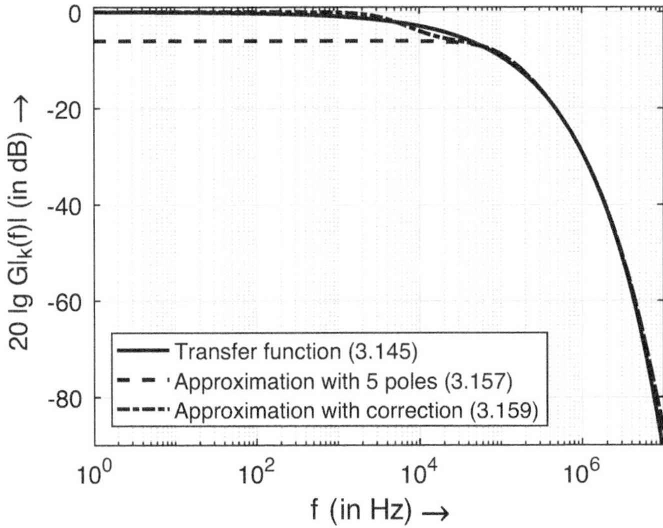

Fig. 3.22 Amplitude response of the transfer function and the rational approximation (without and with correction at low frequencies) of a copper two-wire line (parameters: characteristic frequency $f_0 = 0.178\,\text{MHz} \cdot \text{km}^2$, length $\ell = 2\,\text{km}$)

The cable characteristic frequency is $f_0 = 0.178\,\text{MHz} \cdot \text{km}^2$. For the pole frequencies and time constants of the two cable poles, we obtain:

$$
\begin{aligned}
f_1 &= \frac{\pi^2 f_0}{4\,\ell^2} = & 109.8\,\text{kHz} &\implies & T_1 &= \frac{1}{2\pi f_1} = 0.014\,\text{ms} \\
f_2 &= \frac{\pi^2 f_0}{4\,\ell^2} \cdot 3^2 = 9 f_1 = & 0.988\,\text{MHz} &\implies & T_2 &= \frac{1}{2\pi f_2} = 0.002\,\text{ms}
\end{aligned}
\tag{3.160}
$$

As a transfer function, we get

$$
G_{\text{k}}(p) = \frac{1}{2} \frac{1}{(1 + pT_1)(1 + pT_2)}
\tag{3.161}
$$

and by applying suitable correspondences of the Laplace transformation (see Appendix D or, for example, [11, 13, 74]), the weighting function for $T_1 \neq T_2$ results:

$$
g_{\text{k}}(t) = \mathscr{L}^{-1}\{G_{\text{k}}(p)\} = \frac{1}{2} \cdot \frac{\mathrm{e}^{-t/T_1} - \mathrm{e}^{-t/T_2}}{T_1 - T_2} \cdot 1(t)
\tag{3.162}
$$

□

3.4.4   Transfer Function of the Lossless RF Line

For the lossless or RF line, the propagation constant

$$\gamma = j\omega\sqrt{L'C'} \tag{3.163}$$

is effective according to (3.138). This results in the transfer function

$$G_k(\omega) = e^{-\gamma\,\ell} = e^{-j\omega\ell\sqrt{L'C'}} \quad \text{or} \quad G_k(p) = e^{-p\ell\sqrt{L'C'}} \tag{3.164}$$

and the weighting function

$$g_k(t) = \mathscr{L}^{-1}\{G_k(p)\} = \delta\!\left(t - \ell\sqrt{L'C'}\right) = \delta(t - t_0). \tag{3.165}$$

The lossless RF line causes a pure delay of the signal by $t_0 = \ell\sqrt{L'C'}$. A practical application of this type of line, especially in high-frequency and circuit technology, is its use as a pulse delay line.

3.5   Structural Design of Communication Cables

So far, the electromagnetic wave propagation processes and the resulting transmission properties have been examined with regard to a single electrical two-wire line, i.e., with respect to a *pair* consisting of two insulated copper wires. Practical communication cables often consist of several or a multitude of such electrical lines, which are also referred to as conductor or wire pairs. Communication cables have a cable sheath that mechanically combines the electrical conductors and provides protection against external environmental influences. In this section, some basic terms of the combination of insulated wire pairs into cables will be given to facilitate understanding of the mutual influences of different wire pairs through *crosstalk* discussed in the following section.

To minimize mutual and external electromagnetic influences, insulated wires are twisted together into *pairs* in a helical manner. This combination is referred to as *stranding*. Different levels of stranding are distinguished in communication cables. Some important and common stranding elements in communication cables are (see principle illustration in Fig. 3.23):

- Pair stranding: Two identical insulated single wires are combined into a *pair* by twisting. As a result, the length of the copper wire is greater than the length of the cable (or the sheath length): This factor is also referred to as the stranding extension factor. Its maximum value is approximately 1.02 [56].

Fig. 3.23 Stranding groups in communication cables

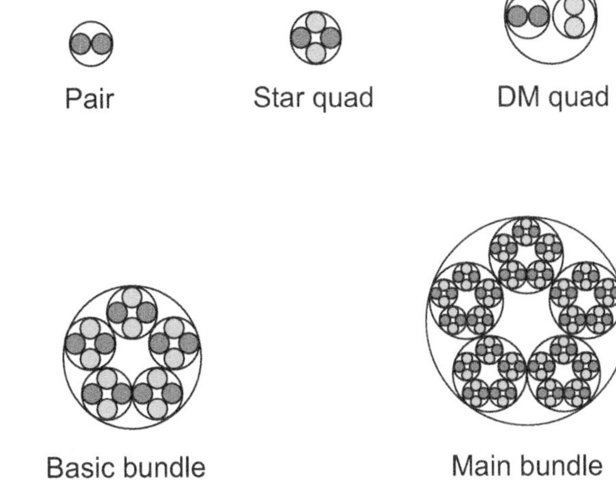

- Four-wire stranding: If two double wires already stranded into a pair are combined into a *four-wire* group, the Dieselhorst-Martin quad[6] (DM quad) is created. If four single wires are stranded in such a way that the single wires belonging to a common pair are opposite each other in the four-wire group, a cross- or star-shaped arrangement is created, the *star quad*. In modern communication cables, star quads are almost exclusively used due to their advantageous mechanical and electrical properties.
- Layer stranding: Cables are constructed from stranding groups (quads) in concentric layers. Layer stranding no longer has any significant practical importance in modern communication cables.
- Bundle stranding: Cables are constructed in stranded bundles with a consistently equal number of stranding groups. Modern communication cables are usually bundle-stranded. For example, five star quads are combined into a *basic bundle* (with a total of 10 double wires) and five basic bundles into a *main bundle* (with 50 double wires).

This brief presentation of the structure of communication cables provides a basic understanding of the relationships presented in the following section, which can occur in the mutual electromagnetic influence of wire pairs in such cables.

Further and detailed information on the structural design of communication cables and the terms used in this context can be found in specialist literature, e.g., [12, 56].

---

[6]This type of stranding was predominantly used in Germany in the more distant past and is now only of historical significance. It was named after its inventors William Dieselhorst and Arthur William Martin.

3.6   Crosstalk in Multi-Pair Communication Cables

3.6.1   Types of Crosstalk

In communication cables, for practical and economic reasons, several or many wire pairs are usually combined. Such cables are referred to as multi-pair cables. Electromagnetic couplings can become effective between these conductor or wire pairs operated in very close spatial proximity, which can lead to signal components being coupled over between adjacent conductor or wire pairs. This phenomenon is referred to as *crosstalk*. The term originates from analog telecommunication technology, where in the low or audio frequency range, e.g., in telephony, such couplings can lead to the reception of—more or less understandable—speech signals from adjacent conductor or wire pairs in a conductor or wire pair, where they appear as interference. Crosstalk usually becomes effective as interference; in modern communications, it can also be partially considered and utilized as a useful signal component.

In Fig. 3.24, the crosstalk occurring in communication cables is schematically represented. Two types of crosstalk are distinguished:

- In *near-end crosstalk*, the transmitter of the near-end crosstalk interference signal and the receiver are at the same line end.
- In *far-end crosstalk*, the transmitter of the far-end crosstalk interference signal and the receiver are at different line ends.

In bidirectionally operated cables, both types of crosstalk occur in principle. There are methods by which near-end crosstalk can be avoided or minimized, e.g., by using different transmission frequency ranges for the two different transmission directions, the wiring of closely adjacent wire pairs with signals that use different frequency ranges, or by using cables in only one transmission direction. However, the latter is not always desirable and practical for economic reasons.

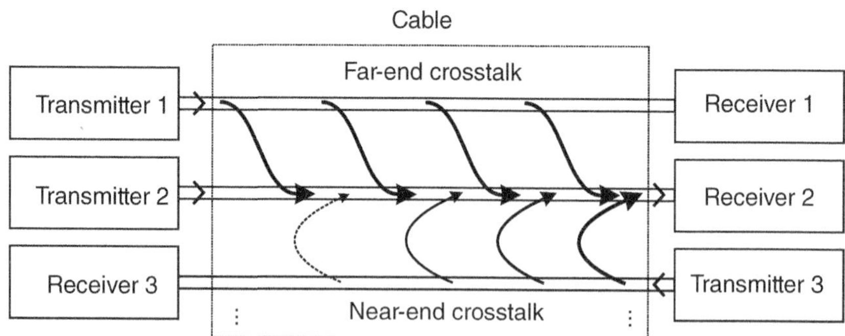

Fig. 3.24 Crosstalk in communication cables

In cables that are operated in only one transmission direction, only far-end crosstalk occurs as a crosstalk effect.

3.6.2   Model of the Cable with Crosstalk

For the description of a communication cable as a transmission channel, it is important to be able to capture the effects of near-end and far-end crosstalk in a system-theoretical model.

Near-End Crosstalk

In near-end crosstalk, the crosstalk signal passes through the coupling between two wire pairs and adds to the useful signal. The near-end crosstalk coupling can be captured by

$$|G_N(f)|^2 = K_N \cdot f^{1.5} \quad \text{with} \quad K_N = n_N^{0.6} \cdot K_{N1} \ . \tag{3.166}$$

The intensity of the total near-end crosstalk coupling is described by the factor K_N. The size n_N denotes the number of interfering wire pairs. The expression $n_N^{0.6}$ takes into account the fact that wire pairs located further away from the considered wire pair in the cable contribute less to the crosstalk interference than directly or closely adjacent ones. The exponent 0.6 is an empirically determined value. The factor K_{N1} describes the near-end crosstalk coupling between two directly adjacent wire pairs in the cable. For numerical calculations, for example, $K_{N1} = 8.536 \cdot 10^{-15} \frac{1}{\text{Hz}^{1.5}}$ can be assumed: This value was empirically determined based on measurements on communication cables [17, 69].

The near-end crosstalk between wire pairs is coupled into an adjacent wire pair without running through much cable length: Therefore, it is largely independent of the length of the cable. It increases with approx. 15 dB/decade with the frequency and is therefore particularly effective at high frequencies. The effect of the near-end crosstalk coupling approximately corresponds to the effect of a differentiating element, so that the interfering near-end crosstalk signal component is an attenuated and differentiated version of the original transmit signal.

Far-End Crosstalk

In far-end crosstalk, the crosstalk signal passes through the coupling between two wire pairs as well as the cable transfer function and is added to the useful signal. The far-end crosstalk coupling can be detected according to

$$|G_F(f)|^2 = K_F \cdot \ell \cdot f^2 \quad \text{with} \quad K_F = n_F^{0.6} \cdot K_{F1} \ . \tag{3.167}$$

The intensity of the total far-end crosstalk coupling is described by the factor K_F, where n_F is the number of interfering wire pairs and the expression $n_F^{0.6}$ models that wire pairs located further away from the considered wire pair in the cable contribute less to the crosstalk interference than directly or closely adjacent ones. The factor K_{F1} describes the far-end crosstalk coupling between two directly adjacent wire pairs in the cable. As an

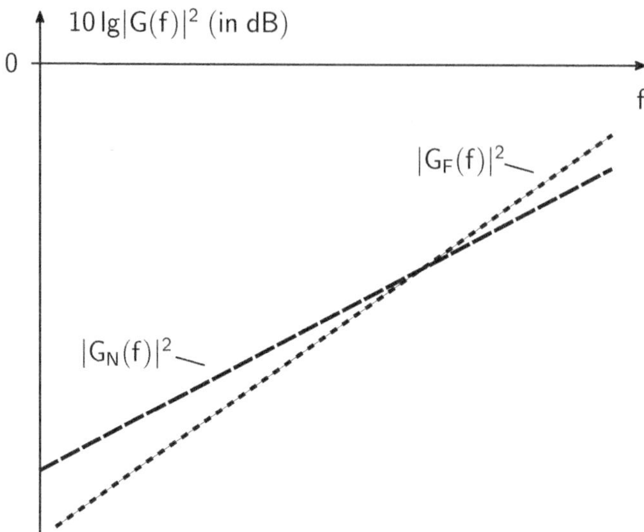

Fig. 3.25 On the frequency dependence of near- and far-end crosstalk coupling

exemplary numerical value, $K_{F1} = 2.6248 \cdot 10^{-17} \frac{1}{\mathrm{Hz}^2 \cdot \mathrm{km}}$ can be used, which has been empirically determined based on measurements on communication cables [17, 69].

The far-end crosstalk passes through the entire cable length and is coupled into an adjacent wire pair distributed over the entire cable length: The far-end crosstalk is dependent on the length of the cable and is also—like the useful signal—attenuated over the entire line length. Therefore, it is particularly pronounced in short cables—due to the then relatively low line attenuation. At low frequencies, the pure far-end crosstalk coupling increases by about 20 dB/decade, but at higher frequencies, the much stronger low-pass evaluation by the cable transfer function predominates.

Frequency Dependence of Near-End and Far-End Crosstalk

In Fig. 3.25, the frequency dependence of near- and far-end crosstalk coupling according to (3.166) and (3.167) is illustrated. It becomes clear that the pure coupling of the far-end crosstalk $|G_F(f)|^2$ increases more with frequency than the near-end crosstalk $|G_N(f)|^2$. However, the far-end crosstalk is additionally attenuated over the entire line length in addition to this coupling, while this is not the case with the near-end crosstalk due to the mechanism of action (see Fig. 3.24). This is illustrated using the diagram in Fig. 3.26: The far-end crosstalk is additionally low-pass evaluated and thus attenuated over the entire cable length ℓ by the cable transfer function $G_k(f)$ in addition to the pure coupling $|G_F(f)|^2$, so that in this case the total for the far-end crosstalk $|G_F(f) \cdot G_k(f)|^2$ is effective. Thus, the far-end crosstalk is dependent on the cable length ℓ: Its effect is more pronounced in short cables (ℓ small) than in long cables (ℓ large).

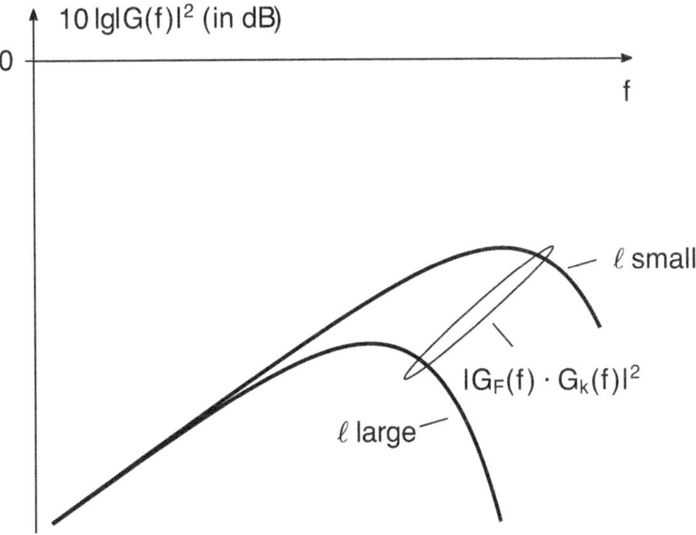

Fig. 3.26 On the frequency dependence of far-end crosstalk in cables in installation operation

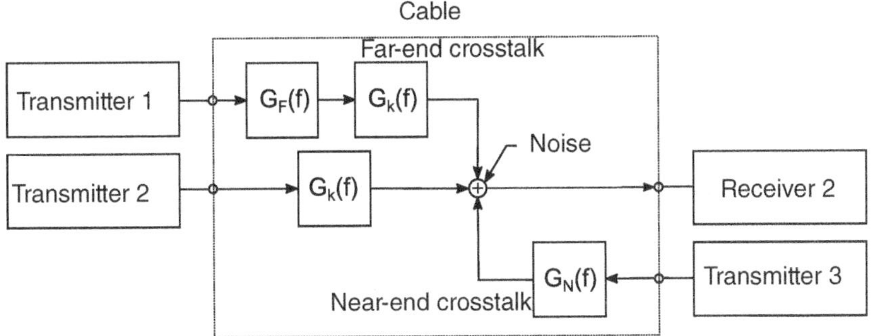

Fig. 3.27 System theoretical model of a communication cable with crosstalk

System Theoretical Model of the Communication Cable

With the previously introduced transmission functions and sizes for influencing the signal transmission in the useful signal path and for the crosstalkfrom adjacent wire pairs, the cable can now be described in system theory. In Fig. 3.27, a corresponding model is shown as a block diagram, which contains all previously discussed elements and additionally the always to be considered disturbance by noise [35].

The transmission path from transmitter 2 to receiver 2 is considered as the path of the useful signal. The signal coupled in by transmitter 1 is transmitted in the same direction as the useful signal and is therefore far-end crosstalk. Transmitter 3 is located at the same line end as receiver 2 and thus the signal coupled in there into the useful signal path is near-end crosstalk.

If there is a near-end crosstalk disturbance due to the type of cable operation, it is generally much larger than the far-end crosstalk disturbance.

3.7   Summary and Bibliographic Notes

The aim of this chapter was to develop a model for describing the behavior of communication cables, starting from the electromagnetic properties of lines, which is suitable for optimizing transmission methods. Starting from the distortion-free communication transmission channel, copper cables with their transmission technical properties were introduced as transmission media. Both the frequency- and length-dependent characteristics of individual symmetric wire pairs and the crosstalk behavior were described. A transfer function—or a weighting function—of the cable was developed in a frequency range used for the transmission of digital signals, which fits into a system theoretical description of transmission systems. Copper cables have a pronounced low-pass character. The cable length has a significant influence: with increasing cable length, the cut-off frequency of lines decreases.

Starting from the theory of lines, propagation processes on metallic lines were described using electromagnetic waves. The theory of lines is very versatile and has diverse fields of application, e.g., in the information and signal transmission over lines, in signal processing in high-frequency circuits, and in the transmission of electrical energy for electrical power supply over long distances. It has not lost any of its relevance over the many decades of its existence—at most, the fields of application have shifted in parts: The transmission of electrical energy over long distances is still an important field of application, telegraphy is no longer—but digital data transmission and processing remains important. A cable transfer function was developed through the theoretical line description, which is applicable for both copper two-wire lines and coaxial cables.

This has created a bridge between the basics of electrical engineering—or theoretical electrical engineering or the theory of electromagnetic fields—via system theory to communications engineering and especially transmission engineering. This approach is also intended to serve as an example of how models for practically relevant transmission channels can be found.

For the design and optimization of transmission systems, this type of modeling is a suitable and sufficiently accurate tool. In practice, there may be deviations from these results based on the transmission line theory, as, for example, effects were not considered that can be caused by inhomogeneities in the connection of cable pieces by soldering or splicing, by complex processes of eddy current formation or current displacement (skin effect at very high frequencies) or by the mismatch of components of the transmitters or receivers to the characteristic impedance of the cable. For this reason, empirical

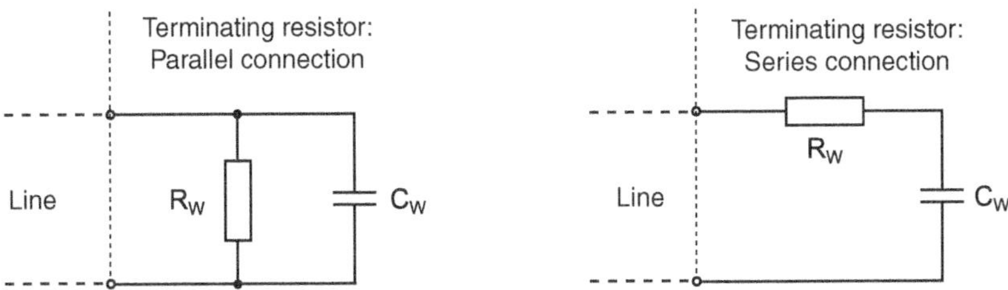

Fig. 3.28 Circuits for terminating the line matched to the line's characteristic impedance

relationships for the attenuation and distortion properties as well as the characteristic impedance of communication cables have been determined in the past based on measurements on real cables in communication networks, e.g., [51]: For optimal dimensioning of transmission systems, these should be consulted for fine-tuning.

Considerations on the transmission line theory can be found in different forms in books on theoretical electrical engineering, e.g., in [33, 41, 50, 61], on electromagnetic field theory, e.g., in [21, 28, 40, 59], or on high-frequency technology, e.g., in [19, 77]. In addition, some works are entirely or largely dedicated to the propagation of signals on lines and especially communication cables, e.g., [12, 46, 56–58]. The transfer function of communication cables is considered, for example, in [29]. All of the references mentioned here are in German.

3.8 Problems

Problem 3.1 (Description of a copper line)

A copper double wire is given, which has the following kilometric line parameters:
$R' = 33\,\Omega/\text{km}$, $L' = 0.7\,\text{mH}/\text{km}$, $G' = 1\,\mu\text{S}/\text{km}$, $C' = 42\,\text{nF}/\text{km}$.

A piece of this type of line with a length of $\ell = 2\,\text{km}$ is to be operated with a harmonic signal at a frequency of $f = 100\,\text{kHz}$. The following are to be calculated:

a) The attenuation constant α and the total attenuation caused by the cable a.
b) The phase constant β and the total phase rotation b at the output with respect to the input.
c) The propagation constant γ.
d) The characteristic impedance Z_W. Which circuit realizes a termination of the cable matched to the line's characteristic impedance?
e) The input impedance Z_0 for short circuit, open circuit and matching at the line output.
f) The wavelength λ of the oscillation.

Problem 3.2 (Transmission and weighting function of a copper line)

A copper two-wire line of length $\ell = 1\,\mathrm{km}$ with the cable characteristic frequency $f_0 = 0.178\,\mathrm{MHz} \cdot \mathrm{km}^2$ is given.

a) Determine the transfer function $G_k(f)$ of this line in the RC range and graphically represent its logarithmic amplitude response!

b) Determine the frequencies of the first five poles of the rational approximation of this cable transfer function!

c) Determine the weighting function $g_k(t)$ of this copper two-wire line in the RC range and graphically represent it in a diagram!

d) Over how many symbol intervals T_s does this weighting function extend, if a binary message signal with the bitrate or with the bit sequence frequency $f_B = 1\,\mathrm{MHz}$ is to be transmitted over it?

3.9   Solutions of the Problems

Solution of Problem 3.1

a) Attenuation constant: The given numerical values are inserted into the relationship (3.65)

$$\alpha = \sqrt{\frac{1}{2}\left(R'G' - \omega^2 L'C'\right) + \frac{1}{2}\sqrt{\left(R'^2 + \omega^2 L'^2\right)\left(G'^2 + \omega^2 C'^2\right)}} \qquad (3.168)$$

and one obtains

$$\alpha = 0.128\,\frac{\mathrm{Np}}{\mathrm{km}}. \qquad (3.169)$$

Note on dimension or unit: From

$$\frac{|U_2|}{|U_1|} = \mathrm{e}^{-\alpha\ell} \qquad (3.170)$$

it follows by applying the natural logarithm

$$\alpha = -\frac{1}{\ell}\ln\frac{|U_2|}{|U_1|} = \frac{1}{\ell}\ln\frac{|U_1|}{|U_2|} \qquad (3.171)$$

This quantity is given in Neper (Np).[7] If one wants to represent the attenuation constant in Dezibel (dB),[8] the decimal logarithm is to be used and one obtains

---

[7] Definition: A (in Np) $= \ln|A|$ (provided A is dimensionless).

[8] Definition: A (in dB) $= 20\,\lg|A|$ (provided A is dimensionless).

$$\frac{|U_2|}{|U_1|} = e^{-\alpha \ell} \quad \Longrightarrow \quad 20 \lg \frac{|U_2|}{|U_1|} = 20 \lg \left(e^{-\alpha \ell} \right) = -\alpha \ell \cdot 20 \lg e, \qquad (3.172)$$

from which

$$-\frac{20 \lg \frac{|U_2|}{|U_1|}}{(20 \lg e) \cdot \ell} = \alpha \quad \Longrightarrow \quad \alpha = \frac{1}{\ell} \cdot \frac{20 \lg \frac{|U_1|}{|U_2|}}{(20 \lg e)} \qquad (3.173)$$

arises. The conversion factor between dB and Np is therefore $20 \lg e = 8.686$. Conversion:

- $1\,\text{dB} = 0.115\,\text{Np}$
- $1\,\text{Np} = 8.686\,\text{dB}$

Total attenuation of the line:

$$a = \alpha \cdot \ell = 0.256\,\text{Np}$$
$$a = \alpha \cdot \ell \cdot 20 \lg e = 2.2\,\text{dB}$$

b) Phase constant:

The given numerical values are inserted into the relationship (3.66)

$$\beta = \sqrt{-\frac{1}{2}\left(R'G' - \omega^2 L'C'\right) + \frac{1}{2}\sqrt{\left(R'^2 + \omega^2 L'^2\right)\left(G'^2 + \omega^2 C'^2\right)}} \qquad (3.174)$$

and one obtains

$$\beta = 3.409 \,\frac{\text{rad}}{\text{km}}. \qquad (3.175)$$

If the phase rotation is to be given in ° (in degrees), the conversion can be done via

$$\beta \,(\text{in}^\circ) = \beta \,(\text{in rad}) \cdot \frac{360^\circ}{2\pi} \qquad (3.176)$$

and one obtains

$$\beta = 195.335^\circ \,\frac{1}{\text{km}}. \qquad (3.177)$$

Total phase rotation of the line:

$$b = \beta \cdot \ell = 6.82\,\text{rad}$$
$$b = \beta \cdot \ell \cdot \frac{180^\circ}{\pi} = 390.67^\circ$$

c) Propagation constant:

The propagation constant is obtained from the relationship (3.64)

$$\gamma = \alpha + j\beta \qquad (3.178)$$

from the previously calculated attenuation and phase constants to

$$\gamma = 0.128 + j\,3.409. \tag{3.179}$$

d) Characteristic impedance:

The characteristic impedance is obtained according to (3.42)

$$Z_{\mathrm{W}} = \sqrt{\frac{R' + j\omega L'}{G' + j\omega C'}} \tag{3.180}$$

for the given numerical values as

$$Z_{\mathrm{W}} = 129.19 - j\,4.837\ \Omega. \tag{3.181}$$

Insert: Calculation Method for the Characteristic Impedance

If the characteristic impedance is to be calculated manually, one must first expand under the root with the conjugate complex of the denominator, calculate the complex root using the representation according to magnitude and phase (or in polar coordinate form), and finally transform the characteristic impedance back into the representation as real and imaginary part. This calculation is shown in this insert.

First, the conjugate complex size of the denominator is extended under the root to achieve a complex number in the representation with real and imaginary part under the root:

$$
\begin{aligned}
Z_{\mathrm{W}} &= \sqrt{\frac{R' + j\omega L'}{G' + j\omega C'}} \\[2mm]
&= \sqrt{\frac{(R' + j\omega L') \cdot (G' - j\omega C')}{(G' + j\omega C') \cdot (G' - j\omega C')}} \\[2mm]
&= \sqrt{\frac{R'G' + j\omega L'G' - j\omega R'C' - j\omega C' \cdot j\omega L'}{G'^2 + \omega^2 C'^2}} \\[2mm]
&= \sqrt{\frac{R'G' + j\omega L'G' - j\omega R'C' + \omega^2 C'L'}{G'^2 + \omega^2 C'^2}} \\[2mm]
&= \sqrt{\frac{R'G' + \omega^2 C'L' + j\omega(L'G' - R'C')}{G'^2 + \omega^2 C'^2}} \\[2mm]
Z_{\mathrm{W}} &= \sqrt{\frac{R'G' + \omega^2 C'L'}{G'^2 + \omega^2 C'^2} + j\,\frac{\omega(L'G' - R'C')}{G'^2 + \omega^2 C'^2}}\,.
\end{aligned}
$$

To calculate the complex root, the complex number under the root is brought into the representation according to magnitude and phase, so that the characteristic impedance can be written as

$$Z_W = \sqrt{A} \cdot e^{j\phi} \qquad (3.182)$$

with

$$A = \sqrt{\left(\frac{R'G' + \omega^2 C'L'}{G'^2 + \omega^2 C'^2}\right)^2 + \left(\frac{\omega(L'G' - R'C')}{G'^2 + \omega^2 C'^2}\right)^2} \qquad (3.183)$$

and

$$\phi = \arctan\left(\frac{\left(\frac{\omega(L'G' - R'C')}{G'^2 + \omega^2 C'^2}\right)}{\left(\frac{R'G' + \omega^2 C'L'}{G'^2 + \omega^2 C'^2}\right)}\right) = \arctan\left(\frac{\omega(L'G' - R'C')}{R'G' + \omega^2 C'L'}\right) . \qquad (3.184)$$

The root of the complex number can then be calculated via

$$Z_W = \sqrt{A \cdot e^{j\phi}} = \sqrt{A} \cdot e^{j\frac{\phi}{2}} = \sqrt{A} \cdot \left(\cos\frac{\phi}{2} + j\sin\frac{\phi}{2}\right) . \qquad (3.185)$$

The transformation into the representation with real and imaginary part results in

$$\mathrm{Re}\{Z_W\} = \sqrt{A} \cdot \cos\frac{\phi}{2} \qquad (3.186)$$

$$\mathrm{Im}\{Z_W\} = \sqrt{A} \cdot \sin\frac{\phi}{2} \qquad (3.187)$$

and with the given numerical values the result is

$$Z_W = 129.19 - j\,4.837\ \Omega . \qquad (3.188)$$

For a termination of a cable matched to its characteristic impedance, the calculated characteristic impedance must be replicated with a circuit. This can essentially be done in two ways:

- Parallel connection of resistance R_W and capacitance C_W: In the extreme case $f = 0$ (direct voltage or current), the capacitance is ineffective and only the resistance remains as an effective element—this implementation is practically favorable.
- Series connection of resistance R_W and capacitance C_W: In the extreme case $f = 0$ (direct voltage or current), the capacitance would have to become infinitely large (short circuit)—therefore, this implementation is unfavorable.

The two possibilities are shown as circuit diagrams in Fig. 3.28.

The calculation of the component values is shown for the assumed numerical values for both circuits in the following insert.

Insert: Calculation of the Circuit Component Values of the Parallel and Series Connection when Terminating the Line Matched to its Characteristic Impedance

1. Parallel connection

In a parallel connection, the conductance values add up to the total conductance. Therefore, the characteristic admittance

$$Y_W = \frac{1}{Z_W} \tag{3.189}$$

is firstly calculated from the characteristic impedance Z_W according to real and imaginary part:

$$
\begin{aligned}
Y_W &= \frac{1}{Z_W} = \frac{1}{\text{Re}\{Z_W\} + j\,\text{Im}\{Z_W\}} \\
&= \frac{(\text{Re}\{Z_W\} - j\,\text{Im}\{Z_W\})}{(\text{Re}\{Z_W\} + j\,\text{Im}\{Z_W\}) \cdot (\text{Re}\{Z_W\} - j\,\text{Im}\{Z_W\})} \\
&= \frac{\text{Re}\{Z_W\} - j\,\text{Im}\{Z_W\}}{(\text{Re}\{Z_W\})^2 + (\text{Im}\{Z_W\})^2} \\
Y_W &= \frac{\text{Re}\{Z_W\}}{(\text{Re}\{Z_W\})^2 + (\text{Im}\{Z_W\})^2} - j\frac{\text{Im}\{Z_W\}}{(\text{Re}\{Z_W\})^2 + (\text{Im}\{Z_W\})^2}.
\end{aligned}
$$

In the parallel connection in Fig. 3.28, the characteristic admittance is composed as follows:

$$Y_W = \frac{1}{R_W} + j\omega C_W = \frac{1}{R_W} + j\,2\pi f\,C_W. \tag{3.190}$$

From the real and imaginary part of the characteristic admittance Y_W, the component values R_W and C_W are now determined with the previously calculated and thus numerically known characteristic impedance Z_W: For the resistance, the result is obtained via

$$\text{Re}\{Y_W\} = \frac{1}{R_W} \quad \Longrightarrow \quad R_W = \frac{1}{\text{Re}\{Y_W\}} \tag{3.191}$$

$$R_W = \frac{(\text{Re}\{Z_W\})^2 + (\text{Im}\{Z_W\})^2}{\text{Re}\{Z_W\}}$$

$$R_W = 129.37\,\Omega,$$

and for the capacitance, the result is obtained according to

$$\text{Im}\{Y_W\} = 2\,\pi f\,C_W \quad \Longrightarrow \quad C_W = \frac{\text{Im}\{Y_W\}}{2\,\pi f} \tag{3.192}$$

$$C_W = -\frac{1}{2\pi f} \cdot \frac{\text{Im}\{Z_W\}}{(\text{Re}\{Z_W\})^2 + (\text{Im}\{Z_W\})^2}$$

$$C_W = 0.46\,\text{nF}.$$

The termination of the line matched to its characteristic impedance can be realized with the considered parallel connection and the component values calculated for it. Strictly speaking, the result is only valid for the operating frequency f, as this is part of the calculations.

2. Series connection

In a series connection, the resistances add up. It can be written

$$Z_W = R_W + \frac{1}{j\omega C_W} = R_W - j\frac{1}{\omega C_W} . \tag{3.193}$$

From the real part of the characteristic impedance, the resistance of the series connection is obtained directly

$$\text{Re}\{Z_W\} = R_W \quad \Longrightarrow \quad R_W = \text{Re}\{Z_W\} \tag{3.194}$$

$$R_W = 129.19\,\Omega, \tag{3.195}$$

and from the imaginary part of the characteristic impedance, the capacitance can be obtained via

$$\text{Im}\{Z_W\} = -\frac{1}{\omega\,C_W} \quad \Longrightarrow \quad C_W = -\frac{1}{2\,\pi f \cdot \text{Im}\{Z_W\}} \tag{3.196}$$

$$C_W = 329\,\text{nF} . \tag{3.197}$$

Here too, this calculation is valid (only) for the given frequency f.

e) Input impedance

The input impedance of a line is generally given by

$$Z_0 = Z_W \cdot \frac{e^{\gamma\ell} + n\,e^{-\gamma\ell}}{e^{\gamma\ell} - n\,e^{-\gamma\ell}} \tag{3.198}$$

according to (3.94). With the calculated quantities characteristic impedance Z_W, propagation constant γ and known length ℓ, the input impedances can be calculated for the three operating states—and thus the known reflection coefficients n. The derived simplified relationships (3.95), (3.96) and (3.97) are used for this. Short circuit ($n = -1$):

$$Z_{0K} = Z_W \tanh \gamma\ell = (45.373 + j\,68.677)\,\Omega. \tag{3.199}$$

Open circuit ($n = +1$):

$$Z_{0L} = Z_W \coth \gamma\ell = (98.946 - j\,177.311)\,\Omega. \tag{3.200}$$

Matching ($n = 0$):

$$Z_{0A} = Z_W = (129.19 - j\,4.837)\ \Omega. \tag{3.201}$$

The input impedances in the calculated example for the three operating states short circuit, open circuit and matching are different: This is the general case, as differently sized reflected waves occur for different line terminations. It can also happen that the input impedances for all three different terminations are equal and equal to the characteristic impedance: Then the line acts as an electrically long line with an input impedance equal to its characteristic impedance, as no reflected wave occurs.

f) Wavelength of the oscillation

The wavelength λ of an oscillation is the distance between two points at which it has the same phase. The angle β is related to 1 km as the angle 2π is to the wavelength λ. Using (3.104) with the given values, we get

$$\frac{\beta}{1\,\text{km}} = \frac{2\pi}{\lambda} \qquad \Longrightarrow \qquad \lambda = \frac{2\pi}{\beta} = 1.843\,\text{km}. \tag{3.202}$$

Solution of Problem 3.2

a) The transfer function of a copper two-wire line in the RC range is given by (3.145) as

$$G_k(f) = e^{-\ell\sqrt{j\frac{f}{f_0}}}. \tag{3.203}$$

For the exemplary numerical values for f_0 and ℓ, its amplitude response is shown in Fig. 3.29: The typical low-pass behavior of such a transmission medium is recognizable.

b) With rational approximation, the transfer function of a copper two-wire line in the RC range can be expressed as

$$G_k(f) \approx \frac{1}{2}\,\frac{1}{\displaystyle\prod_{\nu=1}^{\infty}\left(1 + j\frac{f}{f_\nu}\right)} \qquad \text{with} \qquad f_\nu = \frac{\pi^2\,(2\,\nu - 1)^2}{4\,\ell^2}\,f_0 \tag{3.204}$$

according to (3.157). The pole frequencies of the first five cable poles are at

$$f_1 = \frac{\pi^2 f_0}{4\,\ell^2}\cdot 1^2 = \qquad\quad 0.44\,\text{MHz}$$

$$f_2 = \frac{\pi^2 f_0}{4\,\ell^2}\cdot 3^2 = 9f_1 = \qquad 3.95\,\text{MHz}$$

$$f_3 = \frac{\pi^2 f_0}{4\,\ell^2}\cdot 5^2 = 25f_1 = \quad 10.98\,\text{MHz}$$

$$f_4 = \frac{\pi^2 f_0}{4\,\ell^2}\cdot 7^2 = 49f_1 = \quad 21.52\,\text{MHz}$$

$$f_5 = \frac{\pi^2 f_0}{4\,\ell^2}\cdot 9^2 = 81f_1 = \quad 35.57\,\text{MHz}.$$

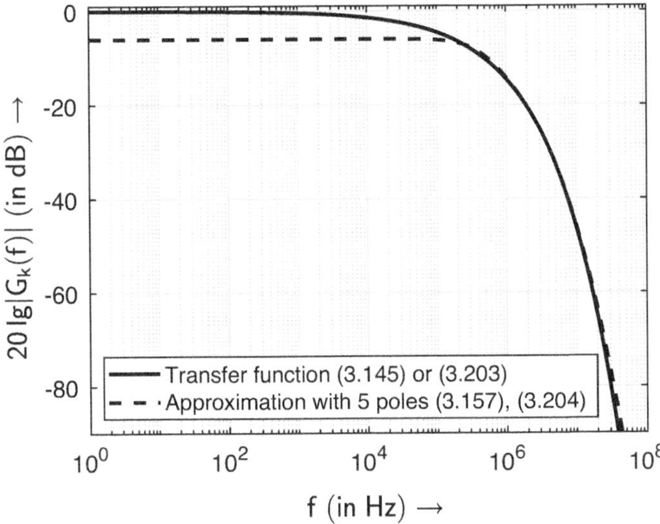

Fig. 3.29 Amplitude response of the transfer function and the rational approximation (with 5 poles) of a copper two-wire line (parameters: characteristic frequency $f_0 = 0.178\,\text{MHz} \cdot \text{km}^2$, length $\ell = 1\,\text{km}$)

The amplitude response of the rational approximation of the cable transfer function with five poles is also shown in Fig. 3.29: Deviations from the original transfer function are particularly noticeable at low frequencies (due to the approach via the cosh function for large x) and at high frequencies (due to the termination of the approximation after the fifth cable pole).

c) The weighting function of a copper two-wire line in the RC range is

$$g_k(t) = \frac{1}{\sqrt{\pi}} \frac{4\pi f_0}{\ell^2} \left[\frac{\ell^2}{4\pi f_0 (t - \tau_0 \ell)} \right]^{\frac{3}{2}} e^{-\frac{\ell^2}{4\pi f_0 (t - \tau_0 \ell)}} \cdot 1(t). \tag{3.205}$$

according to (3.148). It is graphically represented in Fig. 3.30 without considering the delay time τ_0.

d) If a binary, i.e., two-valued, message signal with the bit sequence frequency f_B is transmitted, then the bit sequence frequency is equal to the symbol clock frequency f_T and the symbol duration is

$$T_s = \frac{1}{f_B}. \tag{3.206}$$

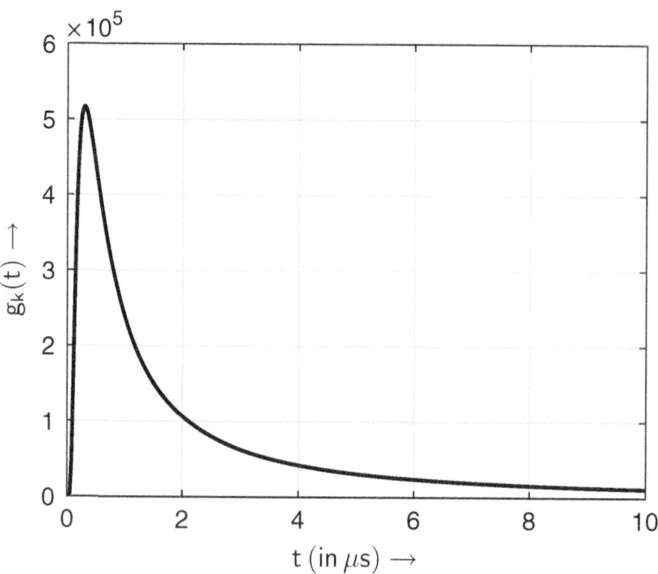

Fig. 3.30 Weighting function of a copper two-wire line (parameters: characteristic frequency $f_0 = 0.178\,\text{MHz} \cdot \text{km}^2$, length $\ell = 1\,\text{km}$)

- For $f_B = 1\,\text{MHz}$, the symbol duration is $T_s = 1/(1\,\text{MHz}) = 1 \cdot 10^{-6}\,\text{s} = 1\,\mu\text{s}$ A look at Fig. 3.30 shows that in this case the weighting function of the copper two-wire line extends with significant proportions over about ten symbol intervals, i.e., intersymbol interferences cause influences over this period. If the bit sequence frequency is further increased, the number of symbol durations would also increase accordingly, assuming a constant transmission channel, over which the transmitted symbols influence each other.

Baseband Transmission over Linearly Distorting Channel

4

Abstract

If a transmission channel causes distortions of the transmitted signal, measures are necessary to compensate for these so that the transmission quality still meets the requirements placed on the transmission system. In this chapter, based on the fundamentals of baseband transmission presented in Chap. 2 and the description of the transmission properties of copper cables developed in Chap. 3, possibilities for compensating for channel distortions are examined and discussed when a copper two-wire line is assumed as a linearly distorting channel for a baseband transmission system. The approach to describing, analyzing, and optimizing such transmission systems from Chap. 2 is essentially maintained and applied to the now present practically relevant case, expanded as needed, and deepened in this way. Detailed numerical examples again serve to illustrate the general insights gained and the principles developed.

4.1 Introduction and General Aspects

In Chap. 2 the transmission over non-distorting channels with noise disturbance was discussed (AWGN channels). When considering, designing, and operating high-bit-rate wired communication systems, the assumption of distortion-free transmission generally cannot be maintained. The real transmission channel increasingly distorts the transmitted signal due to its frequency-dependent properties as the signal bandwidth increases: Intersymbol interferences occur. The considerations in Chap. 2 have shown that the quality of the transmission is strongly influenced by intersymbol interferences: For example, only by clever choice of transmit and receive filters could the first Nyquist criterion be met to avoid intersymbol interferences at the detection point.

When considering real transmission channels, as the bandwidth of the transmitted signal increases, the phenomenon occurs that the spectral components of the transmitted signal are evaluated differently. As the bandwidth of the transmitted signal increases, this effect becomes more pronounced given a certain channel characteristic. Since the symbol clock frequency $f_T = 1/T_s$ is directly proportional to the bandwidth of the transmitted signal, this means that an increase in the symbol rate to be transmitted—the number of bits per symbol is irrelevant for the signal bandwidth—always also implies an increase in the transmission bandwidth.

Therefore, the properties of the transmitted signal—here the frequency range occupied by the transmitted signal—determine whether distortion-free transmission is possible. As the requirements for the data rate to be transmitted increase, it becomes increasingly difficult to achieve this, as the frequency range occupied by the transmitted signal becomes larger. In this case, the individual components of the transmitted signal experience different amplitude and phase distortions, which in principle makes distortion-free transmission impossible and intersymbol interferences occur.

While the frequency-dependent characteristic of the transmission channel can largely be ignored at low bandwidth of the transmitted signal and thus low symbol clock frequency f_T (see Fig. 4.1), at higher symbol clock frequency f_T and thus larger bandwidth of the transmitted signal—and the same given transmission channel—a greater distortion of the transmitted signal occurs (see Fig. 4.2).

In practical transmission technology, distortions are generally caused by real transmission channels. Typical linearly distorting channels can be found in both wired transmission and radio transmission. Using the example of baseband transmission over copper lines, this chapter will explain in detail the basic principles of evaluating the

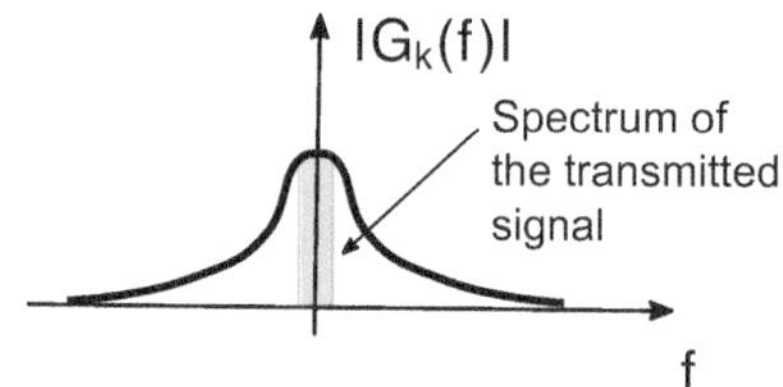

Fig. 4.1 Transmitted signal spectrum and transfer function of the channel in the frequency range at slow data transmission (principle)

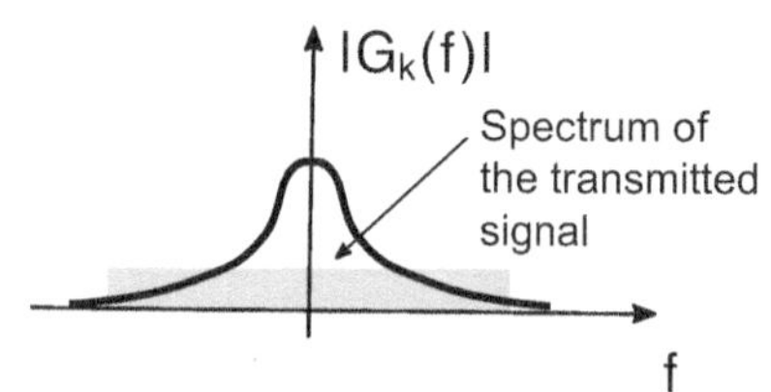

Fig. 4.2 Transmitted signal spectrum and transfer function of the channel in the frequency range at fast data transmission (principle)

useful signal and interference as well as equalization and discuss dependencies. The basis for this are the considerations from Chap. 2 for calculating the signal-to-noise ratio and the bit error probability in baseband transmission systems as well as the system-theoretical model of the wired transmission channel copper two-wire line or copper cable introduced in Chap. 3.

4.2 Compensation of the Distorting Channel Influence

For reliable and high-quality data transmission over distorting transmission channels, where the individual spectral components of the transmitted signal are evaluated differently and thus intersymbol interferences occur, it is necessary in practice to compensate or at least reduce the distortions caused by the respective transmission channel, thereby minimizing their effects on detection and decision-making. This can be done in different ways in practice; however, it is generally always necessary to invert the channel influence in one way or another—at least approximately. The implementation can be done either analogously or with digital signal processing. A functional block in the transmission path, whose main task is to compensate for the signal distortions that have occurred on the transmission channel, is referred to as an *equalizer.*

If the frequency response of the transmission channel has no zeros, one possibility for complete, linear equalization is to invert the channel frequency response and insert such a functional block into the transmission path. In the transfer function $G_k(f)$ determined in Chap. 3 for copper lines, this condition is met, so that such equalization with the inverse cable transfer function $1/G_k(f)$ is possible. Moreover, this type of equalization allows for a relatively clear mathematical-analytical treatment, so that fundamental influences and dependencies in digital information transmission over linearly distorting channels can be well represented and worked out: Therefore, such a transmission system is considered and extensively treated in this chapter. The wired transmission path with inverse cable transfer function as an equalizer in the receiver is shown in Fig. 4.3.

Since only linear components are used in this transmission system, the superposition principle applies, and useful and interfering signals can be considered separately

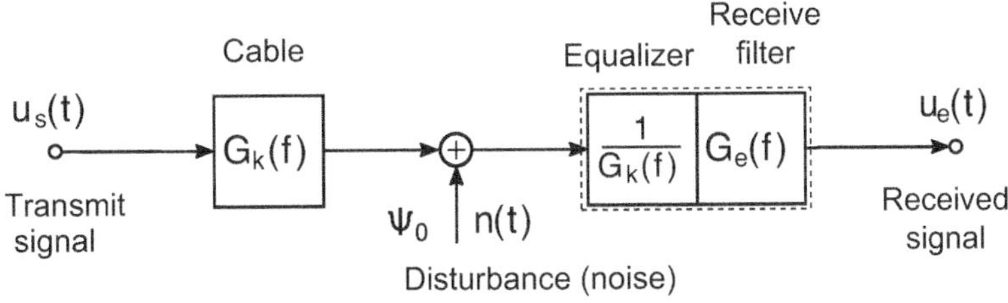

Fig. 4.3 Cable transmission system with distorting channel, linear equalization, and receive filtering

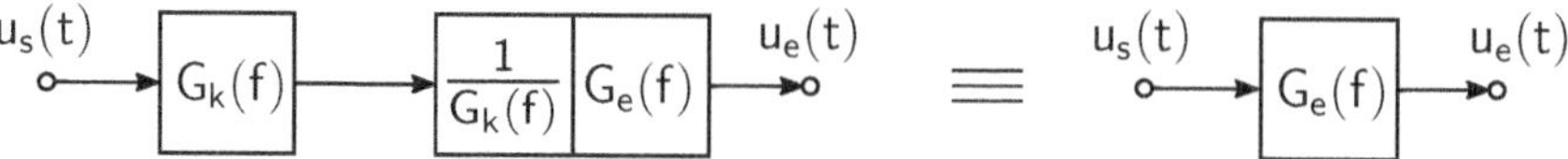

Fig. 4.4 Block diagram for the path of the useful signal

(see Chaps. 1 and 2): Their effect on the quality of data transmission at the detection point can be captured in the receiver using a suitable quality criterion, e.g., the signal-to-noise ratio and the resulting bit error probability.

The transmit signal $u_s(t)$ travels as a useful signal on the transmission path through the cable with the transfer function $G_k(f)$, and subsequently in the receiver through the components equalizer—with the transfer function $1/G_k(f)$—and receive filter—with the transfer function $G_e(f)$. Since the influence of the cable is fully compensated by the equalizer—with $G_k(f) \cdot 1/G_k(f) = 1$ –, the receive filter $G_e(f)$ remains for the effective evaluation of the useful signal: The undisturbed—i.e., without noise influence—receive signal is created by evaluating the transmit signal $u_s(t)$ through the receive filter $G_e(f)$ (see Fig. 4.4). This received useful signal (without disturbance or noise) is referred to as the received useful signal $u_e(t)$. In contrast, the actual output signal consists of the received useful signal $u_e(t)$ and the additive superimposed disturbance evaluated by the equalizer and receive filter: Only this actual—composed of received useful signal and (filtered) disturbance—receive signal can be observed and, for example, measured in real transmission systems. The definition of a received useful signal is, as already mentioned in Chap. 2, useful in linear systems for determining the transmission quality, as the evaluations of the useful signal and disturbance by the transmission system can be recorded separately, thus simplifying the calculations.

The white, Gaussian noise $n(t)$ with the power spectral density Ψ_0 is coupled into the model of the transmission system at the receiver input (see Fig. 4.3). It is evaluated by the transfer functions of the equalizer and the receive filter. Both can be combined into a common transfer function

$$H_e(f) = \frac{G_e(f)}{G_k(f)} \tag{4.1}$$

of an *equalizer-receive filter* (see Fig. 4.5). In practice, such equalizer-receive filters can also be advantageously implemented as a unit when a circuit-technical implementation is necessary.

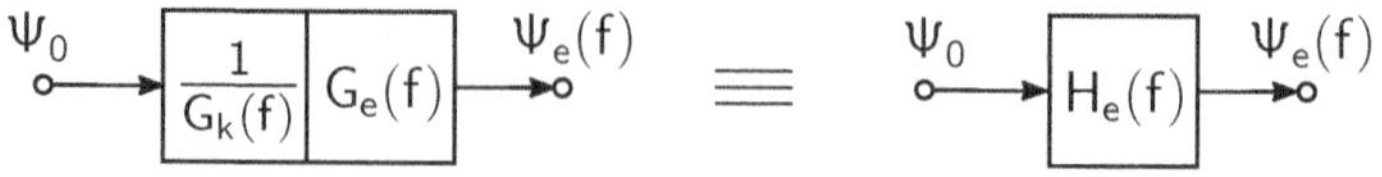

Fig. 4.5 Block diagram for the path of the disturbance

The power spectral density of the noise at the detection point is given by

$$\Psi_e(f) = \Psi_0 |H_e(f)|^2 = \Psi_0 \left| \frac{G_e(f)}{G_k(f)} \right|^2 . \tag{4.2}$$

By equalizing with the inverse cable transfer function in the receiver, not only is the influence of the cable on the useful signal compensated, but the noise signal is also amplified. The amplification caused by the equalizer $1/G_k(f)$ is frequency-dependent and increases with rising frequency. If the cable in a transmission system is operated and utilized up to high frequencies, a stronger increase of the noise can be expected than with lower cable utilization. As explained in Chap. 2, a band limitation of the noise is necessary to limit its power at the detection point. This task is performed by the receive filter $G_e(f)$: It must have a frequency response that allows the noise signal, weighted with the inverse cable transfer function, to reach the detection point with an overall low-pass evaluation. This means that the slope of the receive filter must fall off more steeply than the inverse of the cable transfer function increases according to (3.147): This can be achieved, for example, with the help of a Gaussian low-pass filter with the transfer function

$$G_e(f) = e^{-\ln(2)\cdot\left(\frac{f}{f_G}\right)^2} \quad \bullet\!\!-\!\!\circ \quad g_e(t) = \sqrt{\frac{\pi}{\ln 2}} f_G \cdot e^{-\left(\frac{\pi}{\sqrt{\ln 2}}\right)^2 \cdot t^2 f_G^2} \tag{4.3}$$

see e.g. Chap. 2 or [29].[1]

In Fig. 4.6, exemplary amplitude responses of the cable, the equalizer, and the equalizer-receive filter are shown in the Bode diagram when a Gaussian receive low-pass filter according to (4.3) is used. The pure cable with the transfer function $G_k(f)$ shows a strong low-pass characteristic with—depending on the cable length ℓ and the cable type (captured by the cable characteristic frequency f_0)—relatively low cutoff frequency. The increasing amplification with increasing frequency by the equalizer $1/G_k(f)$ is clearly visible. It becomes clear that the 6-dB cutoff frequency f_G influences how large the area enclosed under $H_e(f)$ (and thus also under $|H_e(f)|^2$) is: This area is a measure of the effective noise power at the detection point (see Chap. 2). In addition, the useful signal is evaluated by the receive filter $G_e(f)$ and its cutoff frequency f_G. These mutual dependencies will be examined in detail in the following.

A receive-side low-pass band limitation, which falls off more steeply than the inverse of the cable transfer function according to (3.147) increases, can also be achieved, for example, by raised cosine or root raised cosine filtering, see Chap. 2. The receive filtering with Gaussian low-pass filter and with root raised cosine low-pass filter will be referred to in a numerical example in the further course of this chapter.

---

[1] This type of description is a non-causal representation: Non-causal systems are those that show a system reaction before the excitation by the input signal, while in causal systems an output signal is observed only when an input signal is present, see e.g. [30, 54].

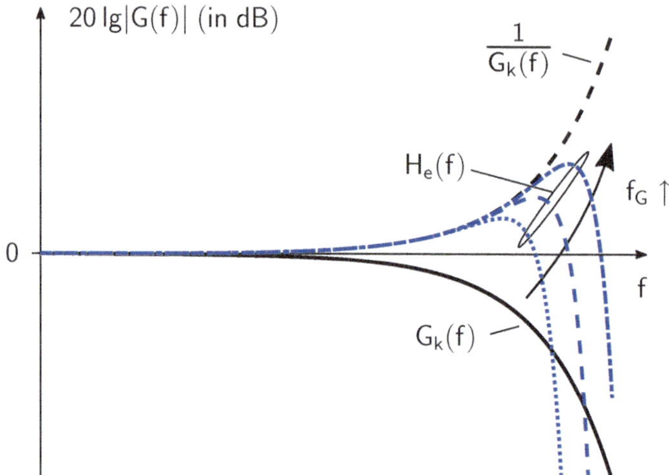

Fig. 4.6 Amplitude responses (in the Bode diagram) of cable, equalizer, and equalizer-receive filter with Gaussian receive low-pass filter with different cutoff frequencies f_G

4.3   Cable Transmission System with Rational Transfer Functions

For the practical implementation of equalizers and, for example, cable replicas, as well as for theoretical investigations, it is often advantageous to use rational transfer functions of the cable and, as a result, also for the equalizer -receive filter: The description with the cable transfer function according to (3.147) and, for example, the Gaussian low-pass filter (4.3) leads to relationships for the useful signal and disturbance at the detection point, which contain integrals that cannot be solved elementarily. These can be evaluated numerically and thus numerical results can be achieved; however, functional dependencies in the theory are not easily recognizable in this way.

Therefore, for the following considerations, the rational approximation of the cable transfer function (3.157) or (3.158) serves as a starting point: It accurately represents the signal distortions caused by the cable and, in the limit case of an infinite product or modeling with infinitely many real poles—except for deviations at very low frequencies—is identical to (3.145). For practical implementations and calculations, the infinite product must be truncated at a finite number m of poles of the cable transfer function. This is done depending on the bandwidth of the signal to be transmitted, so that the influence of the cable is accurately represented for the spectral range to be examined.

The inversion of the rational cable transfer function with m poles results in m zeros for the purely equalizing part of the equalizer receive filter: Such a system would not be realizable separately, as a realizable system may not contain more zeros than poles, i.e., it must have at least as many poles as zeros. The receive low-pass filter is also modeled as a rational transfer function with n poles and it is therefore realized in a unit with the equalizer component. The equalizer receive filter as a system is realizable, provided that at least $n = m$ applies. It becomes clear once again that—in the type of equalization and receive filtering considered here—the integrated realization as an equalizer-receive filter is essential and has practical advantages.

The receive filter is chosen as an RC low-pass filter of nth order:

$$G_{\mathrm{e}}(p) = \frac{1}{\left(1 + pT_{\mathrm{g}}\right)^n} \quad \bullet\!\!-\!\!\circ \quad g_{\mathrm{e}}(t) = \frac{e^{-\frac{t}{T_{\mathrm{g}}}} \cdot t^{n-1}}{T_{\mathrm{g}}^n(n-1)!} \cdot 1(t) \quad \text{with } T_{\mathrm{g}} = \frac{1}{2\pi f_{\mathrm{g}}}. \tag{4.4}$$

The cutoff frequency f_{g} is, for example, for $n = 1$ a 3-dB cutoff frequency, i.e., at the frequency $f = f_{\mathrm{g}}$, the magnitude of the transfer function of (4.4)—at $n = 1$—has dropped to a value of $1/\sqrt{2} \approx 0.7071$ or by $-20\lg(1/\sqrt{2}) = 3\,\mathrm{dB}$ compared to the value at the frequency $f = 0$ (see Chap. 2).[2]

For the rational RC low-pass filter of nth order, one obtains for $\sigma = 0$ and thus $p = \sigma + j\omega \Longrightarrow j\omega = j2\pi f$ as transfer function with the frequency variable f:

$$G_{\mathrm{e}}(p) = \frac{1}{\left(1 + pT_{\mathrm{g}}\right)^n} \quad \Longrightarrow \quad G_{\mathrm{e}}(f) = \frac{1}{\left(1 + j\frac{2\pi f}{2\pi f_{\mathrm{g}}}\right)^n} = \frac{1}{\left(1 + j\frac{f}{f_{\mathrm{g}}}\right)^n}. \tag{4.5}$$

For $n \to \infty$, the transfer function of the RC low-pass filter of nth order (4.5) approximates a Gaussian low-pass filter (4.3), see Chap. 2.

The transfer function of the equalizer receive filter as a whole is using the cable transfer function (3.158)

$$H_{\mathrm{e}}(p) = \frac{2 \prod\limits_{\nu=1}^{m} (1 + pT_\nu)}{\left(1 + pT_{\mathrm{g}}\right)^n}. \tag{4.6}$$

It contains m zeros at the frequencies of the first m cable poles and an n-fold pole at the cutoff frequency f_{g} of the rational RC receive low-pass filter of n-th order. In Fig. 4.7, the transmission system with the components in rational transfer functions is shown.

Figure 4.8 shows for the example $m = 3$ the amplitude responses (line approximations) of the cable, the equalizer receive filter, and the resulting overall system in the Bode diagram. It is assumed that the useful signal to be transmitted occupies such a spectral range that the approximation of the cable properties by a rational transfer function with $m = 3$ poles is sufficient and thus for the cable transfer function (3.157) in this case

---

[2] For a general filter order n, the designation of the cutoff frequency, i.e., by what fraction of the maximum value of the amplitude response at the cutoff frequency f_{g} has dropped compared to the value at $f = 0$, depends on n. Thus, the cutoff frequency at $n = 2$ is a 6-dB cutoff frequency and at $n = 3$ it is a 9-dB cutoff frequency. In general, one could use the designation $n \cdot 3$-dB cutoff frequency for a rational low-pass filter of nth order; however, this designation is not widely used.

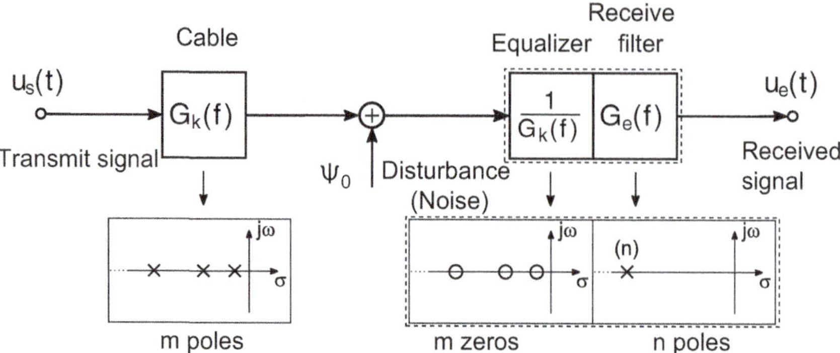

Fig. 4.7 Cable transmission system with equalization at rational transfer functions

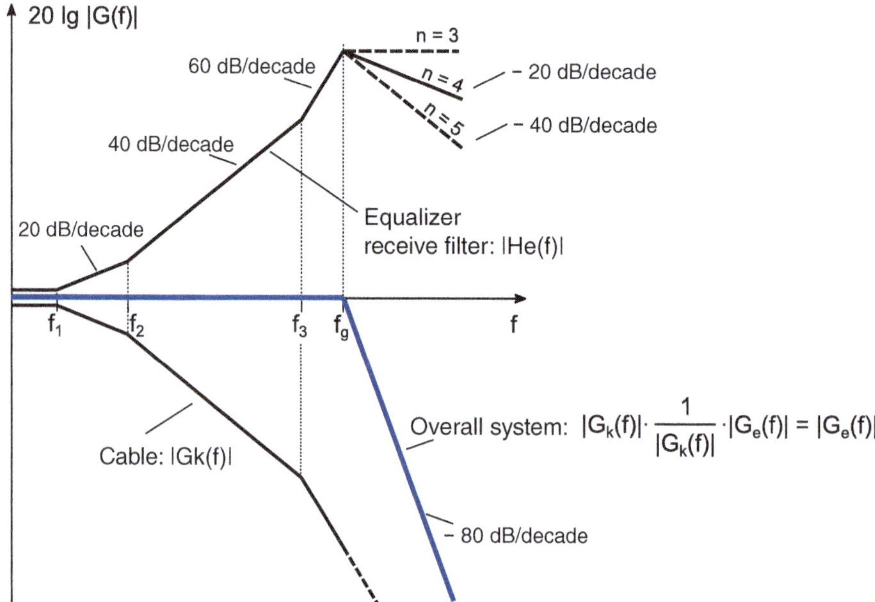

Fig. 4.8 Amplitude responses (straight-line line approximations in the Bode diagram) of cable, equalizer receive filter, and overall system using the example $m = 3$

$$G_k(f) \approx \frac{1}{2} \frac{1}{\prod\limits_{\nu=1}^{m=3}\left(1+j\frac{f}{f_\nu}\right)} \quad \text{with} \quad f_\nu = \frac{\pi^2\,(2\,\nu - 1)^2}{4\,\ell^2} f_0 \tag{4.7}$$

applies. The poles of the cable are located at the frequencies f_1, f_2, and f_3. In the line approximation of the transfer function, the amplitude response of the cable between f_1 and

f_2 shows a drop of 20 dB/decade, between f_2 and f_3 a drop of 40 dB/decade, and above f_3 a drop of 60 dB/decade. The equalizer receive filter has zeros at the frequencies of the cable poles, so that the corresponding increases are recognizable in the respective areas of the line approximation to compensate for the cable frequency response by amplification. The receiver low-pass filter of n-th order with the cutoff frequency f_g must have at least the order $n = m = 3$ so that the equalizer-receive filter is a realizable system. However, the Bode diagram in Fig. 4.8 shows that in this case the equalizer receive filter would not have a band limitation for the noise, so that the noise power at the detection point would be very large—theoretically infinite—and thus the signal-to-noise ratio would be zero: Reliable detection would not be possible. Therefore, a low-pass filter of at least order $n = m + 1 = 4$ is necessary, i.e. generally: $n > m$. With a fourth-order low-pass filter, the noise is band-limited with a falling edge of -20 dB/decade to high frequencies. For $n = 4$, the equalized overall system is also shown in the line approximation: This is the system effectively remaining for the useful signal, i.e. the receive filter $G_e(f)$. For the useful signal, the band limitation with a total of -80 dB/decade is effective.

If the order of the low-pass filter is increased (e.g., $n = 5$), the noise band limitation improves further—here to -40 dB/decade—but the useful signal is also affected—then with -100 dB/decade—and generally distorted more strongly. The degree n of the low-pass filter is a degree of freedom, in addition to the already mentioned effects on the useful signal and disturbance, the implementation effort or complexity increases with increasing n.

The opposing evaluation of the useful signal and the disturbance by the receive filter and its band limitation is of a fundamental nature and finds further expression in the role of the receiving filter cutoff frequency f_g. The cutoff frequency f_g of the receive filter plays a contradictory role (given fixed cable poles):

- If the cutoff frequency f_g is small, the interference is more strongly band-limited and thus the noise power is relatively small: This is favorable for the transmission quality. However, the useful signal is more strongly distorted, generally leaving relatively strong intersymbol interferences, which adversely affects the transmission quality.
- If the cutoff frequency f_g is large, the distortions of the useful signal are less pronounced, leaving relatively minor intersymbol interferences—with a positive effect on the transmission quality. However, the interference is now less strongly band-limited and thus the noise power is larger—with a negative impact on the transmission quality.

There is therefore an opposing dependency of the useful signal and disturbance on the receive filter cutoff frequency f_g—as already discussed in the transmission over a channel with pure noise disturbance in Chap. 2. Therefore, an optimization of the cutoff frequency f_g with respect to the transmission quality, i.e., in terms of a maximum signal-to-noise ratio, is possible and necessary. These relationships will be further analyzed in the course of the following detailed application example and corresponding optimization possibilities will be pointed out.

4.4 Application Example: Rational Linearly Equalized Cable Transmission System

4.4.1 Introduction

To thoroughly investigate the previously discussed properties and dependencies in relation to the evaluation of the useful signal and interference, a comparatively simple but realistic model of a wired transmission system with equalization and receive filtering is considered in an application example. The transmission system is first introduced and then the evaluation of the useful signal and the disturbance are calculated separately in detail. Particular attention is paid to the role of the receive filter cutoff frequency: It can be freely chosen in the design of the transmission system and thus represents an important degree of freedom for the design and optimization of such a transmission system. Finally, the quality of the transmission is evaluated and the receive filter cutoff frequency is optimized based on this.

4.4.2 Cable Transmission System

In Fig. 4.9, the block diagram of the transmission system to be examined is shown: The transmit signal is a unipolar random sequence of rectangular pulses (NRZ—Non Return to Zero) with two amplitude levels (0 V and U_0) with the pulse duration T_s, which passes through the cable at the bit rate or bit sequence frequency $f_B = 1/T_s$ and is disturbed by additive, white, and Gaussian noise with the power spectral density Ψ_0.

The cable transfer function is approximated as a first-order low-pass filter

$$G_k(p) = \frac{1}{2} \cdot \frac{1}{1 + pT_k} \qquad \text{with} \qquad T_k = \frac{1}{2\pi f_k} = \frac{4\,\ell^2}{\pi^2\,\omega_0} \tag{4.8}$$

to achieve good analytical manageability and to be able to work out essential transmission basics—without making the calculations too difficult and endangering the clarity due to too much complexity. This model is created by truncating the infinite product in (3.158) after the first term ($m = 1$) and captures essential distorting properties of the transmission medium copper cable: In Fig. 4.10 the amplitude responses of the chosen cable approximation with one pole and the cable transfer function (3.147) are shown together.

The essential bandwidth limitation that the medium copper cable exerts on the useful signal during transmission is correctly reproduced by the first pole of the approximated transfer function (3.158) of the cable.[3] Towards higher frequencies and thus signal

---

[3] The deviation between the two transfer functions at low frequency was discussed in Chap. 3; this is due to the approach (3.153) in the rational approximation, which gives differences for small frequencies. The approximation quality at high frequencies, which is more significant for the transmission quality, is determined by the number of poles considered in the rational approximation.

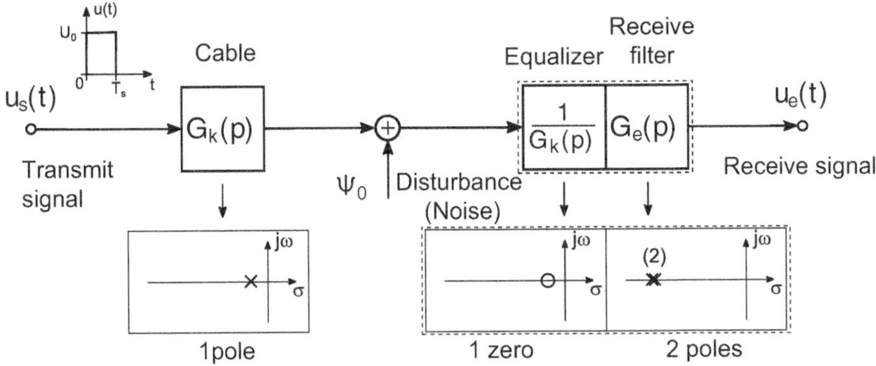

Fig. 4.9 Transmission system with cable and equalizer-receive filter with cable approximation by a single pole

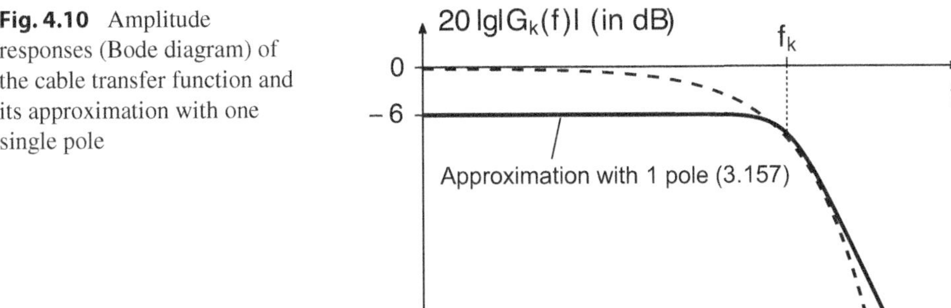

Fig. 4.10 Amplitude responses (Bode diagram) of the cable transfer function and its approximation with one single pole

spectral components, a sufficient number of cable poles must be considered; this results in the transfer function of the equalizer and in connection with the receive filter that of the equalizer-receive filter.

The distorting influence of the cable is—as introduced in the previous section—completely linearly equalized on the receiving side using the inverse cable transfer function. The equalizer has—as the inverse cable transfer function $1/G_k(p)$—a zero at the frequency of the cable pole. The receive filter is a low-pass filter of n-th order. A special feature of this type of equalization and receive filtering is that the zero of the equalizer in the example requires that the receive filter has at least two poles ($n = 2$)—since $n > m$ is required—to achieve a low-pass band limitation of the noise signal. A first-order low-pass filter, as considered in Chap. 2 for the AWGN channel, would not be sufficient here, because the noise would not be band-limited in this case and thus the noise power would become infinite (see Fig. 4.8): Then the signal-to-noise ratio would be zero and reliable transmission and detection would not be possible. The receive filter is therefore exemplarily set as second order with a double-real pole

$$G_e(p) = \frac{1}{\left(1 + pT_g\right)^2} \qquad \text{with} \qquad T_g = \frac{1}{2\pi f_g} \tag{4.9}$$

As a result, the equalizer-receive filter is

$$H_e(p) = \frac{G_e(p)}{G_k(p)} = \frac{2\left(1 + p\,T_k\right)}{\left(1 + p\,T_g\right)^2} \tag{4.10}$$

As described in Chap. 2, the evaluation of the useful signal and disturbance for determining the transmission quality—and thus for optimizing parameters—can be carried out separately in linear systems. In the following, this procedure is carried out step by step for the introduced example of a wired transmission system and explained.

4.4.3   Useful Signal Path and Evaluation of the Useful Signal

A measure of the quality of the useful signal at the detection point is the half vertical eye opening (see Chap. 2). In the general case—with remaining intersymbol interference—it is appropriate to determine this half vertical eye opening for the worst case and to use it for the evaluation of the transmission quality. This worst case, i.e., the smallest half vertical eye opening, exists in two constellations: On the one hand, it occurs when a '1' is sent after a long sequence of '0' pulses and then again a long sequence of '0' pulses and on the other hand, when conversely after a long sequence of '1' pulses a single '0' pulse is sent and then again a long sequence of '1' pulses. The first case is chosen here: To determine how the useful signal is evaluated by the transmission arrangement, the path of a single basic pulse from the transmitter to the output of the receive filter—the detection point—must be considered and its deformation analyzed. The basic transmission pulse $x_s(t)$ is an NRZ-rectangle impulse with the voltage amplitude U_0 and the duration T_s, which passes through the cascade of the transfer functions of cable $G_k(p)$, equalizer $1/G_k(p)$ and receive low-pass filter $G_e(p)$ according to Fig. 4.11.

Since in the model the transfer functions of the cable $(G_k(p))$ and equalizer $(1/G_k(p))$ exactly compensate each other, only the receive filter $G_e(p)$ remains as the element that deforms the transmit signal on its way to the detection point. Therefore, the rectangular

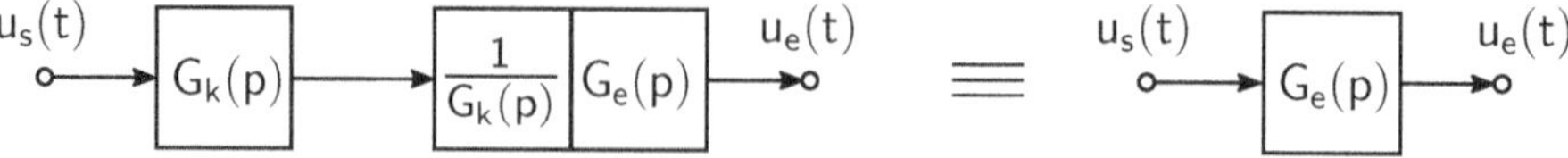

Fig. 4.11 Block diagram for the evaluation of the useful signal

response of the receive filter (4.9) must be calculated. Figure 4.12 shows the amplitude frequency response of the receive filter with the cutoff frequency f_g.

To calculate the rectangular response of the receive filter, the path via the spectral domain is taken (see, for example, Chap. 1 or literature on signal and system theory, e.g., [38]): The Laplace transform of the rectangular NRZ basic transmit pulse of duration T_s with the amplitude U_0 is (using the time shift theorem of Laplace transformation)

$$X_\mathrm{s}(p) = \frac{U_0}{p} - \frac{U_0\,\mathrm{e}^{-pT_\mathrm{s}}}{p}.$$

(4.11)

The Laplace transform of the basic received pulse is thus after multiplication with $G_\mathrm{e}(p)$ according to (4.9)

$$X_\mathrm{e}(p) = X_\mathrm{s}(p) \cdot G_\mathrm{e}(p) = \frac{U_0}{p} \cdot \frac{1 - \mathrm{e}^{-pT_\mathrm{s}}}{\left(1 + pT_\mathrm{g}\right)^2}.$$

(4.12)

By factoring out $1/T_\mathrm{g}^2$ and rearranging, one obtains

$$X_\mathrm{e}(p) = \frac{U_0}{T_\mathrm{g}^2} \cdot \frac{1 - \mathrm{e}^{-pT_\mathrm{s}}}{p\left(\frac{1}{T_\mathrm{g}} + p\right)^2}.$$

(4.13)

The inverse Laplace transform of the Laplace transform of the basic received pulse yields

$$x_\mathrm{e}(t) = \frac{U_0}{T_\mathrm{g}^2} \cdot \mathscr{L}^{-1}\left\{ \frac{1 - \mathrm{e}^{-pT_\mathrm{s}}}{p\left(p + \frac{1}{T_\mathrm{g}}\right)^2} \right\}.$$

(4.14)

With the help of the correspondence (see Appendix D or, for example, [11, 13, 74])

$$\frac{1}{p(p + a)^2} \quad\bullet\!\!-\!\!\circ\quad \frac{1}{a^2}\left(1 - \mathrm{e}^{-at} - at\,\mathrm{e}^{-at}\right)$$

(4.15)

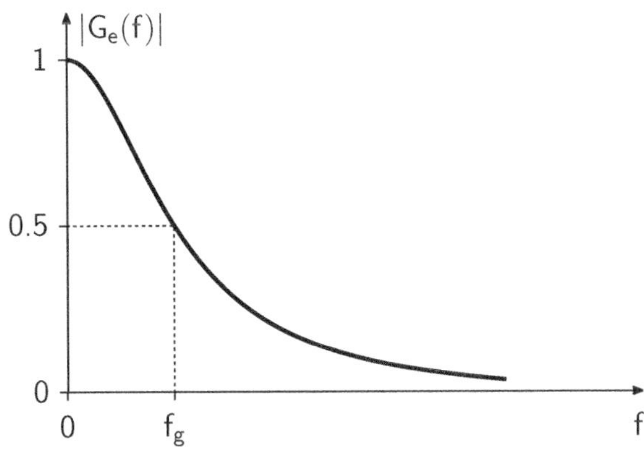

Fig. 4.12 Amplitude response of a second-order receive low-pass filter ($n = 2$) with double-real pole (cutoff frequency f_g)

the time function of the basic receive pulse to

$$x_{\mathrm{e}}(t) = U_0\left[\left(1 - \mathrm{e}^{-\frac{t}{T_{\mathrm{g}}}} - \frac{t}{T_{\mathrm{g}}}\,\mathrm{e}^{-\frac{t}{T_{\mathrm{g}}}}\right)\cdot 1(t) - \left(1 - \mathrm{e}^{-\frac{t-T_{\mathrm{s}}}{T_{\mathrm{g}}}} - \frac{t-T_{\mathrm{s}}}{T_{\mathrm{g}}}\,\mathrm{e}^{-\frac{t-T_{\mathrm{s}}}{T_{\mathrm{g}}}}\right)\cdot 1(t-T_{\mathrm{s}})\right].$$

$$(4.16)$$

Figure 4.13 shows this rectangular response $x_{\mathrm{e}}(t)$ of a low-pass filter with double-real pole at an exemplary receive filter cutoff frequency or receive filter time constant as basic receiving pulse.

In Fig. 4.14, the eye diagram constructed from this basic received pulse—especially with the inner curves for the worst case—is schematically represented, assuming a two-level, unipolar transmission with the transmission amplitudes $0\,\mathrm{V}$ and U_0. At the detection point, it is decided whether a '0' or '1' pulse has been sent, based on the decision threshold at $U_0/2$. The distance between the innermost possible curves of the eye diagram and the decision threshold represents the half vertical eye opening U_{A}.

To achieve the most reliable detection possible, it is necessary to determine the optimal sampling time T_{A}, i.e., to sample the signal at the detection point at the time when a maximum useful signal amplitude is present (see Chap. 2). From Fig. 4.13, it can be seen that the maximum value of the basic received pulse—and thus the maximum vertical eye opening—in the eye diagram (Fig. 4.14) is at times $t \geq T_{\mathrm{s}}$. Therefore, both components from Eq. (4.16) must be taken into account and the time function $x_{\mathrm{e}\,2}(t)$ of the basic received pulse for $t \geq T_{\mathrm{s}}$

$$x_{\mathrm{e}\,2}(t) = U_0\left[\mathrm{e}^{-\frac{t-T_{\mathrm{s}}}{T_{\mathrm{g}}}} + \frac{t-T_{\mathrm{s}}}{T_{\mathrm{g}}}\,\mathrm{e}^{-\frac{t-T_{\mathrm{s}}}{T_{\mathrm{g}}}} - \mathrm{e}^{-\frac{t}{T_{\mathrm{g}}}} - \frac{t}{T_{\mathrm{g}}}\,\mathrm{e}^{-\frac{t}{T_{\mathrm{g}}}}\right] \qquad (4.17)$$

must be used. After multiplying out and regrouping

$$x_{\mathrm{e}\,2}(t) = U_0\left[\mathrm{e}^{-\frac{t}{T_{\mathrm{g}}}}\left(\mathrm{e}^{\frac{T_{\mathrm{s}}}{T_{\mathrm{g}}}} - 1\right) + \mathrm{e}^{-\frac{t}{T_{\mathrm{g}}}}\left[\frac{t}{T_{\mathrm{g}}}\left(\mathrm{e}^{\frac{T_{\mathrm{s}}}{T_{\mathrm{g}}}} - 1\right) - \frac{T_{\mathrm{s}}}{T_{\mathrm{g}}}\,\mathrm{e}^{\frac{T_{\mathrm{s}}}{T_{\mathrm{g}}}}\right]\right] \qquad (4.18)$$

the extreme value problem

$$\frac{\mathrm{d}x_{\mathrm{e}\,2}(t)}{\mathrm{d}t} = 0 \qquad (4.19)$$

Fig. 4.13 Rectangular response of a low-pass filter with double-real pole ($f_{\mathrm{g}} = 0.5/T_{\mathrm{s}}$ or $T_{\mathrm{g}} = T_{\mathrm{s}}/\pi$) as basic receive pulse

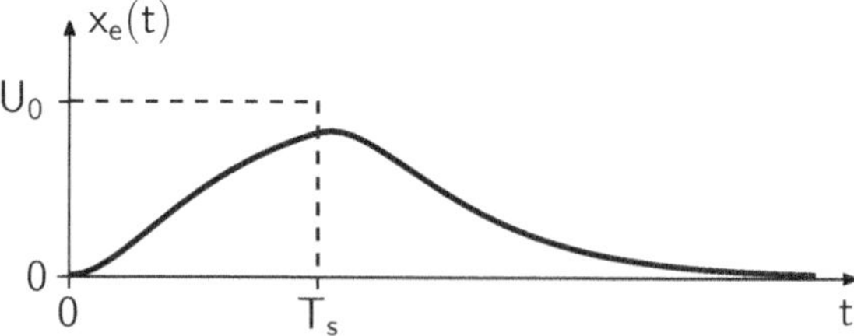

Fig. 4.14 Eye diagram (schematic) at the output of a low-pass filter with double-real pole

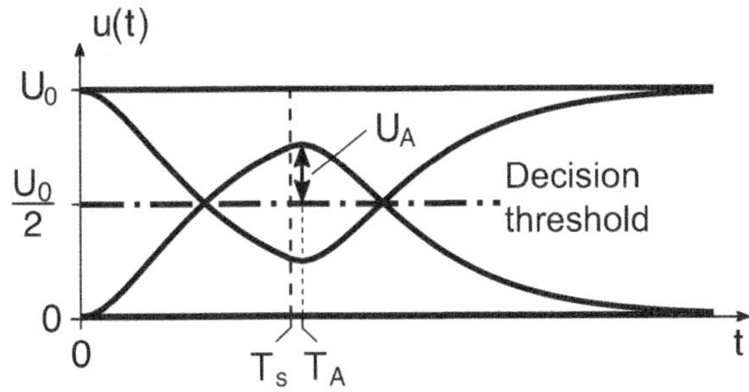

must be solved to determine the maximum of the receive filter rectangular response and its temporal position. To simplify, the variables

$$a = e^{\frac{T_s}{T_g}} - 1 \qquad \text{and} \qquad b = \frac{T_s}{T_g} e^{\frac{T_s}{T_g}} \qquad (4.20)$$

are defined and it results

$$x_{e\,2}(t) = U_0 \left[e^{-\frac{t}{T_g}} \cdot a + e^{-\frac{t}{T_g}} \left(\frac{t}{T_g} \cdot a - b \right) \right]. \qquad (4.21)$$

Using the product rule of calculus (e.g., [11, 74]), the first derivation

$$\begin{aligned}
\frac{dx_{e\,2}(t)}{dt} &= U_0 \left[-\frac{a}{T_g} e^{-\frac{t}{T_g}} - \frac{1}{T_g} e^{-\frac{t}{T_g}} \left(\frac{t}{T_g} a - b \right) + \frac{a}{T_g} e^{-\frac{t}{T_g}} \right] \\
&= -\frac{U_0}{T_g} e^{-\frac{t}{T_g}} \left(\frac{t}{T_g} a - b \right).
\end{aligned} \qquad (4.22)$$

is obtained. The factor containing the e function cannot become zero, leaving

$$\left(\frac{t}{T_g} a - b \right) = 0, \qquad (4.23)$$

and one obtains

$$\frac{t}{T_g} a = b \qquad \Longrightarrow \qquad t = T_A = \frac{b}{a} T_g. \qquad (4.24)$$

From this, the optimal sampling time T_A is obtained—after reinserting a and b:

$$T_A = T_s \frac{e^{\frac{T_s}{T_g}}}{e^{\frac{T_s}{T_g}} - 1}. \qquad (4.25)$$

With $T_g = 1/(2\pi f_g)$, this can also be expressed as

$$T_A = T_s \frac{e^{2\pi f_g T_s}}{e^{2\pi f_g T_s} - 1}. \qquad (4.26)$$

The optimal sampling time T_A thus depends on the given signal characteristic T_s and on the adjustable time constant T_g or cutoff frequency f_g of the receive filter.

The half vertical eye opening U_A for the two-level, unipolar transmission considered here is obtained by subtracting the decision threshold at $U_0/2$ from the amplitude $x_{e2}(T_A)$ achievable at the optimal sampling time T_A, according to

$$U_A = x_{e2}(T_A) - \frac{U_0}{2} \, , \tag{4.27}$$

see also Chap. 2, and it thus results in

$$U_A = U_0 \left[e^{-\frac{T_A}{T_g}} \left(e^{\frac{T_s}{T_g}} - 1 \right) + e^{-\frac{T_A}{T_g}} \left(\frac{T_A}{T_g} \left(e^{\frac{T_s}{T_g}} - 1 \right) - \frac{T_s}{T_g} e^{\frac{T_s}{T_g}} \right) - \frac{1}{2} \right] \tag{4.28}$$

or with $T_g = 1/(2\pi f_g)$, one obtains

$$U_A = U_0 \left[e^{-2\pi f_g T_A} \left(e^{2\pi f_g T_s} - 1 \right) + \dots \right.$$
$$\left. \dots + e^{-2\pi f_g T_A} \left(2\pi f_g T_A \left(e^{2\pi f_g T_s} - 1 \right) - 2\pi f_g T_s \cdot e^{2\pi f_g T_s} \right) - \frac{1}{2} \right] . \tag{4.29}$$

Here, the time variable t must be replaced by the calculated optimal sampling time T_A at each occurrence to obtain the maximum eye opening under the given conditions for the assumed worst case.

Figure 4.15 shows the half vertical eye opening as a function of the receive filter cutoff frequency: There is a range at low cutoff frequencies f_g, where the eye is closed and thus no detection is possible; the intersymbol interferences are too strong. Above the cutoff frequency, where the eye opens, the eye opening increases and tends towards a maximum value, which is given by the half transmission level amplitude $U_0/2$; larger than $U_0/2$ the half vertical eye opening cannot become, even with the receive filter cutoff frequency tending towards infinity.

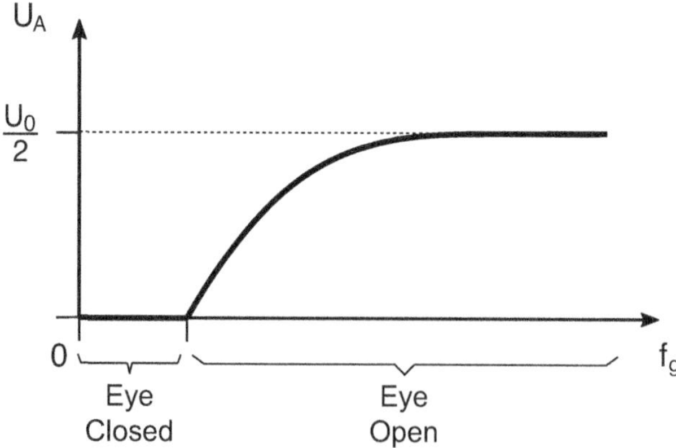

Fig. 4.15 Half vertical eye opening as a function of the receive filter cutoff frequency f_g

4.4.4 Noise Path and Evaluation of the Noise Signal

The noise signal is coupled in at the output of the cable transfer function in the block diagram (Fig. 4.9). It passes through the cascade of transfer functions of equalizer $1/G_k(p)$ and receive low-pass filter $G_e(p)$, and is thus evaluated by the resulting transfer function

$$H_e(p) = \frac{G_e(p)}{G_k(p)} = \frac{2\,(1 + p\,T_k)}{\left(1 + p\,T_g\right)^2} \tag{4.30}$$

The evaluation of the noise signal is shown in the block diagram in Fig. 4.16.

Figure 4.17 shows the amplitude response of the resulting transfer function $H_e(f)$, through which the noise signal is evaluated for a fixed cable pole and for different cutoff frequencies f_g of the receive filter with double-real pole. It can be seen that an amplification is caused by the equalizer towards increasing frequencies and a band limitation exists due to the receive filter. With increasing cutoff frequency, the amplification becomes larger, so that the disturbance is amplified more clearly.

The power spectral density of the noise signal at the detection point is

$$\Psi_e(f) = \Psi_0 \left| \frac{G_e(f)}{G_k(f)} \right|^2 \tag{4.31}$$

and the integration over this received noise power spectral density $\Psi_e(f)$ leads, according to (2.53), to the noise power at the detection point:

$$U_R^2 = \int_{-\infty}^{+\infty} \Psi_e(f)\,\mathrm{d}f = \Psi_0 \int_{-\infty}^{+\infty} \left| \frac{G_e(f)}{G_k(f)} \right|^2 \mathrm{d}f. \tag{4.32}$$

For this, $\left| \frac{G_e(f)}{G_k(f)} \right|^2$ must first be calculated and then the integration must be carried out. Both are shown here in detail in an insert, it can also be skipped with appropriate prior knowledge.

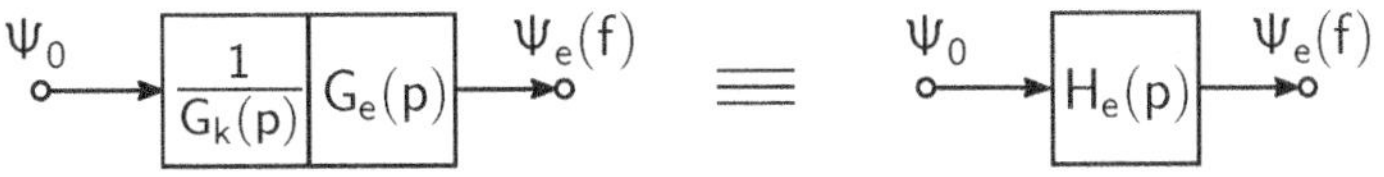

Fig. 4.16 Block diagram for the evaluation of the noise signal

Fig. 4.17 Resulting amplitude responses of the equalizer-receive filter for the evaluation of the interference signal at different receive filter cutoff frequencies f_g

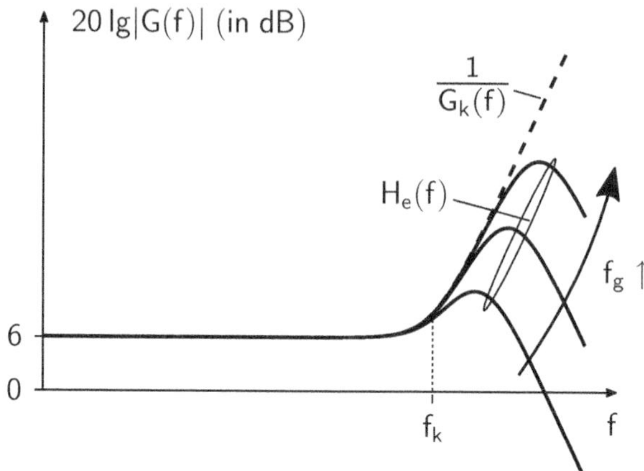

Insert: Calculation of $|H_e(f)|^2 = \left|\dfrac{G_e(f)}{G_k(f)}\right|^2$

To calculate the noise power, the relevant transfer function must first be converted from the Laplace representation to the Fourier representation ($p \to f$): It applies $p = \sigma + j\omega$ and for $\sigma = 0$ one obtains $p = j\omega$ or $p = j2\pi f$, thus making the transition from the frequency variable p to the frequency variable f (see also Chap. 1).

The transfer function relevant for the evaluation of the noise signal, consisting of equalizer and receive low-pass filter, is based on (4.30)—with $T_k = 1/(2\pi f_k)$ and $T_g = 1/(2\pi f_g)$ -

$$H_e(f) = 2\,\frac{1 + j\frac{2\pi f}{2\pi f_k}}{\left(1 + j\frac{2\pi f}{2\pi f_g}\right)^2} = 2\,\frac{1 + j\frac{f}{f_k}}{\left(1 + j\frac{f}{f_g}\right)^2} \tag{4.33}$$

and its magnitude is

$$|H_e(f)| = 2\,\frac{\sqrt{1 + \left(\frac{f}{f_k}\right)^2}}{\left(\sqrt{1 + \left(\frac{f}{f_g}\right)^2}\right)^2}. \tag{4.34}$$

The magnitude square of this transfer function necessary for the calculation of the integral is obtained via the intermediate steps

$$|H_e(f)|^2 = 2^2\,\frac{1 + \left(\frac{f}{f_k}\right)^2}{\left(1 + \left(\frac{f}{f_g}\right)^2\right)^2} = 4\,\frac{\frac{f_k^2}{f_k^2} + \frac{f^2}{f_k^2}}{\left(\frac{f_g^2}{f_g^2} + \frac{f^2}{f_g^2}\right)^2} = 4\,\frac{\frac{1}{f_k^2}\left(f_k^2 + f^2\right)}{\frac{1}{f_g^4}\left(f_g^2 + f^2\right)^2} \tag{4.35}$$

to

$$|H_\mathrm{e}(f)|^2 = 4\,\frac{f_\mathrm{g}^4(f_\mathrm{k}^2 + f^2)}{f_\mathrm{k}^2\left(f_\mathrm{g}^2 + f^2\right)^2}.$$

(4.36)

Insert: Execution of the Integration

With the previously calculated magnitude square $\left|\frac{G_\mathrm{e}(f)}{G_\mathrm{k}(f)}\right|^2$, one can now perform the integration

$$U_\mathrm{R}^2 = \Psi_0 \cdot 4 \cdot \frac{f_\mathrm{g}^4}{f_\mathrm{k}^2} \int\limits_{-\infty}^{+\infty} \frac{f_\mathrm{k}^2 + f^2}{\left(f_\mathrm{g}^2 + f^2\right)^2}\,\mathrm{d}f$$

(4.37)

$$= 4\,\Psi_0 \frac{f_\mathrm{g}^4}{f_\mathrm{k}^2}\left[f_\mathrm{k}^2 \int\limits_{-\infty}^{+\infty} \frac{1}{\left(f_\mathrm{g}^2 + f^2\right)^2}\,\mathrm{d}f + \int\limits_{-\infty}^{+\infty} \frac{f^2}{\left(f_\mathrm{g}^2 + f^2\right)^2}\,\mathrm{d}f\right]$$

(4.38)

The first integral is calculated using

$$\int \frac{1}{\left(a^2 + x^2\right)^2}\,\mathrm{d}x = \frac{x}{2\,a^2\,(a^2 + x^2)} + \frac{1}{2\,a^3}\,\arctan\frac{x}{a} + C$$

(4.39)

from the integral table (see Appendix F.2) or [11, 74] as

$$\int\limits_{-\infty}^{+\infty} \frac{1}{\left(f_\mathrm{g}^2 + f^2\right)^2}\,\mathrm{d}f = \left[\frac{f}{2f_\mathrm{g}^2\,(f_\mathrm{g}^2 + f^2)} + \frac{1}{2f_\mathrm{g}^3}\,\arctan\frac{f}{f_\mathrm{g}}\right]_{-\infty}^{+\infty}$$

$$= \left[\left(0 + \frac{1}{2f_\mathrm{g}^3}\frac{\pi}{2}\right) - \left(0 + \frac{1}{2f_\mathrm{g}^3}\left(-\frac{\pi}{2}\right)\right)\right]$$

(4.40)

$$= \frac{\pi}{4f_\mathrm{g}^3} + \frac{\pi}{4f_\mathrm{g}^3} = \frac{\pi}{2f_\mathrm{g}^3}.$$

For the calculation of the second integral, one uses

$$\int \frac{x^2}{\left(a^2 + x^2\right)^2}\,\mathrm{d}x = -\frac{x}{2\,(a^2 + x^2)} + \frac{1}{2\,a}\,\arctan\frac{x}{a} + C$$

(4.41)

from Appendix F.2 or [11, 74] and obtains

$$
\int_{-\infty}^{+\infty} \frac{f^2}{\left(f_{\mathrm g}^2+f^2\right)^2}\, df = \left[-\frac{f}{2\left(f_{\mathrm g}^2+f^2\right)} + \frac{1}{2f_{\mathrm g}} \arctan \frac{f}{f_{\mathrm g}} \right]_{-\infty}^{+\infty}
$$

$$
= \left[\left(0 + \frac{1}{2f_{\mathrm g}}\frac{\pi}{2} \right) - \left(0 + \frac{1}{2f_{\mathrm g}}\left(-\frac{\pi}{2}\right) \right) \right] \tag{4.42}
$$

$$
= \frac{1}{2f_{\mathrm g}}\frac{\pi}{2} + \frac{1}{2f_{\mathrm g}}\frac{\pi}{2} = \frac{\pi}{2f_{\mathrm g}}.
$$

If one combines both results, the result for the noise power is

$$
U_{\mathrm R}^2 = 4\,\Psi_0\,\frac{f_{\mathrm g}^4}{f_{\mathrm k}^2}\left(f_{\mathrm k}^2\frac{\pi}{2f_{\mathrm g}^3} + \frac{\pi}{2f_{\mathrm g}} \right) = 4\,\Psi_0\left(\frac{\pi}{2}f_{\mathrm g} + \frac{\pi}{2}\frac{f_{\mathrm g}^3}{f_{\mathrm k}^2} \right) \tag{4.43}
$$

$$
U_{\mathrm R}^2 = 4\,\Psi_0\,\frac{\pi}{2}\left(f_{\mathrm g} + \frac{f_{\mathrm g}^3}{f_{\mathrm k}^2} \right) = 2\pi\,\Psi_0\left(f_{\mathrm g} + \frac{f_{\mathrm g}^3}{f_{\mathrm k}^2} \right). \tag{4.44}
$$

The total noise power at the detection point in compact form is

$$
U_{\mathrm R}^2 = 2\pi\,\Psi_0\left(f_{\mathrm g} + \frac{f_{\mathrm g}^3}{f_{\mathrm k}^2} \right). \tag{4.45}
$$

In Fig. 4.18, the noise power as a function of the receive filter cutoff frequency $f_{\mathrm g}$ is shown: The noise power depends on the third power of this cutoff frequency and thus increases very strongly and unlimitedly with increasing cutoff frequency $f_{\mathrm g}$ of the receive filter. In addition, the noise power is a function of the cable cutoff frequency $f_{\mathrm k}$

Fig. 4.18 Noise power at the detection point as a function of the receive filter cutoff frequency $f_{\mathrm g}$

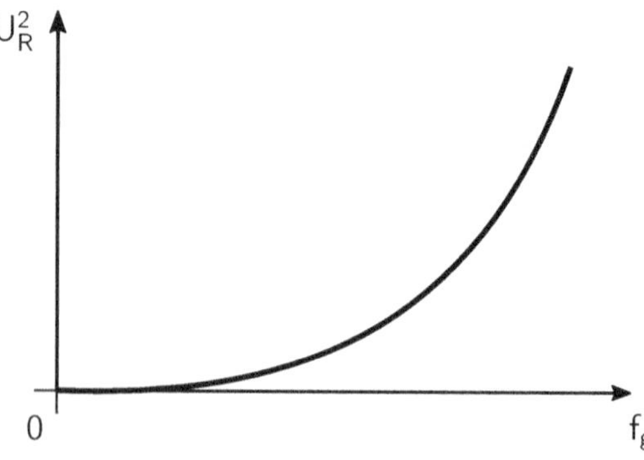

: With a decreasing cable cutoff frequency f_k and thus stronger low-pass effect of the cable, stronger equalization on the receiving side is necessary to compensate for the distorting cable influence on the useful signal. This causes a stronger noise enhancement (or amplification), so that the noise power increases additionally with stronger low-pass effect of the cable to be equalized. This is expressed in (4.45) by the fact that the cable cutoff frequency f_k is in the denominator and thus with decreasing cable cutoff frequency f_k, the noise power U_R^2—under otherwise unchanged boundary conditions—increases.

4.4.5   Quality of Transmission

Signal-to-Noise Ratio

To assess the quality of the examined signal transmission, the characteristics of the useful signal and the disturbance at the detection point must now be summarized in a suitable quantity. Therefore, the signal-to-noise ratio is used to evaluate the quality of the transmission, which is formed according to (2.60) from the half vertical eye opening U_A at the optimal sampling time and the noise power U_R^2 (see Chap. 2):

$$\varrho = \frac{U_A^2}{U_R^2}. \tag{4.46}$$

Figure 4.19 shows this signal-to-noise ratio as a function of the receive filter cutoff frequency f_g. It becomes clear that for the receive filter cutoff frequency f_g, there exists an optimum at which the signal-to-noise ratio exhibits a maximum ϱ_{max}: At very small receive filter cutoff frequencies f_g, the eye is closed and therefore reliable detection and decision about received bits is not possible. After the eye has opened, the signal-to-noise ratio ϱ increases with increasing cutoff frequency f_g, as the half vertical eye opening U_A (see Fig. 4.15) rapidly enlarges in this area, but the noise in its power U_R^2 still remains relatively small (see Fig. 4.18). With further increasing receive filter cutoff frequency f_g, a point is reached at which the increases U_A^2 and U_R^2 in dependence on f_g just balance out: Here, the signal-to-noise ratio reaches its maximum value ϱ_{max}, the receive filter cutoff frequency is therefore optimal at this point—with respect to a maximum signal-to-noise ratio under the given boundary conditions—and is therefore referred to as $f_{g\,opt}$. After this maximum, the signal-to-noise ratio decreases with further increasing receive filter cutoff frequency, as the half vertical eye opening tends towards a maximum value, which is determined by the half-level transmit amplitude $U_0/2$, and the noise now predominates and continues to rise unrestrictedly.

The optimal cutoff frequency can be read from the graphical representation in Fig. 4.19 or determined numerically. It is interesting that the optimal cutoff frequency leads to an eye opening that is not maximum—due to the influence of the noise, it is generally more advantageous for the reliability and thus the quality of the transmission to work at a smaller—than the maximum—eye opening and at a then also smaller noise power, in

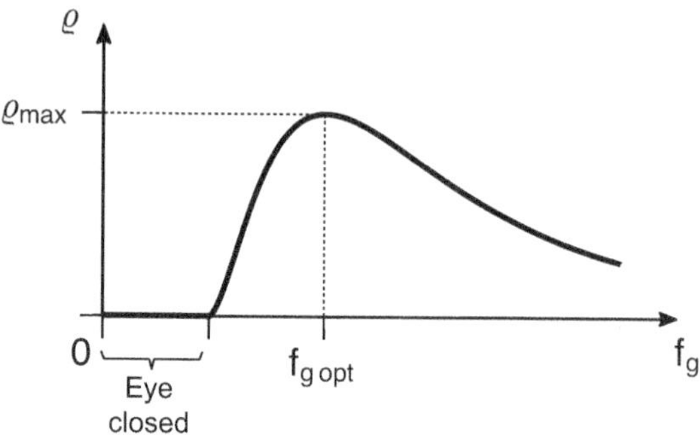

Fig. 4.19 Signal-to-noise ratio ϱ at the detection point as a function of the receive filter cutoff frequency f_g given a (fixed) f_k

order to achieve a maximum signal-to-noise ratio—and thus the best possible quality of the transmission under the given boundary conditions.

Bit Error Probability

Crucial for the quality of a transmission of information encoded in bits over a channel, and thus for the quality of a digital signal transmission, is how many of the bits arriving at the receiving end are correctly detected and how many are incorrectly recognized. A measure for this is the bit error probability P_f—it is therefore the decisive quality criterion for the quality of a digital information or signal transmission. The bit error probability can be calculated directly from the signal-to-noise ratio ϱ according to (2.80) (see Chap. 2)—here for the examined two-level transmission:

$$P_f = \frac{1}{2}\left(1 - \mathrm{erf}\left(\sqrt{\frac{\varrho}{2}}\right)\right) = \frac{1}{2}\,\mathrm{erfc}\left(\sqrt{\frac{\varrho}{2}}\right). \tag{4.47}$$

Figure 4.20 shows the bit error probability P_f as a function of the receive filter cutoff frequency f_g with logarithmic division of the vertical axis: It shows a minimum $P_{f\,\min}$ at the optimal receive filter cutoff frequency $f_{g\,\mathrm{opt}}$, for the same reason given in the previous section for the occurrence of a maximum in the signal-to-noise ratio.

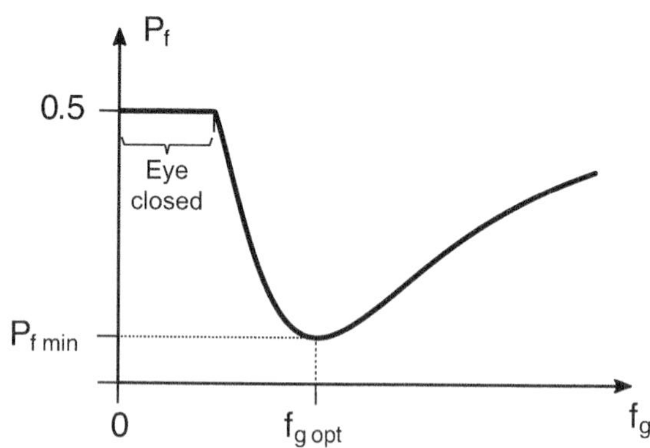

Fig. 4.20 Bit error probability P_f at the detection point as a function of the receive filter cutoff frequency f_g (logarithmic division of the vertical axis) for a given (fixed) f_k

The optimization of the receive filter cutoff frequency f_g can therefore be done with respect to a maximum signal-to-noise ratio ϱ or with respect to a minimum bit error probability P_f: Both approaches are equivalent and lead to the same optimal cutoff frequency $f_{g\,opt}$, as there is a monotonic relationship between the signal-to-noise ratio and the bit error probability via the Gaussian error function $\mathrm{erf}(x)$ (see Appendix E and e.g. [11]).

In practice, therefore, it is often possible to work on the basis of the signal-to-noise ratio, as this is mathematically easier to handle and the additional calculation step with the functions $\mathrm{erf}(x)$ or $\mathrm{erfc}(x)$, which can in principle only be evaluated numerically and are therefore computationally intensive, is omitted.

4.4.6   Numerical Example: Baseband Transmission with a Cable with One Single Pole

Introduction and Task Definition

With the help of a numerical example, the procedure presented in the previous sections for optimizing a baseband transmission system, which is operated over a copper cable with a rational approximation of the transfer functions, is illustrated using a specific task. The transmission system under consideration is shown in Fig. 4.21.

The transmitter of a baseband transmission system sends a unipolar random sequence of rectangular NRZ pulses with the amplitude U_0 and the pulse width T_s. It is assumed that the transfer function of the cable with one pole can be approximated sufficiently accurately according to (4.8). The disturbance is modeled as white, Gaussian noise with the power spectral density Ψ_0. The cable's influence on the useful signal is fully linearly compensated by a zero at the frequency of the cable pole. The receive filter is a second-order low-pass filter ($n = 2$) with a double-real pole at the cutoff frequency f_g. This corresponds to the transmission system that was investigated in the previous sections.

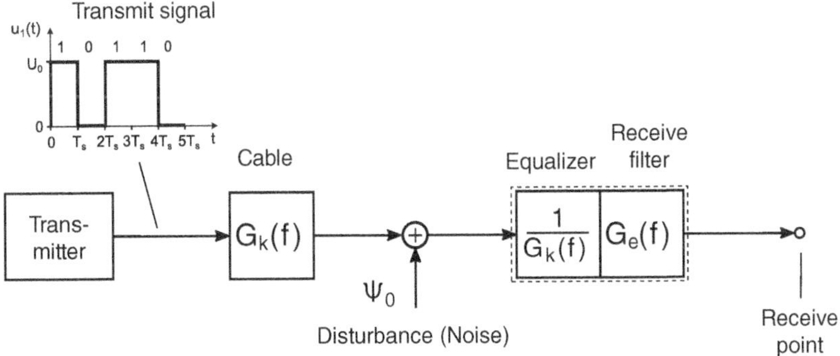

Fig. 4.21 Transmission system for the numerical example with unipolar, binary baseband transmission over a cable with linear equalization

The achievable quality of the transmission, i.e., the signal-to-noise ratio and the bit error probability, is to be calculated. For this purpose, the individual necessary quantities are to be calculated and, if necessary, optimized.

Given Values

The following numerical values are given:

- Signal: Average transmit power $P_s = 1\,\mathrm{V}^2$, pulse duration $T_s = 1\,\mu\mathrm{s}$, unipolar transmission,
- Disturbance: Power spectral density $\Psi_0 = 5 \cdot 10^{-10}\,\frac{\mathrm{V}^2}{\mathrm{Hz}}$,
- Cable: Cable characteristic frequency $f_0 = 0.178\,\mathrm{MHz} \cdot \mathrm{km}^2$, cable length $\ell = 2\,\mathrm{km}$.

Procedure and Results

From the given average transmit power P_s the transmission amplitude is first calculated using (2.7) over

$$P_s = \frac{U_0^2}{2} \qquad \Longrightarrow \qquad U_0 = \sqrt{2\,P_s} \tag{4.48}$$

For $P_s = 1\,\mathrm{V}^2$, the transmission amplitude $U_0 = \sqrt{2}\,\mathrm{V} \approx 1.414\,\mathrm{V}$ is obtained.

Using (4.8), the frequency of the cable pole is calculated as $f_k = 109.8\,\mathrm{kHz}$. The zero point of the equalizer is thus also fixed: It is located at the position of the cable pole.

Figure 4.22 shows the amplitude response of the cable transfer function of the example: The low-pass behavior of the cable with the cutoff frequency f_k is recognizable. In Fig. 4.23, the amplitude response of the equalizer receive filter is shown: The zero at the frequency f_k is fixed; since the cutoff frequency f_g of the receive filter is still to be determined, amplitude responses with various cutoff frequencies f_g of the second-order low-pass are shown to illustrate the influence of the receive filter cutoff frequency in this diagram.

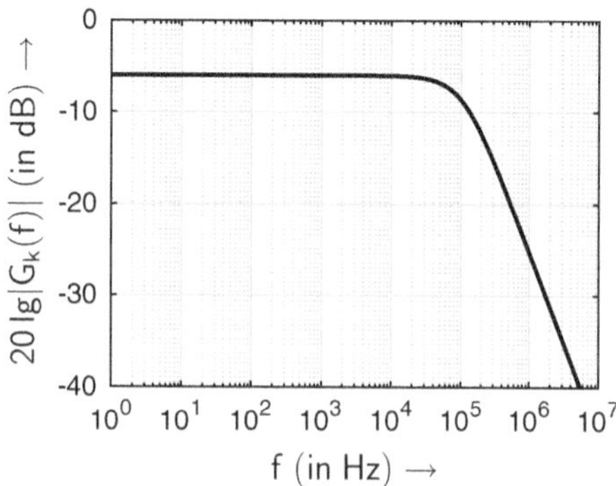

Fig. 4.22 Amplitude response of the cable transfer function approximated with a single pole $G_k(f)$ (Parameters: Cable characteristic frequency $f_0 = 0.178\,\mathrm{MHz} \cdot \mathrm{km}^2$, length $\ell = 2\,\mathrm{km}$)

Fig. 4.23 Amplitude response of the equalizer receive filter $H_e(f)$ with one single equalizer zero and various cutoff frequencies f_g of the receive filter

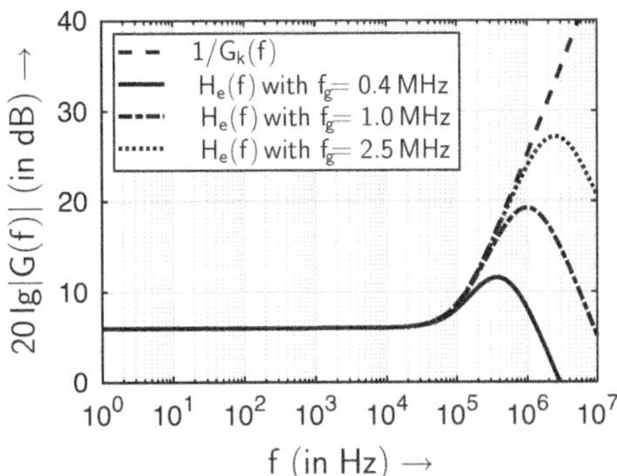

The receive filter is structured as a second-order low-pass filter with a double-real pole, but its cutoff frequency f_g is freely selectable and thus optimizable. Therefore, the half vertical eye opening according to (4.29) and the noise power according to (4.45) are determined as a function of the receive filter cutoff frequency. In Fig. 4.24, the half vertical eye opening and in Fig. 4.25 the noise power at the detection point, each as a function of the receive filter cutoff frequency f_g, are shown. The typical curves are recognizable: The half vertical eye opening is—with a closed eye—initially zero and tends towards a maximum of $U_0/2 = 1/\sqrt{2}\,\mathrm{V} \approx 0.707\,\mathrm{V}$ predetermined by the transmission amplitude, while the noise power increases indefinitely with increasing receive filter cutoff frequency.

If now, according to (4.46), the signal-to-noise ratio and according to (4.47) the bit error probability, also as a function of the receive filter cutoff frequency f_g, are calculated and shown in Figs. 4.26 and 4.27, it is possible to determine the optimal cutoff frequency of the receive filter. It is read from one of the diagrams where the signal-to-noise ratio becomes maximum—or the bit error probability minimum—and $f_{g\,\mathrm{opt}} = 453\,\mathrm{kHz}$ is obtained.

Fig. 4.24 Half vertical eye opening U_A as a function of the receive filter cutoff frequency f_g

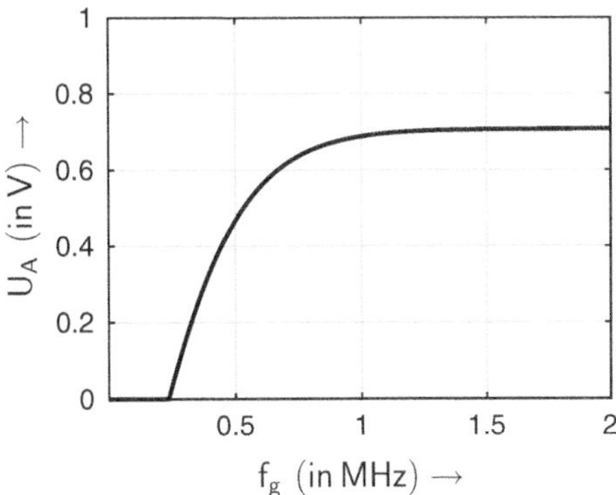

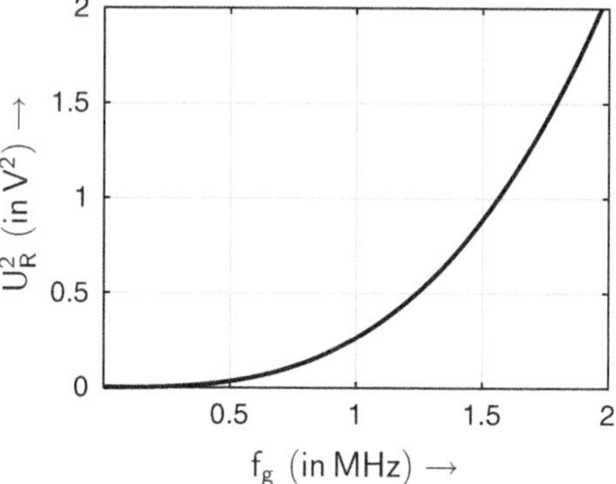

Fig. 4.25 Noise power U_R^2 as a function of the receive filter cutoff frequency f_g

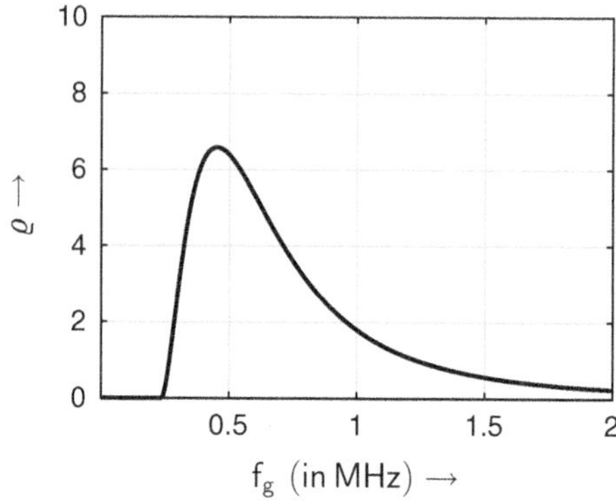

Fig. 4.26 Signal-to-noise ratio ϱ as a function of the receive filter cutoff frequency f_g

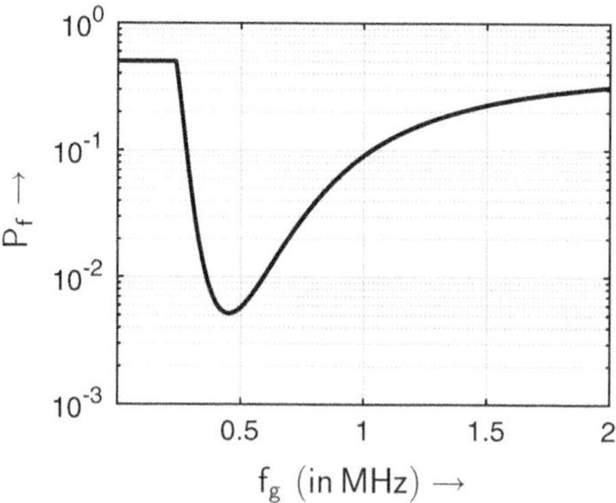

Fig. 4.27 Bit error probability P_f as a function of the receive filter cutoff frequency f_g

Numerical Results for Optimized Transmission System

Once the optimal cutoff frequency $f_{g\,opt}$ of the receive filter has been determined, all other values for signal and disturbance at the detection point can be determined and the quality of transmission can be specified for an optimized transmission system. The individual relationships, which have been formulated as functions of the cutoff frequency f_g of the receive filter or its time constant T_g, are evaluated at the point $f_g = f_{g\,opt}$. The numerical results are compiled in Table 4.1.

The maximum achievable signal-to-noise ratio is $\varrho_{max} = 6.58$ or $\varrho_{max}^* = 10\lg\varrho_{max} = 8.18\,\text{dB}$; it leads to a bit error probability of $P_{f\,min} = 5.2 \cdot 10^{-3}$.

The optimal receive filter cutoff frequency $f_{g\,opt}$ is slightly below half the bit sequence frequency or bit rate $f_B = 1/T_s$. As a result, the half vertical eye opening U_A has a value that is comparatively far from the maximum half vertical eye opening $U_0/2$: This maximum half vertical eye opening would be achieved at a (very) large receive filter cutoff frequency f_g. This would in turn cause a (strong) increase in noise power compared to $f_{g\,opt}$. All these factors together lead to the optimal cutoff frequency in the size that results in its optimization with respect to a maximum signal-to-noise ratio or a minimum bit error probability.

The optimal sampling time T_A is the point in time at which the received signal is sampled at the detection point in the symbol clock T_s. It is interesting that T_A in this type of receive filtering is at times $T_A > T_s$ and thus not within the duration T_s of the considered rectangular transmit pulse. In addition, the optimal sampling time T_A according to (4.26) also depends on the receive filter cutoff frequency f_g.

Some Diagrams for Optimized Transmission System

This section is intended to illustrate some graphical representations of how some time and spectral functions look in an optimized transmission system.

Table 4.1 Given values and numerical results for the linearly equalized cable transmission system with unipolar NRZ rectangular pulses at the optimal receive filter cutoff frequency $f_{g\,opt}$ of the second order low pass filter

Quantity	Numerical Value
Average transmit power	$P_s - 1\,\text{V}^2$
Transmit amplitude	$U_0 = 1.414\,\text{V}$
Noise power spectral density	$\Psi_0 = 5 \cdot 10^{-10}\,\text{V}^2/\text{Hz}$
Pulse duration (bit or symbol duration)	$T_s = 1\,\mu\text{s}$
Optimal receive filter cutoff frequency	$f_{g\,opt} = 453\,\text{kHz}$
Optimal sampling time	$T_A = 1.061\,\mu\text{s}$
Half vertical eye opening	$U_A = 0.411\,\text{V}$
Noise power	$U_R^2 = 0.026\,\text{V}^2$
Signal-to-noise ratio	$\varrho_{max} = 6.58$
Bit error probability	$P_{f\,min} = 5.2 \cdot 10^{-3}$

Fig. 4.28 Amplitude response of the equalizer receive filter at optimal cutoff frequency $f_{g\,opt}$ of the receive filter

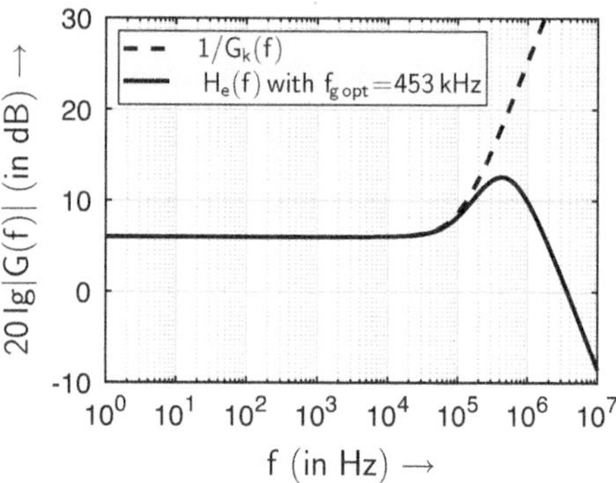

Fig. 4.29 Eye diagram of the useful signal at the detection point at optimal receive filter cutoff frequency $f_{g\,opt}$

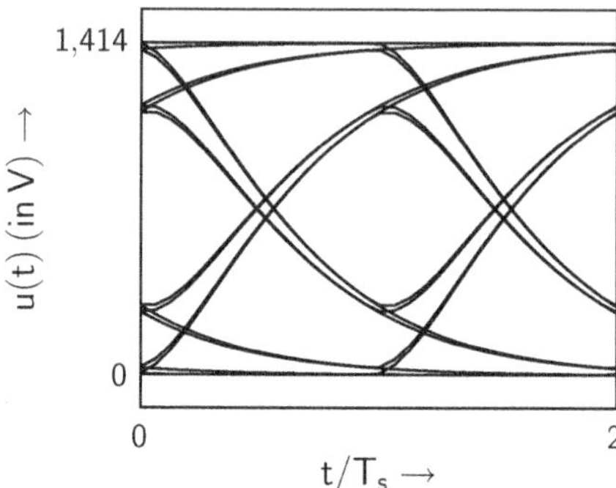

Figure 4.28 shows the amplitude frequency response of the equalizer receive filter $H_e(f)$ and Fig. 4.29 the eye diagram of the useful signal at the detection point—each for the case of the optimized cutoff frequency of the receive low pass filter ($f_{g\,opt} = 453$ kHz). It can be seen that the eye is only moderately open, as the band limitation of the receive filter starts relatively early and thus the optimized transmission system limits the noise in its power on the receiver side relatively well.

4.5　Multi-Level Baseband Transmission Over Cable

The application example with the rational approximation of the cable transfer function with a single pole, linear equalization, and a rational receive low pass filter for unipolar, two-level transmission of an NRZ rectangular pulse sequence over this system was extensively discussed to illustrate the communication engineering relationships and possibilities in optimizing such systems.

In modern telecommunications practice, further pulse shaping and filter functions are used, enabled and supported by largely computer-aided digital signal processing, which have favorable transmission properties. In addition, it can be advantageous to use a multi-level transmission or pulse amplitude modulation. These possibilities will be applied here with reference to Chap. 2 to the wired baseband transmission.

Two important examples will be treated and examined in a comparable manner using the previously described approach:

- A linearly equalized cable transmission system with NRZ rectangular transmit signal and Gaussian low-pass filter according to (2.85) as a receive filter (see Chap. 2): A Gaussian filter is a filter with maximum steepness in the frequency range, leading to pulses without overshoot in the time domain. Moreover, as previously described, it can be considered as a limiting case of the rational RC filter of nth order with $n \to \infty$ and therefore leads to a limiting case of the transmission system previously examined in detail.
- A linearly equalized cable transmission system with root raised cosine transmit and receive filter according to (2.107) and (2.104) (see Chap. 2): Such a baseband transmission system is a Nyquist-1 system, see Chap. 2, and thus free of intersymbol interference at the sampling instants. Moreover, it has a steep band limitation, which proves advantageous for limiting the noise power.

A bipolar signal with the number s of signalling levels according to Chap. 2 is assumed as the transmit signal.[4] The case of s-level transmission allows a general representation and includes the special case $s = 2$.

The transfer function (3.145) of the cable is used, as the two now used receive filter functions (Gaussian low-pass and root raised cosine low-pass filter) drop off more steeply than the inverse of this cable transfer function (3.145) increases with increasing frequency.

Figure 4.30 shows the block diagram of the transmission system, which has been extended by the transfer function $G_s(f)$ of the transmit filter for a general and uniform description, which was also introduced and used in Chap. 2.

4.5.1   Baseband Transmission with NRZ Transmit Signal and Gaussian Receive Filter

We consider an s-level, bipolar baseband transmission system with receiver-side linear equalization of the cable impact. The corresponding block diagram is shown in Fig. 4.30. The transmitter emits a random sequence of NRZ rectangular pulses with the uniformly

---

[4]The previously examined use case represented a special example for the number of signalling levels $s = 2$ and unipolar transmission in the sense and continuation of the basic model of baseband communication considered in Chap. 2 in order to be able to examine and present the basics in detail and comprehensibly.

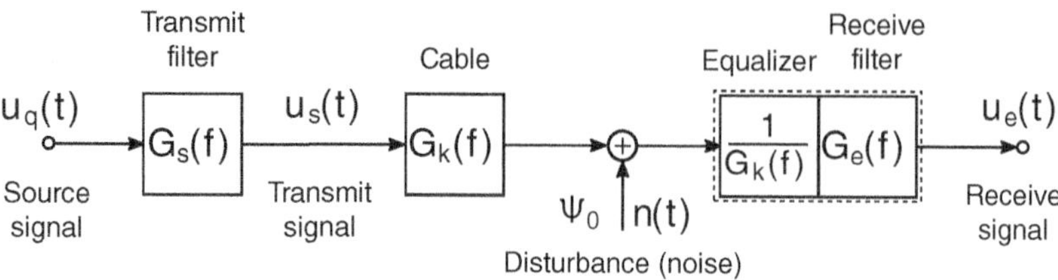

Fig. 4.30 Cable transmission system with transmit and receive filter, channel and linear equalization

distributed amplitudes $\pm U_s, \pm 3\,U_s, \ldots, \pm(s-1)\,U_s$ (see Fig. 4.31). Each of the NRZ rectangular pulses has a duration of T_s. The bit sequence frequency is then $f_B = \mathrm{ld}(s)/T_s$ [34].

This can be equivalently described by a Dirac source

$$u_q(t) = U_s\,T_s \sum_{k=-\infty}^{+\infty} a[k] \cdot \delta(t - k\,T_s) \tag{4.49}$$

with the s-level source symbols $a[k]$ and the half transmission level amplitude U_s following a subsequent transmit filter

$$G_s(f) = \mathrm{si}(\pi f T_s) \tag{4.50}$$

(see Chap. 2). A threshold detection follows the Gaussian receive low-pass filter accordingly (2.85) or (4.3).

Useful Signal Path and Evaluation of the Useful Signal

For the NRZ transmit signal generated by the transmit filtering with the transfer function $G_s(f)$, only the receive filter $G_e(f)$ remains as an element that deforms the transmit signal, based on Fig. 4.4 and the assumed complete linear equalization of the cable influence. After (2.89), the basic received impulse (useful signal) as a response of the Gaussian low-pass to a rectangle impulse of amplitude U_s and duration T_s

$$x_e(t) = \frac{U_s}{2}\left[\mathrm{erf}\left(\frac{\pi f_G}{\sqrt{\ln(2)}}\left(t + \frac{T_s}{2} \right) \right) - \mathrm{erf}\left(\frac{\pi f_G}{\sqrt{\ln(2)}}\left(t - \frac{T_s}{2} \right) \right) \right] \tag{4.51}$$

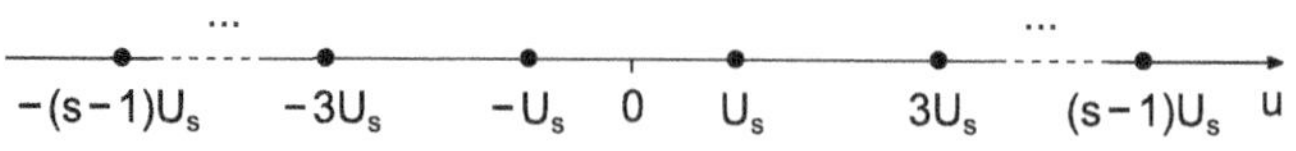

Fig. 4.31 Signal space diagram for s-level, bipolar baseband transmission

is present, see Chap. 2 and e.g. [63]. In the case of the NRZ pulse transmission considered here, the half vertical eye opening for general s-level transmission results in

$$U_\mathrm{A} = U_\mathrm{s}\left[s \cdot \mathrm{erf}\left(\frac{\pi f_\mathrm{G}}{\sqrt{\ln(2)}}\left(\frac{T_\mathrm{s}}{2}\right)\right) - (s-1)\right]. \tag{4.52}$$

The half vertical eye opening as a measure of the quality of the transmission of the useful signal component again depends on the cutoff frequency f_G of the receive filter. The half-level transmit amplitude U_s for s-level transmission can either be directly specified or implicitly via an amplitude or power limitation of the transmitter.

Noise Path and Evaluation of the Noise Signal
The disturbance undergoes an evaluation on the way to the detection point through the equalizer receive filter (compare Fig. 4.5) and one obtains via

$$U_\mathrm{R}^2 = \Psi_0 \int_{-\infty}^{+\infty} \left|\frac{G_\mathrm{e}(f)}{G_\mathrm{k}(f)}\right|^2 \mathrm{d}f = \Psi_0 \int_{-\infty}^{+\infty} \frac{|G_\mathrm{e}(f)|^2}{|G_\mathrm{k}(f)|^2}\,\mathrm{d}f \tag{4.53}$$

with (4.3) and (3.145)

$$U_\mathrm{R}^2 = \Psi_0 \int_{-\infty}^{+\infty} \frac{\left|\mathrm{e}^{-\ln(2)\cdot\left(\frac{f}{f_\mathrm{G}}\right)^2}\right|^2}{\left|\mathrm{e}^{-\ell\sqrt{\mathrm{j}\frac{f}{f_0}}}\right|^2}\,\mathrm{d}f = \Psi_0 \int_{-\infty}^{+\infty} \frac{\mathrm{e}^{-2\ln(2)\cdot\left(\frac{f}{f_\mathrm{G}}\right)^2}}{\mathrm{e}^{-\ell\sqrt{\frac{2f}{f_0}}}}\,\mathrm{d}f. \tag{4.54}$$

Since the integral in (4.54) cannot be solved elementarily, numerical integration methods are used in calculations to achieve results.

Insert: Magnitude of the Cable Transfer Function
The absolute value of the transfer function

$$G_\mathrm{k}(f) = \mathrm{e}^{-\ell\sqrt{\mathrm{j}\frac{f}{f_0}}} \tag{4.55}$$

of the cable can be written as

$$|G_\mathrm{k}(f)| = \left|\mathrm{e}^{-\ell\sqrt{\mathrm{j}\frac{f}{f_0}}}\right| \tag{4.56}$$

$$|G_\mathrm{k}(f)| = \left|\mathrm{e}^{-\ell\sqrt{\mathrm{j}}\sqrt{\frac{f}{f_0}}}\right| \tag{4.57}$$

For this, it is first necessary to explicitly calculate the square root $\sqrt{j}$. One can formulate

$$\sqrt{j} = \sqrt{e^{j\frac{\pi}{2}}} \tag{4.58}$$

$$\sqrt{j} = e^{j\frac{\pi}{4}} \tag{4.59}$$

$$\sqrt{j} = \cos\left(\frac{\pi}{4}\right) + j\sin\left(\frac{\pi}{4}\right) \tag{4.60}$$

where

$$\sin\left(\frac{\pi}{4}\right) = \cos\left(\frac{\pi}{4}\right) = \frac{1}{\sqrt{2}} \tag{4.61}$$

applies. Thus, the result of this intermediate calculation is

$$\sqrt{j} = \frac{1}{\sqrt{2}} + j\frac{1}{\sqrt{2}}, \tag{4.62}$$

which, when inserted into (4.57), leads to

$$|G_k(f)| = \left| e^{-l\left(\frac{1}{\sqrt{2}} + j\frac{1}{\sqrt{2}}\right)\sqrt{\frac{f}{f_0}}} \right| \tag{4.63}$$

$$|G_k(f)| = \left| e^{-l\frac{1}{\sqrt{2}}\sqrt{\frac{f}{f_0}}} \right| \cdot \underbrace{\left| e^{-jl\frac{1}{\sqrt{2}}\sqrt{\frac{f}{f_0}}} \right|}_{=1} \tag{4.64}$$

Since

$$\left| e^{-l\frac{1}{\sqrt{2}}\sqrt{\frac{f}{f_0}}} \right| = e^{-l\frac{1}{\sqrt{2}}\sqrt{\frac{f}{f_0}}} = e^{-l\sqrt{\frac{f}{2f_0}}} \tag{4.65}$$

and

$$\left| e^{-jl\frac{1}{\sqrt{2}}\sqrt{\frac{f}{f_0}}} \right| = 1 \tag{4.66}$$

apply, the result for the magnitude of the cable transfer function is

$$|G_k(f)| = e^{-l\sqrt{\frac{f}{2f_0}}} \tag{4.67}$$

and for the square of this magnitude of the cable transfer function

$$|G_k(f)|^2 = \left[e^{-l\sqrt{\frac{f}{2f_0}}} \right]^2 = e^{-2 \cdot l\sqrt{\frac{f}{2f_0}}} = e^{-\frac{\sqrt{2}\cdot\sqrt{2}}{\sqrt{2}} \cdot l\sqrt{\frac{f}{f_0}}} \tag{4.68}$$

applies resulting in

$$|G_k(f)|^2 = e^{-l\sqrt{\frac{2f}{f_0}}}. \tag{4.69}$$

Quality of Transmission

The quality of transmission is evaluated using the *signal-to-noise ratio* and the resulting *bit error probability*.

Signal-to-Noise Ratio Both eye opening and noise power, and thus the signal-to-noise ratio $\varrho = U_A^2/U_R^2$, are functions of the receive filter cutoff frequency f_G. Therefore, an optimization of the receive filter cutoff frequency is possible and necessary to achieve a maximum signal-to-noise ratio under the given boundary conditions.

Bit Error Probability The bit error probability is obtained according to (2.240), derivation in Chap. 2, via the relationship

$$P_b = \frac{s-1}{s \, \mathrm{ld}(s)} \left[1 - \mathrm{erf}\left(\sqrt{\frac{\varrho}{2}} \right) \right]. \tag{4.70}$$

It shows a minimum at the optimal receive filter cutoff frequency, i.e., where the signal-to-noise ratio is maximum.

Numerical Example

The following numerical values are given as an example:

- Signal: Transmit power $P_s = 1\,\mathrm{V}^2$, bit rate $f_B = 1\,\mathrm{MHz}$, bipolar transmission,
- Disturbance: Power spectral density $\Psi_0 = 5 \cdot 10^{-10}\,\frac{\mathrm{V}^2}{\mathrm{Hz}}$,
- Number of levels: Initially $s = 2$, later variable number of levels s,
- Cable: $f_0 = 0.178\,\mathrm{MHz} \cdot \mathrm{km}^2$, $\ell = 2\,\mathrm{km}$.

Procedure and Results From the given average transmit power (2.230), the half-level transmit amplitude is first calculated using

$$P_s = \frac{U_s}{3}\left(s^2 - 1\right) \quad \Longrightarrow \quad U_s = \sqrt{\frac{3\,P_s}{s^2 - 1}}. \tag{4.71}$$

The half-level transmit amplitude $U_s = 1\,\mathrm{V}$ is obtained for $P_s = 1\,\mathrm{V}^2$ and $s = 2$.

The receive filter is again (as in Chap. 2) predefined in its structure as a Gaussian low-pass, but its cutoff frequency f_G is freely selectable and thus optimizable. Therefore, the half vertical eye opening according to (4.52) and the noise power according to (4.54) are determined as a function of the receive filter cutoff frequency. In Fig. 4.32, the half vertical eye opening U_A is shown for the given numerical values as a function of the cutoff frequency f_G of the receive filter. After the eye has opened, it increases with increasing cutoff frequency and tends towards the maximum value of the half-level transmit amplitude U_s. Figure 4.33 shows the noise power U_R^2 as a function of the receive filter cutoff frequency f_G: It increases indefinitely with increasing receive filter cutoff frequency.

In Fig. 4.34, the signal-to-noise ratio ϱ is shown as a function of the receive filter cutoff frequency f_g: There is again a maximum ϱ_{max} at the optimal receive filter cutoff frequency $f_{G\,opt}$, at which the bit error probability $P_b = P_f$ has a minimum (see Fig. 4.35).

Numerical Results for Optimized Transmission System Once the optimal cutoff frequency $f_{G\,opt}$ of the receive filter is determined, all other values for signal and disturbance at the detection point can be determined and the quality of transmission for an optimized transmission system can be specified. The individual relationships, which have been formulated as functions of the cutoff frequency f_G of the receive filter, are evaluated at the point $f_G = f_{G\,opt}$. The numerical results are compiled in Table 4.2.

With the now optimal receive filter cutoff frequency $f_{G\,opt} = 467\,\text{kHz}$, a maximum signal-to-noise ratio of $\varrho_{max} = 23.93$ or $\varrho_{max}^* = 10\lg\varrho_{max} = 13.79\,\text{dB}$ can be achieved, leading to a minimum bit error probability $P_{b\,min} = 5.01 \cdot 10^{-7}$ (for the case $s = 2$).

Compared to the previously examined numerical example with unipolar, two-level transmission, an improvement in transmission quality can be seen, which is mainly due to the better utilization of the given transmit power in bipolar transmission and the improved receive filter.

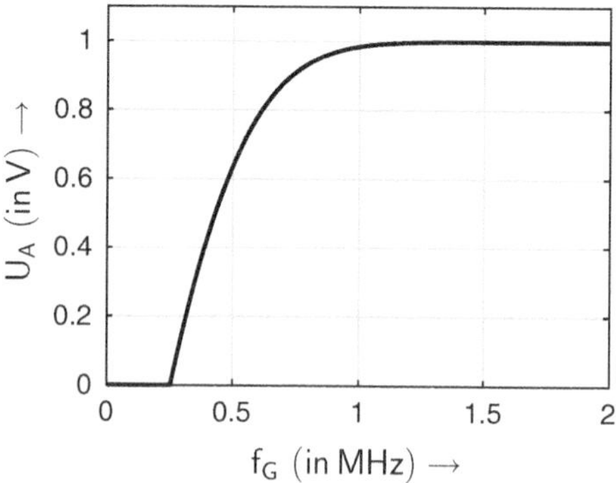

Fig. 4.32 Half vertical eye opening U_A as a function of the receive filter cutoff frequency f_G

Fig. 4.33 Noise power U_R^2 as a function of the receive filter cutoff frequency f_G

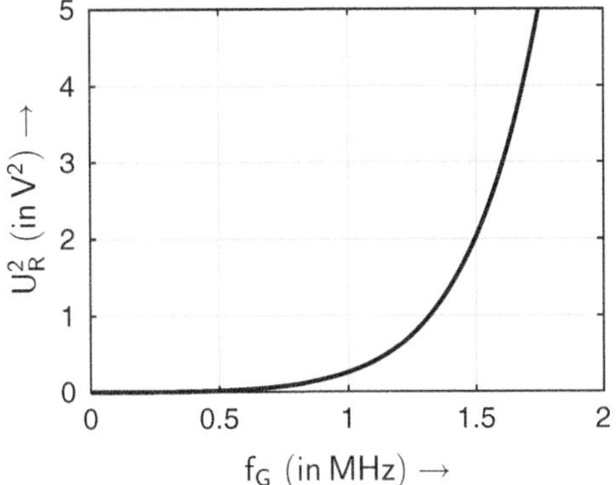

Fig. 4.34 Signal-to-noise ratio ϱ as a function of the receive filter cutoff frequency f_G

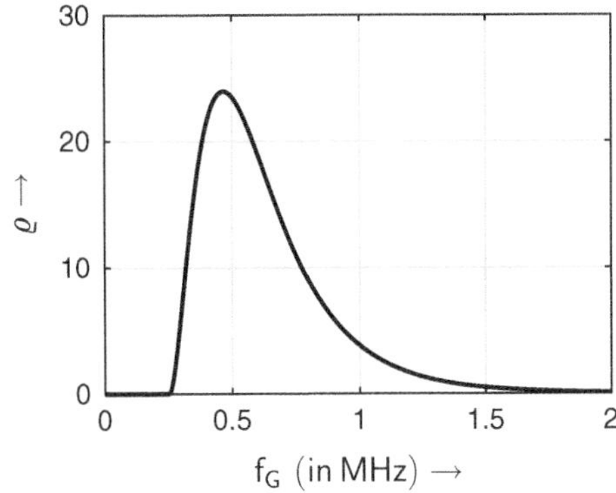

Fig. 4.35 Bit error probability P_b as a function of the receive filter cutoff frequency f_G

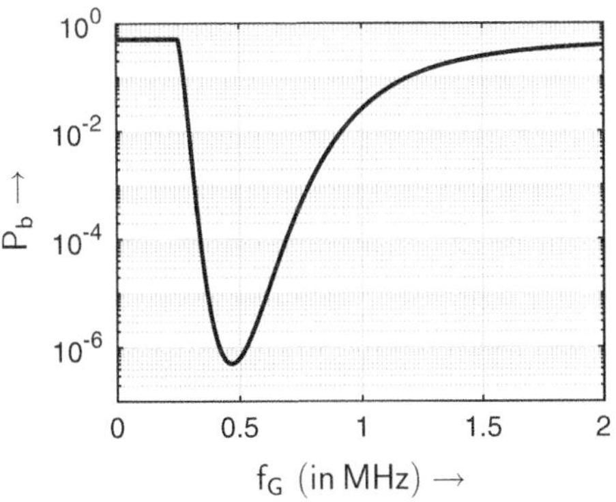

Table 4.2 Given values and numerical results for linearly equalized cable transmission system with bipolar, two-level NRZ rectangular pulses at optimal receive filter cutoff frequency $f_{G\,opt}$ of the Gaussian low-pass filter

Quantity	Numerical Value
Average Transmit Power	$P_s = 1\,\mathrm{V}^2$
Number of Signalling Levels	$s = 2$
Half-level Transmit Amplitude	$U_s = 1\,\mathrm{V}$
Noise Power Spectral Density	$\Psi_0 = 5 \cdot 10^{-10}\,\mathrm{V}^2/\mathrm{Hz}$
Pulse Duration (Symbol Duration)	$T_s = 1\,\mu\mathrm{s}$
Optimal Receive Filter Cutoff Frequency	$f_{G\,opt} = 467\,\mathrm{kHz}$
Half Vertical Eye Opening	$U_A = 0.575\,\mathrm{V}$
Noise Power	$U_R^2 = 0.014\,\mathrm{V}^2$
Signal-to-Noise Ratio	$\varrho_{max} = 23.93$
Bit Error Probability	$P_{b\,min} = 5.01 \cdot 10^{-7}$

Influence of the Number of Signalling Levels If the maximum amplitude range or the average transmit power is limited and thus predetermined in the transmitter, the half-level amplitude U_s decreases with increasing number s of signalling levels: One obtains $(s - 1)$ eye openings, which become smaller with increasing number of levels s (see, for example, Fig. 2.70).

To illustrate the influence of increasing numbers s of signalling levels, the numerical example is now expanded in such a way that different numbers of signalling levels are allowed and considered under otherwise identical specifications regarding signal and interference. With the same predetermined bit rate f_B and average transmit power P_s, the respective half-level transmit amplitudes U_s according to (4.71) and symbol clock frequencies f_T according to (2.195) have to be calculated and thus the previously described calculations for eye opening U_A, noise power U_R^2, signal-to-noise ratio ϱ, and possibly the bit error probability P_b have to be carried out, whereby an optimization of the receive filter cutoff frequency f_G has to be carried out for each number s of signalling levels.

In Fig. 4.36, the maximum achievable signal-to-noise ratio ϱ_{max} as a function of the number of signalling levels s—at a fixed bit sequence frequency of $f_B = 1\,\mathrm{MHz}$ and under otherwise identical boundary conditions—is shown to demonstrate the achievable transmission quality. The progression initially resembles that achieved for an AWGN channel (Fig. 2.106): With increasing number of signalling levels s, the achievable signal-to-noise ratio ϱ_{max} decreases. If the bit rate is increased, e.g., to $f_B = 2\,\mathrm{MHz}$, the bandwidth limitation of the transmission channel becomes more effective and the equalization therefore leads to a stronger noise enhancement: In Fig. 4.37, it becomes apparent that the best performance of the transmission system is now achieved at a number of signalling levels of $s = 4$, and there the highest signal-to-noise ratio ϱ_{max} can be achieved—with a decreasing overall level of the signal-to-noise ratio. If the bit rate is

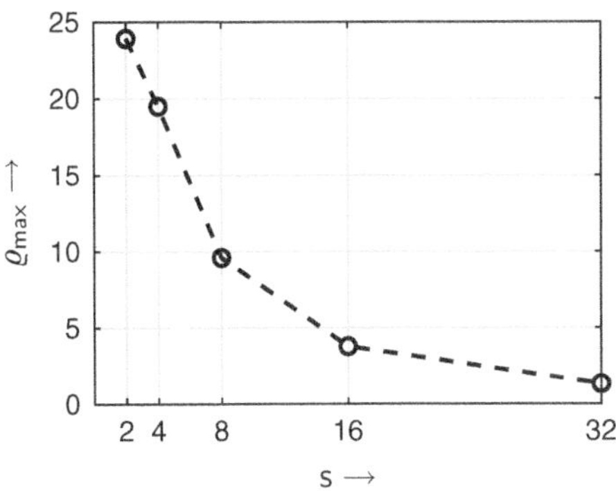

Fig. 4.36 Signal-to-noise ratio ϱ_{max} as a function of the number s of signalling levels for bipolar, multi-level baseband transmission with NRZ transmission pulses and Gaussian receive filtering over linearly equalized cable ($f_B = 1\,\text{MHz}$)

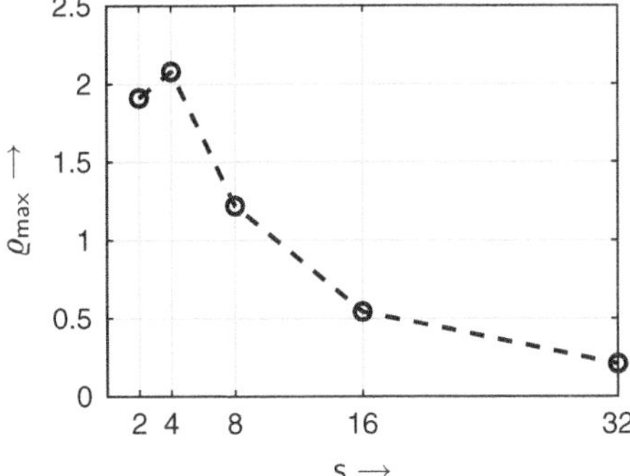

Fig. 4.37 Signal-to-noise ratio ϱ_{max} as a function of the number s of signalling levels for bipolar, multi-level baseband transmission with NRZ transmission pulses and Gaussian receive filtering over linearly equalized cable ($f_B = 2\,\text{MHz}$)

further increased under otherwise unchanged boundary conditions, these tendencies remain: The maximum performance of the transmission system—the highest signal-to-noise ratio—shifts towards higher numbers of signalling levels, with a decreasing overall level of the signal-to-noise ratio.

4.5.2   Baseband Transmission with Root Raised Cosine Transmit and Receive Filters

As before, an s-level, bipolar baseband transmission over the linearly equalized cable transmission system is examined (block diagram according to Fig. 4.30). Transmit and receive filters are chosen as root raised cosine filters (see Chap. 2) with the roll-off factor r, i.e., it applies (with (2.104))

$$G_s(f) = G_e(f) = \sqrt{G_N(f)}, \tag{4.72}$$

so that the first Nyquist criterion is maintained over the cascade of transmit and receive filters. Since the cable impact is completely linearly equalized and thus compensated, this also applies to the entire transmission system from the input of the transmit filter to the output of the receive filter.

Useful Signal Path and Evaluation of the Useful Signal

The first Nyquist criterion is maintained and the useful signal is thus free of intersymbol interference at the detection point. The half vertical eye opening is

$$U_\mathrm{A} = U_\mathrm{s}. \tag{4.73}$$

It is independent of the roll-off factor r. The half-level transmit amplitude U_s can again be given either directly or implicitly via an amplitude or power limitation of the transmitter.

Noise Path and Evaluation of the Noise Signal

The disturbance is evaluated by the equalizer receive filter and the noise power is obtained

$$U_\mathrm{R}^2 = \Psi_0 \int_{-\infty}^{+\infty} \left| \frac{G_\mathrm{e}(f)}{G_\mathrm{k}(f)} \right|^2 \, \mathrm{d}f = \Psi_0 \int_{-\infty}^{+\infty} \frac{|G_\mathrm{e}(f)|^2}{|G_\mathrm{k}(f)|^2} \, \mathrm{d}f \tag{4.74}$$

and

$$U_\mathrm{R}^2 = \Psi_0 \int_{-\infty}^{+\infty} \frac{\left| \sqrt{G_\mathrm{N}(f)} \right|^2}{\left| e^{-\ell \sqrt{\mathrm{j}\frac{f}{f_0}}} \right|^2} \, \mathrm{d}f = \Psi_0 \int_{-\frac{f_\mathrm{T}}{2}(1+r)}^{\frac{f_\mathrm{T}}{2}(1+r)} \frac{G_\mathrm{N}(f)}{e^{-l\sqrt{\frac{2f}{f_0}}}} \, \mathrm{d}f. \tag{4.75}$$

Due to the influence of the cable—via the higher amplification by the equalizer at higher frequencies—on the noise power, the noise power is not independent of the roll-off factor r, unlike with a non-distorting channel (see Chap. 2).

The integral in (4.75) cannot generally be solved elementarily; therefore, numerical integration methods are also used here for calculations.

Quality of the Transmission

The quality of the transmission is again evaluated using the *signal-to-noise ratio* and the resulting *bit error probability*.

Signal-to-Noise Ratio   The signal-to-noise ratio $\varrho = U_\mathrm{A}^2 / U_\mathrm{R}^2$ is formed from the previously achieved results for the evaluation of the useful signal and disturbance. Although it depends—via the noise power U_R^2—on the roll-off factor r, there is no optimization in the sense as before for the cut-off frequency of the receive filter possible and necessary, since the half vertical eye opening does not depend on the roll-off factor and the noise power increases with increasing roll-off factor. It is a question of realization and implementation on the one hand and the occupied bandwidth on the other hand, which roll-off factor r is chosen. Since the eye is always fully open vertically regardless of r, it is advantageous to choose as small roll-off factors as possible due to the then stronger band limitation of the noise, but with very small r detection becomes almost impossible, as the horizontal eye opening becomes very small and a very precise sampling must be realized. For these reasons, roll-off factors in the range $r = 0.3 \ldots 0.5$ are often chosen in practical systems. The discussed dependencies are illustrated in the following numerical example.

Bit Error Probability The bit error probability results from the calculated signal-to-noise ratio according to (4.70) or (2.240).

Numerical Example

The following numerical values are given for this example:

- Signal: Transmit power $P_s = 1\,\mathrm{V}^2$, bit rate $f_B = 1\,\mathrm{MHz}$, bipolar transmission,
- Disturbance: Power spectral density $\Psi_0 = 5 \cdot 10^{-10}\,\frac{\mathrm{V}^2}{\mathrm{Hz}}$,
- Number of levels: Initially $s = 2$, later variable number of levels s,
- Cable: $f_0 = 0.178\,\mathrm{MHz} \cdot \mathrm{km}^2$, $\ell = 2\,\mathrm{km}$.

Procedure and Results From the given average transmit power (2.230), the half-level transmit amplitude is first calculated again via

$$P_s = \frac{U_s}{3}\left(s^2 - 1\right) \qquad \Longrightarrow \qquad U_s = \sqrt{\frac{3\,P_s}{s^2 - 1}}\,. \tag{4.76}$$

For $P_s = 1\,\mathrm{V}^2$ and $s = 2$, the half-level transmit amplitude $U_s = 1\,\mathrm{V}$ is obtained.

In Fig. 4.38, the half vertical eye opening U_A is shown as a function of the roll-off factor r. The overall system is a Nyquist-1 system, and the eye is fully open vertically—

Fig. 4.38 Half vertical eye opening U_A as a function of the roll-off factor r

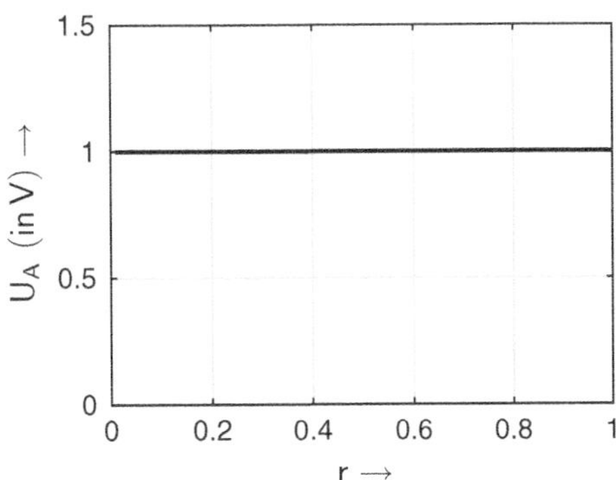

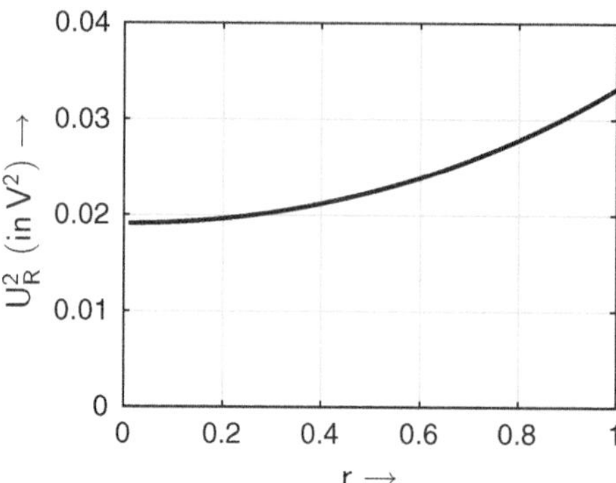

Fig. 4.39 Noise power U_{R}^2 as a function of the roll-off factor r

regardless of the roll-off factor r. The half vertical eye opening is constant and corresponds to the half-level transmit amplitude U_{s}. Figure 4.39 shows the noise power U_{R}^2 as a function of the roll-off factor r: It increases with increasing roll-off factor, as the bandwidth of the receive filter is increased with increasing r, and thus the more pronounced equalizer amplification at higher frequencies leads to a stronger noise enhancement.

In Fig. 4.40, the signal-to-noise ratio ϱ is shown as a function of the roll-off factor r: There is again a maximum $\varrho_{\max}$, but now at the smallest considered roll-off factor r; optimization of the roll-off factor is therefore not possible. The bit error probability P_{b} shows a minimum there (see Fig. 4.41).

Numerical Results for Exemplary Transmission System For example, if a roll-off factor of $r = 0.5$ is chosen as a compromise between performance (signal-to-noise ratio or bit error probability) and feasibility (horizontal eye opening, accuracy of symbol clock sampling), a signal-to-noise ratio of $\varrho = 44.66$ can be achieved. This results in a bit error probability $P_{\mathrm{b}} = 1.17 \cdot 10^{-11}$. The numerical results are compiled in Table 4.3.

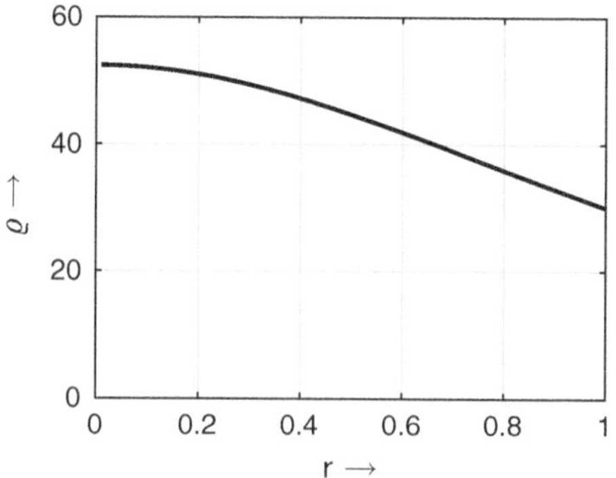

Fig. 4.40 Signal-to-noise ratio ϱ as a function of the roll-off factor r

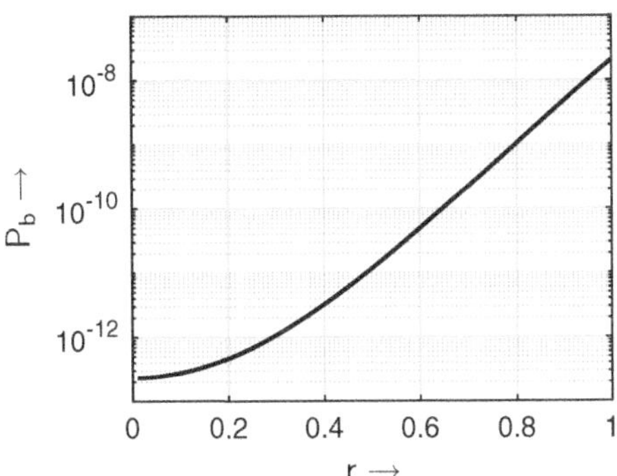

Fig. 4.41 Bit error probability P_b as a function of the roll-off factor r

Table 4.3 Given values and numerical results for a linearly equalized cable transmission system with bipolar, two-level transmit signal at a roll-off factor of $r = 0.5$ of the root raised cosine transmit and receive low-pass filters

Quantity	Numerical Value
Average Transmit Power	$P_s = 1\,\mathrm{V}^2$
Number of Signalling Levels	$s = 2$
Half-level Transmit Amplitude	$U_s = 1\,\mathrm{V}$
Noise Power Spectral Density	$\Psi_0 = 5 \cdot 10^{-10}\,\mathrm{V}^2/\mathrm{Hz}$
Pulse Duration (Symbol Duration)	$T_s = 1\,\mu\mathrm{s}$
Roll-Off Factor	$r = 0.5$
Half Vertical Eye Opening	$U_A = 1\,\mathrm{V}$
Noise Power	$U_R^2 = 0.022\,\mathrm{V}^2$
Signal-to-Noise Ratio	$\varrho_{\max} = 44.66$
Bit Error Probability	$P_{b\,\min} = 1.17 \cdot 10^{-11}$

Influence of the Number of Signalling Levels If the maximum amplitude range or the average transmit power is limited and thus predetermined in the transmitter, the half-level transmit amplitude U_s decreases with increasing number of levels s: One obtains $(s - 1)$ eye openings, which become smaller with increasing number of levels s (see, for example, Fig. 2.70).

To illustrate the influence of increasing step numbers s, the numerical example is now expanded again so that, with otherwise identical specifications regarding signal and interference, different numbers of signalling levels are allowed and considered. Given a predetermined bit rate f_B and average transmit power P_s, the respective half-level transmit amplitudes U_s according to (4.76) and symbol clock frequencies f_T according to (2.195) are to be calculated and thus the previously described calculations for eye opening U_A, noise power U_R^2, signal-to-noise ratio ϱ, and possibly the bit error probability P_b are to be performed.

In Fig. 4.42, the maximum achievable signal-to-noise ratio ϱ as a function of the step number s—at a bit rate of $f_B = 1\,\mathrm{MHz}$ and under otherwise identical boundary conditions—is shown to capture the achievable transmission quality. The signal-to-noise

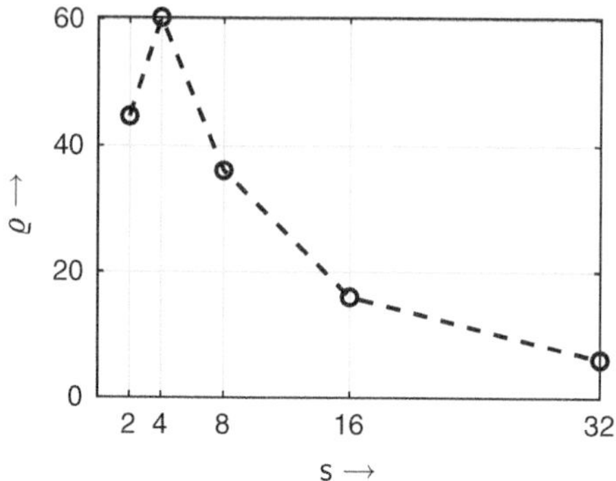

Fig. 4.42 Signal-to-noise ratio ϱ as a function of the number s of signalling levels for bipolar, multi-level baseband transmission with root raised cosine transmit and receive filters with $r = 0.5$ over linearly equalized cable ($f_{\mathrm{B}} = 1\,\mathrm{MHz}$)

ratio now shows—due to the larger eye opening compared to the previously discussed NRZ-Gauss system—a maximum at a bit rate $f_{\mathrm{B}} = 1\,\mathrm{MHz}$ with a step number of $s = 4$ (Fig. 2.106); with further increasing step number s, the achievable signal-to-noise ratio ϱ decreases. If the bit rate is increased, e.g., to $f_{\mathrm{B}} = 2\,\mathrm{MHz}$, the bandwidth limitation of the transmission channel becomes more effective and the equalization therefore leads again to a stronger noise enhancement: In Fig. 4.43, it becomes clear that the greatest performance of the transmission system is again achieved at a step number of $s = 4$, and that the highest signal-to-noise ratio ϱ can be achieved there—with a again decreasing overall level of the signal-to-noise ratio. If the bit rate is further increased under otherwise unchanged boundary conditions, the trends remain the same: The maximum performance of the transmission system—the highest signal-to-noise ratio—shifts towards higher step numbers, with a decreasing overall level of the signal-to-noise ratio.

In comparison of the baseband system with root raised cosine transmit and receive filtering with the system with NRZ transmit signal and Gaussian receive filter, it can be stated that the former has a significantly better performance in terms of signal-to-noise ratio and thus achievable symbol and bit error probability than the latter one. This is

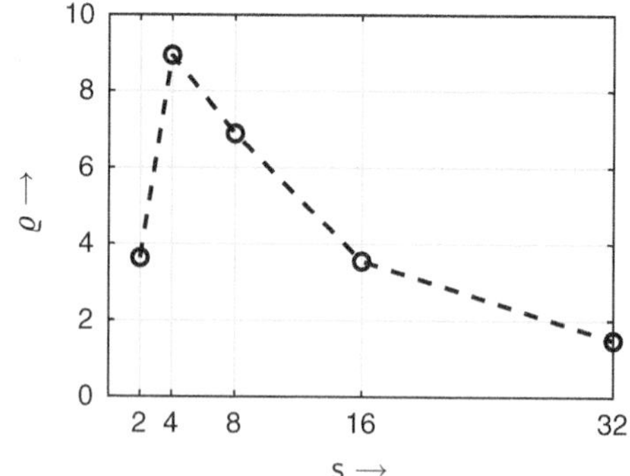

Fig. 4.43 Signal-to-noise ratio ϱ as a function of the number s of signalling levels for bipolar, multi-level baseband transmission with root raised cosine transmit and receive filters with $r = 0.5$ over linearly equalized cable ($f_{\mathrm{B}} = 2\,\mathrm{MHz}$)

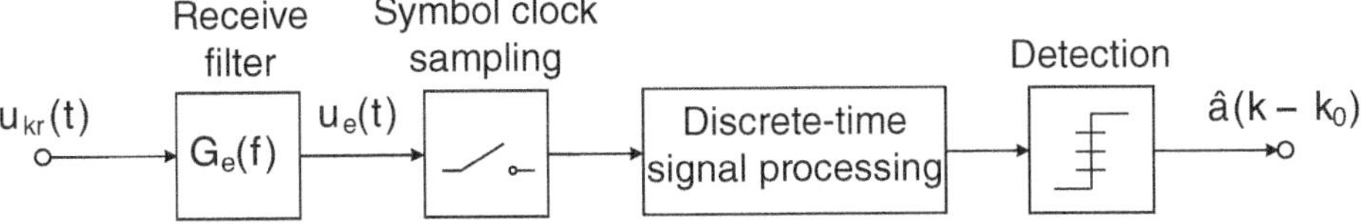

Fig. 4.44 Block diagram of the receiver with time-discrete signal processing

mainly due to the fact that the system with root raised cosine transmit and receive filtering meets the first Nyquist criterion—with solid bandwidth limitation to limit the noise power at the detection point. However, the pulses show—in contrast to the system with NRZ transmit signal and Gaussian receive filter—an overshoot in the time domain, so this must be taken into account in the design of transmit and receive elements.

4.6 Discrete-Time Equalization

4.6.1 Introduction

If a digital information or signal transmission is to be carried out over a distorting transmission channel, the resulting signal distortions must be compensated in order to minimize their effects on the transmission quality. This is necessary, among other reasons, to make the best possible use of given resources—such as bandwidth and transmit power. The compensation of signal distortions is carried out with the help of a unit referred to as an *equalizer*. This was extensively considered in the previous sections using an example, the analog linear equalizer, which is integrated into the receive filter—the so-called *equalizer-receive filter*.

The copper two-wire transmission channel is frequency-dependent and leads to signal distortions depending on the spectral properties of the transmitted signal. Especially at high data transmission rates and high transmission bandwidths, powerful methods for equalization or signal reconstruction are therefore required.

In addition to the discussed analog linear equalization, there are other possibilities and powerful approaches for equalization that compensate for the influence of the channel on signal transmission and that particularly make use of the possibilities of digital signal processing. Optimal methods ensure the best possible transmission quality under given boundary conditions. However, these methods are often very complex and the effort involved in their implementation therefore limits their practical applications. Therefore, equalization methods that do not work optimally but require considerably less implementation effort with typically only minor losses in transmission quality and that are therefore more cost-effective are often of interest.

This section follows the presentations in [2]. With reference to the previous sections of this chapter, it provides an insight into the use of equalizers for signal reconstruction with the aim of reliable symbol decision and extends the considerations to discrete-time equalizer structures and methods for baseband transmission over copper two-wire lines. To this end, the symbol clock model of the underlying transmission system is first introduced. Optimal methods are then discussed and a classification of equalization methods is given. Subsequently, some important linear and nonlinear equalization methods are presented and illustrated with examples. A discussion rounds off the section on equalization methods, with the analog linear equalization also being considered in order to maintain the reference to the previous sections.

4.6.2   Symbol Clock Model of the Baseband Transmission System

With receiver input filters that include equalization—as previously discussed using the example of analog linear equalization (equalizer-receive filter), detection can take place immediately after the symbol clock sampling at the receive filter output. If, on the other hand, signal-adapted receive filters (matched filters) that are adapted to the cascade of transmitter filter and transmission channel, or receive filters with root raised cosine frequency response are used, the linear channel distortions must be compensated by separate signal processing, which can be advantageously and cost-effectively realized discretely in time at the receive filter output before detection. The receiver is to be supplemented by the time-discrete signal processing for sequence estimation or equalization (Fig. 4.44).

The discrete-time equalization methods considered in the following process the symbol clock sample values. Therefore, the transmission system is modeled in the symbol clock (Fig. 4.45). The discrete-time weighting function $h(k) = h(k\,T_s) = \mathcal{Z}^{-1}\{H(z)\}$ of the overall system from the transmitter filter input to the receive filter output is obtained by symbol clock sampling from the time-continuous weighting function $h(t)$ of the overall system. The discrete-time noise signal $w(k)$, which results after receive filtering and symbol clock sampling, is generally colored.[5]

If a distortion-free overall channel $h(k)$ is present in the symbol clock, the cascade of all systems from the input of the transmitter filter to the output of the receive filter fulfills the first Nyquist criterion (see Chap. 2); the discrete-time system model in the symbol clock shown in Fig. 4.46 is obtained. The overall system can also be approximately distortion-free if no significant signal distortions occur when transmitting data at a low rate over a channel with sufficiently large bandwidth. The effects of such a channel on the signal are limited to a frequency-independent attenuation. This is described in the system model by a scaling factor. In a pure AWGN channel, the scaling factor has the value $h[0] = 1$.

---

[5] The special case of white noise (independent noise samples) after receive filtering and symbol clock sampling is obtained when the receive filter $G_e(f)$ (in Fig. 4.44) has a root-Nyquist frequency response (e.g., root raisedcosine filter and e.g., [3, 23]).

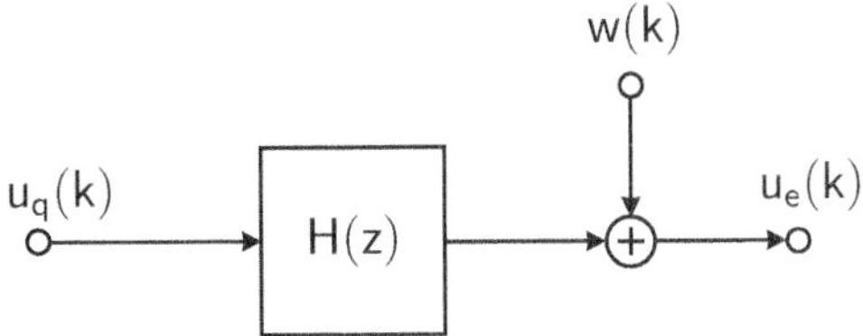

Fig. 4.45 Discrete-time model of the baseband transmission system

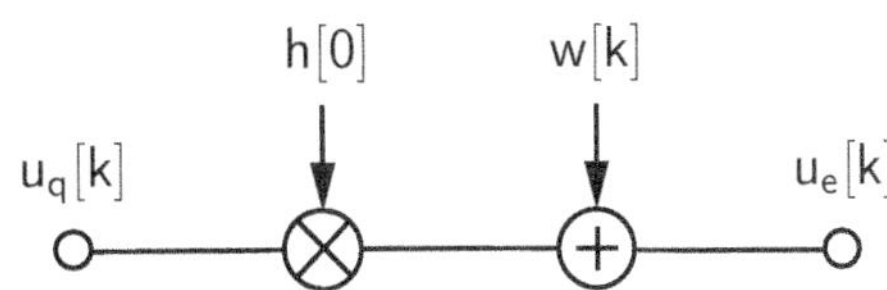

Fig. 4.46 Discrete-time model of a transmission system complying with the first Nyquist criterion (i.e., without intersymbol interference)

If the data transmission rate—while maintaining the same spectral characteristics of the channel—is increased, the individual frequency components of the transmitted signal are evaluated differently, resulting in signal distortions and thus intersymbol interference. In Fig. 4.47, the now resulting discrete-time system model for an example is shown: In addition to the main value $h[0]$, further non-zero components of the overall weighting function $h(k)$ of the transmission system become effective, here exemplarily also the value $h[1]$.

4.6.3   Optimal Methods

The error probability in the transmission of digital information-bearing signals can be minimized if sufficiently long sequences of received symbols are used for the detection of a symbol [60]. Receivers that use this approach and operate according to the principle of sequence estimation (MLSE—Maximum Likelihood Sequence Estimation) use very long sequences of received symbols to make decisions about received symbols [42]. In this way, optimal results are obtained in the detection of received symbols with respect to a minimal error probability [6, 52].

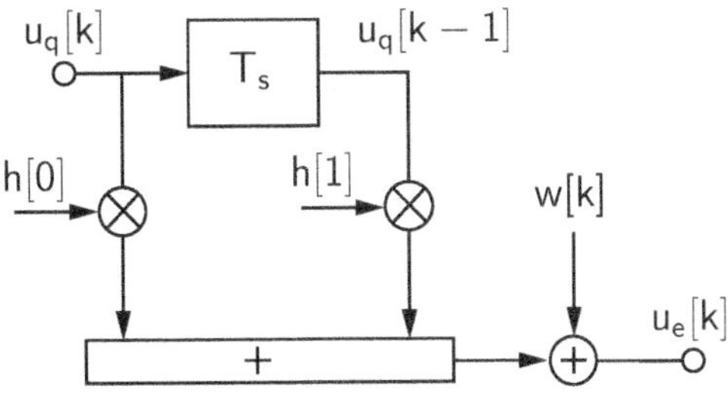

Fig. 4.47 Discrete-time model of an exemplary transmission system with intersymbol interference

The Viterbi algorithm [15] allows for a cost-effective implementation of the MLSE method for sequence estimation. The Viterbi receiver is also referred to as the Viterbi detector. It processes the received signal sampled at the symbol rate $u_e(k)$; it performs equalization and also makes the decision about the received symbol sequence. It is therefore an equalizer with integrated symbol recognition (Fig. 4.48). The Viterbi detector requires an uncorrelated disturbance at its input to achieve its full performance—i.e., the minimum error probability in the decided symbol sequence—i.e., a white noise process must be effective as a disturbance at the output of the receive filter [64].

The complexity of the Viterbi receiver becomes very high with long channel weighting functions $g_k(t)$, as the number of states to be considered increases exponentially with the degree of the channel weight function sampled at the symbol rate—i.e., with the number of non-zero components in the weighting function $h(k)$ of the overall system; a multi-level transmission leads to a further increase in the number of states to be considered in the detection and thus further increases the required effort. Since the weighting function of typical two-wire lines in communication cables extends over many symbol intervals at high-rate transmission, it is often not practical to use Viterbi receivers in real transmission systems. Therefore, in addition to the analog linear equalization, which is integrated into the receive filter (equalizer receive filter), discrete-time equalization methods are often used.

The diagrams in Figs. 4.49, 4.50 and 4.51 show the signal distortions occurring on the transmission path based on the weighting function $h(t) = g_s(t) * g_k(t) * g_e(t)$ of the respective overall system, at different symbol clock frequencies f_T. A copper two-wire line ($l = 2\,\mathrm{km}$, $f_0 = 0.178\,\mathrm{MHz} \cdot \mathrm{km}^2$) is assumed as a channel according to (3.157) considering five poles in the rational approximation. Transmit and receive filters are root raised cosine low-pass filters, each with the roll-off factor $r = 0.5$. It can be seen that with increasing symbol clock frequency f_T and unchanged characteristics of the transmission channel, the number of non-zero symbol clock samples of the resulting weighting function $h(t)$ of the overall system increases.

4.6.4   Classification of Equalization Methods

The linear signal distortions that occur during the transmission of a signal over a distorting channel can, in principle, be corrected in the receiver by linear or non-linear systems [23, 52, 62]. If we consider discrete-time equalizer systems that are used for this pur-

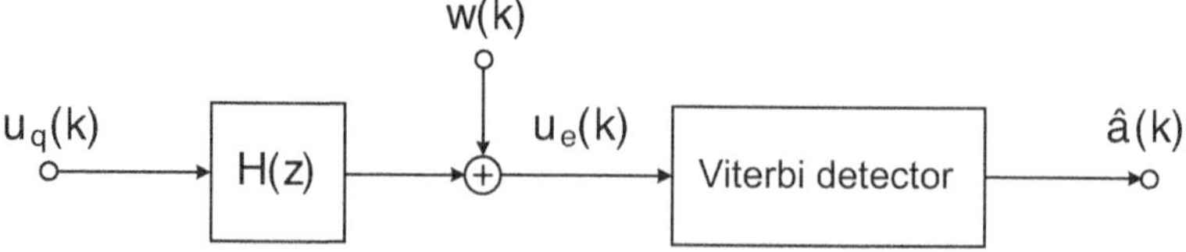

Fig. 4.48 Discrete-time model of a transmission system with Viterbi detector

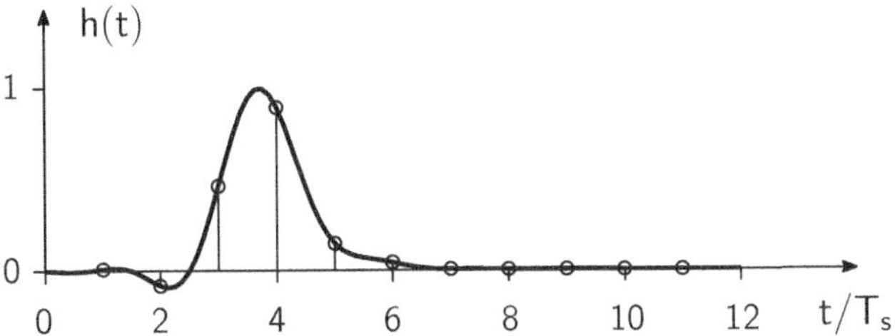

Fig. 4.49 Resulting weighting function of the overall system (normalized to its maximum value) at a symbol clock frequency of $f_T = 0.5\,\text{MHz}$

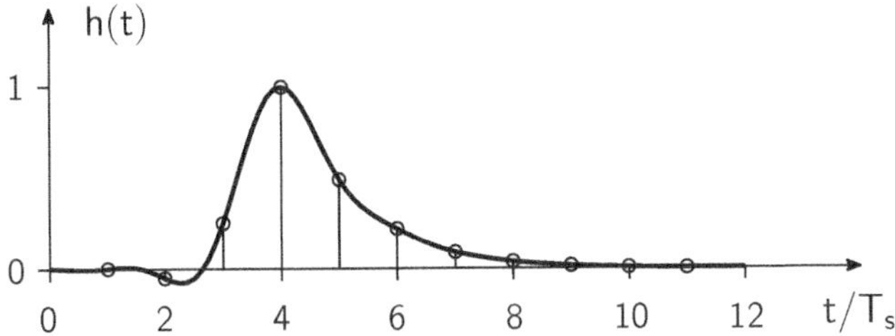

Fig. 4.50 Resulting weighting function of the overall system (normalized to its maximum value) at a symbol clock frequency of $f_T = 1.0\,\text{MHz}$

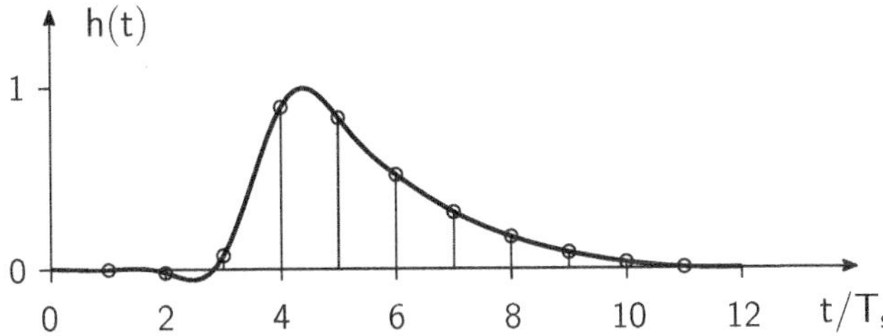

Fig. 4.51 Resulting weighting function of the overall system (normalized to its maximum value) at a symbol clock frequency of $f_T = 2.0\,\text{MHz}$

pose, they can be classified according to the structure of the equalizer and the way in which the equalizer coefficients are determined (see Fig. 4.52).

The classification according to the structure of the equalizing system results in the subdivision into linear and non-linear system structures.

In the classification according to the way in which the equalizer coefficients are determined, a distinction can be made as to whether the weighting function (or impulse response) to be equalized is known, how information about its shape is obtained (via a reference sequence), and whether the determination of the equalizer coefficients can

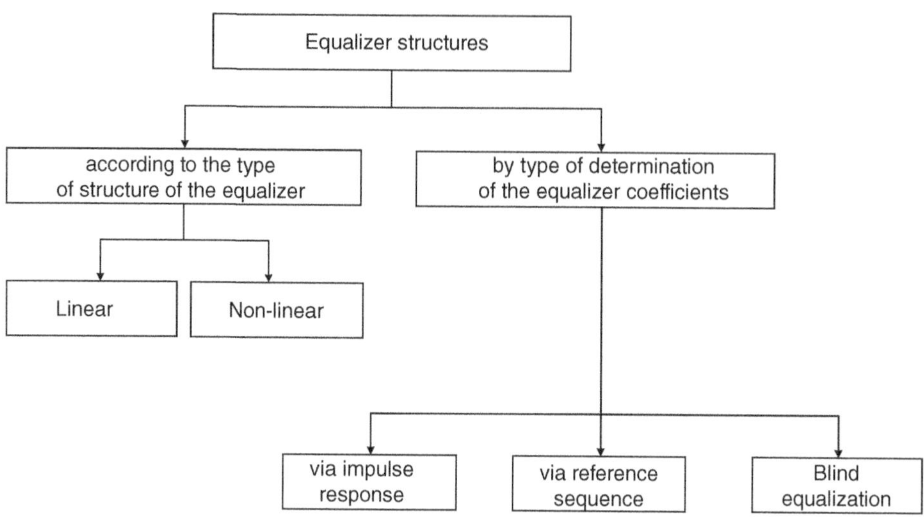

Fig. 4.52 Classification of equalizer structures in time-discrete equalization

even be done without prior knowledge of this weighting function to be equalized (blind equalization).

If we initially assume that the weighting function of the channel to be equalized is known, the equalizer coefficients are often obtained via the inversion of matrices [26], with the sampling values of the weighting function of the overall system being the basis and thus must be known. If the sampling values of the weighting function of the overall system to be equalized are not initially known in the receiver, training data can be used, which are integrated into the user data stream to be transmitted at time intervals as a so-called reference sequence, in order to determine the sampling values of the weighting function of the overall system on this basis, so that they are ultimately available for determining the equalizer coefficients in the equalizer. This approach is often found in radio systems, where the transmission channel typically fluctuates greatly over time. The so-called blind equalization includes methods that determine the equalizer coefficients without knowledge of the weighting function to be equalized—and thus without reference signals or training data [26]. Blind equalization is based on higher-order statistics (e.g., fourth-order moments), on the basis of which a solution can be formulated that allows the required equalizer coefficients to be determined. Blind equalization methods usually require increased implementation effort.

In the following, it is assumed that the weighting function of the channel to be equalized and thus that of the overall system from the transmit filter input to the receive filter output is known as the basis for determining the equalizer coefficients in the receiver.

4.6.5   Linear and Non-linear Equalizers

Linear Analog Equalization

The analog equalizer receive filter has been extensively discussed in the previous sections. It is revisited here and integrated into the systematics of equalizers to relate the basic performance of analog linear equalization to that of discrete-time equalization methods.

The transfer function of the analog, linearly equalizing receive filter (equalizer receive filter) is

$$H_e(f) = \frac{1}{G_k(f)} \cdot G_e(f). \tag{4.77}$$

In this way, the distorting influence of the transmission channel is fully compensated by the inverse channel frequency response. We obtain the baseband system according to Fig. 4.53 [2].

This type of equalization corresponds to the equalization that has been extensively discussed in the previous sections of this chapter. It is mentioned here again for the sake of completeness, to complete the systematics of equalization and to highlight some special features.

In this realization of equalization, whether the first Nyquist criterion is met at the detection point depends on the choice of transmit filter $G_s(f)$ and receive filter $G_e(f)$. If the transmit and receive filter transfer functions are chosen such that their series connection, i.e., the transfer function $G_s(f) \cdot G_e(f)$, fulfills the first Nyquist criterion, then the received useful signal at the symbol clock sampling times $k\,T_s$ has no intersymbol interference. In this case, the eye is vertically fully open at the detection point, regardless of the symbol clock frequency f_T, and the half vertical eye opening results in $U_A = U_s$.

The disturbance or noise signal $n(t)$ is evaluated on its way to the detection point by the equalizer receive filter $G_e(f)/G_k(f)$, and the square of the effective value U_R of the

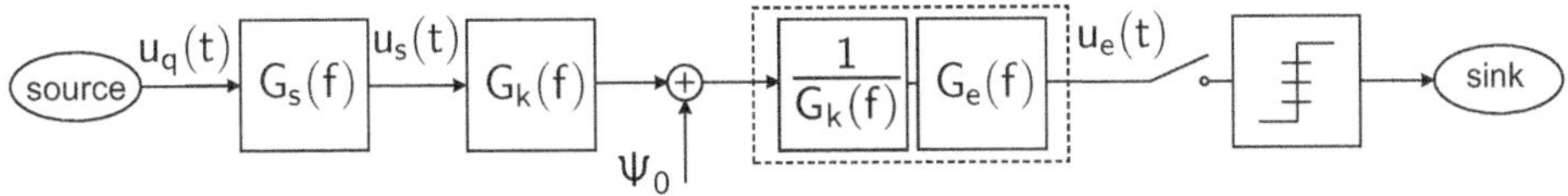

Fig. 4.53 Block diagram of the transmission system with linear equalization of the channel frequency response

noise signal is obtained as noise power at the detection point according to (4.32). The disturbance is amplified frequency-dependently and possibly strongly increased by the linear equalization using the transfer function $1/G_{\mathrm{k}}(f)$. This can lead to high noise power at the detection point.

Example 4.1 (Transmission quality with multi-level transmission and increasing cable utilization). In an example, the behavior of the cable transmission system with variable cable utilization is to be examined, i.e., with fixed properties of the underlying transmission channel, the symbol clock frequency f_{T} is varied and it is determined how individual parameters change that describe the transmission quality. For numerical results, the useful signal $U_{\mathrm{s}} = 1\,\mathrm{V}$, the noise power spectral density $\Psi_0 = 5 \cdot 10^{-10}\,\mathrm{V}^2/\mathrm{Hz}$, and the cable $f_0 = 0.178\,\mathrm{MHz} \cdot \mathrm{km}^2$ as well as $l = 2\,\mathrm{km}$ are assumed, considering five pole in (3.157).

Figure 4.54 exemplarily shows the noise power U_{R}^2 as a function of the clock frequency f_{T}. It becomes clear that with increasing cable utilization—i.e., with increasing symbol clock frequency f_{T} and unchanged cable properties—the noise power increases.

Transmit filter and receive filter transfer functions were each designed as root raised cosine filters with the roll-off factor $r = 0.5$ for $G_{\mathrm{s}}(f)$ and $G_{\mathrm{e}}(f)$, so that the cascade $G_{\mathrm{s}}(f) \cdot G_{\mathrm{e}}(f)$ meets the first Nyquist criterion. This ensures that the eye at the detection point is fully open vertically, regardless of the symbol clock frequency f_{T}, it applies $U_{\mathrm{A}} = U_{\mathrm{s}}$ for all f_{T}. The increasing noise power (see Fig. 4.54) with increasing cable utilization (or increasing clock frequency f_{T}) leads to a decreasing signal-to-noise ratio ϱ with increasing symbol clock frequency f_{T}, as the square of half the vertical eye opening U_{A}^2 is constant regardless of f_{T}. This functional relationship between ϱ and f_{T} is shown in Fig. 4.55.

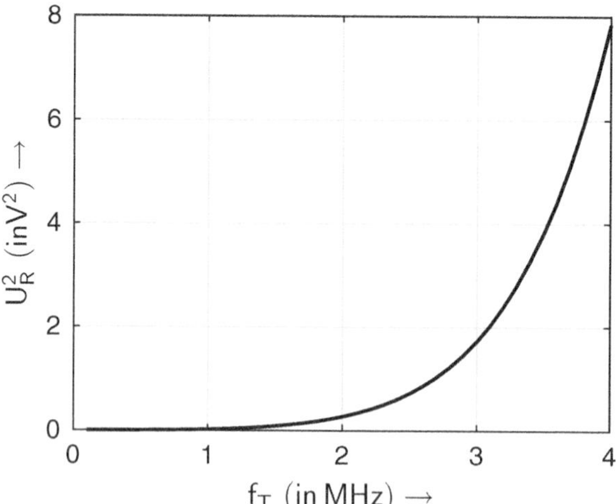

Fig. 4.54 Noise power U_{R}^2 at the detection point with analog, linear equalization of the cable frequency response (parameters: $\ell = 2.0\,\mathrm{km}$, $\Psi_0 = 5 \cdot 10^{-10}\,\mathrm{V}^2/\mathrm{Hz}$)

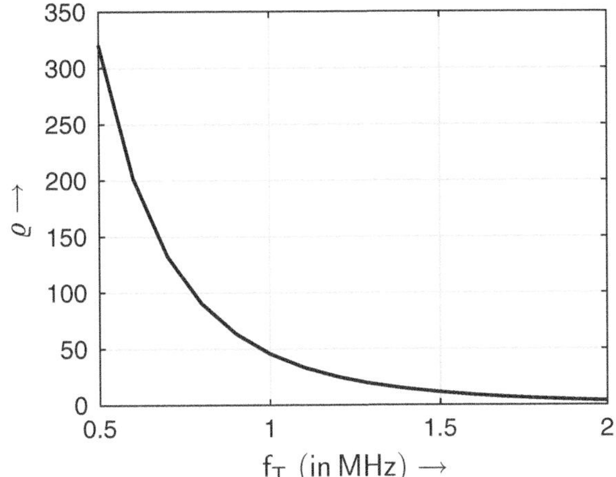

Fig. 4.55 Signal-to-noise ratio ϱ at two-level, bipolar transmission at the detection point with analog, linear equalization of the cable frequency response (parameters: $U_s = 1\,\mathrm{V}$, $\ell = 2.0\,\mathrm{km}$, $\Psi_0 = 5\cdot 10^{-10}\,\mathrm{V}^2/\mathrm{Hz}$)

The roll-off factor r of the root raised cosine filters chosen as transmit and receive filters also has an influence on the achievable transmission quality: The eye is fully open vertically regardless of the roll-off factor; however, the noise power at the detection point increases with increasing roll-off factor, as the bandwidth of the receive filter and thus the area under the curve $|G_e(f)/G_k(f)|^2$ increases with r: The signal-to-noise ratio ϱ is thus a function of the roll-off factor r. To achieve a maximum signal-to-noise ratio at the detection point, a small roll-off factor is advantageous. However, with the smallest possible roll-off factor $r = 0$ (ideal low-pass), reliable detection is no longer guaranteed even with minor deviations from the optimal sampling time [6]. Therefore, in practical implementation, a compromise roll-off factor $r > 0$ should be used, which on the one hand allows reliable detection even with unavoidable clock deviations in practice and on the other hand still causes a bandwidth as small as possible of the receive filter. Therefore, roll-off factors of $r = 0.3$ or $r = 0.5$ are often used, for example. Exemplary amplitude frequency responses of the equalizer receive filter $H_e(f)$ for the two roll-off factors $r = 0.3$ and $r = 0.5$ are shown in Fig. 4.56.

A decreasing signal-to-noise ratio ϱ depending on the symbol clock frequency f_T (see Fig. 4.55) leads directly to an increase in the error probability P_f or P_b due to the monotonic relationship between signal-to-noise ratio and error probability. This will be demonstrated using the bit error probability curves for two selected different symbol clock frequencies, $f_T = 0.6\,\mathrm{MHz}$ and $f_T = 1\,\mathrm{MHz}$.

The average symbol energy at the channel output necessary for the representation of the error probability P_b as a function of E_s/Ψ_0 can be calculated from the received power at the channel output P_e according to

$$E_s = P_e\,T_s = \frac{U_s^2 T_s^2}{3}\left(s^2 - 1\right)\int_{-\infty}^{+\infty} |G_s(f)\,G_k(f)|^2\,\mathrm{d}f. \tag{4.78}$$

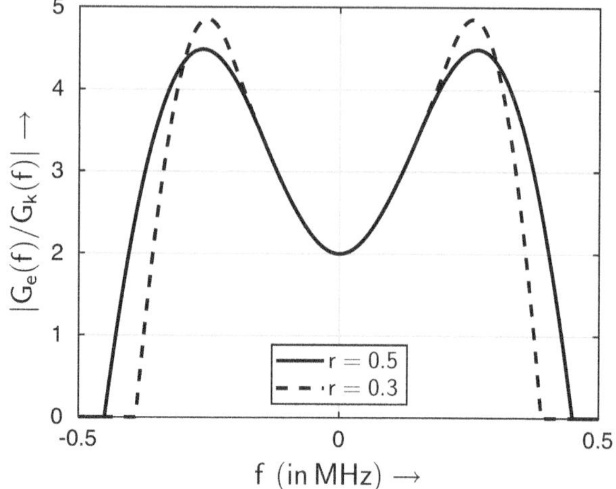

Fig. 4.56 Amplitude frequency response of the equalizer receive filter $H_e(f)$ at different roll-off factors r and linear, analog equalization of the cable frequency response (parameters: $f_T = 0.6\,\mathrm{MHz}$, $\ell = 2.0\,\mathrm{km}$)

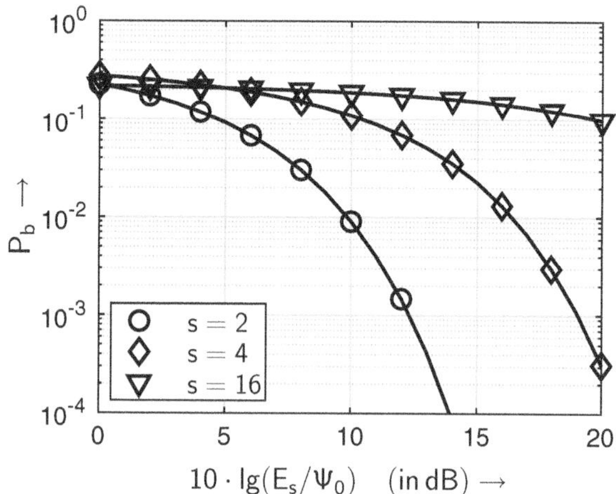

Fig. 4.57 Bit error probability in multi-level transmission and analog, linear equalization of the cable frequency response (Parameters: $\ell = 2.0\,\mathrm{km}$, $f_T = 0.6\,\mathrm{MHz}$, $r = 0.5$)

Figures 4.57 and 4.58 illustrate the resulting curves for the bit error probabilities P_b as a function of $10\lg(E_s/\Psi_0)$ for three different numbers s of signalling levels. It becomes particularly clear that the curves at the lower clock frequency ($f_T = 0.6\,\mathrm{MHz}$) are further to the left in the diagram than at the higher clock frequency ($f_T = 1\,\mathrm{MHz}$). This means that the system with the lower clock frequency allows a higher transmission quality, as a lower bit error probability P_b is achieved at the same E_s/Ψ_0 than at a higher clock frequency (see also Chap. 2).

At a lower data transmission speed, a lower error probability results due to the lower cable utilization and therefore less strong equalization, leading to a lower noise enhancement by the equalizer receive filter. This leads to a larger signal-to-noise ratio and results

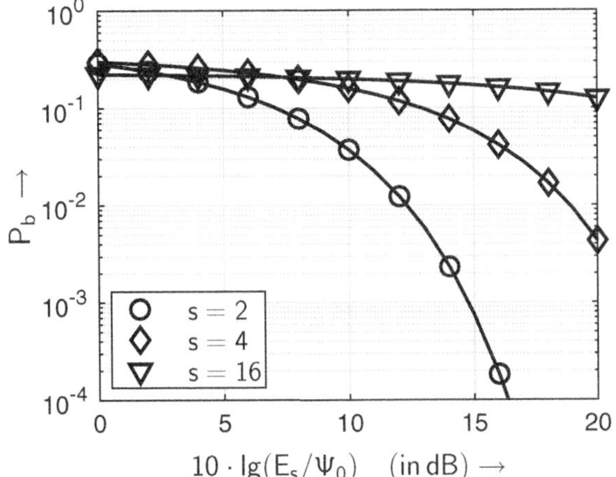

Fig. 4.58 Bit error probability in multi-level transmission and analog, linear equalization of the cable frequency response (Parameters: $\ell = 2.0\,\text{km}$, $f_\text{T} = 1.0\,\text{MHz}$, $r = 0.5$)

in a smaller error probability than at a higher data transmission speed or clock frequency f_T with then stronger noise enhancement. This relationship reflects one of the basic facts of communication engineering: data transmission speed and error probability are interchangeable. This means that an increase in data transmission speed, with all other parameters of the transmission system remaining unchanged, results in an increased error probability and vice versa, a lower error probability with also constant boundary conditions requires a reduction in data transmission speed.$\square$

Linear Discrete-Time Equalization

While previous sections considered analog equalizer structures to introduce basic principles and derive insights from them, now only discrete-time equalizers will be examined. An advantage of such equalizers is that they can be realized discretely and that only the symbol clock sample values need to be considered and processed—not the analog signal. This results in a reduction of complexity.

The goal of linear discrete-time equalization is to generate a—at least approximately—distortion-free overall transfer function

$$H_\text{ges}(z) = H(z) \cdot E(z) \; \bullet\!\!-\!\!\circ \; h_\text{ges}(k) \tag{4.79}$$

for the useful signal transmission (Fig. 4.59) using a time-discrete linear filter $E(z)$. In discrete-time equalization, the practical implementation can advantageously be done using digital signal processing.

Intersymbol interference can be fully compensated if the equalizer transfer function $E(z)$ is chosen to be

$$E(z) = \frac{1}{H(z)} \; . \tag{4.80}$$

In this case, the aim is to achieve the best possible compliance with the first Nyquist condition—without considering the influence of noise.

Example 4.2 (Minimum phase system with linear equalization). For the purpose of illustration, a minimum phase system with the transfer function

$$H(z) = \frac{1}{2} + \frac{1}{4} \cdot z^{-1} = \frac{2z+1}{4z} \tag{4.81}$$

is considered. The resulting pole-zero constellation (pole-zero constellation) in this case is shown in Fig. 4.60.

Minimum phase discrete-time systems have the property that their energy is concentrated in the first coefficients of their weighting function (see also Figs. 4.60 and (4.81)); poles and zeros are located within the unit circle. Figure 4.61 illustrates the PZ constellation of a maximum phase discrete-time system: Here, the zero is located outside the unit circle.

The inversion of $H(z)$ according to (4.80) leads with (4.81) to the result

$$E(z) = \frac{1}{H(z)} = \frac{4z}{2z+1}. \tag{4.82}$$

The pole-zero constellation of the time-discrete equalizer assuming a minimum phase overall impulse response $H(z)$ is shown in Fig. 4.62. Since the pole of the equalizer is not at the origin, the solution of the equalizer problem leads to an IIR filter (Infinite Impulse Response) and thus has an infinitely long impulse response. $\qquad\square$

Exact equalization requires equalizers with an infinitely long impulse response (IIR filter); for practical implementations, equalizers of finite length $m_e + 1$ with

$$E(z) = \sum_{k=0}^{m_e} e[k] \cdot z^{-k} \tag{4.83}$$

(FIR filter, Finite Impulse Response) are important, where in general a compensation of the channel influence is only approximately possible [8]. Figure 4.63 illustrates the resulting structure of the equalizer unit.

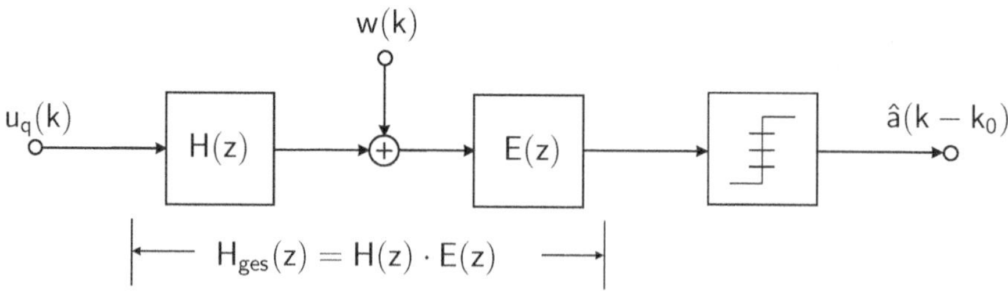

Fig. 4.59 Discrete-time transmission model with linear equalizer

Fig. 4.60 Pole-zero constellation of a minimum phase system $H(z) = \mathcal{Z}\{h(k)\}$

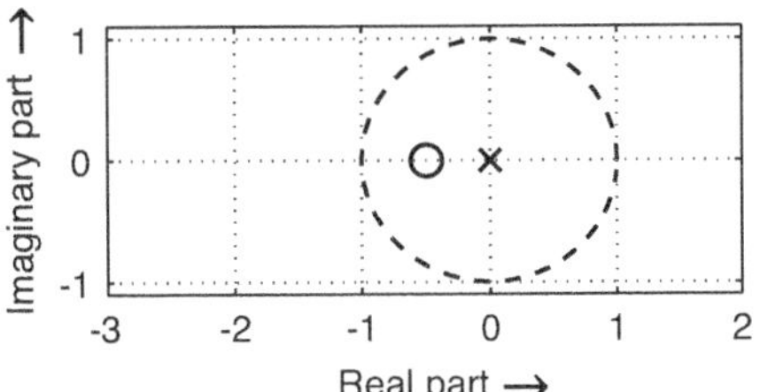

Fig. 4.61 Pole-zero constellation of a maximum phase system $H(z) = \mathcal{Z}\{h(k)\}$

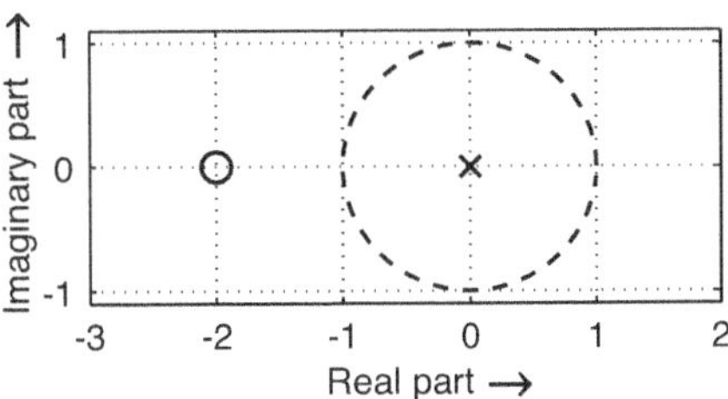

Fig. 4.62 Pole-zero constellation of the equalizer assuming a minimum phase overall impulse response

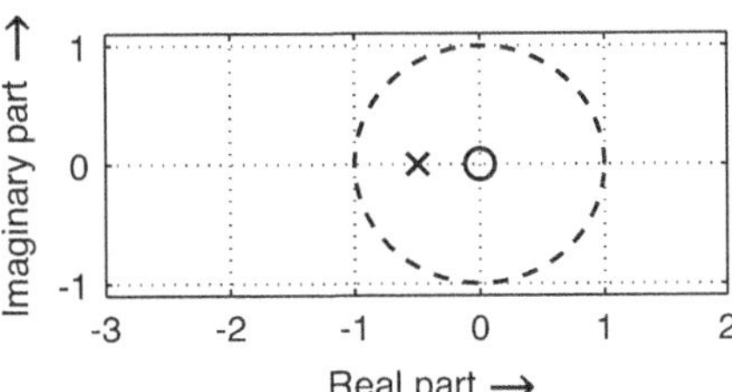

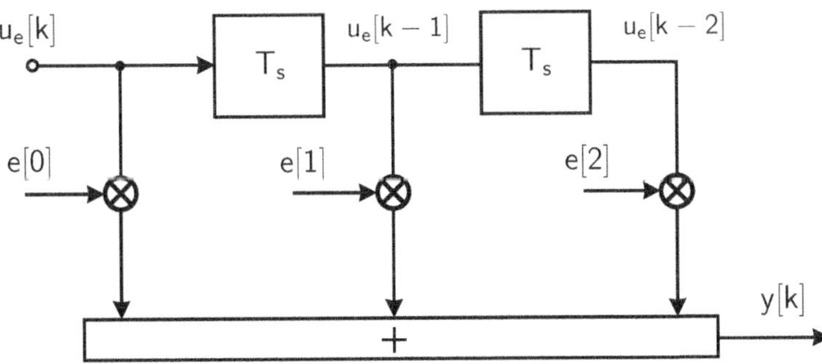

Fig. 4.63 Discrete-time equalizer structure $m_e = 2$

Through threshold decision, the symbol $\hat{a}[k - k_0]$ delayed by k_0 time steps is estimated. The (discrete) runtime k_0 represents an important degree of freedom for system optimization.

As an example, a linear single-sampling equalizer (also referred to as T-equalizer) is considered: Sample values are processed at the symbol duration T_{s}.[6] With the vector of the discrete-time total impulse response (of length $L + 1$)

$$h(k) = \boldsymbol{h} = [h_0, h_1, \ldots, h_L]^{\mathrm{T}},\tag{4.84}$$

the vector of the Nyquist condition

$$\boldsymbol{d} = [0, \ldots, 1, \ldots, 0]^{\mathrm{T}}\tag{4.85}$$

and the resulting channel matrix [26]

$$\boldsymbol{H} = \begin{bmatrix} h_0 & 0 & \cdots & 0 & 0 & 0 \\ h_1 & h_0 & 0 & \cdots & 0 & 0 \\ \vdots & \vdots & \vdots & \vdots & \vdots & \vdots \\ h_L & h_{L-1} & \cdots & h_0 & 0 & 0 \\ 0 & h_L & h_{L-1} & \cdots & h_0 & 0 \\ 0 & 0 & h_L & h_{L-1} & \cdots & h_0 \\ \vdots & \vdots & \vdots & \vdots & \vdots & \vdots \\ 0 & 0 & \cdots & 0 & \cdots & h_L \end{bmatrix}\tag{4.86}$$

the coefficient set for the equalizer is obtained as the vector

$$e = (\boldsymbol{H}^{\mathrm{T}}\boldsymbol{H})^{-1}\boldsymbol{H}^{\mathrm{T}}\boldsymbol{d}.\tag{4.87}$$

Such equalizer structures attempt to compensate for intersymbol interferences that have occurred on the transmission path by connecting an approximately inverse system. Since the equalizer design is independent of the noise signal present at the equalizer input, and an inversion of the channel frequency response is essentially used to determine the equalizer, this type of equalization typically amplifies the noise—or enhances it—similar to the analog linear equalizer. Improvements can be made by equalizer structures that incorporate both the useful and the interfering signal in the equalizer design or equalizers with decision feedback.

Since the symbol rate sampling of the signal at the receive filter output generally does not comply with the sampling theorem (e.g., [38, 48]), a part of the spectrum, which is periodically repeated with the symbol rate, is folded into the useful signal spectrum

---

[6] The term T-equalizer originates from the fact that the symbol duration is often denoted by T. The term *symbol rate sampling* means that a sample value is taken and processed once per symbol interval T.

before equalization, causing aliasing interference [8]. Symbol rate sampling equalizers can then only approximately equalize the received signal due to their principle, i.e., residual intersymbol interferences remain after equalization. The ideal equalizer has infinitely many coefficients.

Fractionally spaced equalizer (e.g., *fractional T/2*-equalizers), which operate at a higher sampling frequency than the symbol rate, avoid this effect [8, 26]; the sampling theorem is then complied with and exact equalization is enabled. After the equalizer, the sampling rate must be reduced to supply the symbol clock samples to the detection circuit.

The system of equations (4.87) for determining the equalizer coefficients is overdetermined and the solution yields the minimum *mean square error* (MSE) between the equalized total impulse response and the vector of the Nyquist condition, defined in (4.85).

A significant disadvantage of linear equalization is the noise enhancement or amplification associated with equalization, which is particularly evident in analog linear equalization, but is also effective in comparable ways in the discrete-time implementations of linear equalization methods. A discrete-time equalization with decision feedback can provide a solution here.

Example 4.3 (Linear time-discrete equalization for three different channel configurations). The three examples presented below are intended to illustrate the adjustment of the position $'1'$ in the vector of the Nyquist condition according to (4.85)—this is a degree of freedom in the design of discrete equalizers. For this purpose, three different channel configurations are considered.

Maximum-Phase Channel Hereafter, a maximum phase channel with the channel impulse response $\mathbf{h} = (0.707, 1.41)^{\mathrm{T}}$ is considered. The pole-zero characteristic resulting from this channel configuration is illustrated in Fig. 4.64.

In this case, the channel matrix results in

$$\mathbf{H} = \begin{pmatrix} 0.7 & 0 & 0 \\ 1.41 & 0.7 & 0 \\ 0 & 1.41 & 0.7 \\ 0 & 0 & 1.41 \end{pmatrix}. \tag{4.88}$$

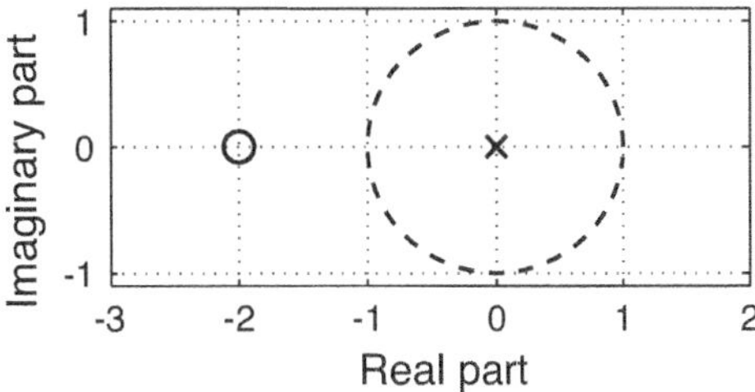

Fig. 4.64 Pole-zero characteristic for a maximum phase channel $H(z) = \mathcal{Z}\{h(k)\}$

With the approach for the equalizer design ($m_\mathrm{e} = 2$) in the following form

$$\mathbf{H} \cdot \mathbf{e} = \mathbf{d} + \boldsymbol{\varepsilon} \quad \rightarrow \quad \text{Solution:} \quad \mathbf{e} = (\mathbf{H}^\mathrm{T} \cdot \mathbf{H})^{-1} \cdot \mathbf{H}^\mathrm{T} \cdot \mathbf{d} \tag{4.89}$$

the equalizer coefficients can be determined. The remaining degree of freedom is the position of the '1' in the vector of the Nyquist condition (4.85), in order to achieve the best possible equalizer result. Table 4.4 illustrates the influence of the position of the '1' in the vector of the Nyquist condition (4.85) on the equalizer result. The error energy in the equalizer result, i.e., $f(k) = e(k) * h(k)$, is analyzed as a criterion for the quality of the equalizer result, which can be calculated as follows

$$E = \sum_{k=-\infty}^{+\infty} |f[k] - d[k - k_0]|^2. \tag{4.90}$$

In the case of perfect equalization, the vector $f(k)$ only has one non-zero coefficient with the amplitude '1', and the error energy in the equalizer vector takes the value zero. It should be noted at this point that this can only be realized for an infinite length of the equalizer impulse response $e(k)$ in the equalizer structure considered here. As a result, it is shown that in this case the '1' in the vector of the Nyquist condition (4.85) should be at the end of the vector to achieve minimal error energy. Furthermore, it is shown that an incorrect choice of the position of the '1' in the vector of the Nyquist condition leads to a strong increase in error energy and thus to poor quality of equalization.

Minimum-Phase Channel Deviating from the maximum-phase channel considered in the previous section, a minimum-phase channel with the characteristic $\mathbf{h} = (1.41, 0.707)^\mathrm{T}$ is now to be analyzed. The pole-zero characteristic resulting from this channel configuration is illustrated in Fig. 4.65.

In this case, the channel matrix results in

$$\mathbf{H} = \begin{pmatrix} 1.41 & 0 & 0 \\ 0.7 & 1.41 & 0 \\ 0 & 0.7 & 1.41 \\ 0 & 0 & 0.7 \end{pmatrix}. \tag{4.91}$$

Table 4.5 again illustrates the influence of the position of the '1' in the vector of the Nyquist condition (4.85) on the equalizer result. It turns out that in this case the '1' in the

Table 4.4 For the choice of the degree of freedom

$d(k)$	$e(k) * h(k)$	E (Error energy)
(1, 0, 0, 0)	(0.2484, 0.3769, −0.1890, 0.0947)	0.7516
(0, 1, 0, 0)	(0.3769, 0.8110, 0.0947, −0.0475)	0.1890
(0, 0, 1, 0)	(−0.1890, 0.0947, 0.9525, 0.0238)	0.0475
(0, 0, 0, 1)	(0.0947, −0.0475, 0.0238, 0.9881)	0.0119

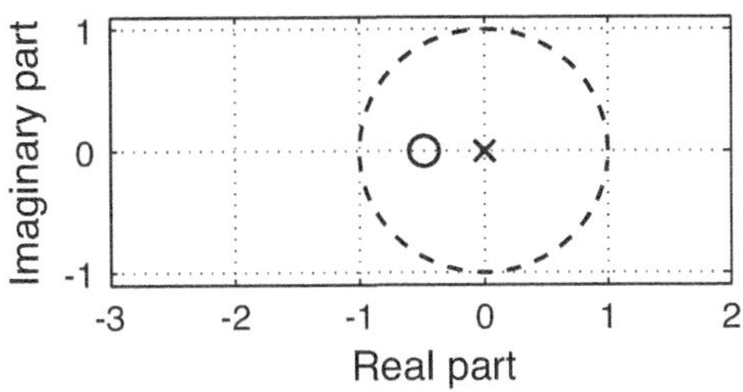

Fig. 4.65 Pole-zero characteristic for a minimum-phase channel $H(z) = \mathcal{Z}\{h(k)\}$

Table 4.5 On the choice of the degree of freedom

$d(k)$	$e(k) * h(k)$	E (Error energy)
(1, 0, 0, 0)	(0.9881, 0.0238, −0.0475, 0.0947)	0.0119
(0, 1, 0, 0)	(0.0238, 0.9525, 0.0947, −0.1890)	0.0475
(0, 0, 1, 0)	(−0.0475, 0.0947, 0.8110, 0.3769)	0.1890
(0, 0, 0, 1)	(0.0947, −0.1890, 0.3769, 0.2484)	0.7516

vector of the Nyquist condition should be at the beginning of the vector to obtain a minimum error energy.

If one wants to further reduce the error energy, the number of equalizer coefficients should be increased.

Unfavorable Channel In addition to the two analyzed channel profiles, which led to a satisfactory equalizer result when the ′1′ in the vector of the Nyquist condition (4.85) were suitably adjusted, there are also channel configurations that cannot be equalized. Such an unfavorable configuration would arise if the coefficients in the channel impulse response had the same amplitude (e.g., $\mathbf{h} = (1.0, 1.0)^{\mathrm{T}}$). The pole-zero characteristic resulting from this channel configuration is illustrated in Fig. 4.66.

For the channel matrix, we now obtain

$$\mathbf{H} = \begin{pmatrix} 1.0 & 0 & 0 \\ 1.0 & 1.0 & 0 \\ 0 & 1.0 & 1.0 \\ 0 & 0 & 1.0 \end{pmatrix}. \tag{4.92}$$

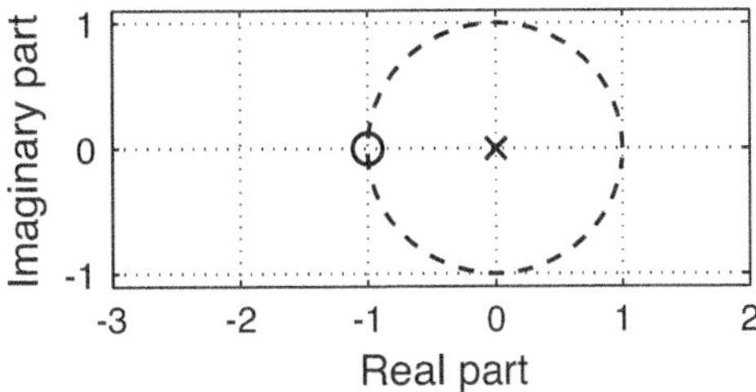

Fig. 4.66 Pole-zero characteristic for a channel difficult to equalize $H(z) = \mathcal{Z}\{h(k)\}$

Table 4.6 again illustrates the influence of the position of the $'1'$ in the vector of the Nyquist condition on the quality of the equalization. The result shows that adjusting the position of the $'1'$ in the vector of the Nyquist condition does not lead to a reduction in error energy. This means that the position of the $'1'$ can be chosen arbitrarily. There is no improvement in the result of the equalization by changing the position of the $'1'$. $\square$

Equalization with Decision Feedback

Equalization with Quantized Feedback In channels with strong intersymbol interference, the application of linear equalization methods leads to a significant increase in noise [64, 65]; this effect can be avoided by using decision feedback equalization (DFE). However, possibly incorrectly decided symbols now reduce the performance, as they affect the detection of subsequent symbols via the feedback and can thus lead to error propagation.

In the decision feedback equalization according to Fig. 4.67, the linear filtering with the forward filter $H_\mathrm{v}(z)$ provides an impulse response $h_\mathrm{ges}(k)$ that has no (or at least negligible) precursors before the detection value $k = k_0$. The main goal of the entire receive filtering in the forward branch in an equalization with decision feedback is—in addition to noise band limitation—to generate a resulting basic pulse $h_\mathrm{ges}(k)$ that only has pulse followers, since compensation of the pulse precursors by decision feedback is not possible for reasons of causality [63]. The followers of $h_\mathrm{ges}(k)$ are eliminated by the feedback of already decided symbols via a filter

$$H_\mathrm{r}(z) - 1 = \sum_{k=1}^{m_\mathrm{r}} h_\mathrm{r}[k] \cdot z^{-k} \tag{4.93}$$

assuming a correct decision of previous symbols (see Fig. 4.67).

Example 4.4 (Minimum-phase system and nonlinear equalization). For the purpose of illustration, a minimum-phase system with the transfer function

$$H(z) = \frac{\sqrt{2}}{2} + \frac{\sqrt{2}}{4} \cdot z^{-1} \tag{4.94}$$

Table 4.6 Choice of degree of freedom

$d(k)$	$e(k) * h(k)$	E (Error energy)
(1, 0, 0, 0)	(0.7500, 0.2500, −0.2500, 0.2500)	0.2500
(0, 1, 0, 0)	(0.2500, 0.7500, 0.2500, −0.2500)	0.2500
(0, 0, 1, 0)	(−0.2500, 0.2500, 0.7500, 0.2500)	0.2500
(0, 0, 0, 1)	(0.2500, −0.2500, 0.2500, 0.7500)	0.2500

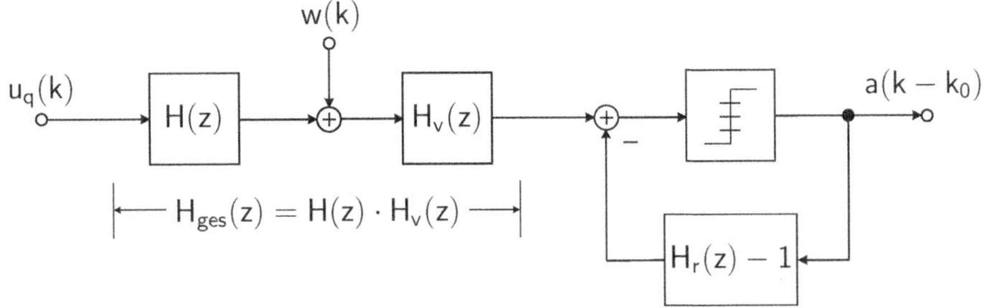

Fig. 4.67 Model of a transmission system with decision feedback

is considered. Since the total impulse response present at the input of the equalizer is minimum-phase, the solution for $H_\mathrm{v}(z)$ results in

$$H_\mathrm{v}(z) = \frac{1}{h[0]} = \frac{2}{\sqrt{2}}. \tag{4.95}$$

In this case, the forward filter $H_\mathrm{v}(z)$ only has to amplify (normalize) the received signal so that $h_\mathrm{ges}[0] = 1$ becomes (normalization of the impulse response $h_\mathrm{ges}(k) = h(k) * h_\mathrm{v}(k)$).

The feedback filter

$$H_\mathrm{r}(z) - 1 = H(z) \cdot H_\mathrm{v}(z) - 1 = \frac{H(z)}{h[0]} - 1 = \frac{1}{2}\, z^{-1} \tag{4.96}$$

then has the task of eliminating the postcursors of $h_\mathrm{ges}(k)$ and can be determined via the approach

$$H_\mathrm{r}(z) - 1 = H_\mathrm{ges}(z) - 1 \tag{4.97}$$

(Fig. 4.68).   □

A disadvantage of an equalizer strategy with quantized feedback is a possible error propagation, which is caused by the nonlinear part of the equalizer network. The error propagation will significantly reduce the performance efficiency, especially at low or medium signal-to-noise ratios. Another problem arises when the DFE is to be combined with channel coding. If the threshold decision is replaced by a channel decoder, it must immediately make preliminary decisions about the transmitted symbols for the feedback branch, which will generally lead to conflicts with the principle of channel coding.

Tomlinson-Harashima Precoding As a remedy in cases where the application of an equalizer with quantized feedback does not lead to a usable performance of the transmission system a Tomlinson-Harashima precoding (THP) can be applied [23]. This can

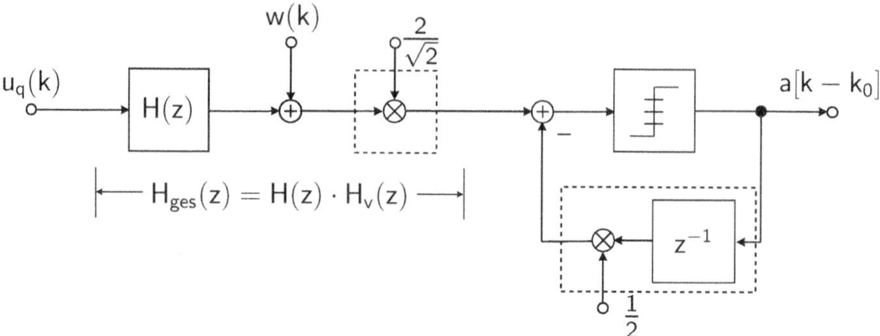

Fig. 4.68 Structure of a transmission system with decision feedback assuming channel characteristics according to (4.94)

be useful, for example, in systems with quantized feedback at low input signal-to-noise ratios, where the performance of the transmission system is dominated by error propagation resulting from the feedback.

The basic idea is to move the feedback filter to the transmitter, which leaves the overall transmission behavior unchanged if an equalization with decision feedback without decision errors in the feedback branch is present beforehand. Figure 4.69 illustrates the resulting structure of the transmission system. Practically, by moving the feedback filter to the transmitter, error propagation can be avoided at the expense of an expected increase in transmit power.

It is therefore necessary that the increase in transmit power can be suitably limited. This problem was solved by introducing a modulo operation, which leads to a limitation of the transmission amplitudes and largely leaves the transmission spectrum unchanged [23]. This method is referred to as *Tomlinson-Harashima precoding*. Figure 4.70 shows an exemplary structure of such a transmission system.

The principle of precoding is that the precoder suitably limits the increase in transmit power by introducing a modulo operation. However, a transmission with optimal Tomlinson-Harashima precoding requires the transmission of channel state information to the transmitting side, which requires a return channel. A fixed precoder can provide

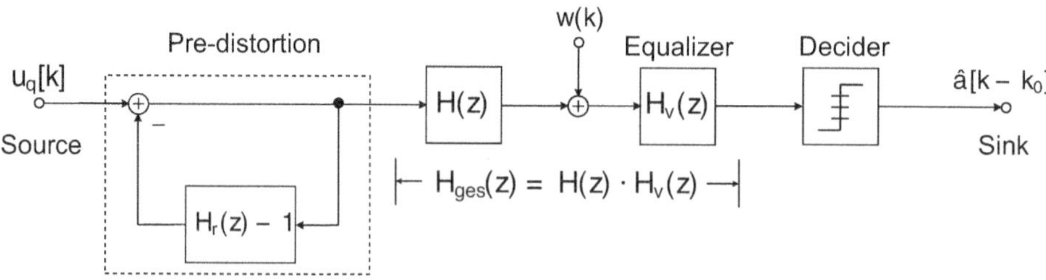

Fig. 4.69 Model of a transmission system with transmitter-side pre-distortion

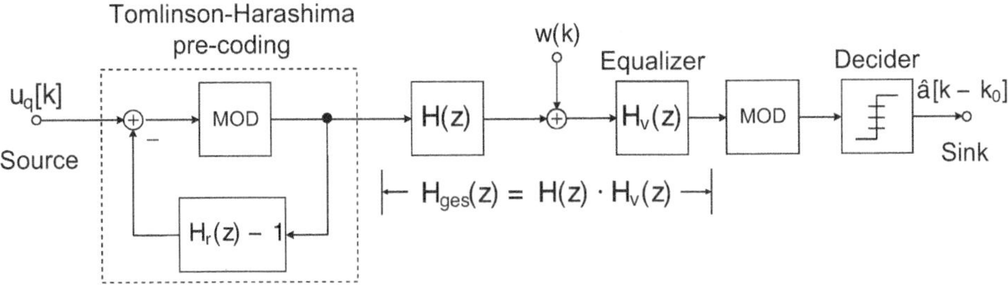

Fig. 4.70 Model of a transmission system with Tomlinson-Harashima precoding

a solution here, which was designed for a fixed reference application, i.e., for a typical transmission channel, and whose parameters are close to those of the actual application or the actual transmission channel.

Such an approach could be of interest, for example, for wired transmission methods, where the properties of the transmission medium change only slowly over time (essentially due to temperature changes or signs of aging).

The advantage of Tomlinson-Harashima precoding is that the quantized feedback can be implemented without error propagation. The transmission spectrum remains largely unchanged despite the transmitter-side preprocessing and the amplitude is limited, which is significant for the realization of transmission units. A disadvantage of this method is that the resulting overall weighting function $h(k)$ must be known to the transmitter, which may require a reverse channel and thus represents an additional implementation effort.

4.6.6   Discussion and Conclusion on Equalization

The performance of the considered approaches and methods for equalization in the transmission of digital signals over symmetric two-wire lines can be determined based on the estimation of the respective bit error probability through a discrete-time numerical simulation [24, 64]. The transmission is simulated and the number of determined bit errors is set in relation to the total number of transmitted bits, resulting in a relative average bit error frequency, which is used as an estimate for the bit error probability when a sufficiently large number of bits are transmitted in the simulation. In this way, the various discussed equalization methods can be compared with each other.

For numerical results, a bipolar, two-level baseband transmission with a symbol clock frequency of $f_T = 1/T_s = 1\,\text{MHz}$ and root raised cosine transmit and receive filters, each with the roll-off factor $r = 0.5$, over a symmetric copper line with the transfer function according to (3.157) was assumed (length $l = 2\,\text{km}$, cable characteristic frequency $f_0 = 0.178\,\text{MHz} \cdot \text{km}^2$). The signal was normalized so that the average symbol energy E_s of the useful signal $u_k(t)$ at the receiver input is the same in each of the considered cases (see Fig. 4.71). The symbol clock frequency was set—in relation to the properties

Fig. 4.71 On the definition of the ratio E_s/Ψ_0 of average symbol energy to noise power spectral density at the receiver input

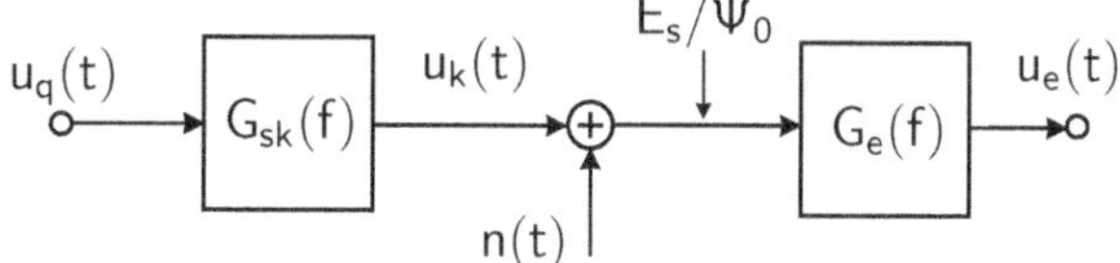

of the transmission channel—as indicated, so that in particular the Viterbi detector can be implemented with reasonable effort: Here with a memory over four symbols. Additionally, the chosen number of signalling levels ($s = 2$) of this example system limits the number of possible states that are to be processed by the Viterbi detector: This would be increased with multistage, see e.g. [42]. In Fig. 4.72, the estimated bit error probabilities are shown depending on the ratio E_s/Ψ_0 of average symbol energy at the receiver input to noise power spectral density. The estimated bit error probabilities for transmission over a distortion-free channel with interference by additive white Gaussian noise and for transmission over the given exemplary channel without equalization are shown as limit cases. The linear equalization by the inverse cable transfer function (analog equalization) and the discrete-time linear equalization with simple sampling (*T*-equalization) yield very similar performances for the chosen example system. With the help of decision feedback equalization (DFE), an improvement in transmission quality can be achieved compared to the linear methods under otherwise identical boundary conditions. The sequence estimation by Viterbi detection is the most powerful method among the considered variants.

From a practical point of view, very useful results can be achieved with linear equalization methods in terms of the achievable transmission quality. Improvements can be achieved, in particular, by non-linear equalization methods, e.g. with decision feedback. The sequence estimation using Viterbi detection is the most powerful, but often requires

Fig. 4.72 Bit error probabilities for some equalization methods depending on the ratio of average symbol energy to noise power spectral density at the receiver input for exemplary cable transmission

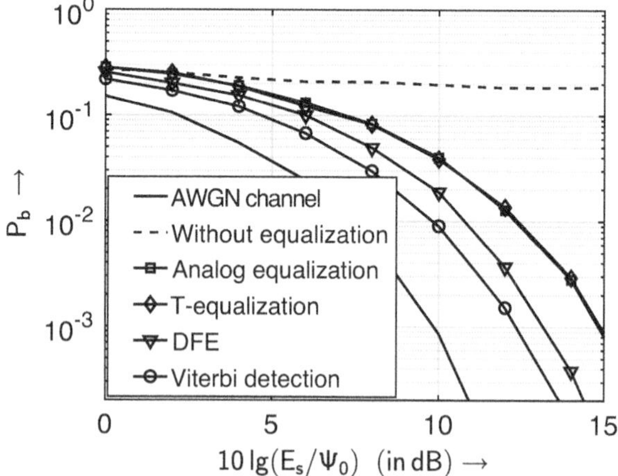

a high implementation effort in wired transmission, as the weighting function of a copper line can extend over a large number of symbol intervals at high-rate transmission.

Detailed presentations, especially on the discrete-time implemented equalization methods, can be found e.g. in [14, 23, 26]

4.7 Summary and Bibliographic Notes

The aim of this chapter was to demonstrate principles in the transmission of digital signals over linearly distorting channels. Using the example of the copper cable transmission channel, the evaluation of the useful signal and the disturbance was extensively presented and supported with numerical examples. It became clear how transmission systems can be analyzed and optimized using the tools of signal and system theory. An application example with rational transfer functions was thoroughly examined and all parameters were derived and analyzed: This consideration served to develop and consolidate understanding of basic communication technology relationships. To round off these considerations, important filter functions were considered in practice and examined for multi-level baseband transmission when equalizing the cable influence. The results achieved in the numerical examples were compared and it became apparent which filter functions provide advantages in terms of the quality of the transmission.

In a subsequent section, the consideration of equalization was expanded and a classification of equalizer structures was made. In particular, some important discrete-time equalizer structures were specified, which can be practically implemented advantageously with the help of digital signal processing.

The basics of baseband transmission are presented in many textbooks on information, communication, or signal transmission, e.g. in [26, 42, 48, 55, 72] (in German). The focus is on transmission over AWGN channels. Equalization methods are also discussed in these textbooks. Applications are often presented in the field of radio transmission. The connection with transmission over copper cable is often only hinted at and treated quite briefly: This topic was therefore described in detail in this chapter in order to work out and clarify important relationships in the transmission and evaluation of the useful signal and the disturbance over an exemplary—but typical and widely used—transmission channel. Representations of this in English-language technical literature can be found e.g. in [8, 9, 52, 62, 70].

4.8 Problems

Problem 4.1 (Output signal of a cable)

Calculate the output signal at the end of a cable, assuming a rectangle impulse of amplitude U_0 and duration T_s as the transmit signal and the cable's transfer function is approximated with the first two poles of the rational approximation (3.158)!

Problem 4.2 (Baseband transmission over a cable with NRZ pulses)

A $\ell = 2\,\text{km}$ long copper cable with the characteristic frequency $f_0 = 0.178\,\text{MHz} \cdot \text{km}^2$ is used for data transmission with a random, two-level, and unipolar NRZ pulse sequence of amplitude $U_0 = 5\,\text{V}$ without equalization. The cable transfer function should be approximated by the first pole of the rational approximation. The noise disturbances are negligible, i.e., very good transmission conditions exist on the cable.

a) At which bit sequence frequency (bit rate) $f_B = 1/T_s$ does the eye at the cable output just close?
b) What bit sequence frequency (bit rate) can be achieved if a vertical eye opening of 30 % or 50 % is required under real operational circumstances?

Problem 4.3 (Baseband transmission with NRZ pulses over linearly equalized cable and second-order low-pass receive filter)

The transmitter of a baseband transmission system sends a unipolar, rectangular NRZ binary signal $u_0(t)$ with the amplitude U_0 and the pulse width T_s. The useful signal is transmitted over a cable of length $\ell = 2\,\text{km}$, which is modeled sufficiently accurately by the first pole of the cable transfer function (3.157) at the frequency $f_k = f_1$ (cable characteristic frequency: $f_0 = 0.178\,\text{MHz} \cdot \text{km}^2$). On the transmission path, the useful signal is disturbed by white, Gaussian noise with the constant noise power spectral density Ψ_0 (in V^2/Hz). In the receive, the received signal is completely linearly equalized by the inverse cable transfer function. The receive filter is a second-order RC low-pass filter (a double-real pole at the cutoff frequency ω_g or f_g).

To be determined are:

a) the basic receive impulse $x_e(t)$ at the detection point and the eye diagram resulting from the function of the received useful signal $u_e(t)$,
b) the half vertical eye opening U_A with redundancy-free coding as a function of the receive filter cutoff frequency f_g,
c) the function of the effective square value U_R^2 of the noise at the detection point as a function of the receive filter cutoff frequency f_g,
d) the signal-to-noise ratio ϱ at the detection point, also as a function of the receive filter cutoff frequency f_g,
e) the signal-to-noise ratio ϱ at the optimal receive filter cutoff frequency $f_{g\,\text{opt}}$ with the given values $U_0 = 5\,\text{V}$, $T_s = 1\,\mu\text{s}$ and $\Psi_0 = 10^{-9}\,\text{V}^2/\text{Hz}$ as well as
f) the bit error probability P_b.

4.9   Solutions of the Problems

Solution of Problem 4.1

The output signal at the end of a cable can advantageously be calculated via the image or spectral domain using the Laplace transformation. For this, the Laplace transform of the transmit signal and the transfer function of the cable must be known.

The Laplace transform of a rectangular pulse of amplitude U_0 and duration T_s as a transmit signal is according to (4.11)

$$X_s(p) = \frac{U_0}{p} - \frac{U_0\,e^{-pT_s}}{p} = \frac{U_0}{p}\left(1 - e^{-pT_s}\right). \tag{4.98}$$

The cable transfer function according to (3.158) considering two cable poles is given as

$$G_k(p) = \frac{1}{2}\frac{1}{(1+pT_1)(1+pT_2)} \quad \text{with} \quad T_1 = 9\,T_2. \tag{4.99}$$

A transformation with regard to the inverse Laplace transformation results in

$$G_k(p) = \frac{1}{2}\frac{1}{T_1\left(p + \frac{1}{T_1}\right)T_2\left(p + \frac{1}{T_2}\right)} = \frac{1}{2\,T_1\,T_2}\cdot\frac{1}{\left(p + \frac{1}{T_1}\right)\left(p + \frac{1}{T_2}\right)}. \tag{4.100}$$

Taking into account the coupling of the cable poles according to (3.158) applies $T_1 = 3^2\,T_2 = 9\,T_2$ and it further results in

$$G_k(p) = \frac{9}{2\,T_1^2}\cdot\frac{1}{\left(p + \frac{1}{T_1}\right)\left(p + \frac{9}{T_1}\right)}. \tag{4.101}$$

Now, using the inverse Laplace transformation

$$x_k(t) = U_0 \cdot \mathscr{L}^{-1}\left\{G_k(p)\cdot\frac{1 - e^{-pT_s}}{p}\right\}$$

$$x_k(t) = \frac{U_0}{2}\frac{9}{T_1^2}\cdot\mathscr{L}^{-1}\left\{\frac{1}{p\left(p + \frac{1}{T_1}\right)\left(p + \frac{9}{T_1}\right)}\cdot\left(1 - e^{-pT_s}\right)\right\}$$

with the help of the correspondence (see Appendix D or e.g. [11, 13, 74])

$$\frac{1}{p(p+a)(p+b)} \quad\bullet\!\!-\!\!\circ\quad \frac{1}{ab(a-b)}\left((a-b) + b\,e^{-at} - a\,e^{-bt}\right) \tag{4.102}$$

and the shift theorem of the Laplace transformation, the time function of the received pulse is calculated to[7]

$$
x_k(t) = \frac{U_0}{2} \frac{9}{T_1^2} \frac{T_1^2}{9} \frac{T_1}{8} \left[\left(\frac{8}{T_1} + \frac{1}{T_1} e^{-9\frac{t}{T_1}} - \frac{9}{T_1} e^{-\frac{t}{T_1}} \right) \cdot 1(t) - \ldots \right.
$$
$$
\left. \ldots - \left(\frac{8}{T_1} + \frac{1}{T_1} e^{-9\frac{t-T_s}{T_1}} - \frac{9}{T_1} e^{-\frac{t-T_s}{T_1}} \right) \cdot 1(t - T_s) \right].
$$

calculated. After simplifying, the result is

$$
x_k(t) = \frac{U_0}{2} \left[\left(1 + \frac{1}{8} e^{-9\frac{t}{T_1}} - \frac{9}{8} e^{-\frac{t}{T_1}} \right) \cdot 1(t) - \ldots \right.
$$
$$
\left. \ldots - \left(1 + \frac{1}{8} e^{-9\frac{t-T_s}{T_1}} - \frac{9}{8} e^{-\frac{t-T_s}{T_1}} \right) \cdot 1(t - T_s) \right]. \tag{4.103}
$$

In Fig. 4.73, the determined rectangular response of the cable for various ratios of symbol duration T_s to the time constant of the first cable pole T_1 is shown: If the symbol duration T_s becomes smaller in relation to the time constant T_1 of the first cable pole, the pulse $x_k(t)$ at the cable output is more strongly deformed and intersymbol interference occurs at the output of the cable. The maximum amplitude of the pulse at the cable output is $U_0/2$: This is due to the rational cable approximation (3.158) and can technically be interpreted as being caused by matching the receiver input impedance to the characteristic impedance of the cable.

Solution of Problem 4.2

The Laplace transform of a rectangular pulse with amplitude U_0 and duration T_s as a transmit signal is again

$$
X_s(p) = \frac{U_0}{p} - \frac{U_0\, e^{-pT_s}}{p} = \frac{U_0}{p} \left(1 - e^{-pT_s} \right). \tag{4.104}
$$

according to (4.11). The cable transfer function, considering the first cable pole, is

$$
G_k(p) = \frac{1}{2} \frac{1}{(1 + pT_1)} \quad \text{with} \quad T_1 = \frac{2\,\ell^2}{\pi^3 f_0} \tag{4.105}
$$

according to (3.158) or after transformation in preparation for the inverse Laplace transformation

$$
G_k(p) = \frac{1}{2} \frac{1}{T_1} \frac{1}{\left(p + \frac{1}{T_1} \right)}. \tag{4.106}
$$

---

[7]For the practical execution of the inverse transformation, $a = 9/T_1$ and $b = 1/T_1$ are advantageously chosen.

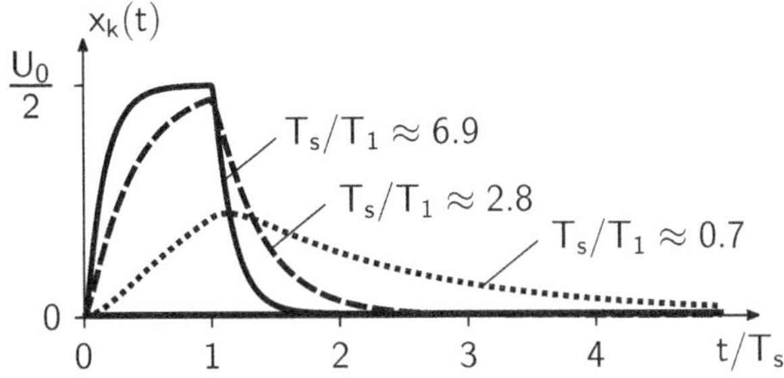

Fig. 4.73 Rectangular response of the cable at various ratios of symbol duration T_s to the time constant of the first cable pole T_1

Thus, the pulse present at the output of the cable can be written as

$$x_k(t) = U_0 \cdot \mathcal{L}^{-1}\left\{ G_k(p) \cdot \frac{1 - e^{-pT_s}}{p} \right\}$$

$$x_k(t) = \frac{U_0}{2}\frac{1}{T_1} \cdot \mathcal{L}^{-1}\left\{ \frac{1}{p\left(p + \frac{1}{T_1}\right)} \cdot \left(1 - e^{-pT_s}\right) \right\}.$$

Using the correspondence (see Appendix D or e.g. [11, 13, 74])

$$\frac{1}{p(p+a)} \quad \bullet\!\!-\!\!\circ \quad \frac{1}{a}\left(1 - e^{-at}\right) \tag{4.107}$$

and the shift theorem of the Laplace transformation, the time function of the received pulse results in

$$x_k(t) = \frac{U_0}{2}\frac{1}{T_1} \cdot T_1\left[\left(1 - e^{-\frac{t}{T_1}}\right) \cdot 1(t) - \left(1 - e^{-\frac{t-T_s}{T_1}}\right) \cdot 1(t - T_s)\right]$$

$$x_k(t) = \frac{U_0}{2}\left[\left(1 - e^{-\frac{t}{T_1}}\right) \cdot 1(t) - \left(1 - e^{-\frac{t-T_s}{T_1}}\right) \cdot 1(t - T_s)\right].$$

The maximum pulse amplitude at the cable output is $U_0/2$: This is due to the rational cable approximation (3.158) and can technically be interpreted as being caused by matching the receiver input impedance to the characteristic impedance of the cable.

The half vertical eye opening at the optimal sampling time $T_A = T_s$ is, using (2.34)—now with the maximum achievable amplitude $U_0/2$, the time constant T_1 of the first cable pole and the decision threshold at $U_0/4$—

$$U_A = \frac{U_0}{2}\left(1 - e^{-\frac{T_s}{T_1}}\right) - \frac{U_0}{4} = \frac{U_0}{2}\left(\frac{1}{2} - e^{-\frac{T_s}{T_1}}\right). \tag{4.108}$$

a) The bit rate $f_B = 1/T_s$ is sought, at which the eye just closes, i.e. $U_A = 0$ applies and one obtains

$$0 = \frac{U_0}{2}\left(\frac{1}{2} - e^{-\frac{T_s}{T_1}}\right)$$

$$0 = \frac{1}{2} - e^{-\frac{T_s}{T_1}}$$

$$e^{-\frac{T_s}{T_1}} = \frac{1}{2} \qquad\qquad \Big|\ \ln(\ldots)$$

$$-\frac{T_s}{T_1} = \ln\left(\frac{1}{2}\right)$$

$$T_s = -T_1 \cdot \ln 0.5.$$

With the given numerical values, one obtains $T_1 = 1.45\,\mu s$ and thus $T_s = 1.005\,\mu s$. The maximum bit sequence frequency at which the eye vertically just closes is in this case

$$f_B = \frac{1}{T_s} = \frac{1}{1.005 \cdot 10^{-6}\,s} = 0.994\,\text{MHz.} \tag{4.109}$$

b) In real operation, the eye must be at least partially open for reliable detection. The maximum *half* vertical eye opening is $U_0/4$.

If the eye is to be vertically open to $30\,\%$, at least $0.3 \cdot U_0/4$ must be available. The calculation is then

$$0.3 \cdot \frac{U_0}{4} = \frac{U_0}{2}\left(\frac{1}{2} - e^{-\frac{T_s}{T_1}}\right)$$

$$\frac{0.3}{2} = \frac{1}{2} - e^{-\frac{T_s}{T_1}}$$

$$e^{-\frac{T_s}{T_1}} = \frac{1}{2} - \frac{0.3}{2} = \frac{0.7}{2} \qquad\qquad \Big|\ \ln(\ldots)$$

$$-\frac{T_s}{T_1} = \ln\left(\frac{0.7}{2}\right)$$

$$T_s = -T_1 \cdot \ln 0.35.$$

and one obtains for the achievable bit sequence frequency

$$f_B = \frac{1}{T_s} = \frac{1}{1.522 \cdot 10^{-6}\,s} = 0.657\,\text{MHz.} \tag{4.110}$$

If a vertical eye opening of 50 % is required, at least $0.5 \cdot U_0/4$ must be available. The calculation proceeds analogously to before

$$0.5 \cdot \frac{U_0}{4} = \frac{U_0}{2} \left(\frac{1}{2} - e^{-\frac{T_s}{T_1}} \right)$$

$$\frac{0.5}{2} = \frac{1}{2} - e^{-\frac{T_s}{T_1}}$$

$$e^{-\frac{T_s}{T_1}} = \frac{1}{2} - \frac{0.5}{2} = \frac{0.5}{2} \qquad \Big| \quad \ln(\ldots)$$

$$-\frac{T_s}{T_1} = \ln\left(\frac{0.5}{2}\right)$$

$$T_s = -T_1 \cdot \ln 0.25.$$

and one obtains for the achievable bit sequence frequency (bit rate)

$$f_B = \frac{1}{T_s} = \frac{1}{2.009 \cdot 10^{-6}\,\text{s}} = 0.497\,\text{MHz}. \tag{4.111}$$

It can be seen that a larger vertical eye opening can be achieved by reducing the bit sequence frequency f_B: With a lower bit sequence frequency f_B, the transmit signal is less strongly deformed and a larger vertical eye opening remains for signal detection.

Solution of Problem 4.3 This task can be advantageously solved using the extensively derived relationships and results for baseband transmission of NRZ rectangular pulses over a linearly equalized cable with receive filtering with RC low-pass filter of second order in Chapter 4 and especially in reference to the section *Numerical Example: Baseband Transmission with a Cable with One Pole*.

a) The time function $x_e(t)$ of the received basic pulse at the detection point results according to the extensively derived result (4.16) to

$$x_e(t) = U_0 \left[\left(1 - e^{-\frac{t}{T_g}} - \frac{t}{T_g} e^{-\frac{t}{T_g}} \right) \cdot 1(t) - \left(1 - e^{-\frac{t-T_s}{T_g}} - \frac{t - T_s}{T_g} e^{-\frac{t-T_s}{T_g}} \right) \cdot 1(t - T_s) \right]. \tag{4.112}$$

It is shown in Fig. 4.13. An example of an eye diagram of the received useful signal $u_e(t)$ is shown in Fig. 4.14.

b) The half vertical eye opening is obtained according to (4.28) or (4.29), there with derivation, as a function of the receive filter cutoff frequency f_g (or the corresponding time constant $T_g = 1/(2\pi f_g)$) to

$$U_A = U_0 \left[e^{-\frac{T_A}{T_g}} \left(e^{\frac{T_s}{T_g}} - 1 \right) + e^{-\frac{T_A}{T_g}} \left(\frac{T_A}{T_g} \left(e^{\frac{T_s}{T_g}} - 1 \right) - \frac{T_s}{T_g} e^{\frac{T_s}{T_g}} \right) - \frac{1}{2} \right]. \tag{4.113}$$

The optimal sampling time is according to (4.25) or (4.26)

$$T_A = T_s \frac{e^{\frac{T_s}{T_g}}}{e^{\frac{T_s}{T_g}} - 1} = T_s \frac{e^{2\pi f_g T_s}}{e^{2\pi f_g T_s} - 1}. \tag{4.114}$$

c) The function of the noise effective square value (noise power) U_R^2 at the detection point as a function of the receive filter cutoff frequency f_g is according to (4.45), there with derivation,

$$U_R^2 = 2\pi \ \Psi_0 \left(f_g + \frac{f_g^3}{f_k^2} \right). \tag{4.115}$$

In addition to the receive filter cutoff frequency f_g, the frequency f_k of the cable pole considered in the equalization also influences the noise power.

d) For the signal-to-noise ratio ϱ at the detection point, also as a function of the receive filter cutoff frequency f_g (or the corresponding time constant $T_g = 1/(2\pi f_g)$), one obtains according to (2.60) with the results for the half vertical eye opening U_A and the noise power U_R^2

$$\varrho = \frac{U_A^2}{U_R^2} = \frac{U_0^2 \left[e^{-\frac{T_A}{T_g}} \left(e^{\frac{T_s}{T_g}} - 1 \right) + e^{-\frac{T_A}{T_g}} \left(\frac{T_A}{T_g} \left(e^{\frac{T_s}{T_g}} - 1 \right) - \frac{T_s}{T_g} e^{\frac{T_s}{T_g}} \right) - \frac{1}{2} \right]^2}{2\pi \ \Psi_0 \left(f_g + \frac{f_g^3}{f_k^2} \right)}, \tag{4.116}$$

where practically usable results are only to be expected with a vertically open eye. The signal-to-noise ratio ϱ is shown for the given numerical values in Fig. 4.74 as a function of the receive filter cutoff frequency f_g.

With the given numerical values, the numerical results are calculated below.

Optimal receive filter cutoff frequency (via representation of the signal-to-noise ratio ϱ as a function of f_g and reading off the optimal cutoff frequency $f_{g\,\mathrm{opt}}$ where the signal-to-noise ratio is maximum, see Fig. 4.74): $f_{g\,\mathrm{opt}} = 453\,\mathrm{kHz}$.

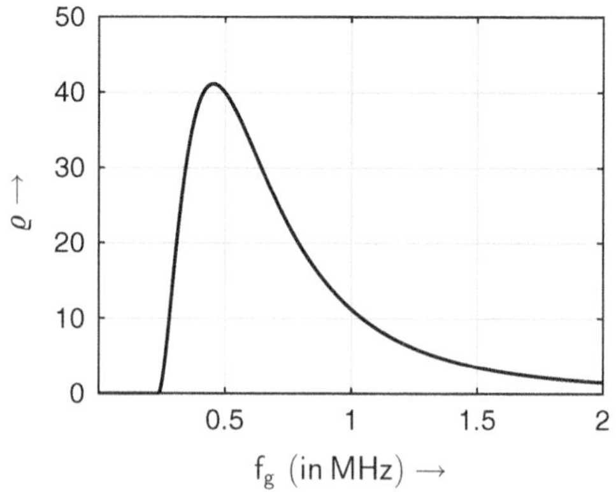

Fig. 4.74 Signal-to-noise ratio ϱ as a function of the receive filter cutoff frequency f_g with the values given in the task

Frequency of the first cable pole considered in the equalizer according to (3.157): $f_k = 109.8\,\text{kHz}$

If one sets as the receive filter cutoff frequency f_g the optimal cutoff frequency $f_{g\,\text{opt}} = 453\,\text{kHz}$ and as the frequency of the cable pole considered in the equalizer $f_k = 109.8\,\text{kHz}$ in the obtained equations, one obtains the further numerical results:

$U_A = 1.4519\,\text{V}, U_R^2 = 0.0513\,\text{V}^2, \varrho = 41.0983 \mathrel{\hat{=}} 16.1378\,\text{dB}$

e) The error probability for binary transmission is obtained according to (2.80), there with derivation, to

$$P_f = \frac{1}{2}\left[1 - \text{erf}\left(\sqrt{\frac{\varrho}{2}}\right)\right] = \frac{1}{2}\,\text{erfc}\left(\sqrt{\frac{\varrho}{2}}\right). \tag{4.117}$$

In binary or two-valued transmission, this probability P_f that a received symbol is decided incorrectly with respect to the decision threshold, coincides with the bit error probability P_b.

As a numerical result, one obtains (e.g., when using a computer algebra or spread-sheet program) $P_f = 7.2549 \cdot 10^{-11}$.

If one uses the table of the erfc function in Appendix E, the argument $x = \sqrt{\varrho/2} = 4.5329$ must first be calculated. As a (rough) value, one obtains by reading off for $x = 4.5 \longrightarrow \text{erfc}(x) = 1.97 \cdot 10^{-10}$ and finally by division by 2 the approximate value $P_f \approx 9.85 \cdot 10^{-11}$ for the bit error probability. An improvement in accuracy when using the table of the erfc function can be achieved if the more accurate tables available in reference works, worked out to several places after the decimal point, are used.

Summary and Outlook 5

5.1 Summary

In the information society, communication networks represent an indispensable infrastructure on which various processes in people's lives are based. In the current age, where information exchange and communication play an important and increasingly significant role, extensive amounts of data or information must be transmitted over these networks—and thus over the transmission channels they provide. The information can already be in digital form, or it may originally be of an analog nature—e.g., voice, music, as well as images and videos—and is then converted into a digital form, as further processing, storage, and transmission are predominantly efficient in digital form. Therefore, the transmission of digital signals is of paramount importance. The task in designing a communication system is then to process a digital signal—which may have originated from an analog signal through digitization—for example, a sequence of bits, in such a way that it can be transmitted as a continuous signal with the highest possible or required transmission rate, the highest possible transmission quality, and the lowest possible energy demand over a given transmission channel. The transmission medium causes the transmitted signal to be influenced in its form and properties by a multitude of disturbance effects acting on this signal during transmission. Despite the distortions and disturbances affecting the signal, the receiver must be able to reconstruct the digital signal with the smallest possible error probability so that high-quality data detection is possible—and in this way, high transmission quality can be ensured.

This textbook addresses this fundamental problem of communication technology, using baseband transmission, to transmit information over a given transmission medium with the highest possible transmission rate and the lowest possible energy consumption at a minimal probability of an erroneous decision in the receiver. The approaches and

© The Editor(s) (if applicable) and The Author(s), under exclusive license to Springer
Fachmedien Wiesbaden GmbH, part of Springer Nature 2025
C. Lange and A. Ahrens, *Digital Transmission Engineering*,
https://doi.org/10.1007/978-3-658-46789-0_5

possibilities for optimization have been described in detail and as clearly as possible, while being theoretically sound. After a repetitive description of important signal and system theoretical relationships, the basics of transmitting digital signals over a distortion-free channel with noise disturbances were introduced. For the transmission channel copper cable, an appropriate description form suitable for the analysis and optimization of transmission systems was developed using electrical engineering and system theoretical considerations, which allows capturing the signal distortions. Subsequently, possibilities were extensively presented and calculated to compensate for the signal distortions caused by such copper cables. The presentations are designed to be largely independent of current systems—and thus their potential short lifespan—with the focus on fundamental communication technology relationships.

5.2   Outlook

Trends in the development of communication or message networks show that more and more devices are being networked and the transmission of ever larger amounts of data is becoming necessary. This places ever higher demands on transmission speeds and the energy efficiency of messaging systems, and thus overall on the performance of transmission systems.

Transmission with a high transmission rate and low energy consumption while maintaining the highest possible transmission quality is therefore of great importance now and in the future. For this purpose, a large number of very powerful and efficient transmission methods have been and are being developed. Examples of advanced transmission techniques include a variety of equalization and detection methods, digital transmission methods that also use multicarrier and multichannel concepts, as well as multiplexing techniques for multiple use of transmission channels.

This book provides a solid basis for understanding and applying the basics of digital signal transmission. With the help of further technical literature, one is then able to delve into modern, advanced and future messaging systems and transmission methods and further develop and optimize them.

The baseband transmission methods discussed in this book form the technical basis for many currently—and future—used transmission methods. While the methods considered here only allow the arrangement of signal points on the real axis, the shift of such a baseband signal into the bandpass range by modulation allows complex symbols to be used already in the baseband. Such complex signal space diagrams enable better utilization of the transmission channel resource and they form the basis for the widely used methods of radio transmission, for example.

Thus, the basics of communication technology discussed in this book form the basis for modern and advanced transmission methods of multicarrier and multichannel transmission, which find practical applications in wired transmission in the various DSL systems for copper cables (Asymmetric Digital Subscriber Line (ADSL), Very High-Speed

Digital Subscriber Line (VDSL) up to vectoring and DSL variants in standardization), in optical message transmission over fiber optic lines with e.g. wavelength division multiplexing methods (Wavelength Division Multiplex (WDM)) as well as in radio transmission with multiple antenna technology (Multiple Input Multiple Output (MIMO)).

The understanding of the utilization of the transmission channel has changed in recent years and decades, particularly with the introduction of MIMO technology. Based on the results achieved in the 1970s in multichannel technology, which was originally developed to exploit crosstalk in multi-pair copper cables, the application of the MIMO principle was able to make further degrees of freedom usable for transmission—since the 1990s especially in radio technology and in advanced DSL systems. The use of these degrees of freedom allows for improved utilization of transmission channels and therefore MIMO technology has been an integral part of many current and future transmission standards since this time. It has been recognized that interference does not always have to be a disturbing component in data transmission, but can be advantageously used to improve transmission quality or increase transmission speed through advanced signal processing within a MIMO system.

A large number of communication engineering areas can be connected to the topics covered in this textbook—also with a view to the trends—e.g. modulation to flexibly use further frequency ranges of the cable or to apply to the large complex of radio transmission, the simulation of communication systems with the necessary digital signal processing up to the design, organization and optimization of communication networks. There are a number of textbooks in the communication technology literature for whose understanding this textbook forms the foundation.

Appendix A: Overview of System Theoretical Relationships

This chapter of the appendix presents important relationships of signal and system theory in a compact manner. First, signals and important characteristics and properties are specified. The signals are classified according to deterministic and random time processes. Subsequently, some aspects of the evaluation of these signals by linear time-invariant systems are presented.

The overviews are intended as a compact supplement to the fundamentals of signal and system theory considered in Chap. 1, and thus, for example, also for reference.

A.1 Signals

See Tables A.1 and A.2.

A.2 Evaluation by Systems

See Tables A.3 and A.4.

Table A.1 System theoretical relationships (Assumption: Power signals)

Determined Time Processes	Designation	Random Time Processes
$U(f) = \mathcal{F}\{u(t)\}$	Fourier Transformation	$\Psi(f) = \mathcal{F}\{\psi(t)\}$
$u(t) = \mathcal{F}^{-1}\{U(f)\}$	Inverse Fourier Transformation	$\psi(\tau) = \mathcal{F}^{-1}\{\Psi(f)\}$
$u(t)$	Time function (in V) (Power signal)	Defined, but not mathematically describable by formula
$U(f) = \mathcal{F}\{u(t)\} = \int\limits_{-\infty}^{+\infty} u(t)\, e^{-j2\pi f t}\, dt$	amplitude spectral density $\equiv$ voltage per Hz bandwidth in $\frac{V}{Hz}$	Not defined
$\psi(\tau) = \frac{1}{T} \int\limits_{-T/2}^{+T/2} u(t) \cdot u(t+\tau)\, dt$	ACF = Autocorrelation function (in V^2), describes temporal dependencies within a function $u(t)$	$\psi(\tau) = \lim\limits_{T\to\infty} \frac{1}{T} \int\limits_{-T/2}^{+T/2} u(t) \cdot u(t+\tau)\, dt$
$\Psi(f) = \frac{1}{T} U(f) \cdot U^*(f) = \frac{1}{T} |U(f)|^2$	power spectral density (at $R = 1\,\Omega$) $\equiv$ (effective value)^2 per Hz bandwidth in $\frac{V^2}{Hz}$	$\Psi(f) = \int\limits_{-\infty}^{+\infty} \psi(\tau)\, e^{-j2\pi f \tau}\, d\tau \quad \psi(\tau) = \mathcal{F}^{-1}\{\Psi(f)\}$ Wiener-Chintschin theorem
$\psi(0) = u_{\mathrm{eff}}^2 = \frac{1}{T} \int\limits_{-T/2}^{+T/2} u^2(t)\, dt = \int\limits_{-\infty}^{+\infty} \Psi(f)\, df$ (from Parseval Theorem)	(effective value)^2 (in V^2) $\equiv$ (root mean square)$^2 \equiv$ power (at $R = 1\,\Omega$)	$\psi(0) = u_{\mathrm{eff}}^2 = \lim\limits_{T\to\infty} \frac{1}{T} \int\limits_{-T/2}^{+T/2} u^2(t)\, dt = \int\limits_{-\infty}^{+\infty} \Psi(f)\, df$ (from Parseval theorem)
$\psi_{1,2}(\tau) = \frac{1}{T} \int\limits_{-T/2}^{+T/2} u_1(t) \cdot u_2(t+\tau)\, dt$	CCF = Cross-correlation function (in V^2), describes temporal dependencies between two functions $u_1(t)$ and $u_2(t)$	$\psi_{1,2}(\tau) = \lim\limits_{T\to\infty} \frac{1}{T} \int\limits_{-T/2}^{+T/2} u_1(t) \cdot u_2(t+\tau)\, dt$

Table A.2 System theoretical relationships (Assumption: Energy signals)

Determined Time Processes	Designation	**Random Time Processes**
$U(f) = \mathcal{F}\{u(t)\}$	Fourier Transformation	Not defined
$u(t) = \mathcal{F}^{-1}\{U(f)\}$	Inverse Fourier Transformation	Not defined
$u(t)$	Time function (in V) (Energy signal)	Not defined
$U(f) = \mathcal{F}\{u(t)\} = \int\limits_{-\infty}^{+\infty} u(t)\,e^{-j2\pi f t}\,dt$	amplitude spectral density $\equiv$ voltage per Hz bandwidth in $\frac{V}{Hz}$	Not defined
$\psi(\tau) = \int\limits_{-\infty}^{+\infty} u(t)\cdot u(t+\tau)\,dt$	ACF = Autocorrelation function (in $V^2 s$), describes temporal dependencies within a function $u(t)$	Not defined
$\Psi(f) = U(f)\cdot U^*(f) = \lvert U(f)\rvert^2$	Energy density spectrum (at $R = 1\,\Omega) \equiv$ Energy per Hz bandwidth in $\frac{(Vs)^2}{Hz}$	Not defined
$\psi(0) = \int\limits_{-\infty}^{+\infty} u^2(t)\,dt = \int\limits_{-\infty}^{+\infty} \Psi(f)\,df$ (from Parseval theorem)	(Impulse-) Energy (at $R = 1\,\Omega$) (in $V^2 s$)	Not defined
$\psi_{1,2}(\tau) = \int\limits_{-\infty}^{+\infty} u_1(t)\cdot u_2(t+\tau)\,dt$	CCF = Cross-correlation function (in $V^2 s$), describes temporal dependencies between two functions $u_1(t)$ and $u_2(t)$	Not defined

Table A.3 Evaluation of deterministic and random signals through systems

Deterministic Time Processes	Random Time Processes
$U_1(\omega)$ — $G(\omega)$, $g(t)$ — $U_2(\omega)$; $u_1(t)$ — $u_2(t)$	$\Psi_1(\omega)$ — $G(\omega)$, $g(t)$ — $\Psi_2(\omega)$; $\psi_1(\tau)$ — $\psi_2(\tau)$
$U_2(\omega) = U_1(\omega)\cdot G(\omega)$	$\Psi_2(\omega) = \Psi_1(\omega)\cdot \lvert G(\omega)\rvert^2$
$u_2(t) = u_1(t) * g(t)$	$\psi_2(\tau) = \psi_1(\tau) * \psi_g(\tau)$
$g(t) = \mathcal{F}^{-1}\{G(\omega)\}$ weighting function (in 1/s or Hz)	Autocorrelation function (ACF) of the weighting function $g(t)$: $\psi_g(\tau) = \int\limits_{-\infty}^{+\infty} g(t)\cdot g(t+\tau)\,dt$

Table A.4 Relationships regarding Multiplication in the Frequency and Time Domain

Multiplication in the Frequency Domain	Multiplication in the Time Domain
$U_2(\omega) = U_1(\omega) \cdot G(\omega)$	$u_3(t) = u_1(t) \cdot u_2(t)$
$u_2(t) = u_1(t) * g(t)$	$U_3(\omega) = \frac{1}{2\pi} U_1(\omega) * U_2(\omega)$
$\Psi_2(\omega) = \Psi_1(\omega) \cdot |G(\omega)|^2$	or
$\psi_2(\tau) = \psi_1(\tau) * \psi_g(\tau)$	$U_3(f) = U_1(f) * U_2(f)$

A.3 Characteristic Time Functions

A.3.1 Some Basic Time Functions

$$
\text{Dirac impulse} \qquad \delta(t) \quad \circ\!\!-\!\!\bullet \quad 1\,\frac{\text{V}}{\text{Hz}} \qquad \text{with} \quad \int_{-\infty}^{+\infty} \delta(t)\,\mathrm{d}t = 1
$$

$$
\text{DC voltage} \qquad 1\,\text{V} \quad \circ\!\!-\!\!\bullet \quad 2\pi\,\delta(\omega) \qquad \text{with} \quad \frac{1}{2\pi}\int_{-\infty}^{+\infty} 2\pi\,\delta(\omega)\,\mathrm{d}\omega = 1
$$

$$
\text{Step function} \qquad 1(t) \quad \circ\!\!-\!\!\bullet \quad \frac{1}{j\omega} + \pi\,\delta(\omega) \qquad \text{with} \quad 1(t) = \int_{-\infty}^{t} \delta(\tau)\,\mathrm{d}\tau
$$

A.3.2 Some Further Common Time Functions

Rectangular Impulse

$$
u(t) = \begin{cases} U_0 & \text{for} \ \ |t| \le \frac{T_R}{2} \\ 0 & \text{for} \ \ |t| > \frac{T_R}{2} \end{cases} \quad \circ\!\!-\!\!\bullet \quad U(\omega) = U_0 \cdot T_R \cdot \text{si}\left(\frac{\omega\,T_R}{2}\right)
$$

Gaussian Impulse

$$
u(t) = U_0\,e^{-\ln(2)\left(\frac{t}{\tau}\right)^2} \quad \circ\!\!-\!\!\bullet \quad U(\omega) = U_0\frac{2\sqrt{\pi\,\ln(2)}}{\omega_G}e^{-\ln(2)\left(\frac{\omega}{\omega_G}\right)^2}
$$

with $\tau = \frac{2\ln 2}{\omega_{\mathrm{G}}}$

Periodic Impulse Sequence

$$u(t) = C \cdot \sum_{m=-\infty}^{+\infty} \delta(t - m\,T_0) \quad \circ\!\!-\!\!\bullet \qquad U(\omega) = C \cdot \frac{\omega_0}{2\pi} \sum_{n=-\infty}^{+\infty} \delta(\omega - n\,\omega_0)$$

or

$$U(f) = C \cdot f_0 \sum_{n=-\infty}^{+\infty} \delta(f - n\,f_0)$$

with period duration $T_0 = \frac{1}{f_0} = \frac{2\pi}{\omega_0}$

Appendix B: Dimensions in Signal and System Theory

B.1 Dimensions in Signal Theory

B.1.1 Some Basics

To come as close as possible to physical reality, all determined signals in this book are understood as voltages $u(t)$ (in V). Through the Fourier transformation, one obtains from this the amplitude spectral density

$$U(f) = \int_{-\infty}^{+\infty} u(t)\,e^{-j2\pi ft}\,dt \quad \left(\text{in } \frac{V}{Hz} \text{ or } V\,s\right). \tag{B.1}$$

Random time processes are described with the help of the autocorrelation function $\psi(\tau)$ (in V^2), which is linked via the Fourier transformation

$$\Psi(f) = \int_{-\infty}^{+\infty} \psi(\tau)\,e^{-j2\pi f\tau}\,d\tau \quad \left(\text{in } \frac{V^2}{Hz} \text{ or } V^2 s\right) \tag{B.2}$$

with the power spectral density $\Psi(f)$. In this book, a signal-theoretic power of dimension (voltage)^2 with the unit (Volt)^2 is used. At a real, constant resistance, this quantity is proportional to the physical power with the unit Watt. Accordingly, power spectral density refers to the squared amplitude spectral density (in V^2/Hz).

The following briefly introduces two characteristic functions.

Dirac or Delta Impulse in the Time Domain
The function

$$u(t) = A \cdot \delta(t) \tag{B.3}$$

is referred to as a Dirac impulse or a Delta function in the time domain. Here, A is the area of the impulse (the impulse integral [31]). With the definition

© The Editor(s) (if applicable) and The Author(s), under exclusive license to Springer
Fachmedien Wiesbaden GmbH, part of Springer Nature 2025
C. Lange and A. Ahrens, *Digital Transmission Engineering*,
https://doi.org/10.1007/978-3-658-46789-0

$$\int\limits_{-\infty}^{+\infty} \delta(t)\, dt = 1 \quad \longrightarrow \quad \delta(t) \quad \left(\text{in } \frac{1}{s} \text{ or Hz} \right) \tag{B.4}$$

it follows that the time impulse has the unit (1/s). The value A (interpretable as area) must then have the unit V s or V/Hz so that $u(t)$ in (B.3) has the required dimension of a voltage (in the unit V). The spectrum of an exemplary time impulse from

$$u(t) = A \cdot \delta(t) = 1\, \frac{V}{Hz} \cdot \delta(t) \tag{B.5}$$

is thus a constant ($A = 1\,\text{V/Hz}$) in the frequency domain of

$$U(f) = 1\, \frac{V}{Hz}. \tag{B.6}$$

Dirac Impulse in the Frequency Domain

A Dirac impulse in the frequency domain is referred to (analogous to the Dirac impulse in the time domain) as the function

$$U(f) = B \cdot \delta(f) \ . \tag{B.7}$$

Here, B is the area of the Dirac impulse (the impulse integral [31]). With the definition

$$\int\limits_{-\infty}^{+\infty} \delta(f)\, df = 1 \quad \longrightarrow \quad \delta(f) \quad \left(\text{in } \frac{1}{Hz} \text{ or s} \right) \tag{B.8}$$

it follows that the frequency-domain Dirac impulse has the unit (1/Hz). The value B (interpretable as the area of the Dirac impulse in the frequency domain) must then have the unit V = V/Hz $\cdot$ Hz so that $U(f)$ in (B.7) has the required dimension of an amplitude density (in the unit V/Hz). The spectrum of an exemplary direct voltage of

$$u(t) = 1\,\text{V} \tag{B.9}$$

is thus a Dirac impulse in the frequency domain:

$$U(f) = B \cdot \delta(f) = 1\,\text{V} \cdot \delta(f). \tag{B.10}$$

B.1.2 Overview

Definition: $1\,s = \frac{1}{Hz}$

$$u(t) \text{ in V} \ \circ\!\!-\!\!\bullet\ U(\omega) \text{ in } \frac{V}{Hz} \text{ or Vs, since } U(\omega) = \int\limits_{-\infty}^{+\infty} u(t)\, e^{-j\omega t}\, dt$$

Time-domain Dirac Impulse

$$u(t) = A \cdot \delta(t)$$

$$(\mathrm{V}) = (\mathrm{Vs}) \cdot \left(\tfrac{1}{\mathrm{s}}\right)$$

Definition: $\int\limits_{-\infty}^{+\infty} \delta(t)\, \mathrm{d}t = 1 \implies \delta(t)$ in $\tfrac{1}{\mathrm{s}}$

$A = $ Area of the Dirac impulse (in the time domain)

$\implies$ Spectrum of a Dirac impulse function:

$$u(t) = 1\,\tfrac{\mathrm{V}}{\mathrm{Hz}} \cdot \delta(t) \; \circ\!\!-\!\!\bullet \; 1\,\tfrac{\mathrm{V}}{\mathrm{Hz}} = U(f)$$

Frequency-domain Dirac impulse

$$U(f) = B \cdot \delta(f)$$

$$\left(\tfrac{\mathrm{V}}{\mathrm{Hz}}\right) = \left(\tfrac{\mathrm{V}}{\mathrm{Hz}} \cdot \mathrm{Hz}\right) \cdot \left(\tfrac{1}{\mathrm{Hz}}\right)$$

Definition: $\int\limits_{-\infty}^{+\infty} \delta(f)\, \mathrm{d}f = 1 \implies \delta(f)$ in $\tfrac{1}{\mathrm{Hz}}$ or s

$B = $ Area of the Dirac impulse (in the frequency domain)

$$\implies \int\limits_{-\infty}^{+\infty} \delta(f)\, \mathrm{d}f = 1$$

Spectrum of a DC voltage of 1 V:

$$U(f) = 1\,\mathrm{V} \cdot \delta(f) = 1\,\tfrac{\mathrm{V}}{\mathrm{Hz}}$$

Convolution

$$u_1(t) * u_2(t) = \int\limits_{-\infty}^{+\infty} u_1(\tau) \cdot u_2(t - \tau)\, \mathrm{d}\tau$$

$$\left(\mathrm{V}^2\right) = (\mathrm{V}) \cdot (\mathrm{V}) \cdot (\mathrm{s})$$

Transfer function

$$G(\omega) = \frac{U_2(\omega)}{U_1(\omega)}$$

$$(1) = \left(\tfrac{\mathrm{V}}{\mathrm{Hz}}\right) \div \left(\tfrac{\mathrm{V}}{\mathrm{Hz}}\right)$$

System response (weighting function)

$$g(t) = \frac{1}{2\pi} \int\limits_{-\infty}^{+\infty} G(\omega)\, \mathrm{e}^{\mathrm{j}\omega t}\, \mathrm{d}\omega$$

$$(\mathrm{Hz}) \text{ or } \left(\tfrac{1}{\mathrm{s}}\right)$$

Dirac impulse response: $u(t) = 1\,\tfrac{\mathrm{V}}{\mathrm{Hz}} \cdot g(t)$

B.2 Dimensions in System Theory

B.2.1 Some Basics

Deterministic Signals

In Fig. B.1, a system with determined input and output variables is shown in both the time and frequency domain. The transfer function is represented by

$$G(f) = \frac{U_2(f)}{U_1(f)}\,. \tag{B.11}$$

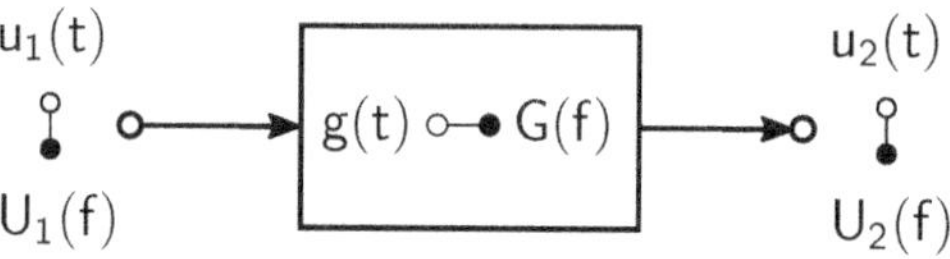

Fig. B.1 On the definition of a system with determined input and output sizes

It is therefore dimensionless. Through the Fourier transformation, the transfer function is linked with the weighting function:

$$g(t) = \int_{-\infty}^{+\infty} G(f)\, e^{j2\pi ft}\, df \quad \left(\text{in Hz or } \frac{1}{s}\right). \tag{B.12}$$

The input-output relationship of the system shown in Fig. B.1 is

$$U_2(f) = G(f) \cdot U_1(f). \tag{B.13}$$

Since $G(f)$ is dimensionless, the amplitude density present at the system's input $U_1(f)$ (in V/Hz) invokes also an amplitude density $U_2(f)$ (in V/Hz) at the output. In the time domain, the input-output relationship for linear, time-invariant systems is given by the convolution

$$u_2(t) = g(t) * u_1(t) = \int_{-\infty}^{+\infty} g(\tau) \cdot u_1(t - \tau)\, d\tau. \tag{B.14}$$

Since the convolution operation is dimension-bearing, the voltage present at the input $u_1(t)$ (in V) leads to an output voltage $u_2(t)$ (in V) as follows: $g(t)$ (in 1/s) multiplied by the unit (s) from the convolution integral cancels out and what remains is the dimension of $u_1(t)$ (in V) for the result $u_2(t)$ (in V).

If the system $g(t)$ is subjected to a Dirac impulse

$$u_1(t) = A\,\delta(t) \tag{B.15}$$

as an input signal, the Dirac impulse response

$$u_2(t) = u_1(t) * g(t) \tag{B.16}$$

$$u_2(t) = A\,\delta(t) * g(t) \tag{B.17}$$

$$u_2(t) = A \cdot g(t) \tag{B.18}$$

is obtained as the signal at the output of the system. The weighting function and the impulse response differ only by the factor A; however, this is dimensionally loaded. It is therefore important to note that the weighting function $g(t)$ as a system characteristic, with dimensionless transfer function $G(f)$, has the dimension $(\text{time})^{-1}$; on the other hand, the impulse response as the system's reaction to an impulse at the input has the dimension of the input signal (here in the unit V). In this respect, the terms *weighting function* and *impulse response* must be distinguished and not used synonymously.

Random Signals

The input-output relationship for random variables (see Fig. B.2) is in the frequency domain

$$\Psi_2(f) = |G(f)|^2 \cdot \Psi_1(f). \tag{B.19}$$

Since $G(f)$ is dimensionless (and thus also $|G(f)|^2$), the power density (or squared amplitude density) present at the system's entrance $\Psi_1(f)$ (in V^2/Hz) induces also a power density $\Psi_2(f)$ (in V^2/Hz) at the output. In the time domain, the input-output relationship is given by the convolution

$$\psi_2(\tau) = \psi_g(\tau) * \psi_1(\tau) = \int_{-\infty}^{+\infty} \psi_g(t) \cdot \psi_1(t - \tau)\, dt \tag{B.20}$$

with

$$\psi_g(\tau) = \int_{-\infty}^{+\infty} g(t) \cdot g(t + \tau)\, dt \tag{B.21}$$

as the autocorrelation function of the weighting function $g(t)$. In (B.21) the unit of $g(t)$ (in $1/s$) cancels out against the unit (s) of the dt from the convolution integral and the dimension remains $g(t)$ (in $1/s$) for the result $\psi_g(\tau)$ (in $1/s$). Thus, in (B.20) $\psi_g(\tau)$ has the unit $(1/s)$, which cancels out against the unit (s) of the dt from the convolution integral and it remains as the unit for the result $\psi_2(\tau)$ the unit of the autocorrelation function $\psi_1(\tau)$ of the input signal (V^2).

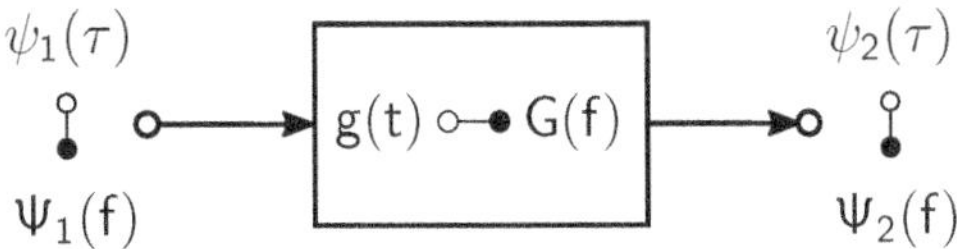

Fig. B.2 On the definition of a system with random input and output sizes

B.2.2 Overview

$$u_2(t) = u_1(t) * g(t)$$

$$(\mathrm{V}) = (\mathrm{V}) \cdot \left(\tfrac{1}{\mathrm{s}}\right) \cdot (\mathrm{s}) \quad (\ddagger)$$

$$\Psi_2(\omega) = \Psi_1(\omega) \cdot |G(\omega)|^2$$

$$\left(\tfrac{\mathrm{V}^2}{\mathrm{Hz}}\right) = \left(\tfrac{\mathrm{V}^2}{\mathrm{Hz}}\right) \cdot (1^2)$$

$$\bullet\!\!-\!\!\circ$$

$$\psi(\tau) = \tfrac{1}{2\pi} \int\limits_{-\infty}^{+\infty} \Psi(\omega)\, e^{j\omega\tau}\, d\omega$$

$$(\mathrm{V}^2) = \left(\tfrac{\mathrm{V}^2}{\mathrm{Hz}}\right) \cdot (\mathrm{Hz})$$

$$\psi_2(\tau) = \psi_1(\tau) * \psi_g(\tau)$$

$$(\mathrm{V}^2) = (\mathrm{V}^2) \cdot \left(\tfrac{1}{\mathrm{s}}\right) \cdot (\mathrm{s}) \quad (\ddagger)$$

$$\text{with } \psi_g(\tau) = \int\limits_{-\infty}^{+\infty} g(t) \cdot g(t + \tau)\, dt$$

$$\left(\tfrac{1}{\mathrm{s}}\right) = \left(\tfrac{1}{\mathrm{s}}\right) \cdot \left(\tfrac{1}{\mathrm{s}}\right) \cdot (\mathrm{s})$$

(ACF of the system response)

$(\ddagger)$ Note: The last unit (s) originates from the convolution integral.

Appendix C: Fourier Transformation

C.1 Definition: Fourier Transformation

The Fourier transform of a time function $u(t)$ is defined as

$$U(\omega) = \mathcal{F}\{u(t)\} = \int\limits_{-\infty}^{+\infty} u(t)\,\mathrm{e}^{-\mathrm{j}\omega t}\,\mathrm{d}t \;. \tag{C.1}$$

The inverse Fourier transformation is

$$u(t) = \mathcal{F}^{-1}\{U(\omega)\} = \frac{1}{2\pi}\int\limits_{-\infty}^{+\infty} U(\omega)\,\mathrm{e}^{\mathrm{j}\omega t}\,\mathrm{d}\omega. \tag{C.2}$$

If the frequency f instead of the angular frequency $\omega = 2\pi f$ is used, one obtains a symmetric form of the Fourier transformation with

$$U(f) = \mathcal{F}\{u(t)\} = \int\limits_{-\infty}^{+\infty} u(t)\,\mathrm{e}^{-\mathrm{j}2\pi f t}\,\mathrm{d}t \tag{C.3}$$

and

$$u(t) = \mathcal{F}^{-1}\{U(f)\} = \int\limits_{-\infty}^{+\infty} U(f)\,\mathrm{e}^{\mathrm{j}2\pi f t}\,\mathrm{d}f. \tag{C.4}$$

© The Editor(s) (if applicable) and The Author(s), under exclusive license to Springer
Fachmedien Wiesbaden GmbH, part of Springer Nature 2025
C. Lange and A. Ahrens, *Digital Transmission Engineering*,
https://doi.org/10.1007/978-3-658-46789-0

C.2 Laws and Calculation Rules of Fourier Transformation

Modulation	$U(\omega - \omega_0) \;\bullet\!\!-\!\!\circ\; u(t) \cdot e^{j\omega_0 t}$
Time Shift theorem	$u(t - t_0) \;\circ\!\!-\!\!\bullet\; U(\omega) \cdot e^{-j\omega t_0}$
Convolution theorem	$U_1(\omega) \;\bullet\!\!-\!\!\circ\; u_1(t); \quad U_2(\omega) \;\bullet\!\!-\!\!\circ\; u_2(t)$
	$u_1(t) \cdot u_2(t) \;\circ\!\!-\!\!\bullet\; \frac{1}{2\pi} U_1(\omega) * U_2(\omega)$
	$U_1(\omega) \cdot U_2(\omega) \;\bullet\!\!-\!\!\circ\; u_1(t) * u_2(t)$
Differentiation theorem	$\frac{\mathrm{d}}{\mathrm{d}t} u(t) \;\circ\!\!-\!\!\bullet\; j\omega \cdot U(\omega)$
Integration theorem	$\displaystyle\int_{-\infty}^{t} u(\tau)\,\mathrm{d}\tau \;\circ\!\!-\!\!\bullet\; U(\omega)\left[\frac{1}{j\omega} + \pi\,\delta(\omega)\right]$

C.3 Correspondences of the Fourier Transformation

In Tables C.1 and C.2, selected correspondences of the Fourier transformation are compiled. The focus is on the correspondences that were used in this book; some other important correspondences have also been included. They—and many others—can be found, for example, in [11, 74] or other comparable works.

Table C.1 Selection of some correspondences of the Fourier transformation at frequency variable f (with U_0 in V and T in s)

$u(t)$	$U(f) = \mathcal{F}\{u(t)\}$
U_0	$U_0\,\delta(f)$
$U_0\,T\,\delta(t)$	$U_0\,T$
$U_0\,\cos(2\pi f_0 t)$	$\frac{U_0}{2}(\delta(f - f_0) + \delta(f + f_0))$
$U_0\,\sin(2\pi f_0 t)$	$\frac{U_0}{2j}(\delta(f - f_0) - \delta(f + f_0))$
$U_0\,e^{j2\pi f_0 t}$	$U_0\,\delta(f - f_0)$
$U_0\,\mathrm{rect}\!\left(\frac{t}{T}\right)$	$U_0\,T\,\mathrm{si}(\pi f T)$

Table C.2 Selection of some correspondences of the Fourier transformation at frequency variable ω (with U_0 in V and T in s)

$u(t)$	$U(\omega) = \mathcal{F}\{u(t)\}$
U_0	$U_0\,2\pi\,\delta(\omega)$
$U_0\,T\,\delta(t)$	$U_0\,T$
$U_0\,\cos(\omega_0 t)$	$U_0\,\pi\,(\delta(\omega - \omega_0) + \delta(\omega + \omega_0))$
$U_0\,\sin(\omega_0 t)$	$\frac{U_0}{j}\,\pi\,(\delta(\omega - \omega_0) - \delta(\omega + \omega_0))$
$U_0\,e^{j\omega_0 t}$	$U_0\,2\pi\,\delta(\omega - \omega_0)$
$U_0\,\mathrm{rect}\!\left(\frac{t}{T}\right)$	$U_0\,T\,\mathrm{si}\!\left(\frac{\omega T}{2}\right)$

Appendix D: Laplace Transformation

D.1 Definition: Laplace Transformation

The Laplace transform of a time function $u(t)$ is defined as

$$U(p) = \mathcal{L}\{u(t)\} = \int_0^\infty u(t)\, e^{-pt}\, dt \, . \tag{D.1}$$

The Laplace inverse transformation is

$$u(t) = \mathcal{L}^{-1}\{U(p)\} = \frac{1}{j2\pi} \int_{\sigma-j\infty}^{\sigma+j\infty} U(p)\, e^{pt}\, dp \, . \tag{D.2}$$

In the Laplace transformation—unlike the Fourier transformation—increasing time functions are also allowed. However, for times $t < 0$ the condition $u(t) = 0$ must be met (one-sided Laplace transformation).

Since in this book in connection with the Laplace transformation exclusively one-sided time functions with the property

$$x(t) = 0 \qquad \text{for} \quad t < 0$$

are considered, the addition $1(t)$ is omitted, provided there is no risk of confusion. In this respect, the notations

$$x(t) \qquad \text{and} \qquad x(t) \cdot 1(t)$$

and

$$x(t - t_0) \qquad \text{and} \qquad x(t - t_0) \cdot 1(t - t_0)$$

are used interchangeably.

© The Editor(s) (if applicable) and The Author(s), under exclusive license to Springer Fachmedien Wiesbaden GmbH, part of Springer Nature 2025
C. Lange and A. Ahrens, *Digital Transmission Engineering,*
https://doi.org/10.1007/978-3-658-46789-0

D.2 Laws and Calculation Rules of Laplace Transformation

The following theorems or calculation rules can often simplify practical calculations in the Laplace transformation as well as the handling of Laplace transforms. The designations of the theorems refer to the mathematical operations that are performed on the time (or original) functions.

Addition Theorem

$$\mathcal{L}\{a_1 u_1(t) + a_2 u_2(t)\} = a_1 \mathcal{L}\{u_1(t)\} + a_2\{\mathcal{L}u_2(t)\} \tag{D.3}$$

The transformation is linear.

Convolution Theorem

$$\mathcal{L}\left\{\int_0^t u_1(t-\tau)\, u_2(\tau)\, d\tau\right\} = \mathcal{L}\{u_1(t)\} \cdot \{\mathcal{L}u_2(t)\} \tag{D.4}$$

The convolution product in the original domain corresponds to the (ordinary) product in the spectral domain.

Integration Theorem

$$\mathcal{L}\left\{\int_0^t u(\tau)\, d\tau\right\} = \frac{1}{p}\mathcal{L}\{u(t)\} = \frac{1}{p}U(p) \tag{D.5}$$

The integration in the original domain corresponds to the division by p in the spectral domain.

Derivative Theorem

$$\mathcal{L}\{u^{(n)}(t)\} = p^n\, U(p) - p^{n-1}u_0 - \ldots - p\, u_0^{(n-2)} - u_0^{(n-1)} \tag{D.6}$$

The following applies $u_0^{(v)} = \lim_{t \to +0} \frac{d^v u(t)}{dt^v}$.

Time-Shifting Theorem

$$\mathcal{L}\{u(t-b)\} = e^{-bp}\, \mathcal{L}\{u(t)\} \tag{D.7}$$

Similarity Theorem

$$\mathcal{L}\{u(at)\} = \frac{1}{a}\, U\left(\frac{p}{a}\right) \qquad \text{for} \qquad a > 0 \tag{D.8}$$

Frequency-Shifting Theorem

$$\mathcal{L}\left\{e^{-at}u(t)\right\} = U(p+a) \tag{D.9}$$

Multiplication Theorem

$$\mathcal{L}\{t^n u(t)\} = (-1)^n\, U^{(n)}(p) \qquad \text{for} \qquad n = 1, 2, \ldots \tag{D.10}$$

D.3 Correspondences of the Laplace Transformation

Selected correspondences of the Laplace transformation are compiled in Table D.1. The focus is on the correspondences that were used in this book; some other important correspondences have also been included. They—and many others—can be found in e.g. [11, 13, 74] or other comparable works.

Table D.1 Selection of some correspondences when applying the one-sided Laplace transformation

$u(t)$	$U(p) = \mathcal{L}\{u(t)\}$
$1(t)$	$\frac{1}{p}$
e^{-at}	$\frac{1}{p+a}$
t	$\frac{1}{p^2}$
t^n	$\frac{n!}{p^{n+1}}$
$\frac{1}{T}\,e^{-\frac{t}{T}}$	$\frac{1}{1+pT}$
$\frac{1}{a}\left(1 - e^{-at}\right)$	$\frac{1}{p(p+a)}$
$1 - e^{-\frac{t}{T}}$	$\frac{1}{p(1+pT)}$
$t\,e^{-at}$	$\frac{1}{(p+a)^2}$
$\frac{1}{a^2}\left(1 - e^{-at} - at\,e^{-at}\right)$	$\frac{1}{p(p+a)^2}$
$\frac{1}{b-a}\left(e^{-at} - e^{-bt}\right)$	$\frac{1}{(p+a)(p+b)}$
$\frac{1}{a-b}\left(a\,e^{-at} - b\,e^{-bt}\right)$	$\frac{p}{(p+a)(p+b)}$
$\frac{1}{ab(a-b)}\left((a-b) + b\,e^{-at} - a\,e^{-bt}\right)$	$\frac{1}{p(p+a)(p+b)}$
$\frac{e^{-t/T_1} - e^{-t/T_2}}{T_2 - T_1}$	$\frac{1}{(1+pT_1)(1+pT_2)}$
$\frac{1}{a}\sin(at)$	$\frac{1}{p^2+a^2}$
$\cos(at)$	$\frac{p}{p^2+a^2}$
$\frac{1}{a}\sinh(at)$	$\frac{1}{p^2-a^2}$
$\cosh(at)$	$\frac{p}{p^2-a^2}$
$\delta(t)$	1
$\delta(t-T)$	e^{-pt}
$\delta^{(n)}(t)$	p^n

Appendix E: Error Function erf(*x*) and Complementary Error Function erfc(*x*)

E.1 Definitions and Properties

The non-closed solvable integral

$$\operatorname{erf}(x) = \frac{2}{\sqrt{\pi}} \int\limits_{0}^{x} \mathrm{e}^{-t^2}\, \mathrm{d}t \tag{E.1}$$

is referred to as the *Gaussian error function* (error function). Some properties of this function are

$$\operatorname{erf}(-x) = -\operatorname{erf}(x) \qquad \text{(odd function)}$$
$$\operatorname{erf}(-\infty) = -1$$
$$\operatorname{erf}(\infty) = 1$$
$$\operatorname{erf}(0) = 0$$
$$\frac{\mathrm{d}}{\mathrm{d}x}\operatorname{erf}(x) = \frac{2}{\sqrt{\pi}}\mathrm{e}^{-x^2} \, .$$

Often in this context, the *complementary error function*

$$\operatorname{erfc}(x) = 1 - \operatorname{erf}(x) = \frac{2}{\sqrt{\pi}} \int\limits_{x}^{\infty} \mathrm{e}^{-t^2}\, \mathrm{d}t \tag{E.2}$$

with the properties

$$\operatorname{erfc}(-x) = 2 - \operatorname{erfc}(x)$$
$$\operatorname{erfc}(-\infty) = 2$$
$$\operatorname{erfc}(\infty) = 0$$

is used.

© The Editor(s) (if applicable) and The Author(s), under exclusive license to Springer Fachmedien Wiesbaden GmbH, part of Springer Nature 2025
C. Lange and A. Ahrens, *Digital Transmission Engineering*,
https://doi.org/10.1007/978-3-658-46789-0

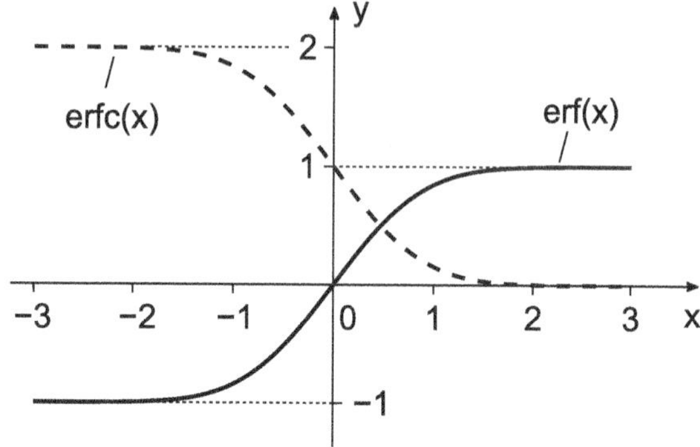

Fig. E.1 Course of the functions $y = \mathrm{erf}(x)$ and $y = \mathrm{erfc}(x)$

Figure E.1 illustrates the course of the Gaussian error function $\mathrm{erf}(x)$ and the complementary Gaussian error function $\mathrm{erfc}(x)$.

E.2 Calculation Possibilities and Approximations

The Gaussian error function $\mathrm{erf}(x)$ – as well as its complementary function $\mathrm{erfc}(x)$ – can practically be calculated

- with the help of tables or
- through approximations, which are mostly created by series expansions.

Some numerical values for the function $\mathrm{erfc}(x)$ are in Table E.1.

The function $\mathrm{erf}(x)$ can be developed into the series

$$\mathrm{erf}(x) = \frac{2}{\sqrt{\pi}}\, \mathrm{e}^{-x^2} \sum_{k=0}^{\infty} \frac{2^k\, x^{2k+1}}{(2k+1)!!}\,. \tag{E.3}$$

Some important approximations for the practical calculation of the function $\mathrm{erf}(x)$ are:

$$\mathrm{erf}(x) \approx 1 - \frac{1}{\sqrt{\pi}\, x}\mathrm{e}^{-x^2} \qquad\qquad \text{for} \quad x > 4$$

$$\mathrm{erf}(x) \approx -1 + \frac{1}{\sqrt{\pi}\, x}\mathrm{e}^{-x^2} \qquad\qquad \text{for} \quad x < -4$$

$$\mathrm{erf}(x) \approx \frac{2}{\sqrt{\pi}}\, x \qquad\qquad\qquad\qquad \text{for} \quad x \to 0$$

$$\mathrm{erf}(x) \approx 1 - \frac{1}{\sqrt{\pi}}\left(\frac{1}{x} - \frac{1}{2x^3} + \frac{3}{4x^5}\right)\mathrm{e}^{-x^2} \quad \text{for} \quad |x| \geq 5$$

Table E.1 Some numerical values of the function $\text{erfc}(x)$

x	$\text{erfc}(x)$	x	$\text{erfc}(x)$	x	$\text{erfc}(x)$
0	1	2.0	0.004678	4.0	$1.54 \cdot 10^{-8}$
0.1	0.8875	2.1	0.002979	4.1	$6.70 \cdot 10^{-9}$
0.2	0.7773	2.2	0.001863	4.2	$2.86 \cdot 10^{-9}$
0.3	0.6714	2.3	0.001143	4.3	$1.19 \cdot 10^{-9}$
0.4	0.5716	2.4	0.000688	4.4	$4.89 \cdot 10^{-10}$
0.5	0.4795	2.5	0.000407	4.5	$1.97 \cdot 10^{-10}$
0.6	0.3961	2.6	0.000236	4.6	$7.75 \cdot 10^{-11}$
0.7	0.3222	2.7	0.000134	4.7	$2.99 \cdot 10^{-11}$
0.8	0.2579	2.8	$7.50 \cdot 10^{-5}$	4.8	$1.14 \cdot 10^{-11}$
0.9	0.2031	2.9	$4.11 \cdot 10^{-5}$	4.9	$4.22 \cdot 10^{-12}$
1.0	0.1573	3.0	$2.21 \cdot 10^{-5}$	5.0	$1.54 \cdot 10^{-12}$
1.1	0.1198	3.1	$1.65 \cdot 10^{-5}$	5.1	$5.49 \cdot 10^{-13}$
1.2	0.0897	3.2	$6.03 \cdot 10^{-6}$	5.2	$1.93 \cdot 10^{-13}$
1.3	0.0659	3.3	$3.06 \cdot 10^{-6}$	5.3	$6.62 \cdot 10^{-14}$
1.4	0.0477	3.4	$1.52 \cdot 10^{-6}$	5.4	$2.23 \cdot 10^{-14}$
1.5	0.0339	3.5	$7.43 \cdot 10^{-7}$	5.5	$7.33 \cdot 10^{-15}$
1.6	0.0237	3.6	$3.56 \cdot 10^{-7}$	5.6	$2.33 \cdot 10^{-15}$
1.7	0.0162	3.7	$1.67 \cdot 10^{-7}$	5.7	$7.77 \cdot 10^{-16}$
1.8	0.0109	3.8	$7.70 \cdot 10^{-8}$	5.8	$2.22 \cdot 10^{-16}$
1.9	0.0072	3.9	$3.48 \cdot 10^{-8}$	5.9	$1.11 \cdot 10^{-16}$

A power series development is

$$\text{erf}(x) = \frac{2}{\sqrt{\pi}} \sum_{k=0}^{\infty} \frac{(-1)^k x^{2k+1}}{k!\,(2k+1)}$$

$$\text{erf}(x) = \frac{2}{\sqrt{\pi}} \left[x - \frac{x^3}{3} + \frac{x^5}{10} - \frac{x^7}{42} + \frac{x^9}{216} - \cdots \right].$$

E.3 Relationships to Other Functions

The two functions $\text{erf}(x)$ and $\text{erfc}(x)$ are closely related to the function

$$Q(x) = \frac{1}{\sqrt{2\pi}} \int_{x}^{\infty} e^{-\frac{t^2}{2}}\, dt \tag{E.4}$$

and to the Gaussian error integral (distribution function of the standard normal distribution)

$$\Phi(x) = \frac{1}{\sqrt{2\pi}} \int_{-\infty}^{x} e^{-\frac{t^2}{2}}\, dt. \tag{E.5}$$

The mutual conversion relationships are

$$Q(x) = \frac{1}{2}\, \mathrm{erfc}\left(\frac{x}{\sqrt{2}}\right)$$

$$\mathrm{erfc}(x) = 2\, Q\left(x \cdot \sqrt{2}\right)$$

and

$$\Phi(x) = \frac{1}{2}\left(1 + \mathrm{erf}\left(\frac{x}{\sqrt{2}}\right)\right)$$

$$\mathrm{erf}(x) = 2\, \Phi\left(x \cdot \sqrt{2}\right) - 1.$$

Appendix F: Mathematical Formulas and Relationships

F.1 Hyperbolic Functions

Hyperbolic sine

$$\sinh x = \frac{e^x - e^{-x}}{2} \tag{F.1}$$

Hyperbolic cosine

$$\cosh x = \frac{e^x + e^{-x}}{2} \tag{F.2}$$

Relationship between hyperbolic sine and hyperbolic cosine

$$\cosh^2 x - \sinh^2 x = 1 \tag{F.3}$$

Hyperbolic tangent

$$\tanh x = \frac{\sinh x}{\cosh x} \tag{F.4}$$

Hyperbolic cotangent

$$\coth x = \frac{\cosh x}{\sinh x} = \frac{1}{\tanh x} \tag{F.5}$$

The hyperbolic cosine function $\cosh x$ corresponds to the *catenary* (also: *chain curve* or *catenoid*) and mathematically describes the curve that a chain (or a rope) suspended at its ends exhibits under the influence of gravity.

F.2 Integrals

This section compiles integrals used in the book. They—and many others—can be found, for example, in [11, 74] or other comparable works.

$$\int \frac{1}{a^2 + x^2}\, dx = \frac{1}{a} \arctan \frac{x}{a} + C \tag{F.6}$$

$$\int \frac{1}{\left(a^2 + x^2\right)^2}\, dx = \frac{x}{2\,a^2\,(a^2 + x^2)} + \frac{1}{2\,a^3} \arctan \frac{x}{a} + C \tag{F.7}$$

$$\int \frac{x^2}{\left(a^2 + x^2\right)^2}\, dx = -\frac{x}{2\,(a^2 + x^2)} + \frac{1}{2\,a} \arctan \frac{x}{a} + C \tag{F.8}$$

$$\int_0^\infty e^{-a^2 x^2}\, dx = \frac{\sqrt{\pi}}{2\,a} \qquad \text{für} \qquad a > 0 \tag{F.9}$$

$$\int \cos ax\, dx = \frac{1}{a} \sin ax + C \tag{F.10}$$

F.3 Trigonometric Relationships

Sine

$$\sin x = \frac{1}{2\mathrm{j}} \left(e^{\mathrm{j}x} - e^{-\mathrm{j}x}\right) \tag{F.11}$$

Cosine

$$\cos x = \frac{1}{2} \left(e^{\mathrm{j}x} + e^{-\mathrm{j}x}\right) \tag{F.12}$$

Relationship between Sine and Cosine

$$\cos^2 x + \sin^2 x = 1 \tag{F.13}$$

$$\sin x = \cos\left(x - \frac{\pi}{2}\right) \quad \text{(radian measure)} \tag{F.14}$$

Some Addition and Multiplication Theorems

$$\cos x \cdot \cos y = \frac{1}{2}(\cos(x - y) + \cos(x + y)) \tag{F.15}$$

$$\sin x \cdot \sin y = \frac{1}{2}(\cos(x - y) - \cos(x + y)) \tag{F.16}$$

$$\sin x + \sin y = 2 \sin\left(\frac{x + y}{2}\right) \cos\left(\frac{x - y}{2}\right) \tag{F.17}$$

$$\cos(\alpha - \beta) = \cos(\alpha) \cos(\beta) + \sin(\alpha) \sin(\beta) \tag{F.18}$$

Appendix G: On the Definition of Frequency

A cosine oscillation can accordingly

$$u(t) = U_0 \cos(\omega_0 t) = U_0 \cos(2\pi f_0 t) = \frac{U_0}{2}\left(e^{j\omega_0 t} + e^{-j\omega_0 t}\right) \qquad \text{(G.1)}$$

be represented by two rotating vectors with opposite directions of rotation (see Fig. G.1).

Definition: The angular frequency

$$\omega_0 = 2\pi f_0 = \frac{d\varphi}{dt} \qquad \text{(G.2)}$$

is the angular velocity of rotating vectors.

Thus, a cosine oscillation always involves *two* frequencies, the positive frequency $+f_0$ (or $+\omega_0$) and the negative frequency $-f_0$ (or $-\omega_0$). The resultant of both rotating vectors is real.

Note: The relationship

$$|f| = \frac{1}{\text{period } T} \qquad \text{(G.3)}$$

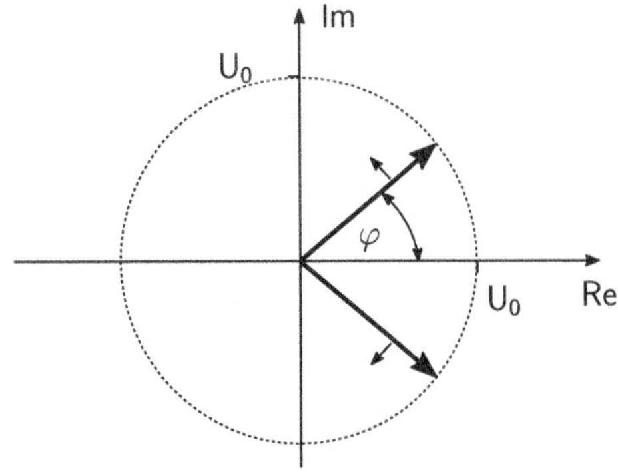

Fig. G.1 On the definition of frequency via two rotating vectors with opposite directions of rotation

is *not* a definition of frequency, but merely a way to measure the *magnitude* of a frequency.

If, in addition to the rotation of the vectors, a temporal change in the length of the vectors, i.e., the amplitude, is allowed, the complex frequency is obtained

$$p = \sigma + j\omega. \tag{G.4}$$

Here, ω describes the angular frequency and σ the change in length of the vectors during exponential increase or decrease.

References

1. AHRENS, A. ; LANGE, C.: Kriterien zur Bewertung der Qualität digitaler Signal übertragung. In: *Frequenz* 58 (2004), January/February, No. 1/2, pp. 3–10
2. AHRENS, A. ; LANGE, C.: *Signal-Rausch-Verhältnis und Fehlerrate im Kabel*. Berlin: Mensch & Buch Verlag, 2012
3. ANDERSON, J. B.: *Digital Transmission Engineering*. 2nd ed. Piscataway, NJ: IEEE Press; John Wiley & Sons, 2005
4. ARFKEN, G. B. ; Weber, H. J.: Mathematical Methods for Physicists. 5th ed. San Diego, San Francisco, New York: Harcourt/Academic, 2001
5. BARRY, J. R. ; LEE, E. A. ; MESSERSCHMITT, D. G.: *Digital Communication*. 3rd ed. Boston: Kluwer Academic Publishers, 2004
6. BENEDETTO, S. ; BIGLIERI, E.: *Principles of Digital Transmission – With Wireless Applications*. New York: Kluwer Academic/Plenum Publishers, 1999
7. BENING, F.: *Z-Transformation f ür Ingenieure*. Stuttgart: B. G. Teubner, 1995
8. BINGHAM, J. A. C.: *The Theory and Practice of Modem Design*. New York: Wiley, 1988
9. BINGHAM, J. A. C.: *ADSL, VDSL, and Multicarrier Modulation*. New York: Wiley, 2000
10. BÖHME, J. F.: *Stochastische Signale*. 2nd ed. Stuttgart: B. G. Teubner, 1998
11. BRONSTEIN, I. N. ; SEMENDJAJEW, K. A.: *Taschenbuch der Mathematik*. 20th ed. Moskau; Leipzig: Nauka; BSB B. G. Teubner, 1981
12. DEUTSCH, B. ; MOHR, S. ; ROLLER, A. ; ROST, H.: *Elektrische Nachrichtenkabel*. Erlangen, München: Publicis MCD Verlag, 1998
13. DOBESCH, H. ; SULANKE, H.: *Zeitfunktionen*. 3rd ed. Berlin: Verlag Technik, 1970
14. FISCHER, R. F. H.: *Precoding and Signal Shaping for Digital Transmission*. New York: Wiley, 2002
15. FORNEY, G. D.: The Viterbi Algorithm. In: *Proceedings of the IEEE* 61 (1973), March, No. 3, pp. 268–278
16. FRITZSCHE, G.: *Signale und Funktionaltransformationen*. Berlin: Verlag Technik, 1985
17. GALLI, S. ; KERPEZ, K. J.: Methods of Summing Crosstalk From Mixed Sources – Part I. In: *IEEE Transactions on Communications* 50 (2002), March, No. 3, pp. 453–461
18. GIROD, B. ; RABENSTEIN, R. ; STENGER, A. K. E.: *Einführung in die Systemtheorie*. 4th ed. Wiesbaden: Vieweg+Teubner, 2007
19. GUSTRAU, F.: *Hochfrequenztechnik*. 3rd, updated ed. München: Hanser, 2019
20. HÄNSLER, E.: *Statistische Signale*. 3rd ed. Berlin, Heidelberg, New York: Springer, 2001
21. HENKE, H.: *Elektromagnetische Felder*. 6th ed. Berlin: Springer Vieweg, 2020

22. HÖHER, P. A.: *Grundlagen der digitalen Informationsübertragung*. 2nd ed. Wiesbaden: Springer Vieweg, 2013

23. HUBER, J.: *Trelliscodierung*. Berlin, Heidelberg: Springer, 1992

24. JERUCHIM, M. C. ; BALABAN, P. ; SHANMUGAN, K. S.: *Simulation of Communication Systems – Modeling, Methodology, and Techniques*. 2nd ed. New York, Dordrecht, London: Kluwer Academic/Plenum Publishers, 2000

25. JONDRAL, F. ; WIESLER, A.: *Wahrscheinlichkeitsrechnung und stochastische Prozesse*. 2nd ed. Stuttgart, Leipzig, Wiesbaden: B. G. Teubner, 2002

26. KAMMEYER, K.-D. ; DEKORSY, A.: *Nachrichtenübertragung*. 6th ed. Wiesbaden: Springer Vieweg, 2018

27. KARRENBERG, U.: *Signale – Prozesse – Systeme*. 7th ed. Wiesbaden: Springer Vieweg, 2017

28. KLINGBEIL, H.: *Grundlagen der elektromagnetischen Feldtheorie*. 3rd ed. Berlin: Springer Spektrum, 2018

29. KRESS, D.: *Theoretische Grundlagen der Übertragung digitaler Signale*. Berlin: Akademie-Verlag, 1979

30. KRESS, D. ; B, KAUFHOLD.: *Signale und Systeme verstehen und vertiefen*. Wiesbaden: Vieweg+Teubner, 2010

31. KRESS, D. ; IRMER, R.: *Angewandte Systemtheorie*. Berlin: Verlag Technik, 1989

32. KRESS, D. ; KRIEGHOFF, M. ; GRÄFE, W.-R.: Gütekriterien bei der Übertragung digitaler Signale. In: XX. *Internationales Wissenschaftliches Kolloquium*. Ilmenau: Technische Hochschule, 1975 (series of lectures „Nachrichtentechnik"), pp. 159–162

33. KÜPFMÜLLER, K. ; MATHIS, W. ; REIBIGER, A.: *Theoretische Elektrotechnik*. 17th, revised ed. Berlin; Heidelberg: Springer, 2006

34. LANGE, C.: Optimierung eines Mehrträgerverfahrens zur leitungsgebundenen Übertragung digitaler Signale. In: *Frequenz* 58 (2004), January/February, No. 1/2, pp. 11–19

35. LANGE, C. ; AHRENS, A.: *Optimierung leitungsgebundener Übertragungssysteme*. Berlin: Mensch & Buch Verlag, 2006

36. LANGE, F. H.: *Signale und Systeme*. Bd. 2: Gesteuerte elektronische Systeme. Berlin: Verlag Technik, 1968

37. LANGE, F. H.: *Signale und Systeme*. Bd. 3: Regellose Vorgänge. Verlag Technik, 1968

38. LANGE, F. H.: *Signale und Systeme*. Bd. 1: Spektrale Darstellung. 2nd, extensively revised ed. Berlin: Verlag Technik, 1975

39. LANGE, F. H.: *Signale und Systeme*. Bd. 1: Spektrale Darstellung; 2: Gesteuerte elektronische Systeme; 3: Regellose Vorgänge. Berlin: Verlag Technik, since 1965

40. LEHNER, G. ; KURZ, S.: *Elektromagnetische Feldtheorie*. 9th ed. Berlin: Springer Vieweg, 2021

41. LEONE, M.: *Theoretische Elektrotechnik*. Wiesbaden: Springer Vieweg, 2018

42. LINDNER, J.: *Informationsübertragung*. Berlin; Heidelberg: Springer, 2005

43. LUNZE, K.: *Einführung in die Elektrotechnik*. 13th ed. Berlin: Verlag Technik, 1991

44. MARKO, H.: Optimale und fast optimale binäre und mehrstufige digitale Übertragungssysteme. In: *Archiv für Elektronik und Übertragungstechnik (AEÜ)* 28 (1974), October, No. 10, pp. 402–414

45. MARKO, H.: *Methoden der Systemtheorie*. Berlin, Heidelberg, New York: Springer, 1977

46. MARTINI, H.: *Theorie der Übertragung auf elektrischen Leitungen*. Heidelberg: Hüthig, 1974

47. MERTINS, A.: Signaltheorie. 4th ed. Wiesbaden: Springer Vieweg, 2020

48. OHM, J.-R. ; LÜKE, H. D.: *Signalübertragung*. 12th ed. Berlin; Heidelberg: Springer Vieweg, 2014

49. OPPENHEIM, A. V. ; SCHAFER, R. W.: *Zeitdiskrete Signalverarbeitung*. 3rd ed. München, Wien: Oldenbourg, 1999

50. PHILIPPOW, E. ; BONFIG, K.W. (Ed.) ; BECKER, W.-J. (Hrsg.): *Grundlagen der Elektrotechnik*. 10th, revised ed. Berlin: Verlag Technik, 2000

51. POLLAKOWSKI, M. ; WELLHAUSEN, H.-W.: Eigenschaften symmetrischer Ortsanschlußkabel im Frequenzbereich bis 30 MHz. In: *Der Fernmeldeingenieur* 49 (1995), September/October, No. 9/10, pp. 1–58

52. PROAKIS, J. G. ; SALEHI, M.: *Digital Communications*. 5th ed. Boston; New York; San Francisco: McGraw-Hill, 2008

53. PUN, L.: *Abriss der Optimierungspraxis*. Berlin: Akademie-Verlag, 1974

54. RENNERT, I. ; BUNDSCHUH, B.: *Signale und Systeme*. 2nd ed. München: Fachbuchverlag Leipzig im Carl Hanser Verlag, 2013

55. ROPPEL, C.: *Grundlagen der Nachrichtentechnik*. München: Hanser, 2018

56. SCHMID, H.: *Theorie und Technik der Nachrichtenkabel*. Heidelberg: Hüthig, 1976

57. SCHRÖDER, H.: *Elektrische Nachrichtentechnik – Band 1: Grundlagen, Theorie und Berechnung passiver Übertragungswerte*. Berlin-Borsigwalde: Verlag für Radio-Foto-Kinotechnik, 1967

58. SCHUBERT, W.: *Nachrichtenkabel und Übertragungssysteme*. 3rd ed. Berlin; München: Siemens Aktiengesellschaft, 1986

59. SCHWAB, A. J.: *Begriffswelt der Feldtheorie*. 7th, revised and extended ed. Berlin; Heidelberg: Springer Vieweg, 2013

60. SHANNON, C. E.: A Mathematical Theory of Communication. In: *Bell System Technical Journal* 27 (1948), July and October, No. 1, pp. 379–423 and 623–656

61. SIMONYI, K.: *Theoretische Elektrotechnik*. 10th ed. Leipzig: Barth, 1993

62. SKLAR, B. ; HARRIS, F.: *Digital Communications – Fundamentals and Applications*. 3rd ed. New York: Pearson, 2021

63. SÖDER, G.: Optimierung und Vergleich digitaler Koaxialkabelsysteme. In: *Archiv für Elektronik und Übertragungstechnik (AEÜ)* 36 (1982), No. 11/12, pp. 451–464

64. SÖDER, G.: *Modellierung, Simulation und Optimierung von Nachrichtensystemen*. Berlin, Heidelberg: Springer, 1993

65. SÖDER, G. ; TRÖNDLE, K.: *Digitale Übertragungssysteme – Theorie, Optimierung und Dimensionierung der Basisbandsysteme*. Berlin, Heidelberg: Springer, 1985

66. SPANIER, J. ; OLDHAM, K. B.: *An Atlas of Functions*. Washington, New York, London: Hemisphere Publishing, 1987

67. STEINBUCH, K. ; RUPPRECHT, W.: *Nachrichtentechnik – Band II: Nachrichten übertragung*. 3rd revised ed. Berlin, Heidelberg, New York: Springer, 1982

68. ULRICH, H. ; WEBER, H.: *Laplace-, Fourier- und z-Transformation*. 10th ed. Wiesbaden: Springer Vieweg, 2017

69. VALENTI, C.: NEXT and FEXT Models for Twisted-Pair North American Loop Plant. In: *IEEE Journal on Selected Areas in Communications* 20 (2002), June, No. 5, pp. 893–900

70. WAGGENER, W. N.: *Pulse Code Modulation Techniques with Applications in Communications and Data Recording*. New York: Van Nostrand Reinhold, 1995

71. WELLHAUSEN, H.-W.: Neue Nutzungsmöglichkeiten vorhandener Kupferanschlußleitungsnetze. In: *Nachrichtentechnische Zeitschrift (NTZ)* 17 (1995), No. 4, pp. 18–27

72. WERNER, M.: *Nachrichtentechnik*. 8th ed. Wiesbaden: Springer Vieweg, 2017

73. WUNSCH, G. ; SCHREIBER, H.: *Stochastische Systeme*. 4th ed. Berlin, Heidelberg: Springer, 2006

74. ZEIDLER, E. (Ed.): *Springer-Handbuch der Mathematik I*. Wiesbaden: Springer Spektrum, 2013

75. ZEIDLER, E. (Ed.): *Springer-Handbuch der Mathematik II*. Wiesbaden: Springer Spektrum, 2013

76. ZEIDLER, E. (Ed.): *Springer-Handbuch der Mathematik III*. Wiesbaden: Springer Spektrum, 2013

77. ZINKE, O. ; BRUNSWIG, H. ; VLCEK, A. (Ed.) ; HARTNAGEL, H. (Ed.) ; MAYER, K. (Ed.): *Hochfrequenztechnik 1*. 6th, revised ed. Berlin; Heidelberg: Springer, 2000

GPSR Compliance

The European Union's (EU) General Product Safety Regulation (GPSR) is a set of rules that requires consumer products to be safe and our obligations to ensure this.

If you have any concerns about our products, you can contact us on ProductSafety@springernature.com

In case Publisher is established outside the EU, the EU authorized representative is:

Springer Nature Customer Service Center GmbH
Europaplatz 3
69115 Heidelberg, Germany

Batch number: 09026430

Printed by Printforce, the Netherlands